Bioactives and Functional Ingredients in Foods

Bioactives and Functional Ingredients in Foods

Editors

Severina Pacifico
Simona Piccolella

MDPI • Basel • Beijing • Wuhan • Barcelona • Belgrade • Manchester • Tokyo • Cluj • Tianjin

Editors

Severina Pacifico
Department of
Environmental, Biological
and Pharmaceutical Sciences
and Technologies
University of Campania
"Luigi Vanvitelli"
Caserta
Italy

Simona Piccolella
Department of
Environmental, Biological
and Pharmaceutical Sciences
and Technologies
University of Campania
"Luigi Vanvitelli"
Caserta
Italy

Editorial Office
MDPI
St. Alban-Anlage 66
4052 Basel, Switzerland

This is a reprint of articles from the Special Issue published online in the open access journal *Molecules* (ISSN 1420-3049) (available at: www.mdpi.com/journal/molecules/special_issues/bioactive_ingredient_foods).

For citation purposes, cite each article independently as indicated on the article page online and as indicated below:

LastName, A.A.; LastName, B.B.; LastName, C.C. Article Title. *Journal Name* **Year**, *Volume Number*, Page Range.

ISBN 978-3-0365-5158-6 (Hbk)
ISBN 978-3-0365-5157-9 (PDF)

© 2022 by the authors. Articles in this book are Open Access and distributed under the Creative Commons Attribution (CC BY) license, which allows users to download, copy and build upon published articles, as long as the author and publisher are properly credited, which ensures maximum dissemination and a wider impact of our publications.

The book as a whole is distributed by MDPI under the terms and conditions of the Creative Commons license CC BY-NC-ND.

Contents

Lilian Alnsour, Reem Issa, Shady Awwad, Dima Albals and Idrees Al-Momani
Quantification of Total Phenols and Antioxidants in Coffee Samples of Different Origins and Evaluation of the Effect of Degree of Roasting on Their Levels
Reprinted from: *Molecules* **2022**, *27*, 1591, doi:10.3390/molecules27051591 1

Betül Gıdık
Antioxidant, Antimicrobial Activities and Fatty Acid Compositions of Wild *Berberis* spp. by Different Techniques Combined with Chemometrics (PCA and HCA)
Reprinted from: *Molecules* **2021**, *26*, 7448, doi:10.3390/molecules26247448 11

Jolanta Gawałek
Effect of Spray Dryer Scale Size on the Properties of Dried Beetroot Juice
Reprinted from: *Molecules* **2021**, *26*, 6700, doi:10.3390/molecules26216700 25

Syed Amir Ashraf, Abd Elmoneim O. Elkhalifa, Khalid Mehmood, Mohd Adnan, Mushtaq Ahmad Khan and Nagat Elzein Eltoum et al.
Multi-Targeted Molecular Docking, Pharmacokinetics, and Drug-Likeness Evaluation of Okra-Derived Ligand Abscisic Acid Targeting Signaling Proteins Involved in the Development of Diabetes
Reprinted from: *Molecules* **2021**, *26*, 5957, doi:10.3390/molecules26195957 41

Yuji Iwaoka, Shoichi Suzuki, Nana Kato, Chisa Hayakawa, Satoko Kawabe and Natsuki Ganeko et al.
Characterization and Identification of Bioactive Polyphenols in the *Trapa bispinosa* Roxb. Pericarp Extract
Reprinted from: *Molecules* **2021**, *26*, 5802, doi:10.3390/molecules26195802 65

Cristina Arteaga, Nuria Boix, Elisabet Teixido, Fernanda Marizande, Santiago Cadena and Alberto Bustillos
The Zebrafish Embryo as a Model to Test Protective Effects of Food Antioxidant Compounds
Reprinted from: *Molecules* **2021**, *26*, 5786, doi:10.3390/molecules26195786 81

Shadab Faramarzi, Simona Piccolella, Lorenzo Manti and Severina Pacifico
Could Polyphenols Really Be a Good Radioprotective Strategy?
Reprinted from: *Molecules* **2021**, *26*, 4969, doi:10.3390/molecules26164969 93

Martha A. Flores-Mancha, Martha G. Ruíz-Gutiérrez, Rogelio Sánchez-Vega, Eduardo Santellano-Estrada and América Chávez-Martínez
Effect of Encapsulated Beet Extracts (*Beta vulgaris*) Added to Yogurt on the Physicochemical Characteristics and Antioxidant Activity
Reprinted from: *Molecules* **2021**, *26*, 4768, doi:10.3390/molecules26164768 113

Jolanta Sarowska, Dorota Wojnicz, Agnieszka Jama-Kmiecik, Magdalena Frej-Madrzak and Irena Choroszy-Król
Antiviral Potential of Plants against Noroviruses
Reprinted from: *Molecules* **2021**, *26*, 4669, doi:10.3390/molecules26154669 131

Nuntouchaporn Hutachok, Pimpisid Koonyosying, Tanachai Pankasemsuk, Pongsak Angkasith, Chaiwat Chumpun and Suthat Fucharoen et al.
Chemical Analysis, Toxicity Study, and Free-Radical Scavenging and Iron-Binding Assays Involving Coffee (*Coffea arabica*) Extracts
Reprinted from: *Molecules* **2021**, *26*, 4169, doi:10.3390/molecules26144169 153

Surakshi Wimangika Rajapaksha and Naoto Shimizu
Development and Characterization of Functional Starch-Based Films Incorporating Free or Microencapsulated Spent Black Tea Extract
Reprinted from: *Molecules* **2021**, *26*, 3898, doi:10.3390/molecules26133898 . . . . **187**

Abd Elmoneim O. Elkhalifa, Eyad Al-Shammari, Mohd Adnan, Jerold C. Alcantara, Khalid Mehmood and Nagat Elzein Eltoum et al.
Development and Characterization of Novel Biopolymer Derived from *Abelmoschus esculentus* L. Extract and Its Antidiabetic Potential
Reprinted from: *Molecules* **2021**, *26*, 3609, doi:10.3390/molecules26123609 . . . . **203**

Andri Frediansyah, Fitrio Romadhoni, Suryani, Rifa Nurhayati and Anjar Tri Wibowo
Fermentation of Jamaican Cherries Juice Using *Lactobacillus plantarum* Elevates Antioxidant Potential and Inhibitory Activity against Type II Diabetes-Related Enzymes
Reprinted from: *Molecules* **2021**, *26*, 2868, doi:10.3390/molecules26102868 . . . . **215**

Abd Elmoneim O. Elkhalifa, Eyad Alshammari, Mohd Adnan, Jerold C. Alcantara, Amir Mahgoub Awadelkareem and Nagat Elzein Eltoum et al.
Okra (*Abelmoschus Esculentus*) as a Potential Dietary Medicine with Nutraceutical Importance for Sustainable Health Applications
Reprinted from: *Molecules* **2021**, *26*, 696, doi:10.3390/molecules26030696 . . . . **229**

Mauren Estupiñan-Amaya, Carlos Alberto Fuenmayor and Alex López-Córdoba
New Freeze-Dried Andean Blueberry Juice Powders for Potential Application as Functional Food Ingredients: Effect of Maltodextrin on Bioactive and Morphological Features
Reprinted from: *Molecules* **2020**, *25*, 5635, doi:10.3390/molecules25235635 . . . . **251**

Cristhian J. Yarce, Maria J. Alhajj, Julieth D. Sanchez, Jose Oñate-Garzón and Constain H. Salamanca
Development of Antioxidant-Loaded Nanoliposomes Employing Lecithins with Different Purity Grades
Reprinted from: *Molecules* **2020**, *25*, 5344, doi:10.3390/molecules25225344 . . . . **267**

Article

Quantification of Total Phenols and Antioxidants in Coffee Samples of Different Origins and Evaluation of the Effect of Degree of Roasting on Their Levels

Lilian Alnsour [1], Reem Issa [1,*], Shady Awwad [2], Dima Albals [3] and Idrees Al-Momani [4]

[1] Department of Pharmaceutical Sciences, Pharmacological and Diagnostic Research Center (PDRC), Faculty of Pharmacy, Al-Ahliyya Amman University, Amman 19328, Jordan; l.alnsour@ammanu.edu.jo
[2] Department of Pharmaceutical Chemistry & Pharmacognosy, Applied Science Private University, Amman 11931, Jordan; sh_awwad@asu.edu.jo
[3] Department of Medicinal Chemistry and Pharmacognosy, Faculty of Pharmacy, Yarmouk University, Irbid 21163, Jordan; dimabals@yu.edu.jo
[4] Department of Chemistry, Faculty of Science, Yarmouk University, Irbid 21163, Jordan; imomani@yu.edu.jo
* Correspondence: r.issa@ammanu.edu.jo

Abstract: Phenolic and antioxidant compounds have received considerable attention due to their beneficial effects on human health. The aim of this study is to determine the content of total phenols and antioxidants in fifty-two coffee samples of different origins, purchased from the Jordanian local market, and investigate the effect of the degree of roasting on the levels of these compounds. The coffee samples were extracted using the hot water extraction method, while Folin–Ciocalteu (FC) and 1,1-diphenyl-2-picrylhydrazyl (DPPH) assay methods were used to analyze these compounds. The results showed that the highest content of total phenol (16.55 mg/g equivalent to GAE) was found in the medium roasted coffee, and the highest content of antioxidants (1.07 mg/g equivalent to TEAC) content was found in the green coffee. Only light and medium roasted coffee showed a significant correlation ($p < 0.05$, $R^2 > 0.95$) between the average of total phenolic and antioxidant content. A negative correlation between the antioxidant content and the degree of roasting ($p < 0.05$, $R^2 > 0.95$) were shown, while it did not correlate with phenolic contents. Previously, a positive correlation between antioxidant and chlorogenic acids content was observed, with no correlation between the origin of coffee samples nor heavy metal content, which was previously determined for the same coffee samples. These findings suggest that the antioxidant content for coffee extracts is largely determined by its chlorogenic acid content, rather than the coffee origin or total phenolic and heavy metals content.

Keywords: coffee; phenols; antioxidant; TEAC; roasting; GAE

1. Introduction

Coffee is one of the most popular beverages in the world. It was introduced to the New World in the mid-17th century, although its history dates to the 15th century when coffee plants were supposedly cultivated in Southern Arabia and taken originally from Ethiopia. It was not until the 1950s that instant coffee was produced. Nowadays, Brazil is the top coffee-producing country in the world, followed by Vietnam and Colombia [1].

There are over 120 species of Coffea plant, with *Coffea arabica* and *Coffea canephora* (also known as "Robusta") being the most popular commercially. The former contributes to 70% of the world's coffee consumption. The latter contains more caffeine and lower lipid content, which is why it tastes more bitter [2]. It is also cheaper to produce compared with *Coffea arabica* [3]. Green coffee beans are the seeds from the coffee tree fruits. The same species of coffee can be cultivated differently to produce a wide variety of coffee beans having different flavors and aromas, depending on the soil, climate, and altitude of their growing areas, which means that coffee is affected by its geographical origin [4].

Roasting is also an important determinant of taste and aroma in brewed coffee. Green coffee beans are heated at 200–240° for 10–15 min, depending on the degree of roasting required. This may considerably alter their chemical composition; lowering the sugars, water, and chlorogenic acids while forming new compounds, such as melanoidins due to the Maillard reaction [5]. In a study by Mayer and others [6], it was shown that the concentration of certain compounds in roasted coffee beans was greatly affected by the degree of roast. For example, Colombian and Kenyan coffees have increasing amounts of phenolic compounds, such as guaiacol, with an increased roasting degree.

The quality of coffee can be affected by levels of chemical fertilizers and pesticides used in soil, which contribute to heavy metal contamination. The preparation method and degree of roasting also affect its final heavy metal composition. Albals et al. [7] determined the heavy metal content in different green and roasted coffee samples consumed in Jordan, taken from five origins: Brazil, Ethiopia, Kenya, Colombia, and India. According to the results, there was a significant difference in the levels of Zn, Cr, and Co in green and roasted coffee beans. All levels were below the tolerable upper limit of daily intake (TULD) of metals determined by the World Health Organization (WHO) and thus were safe for consumption.

Coffee is known to be an essential source of antioxidants due to the presence of alkaloids, flavonoids, and phenolic compounds. Consumption of coffee is therefore attributed to improving health [4]. Moreover, it is the main antioxidant found in the diets of Americans, Japanese, Danish, and Brazilians [8].

Phenolic compounds are widely abundant in fruits, vegetables, dry legumes, chocolate, and beverages like coffee, tea, and cocoa [9]. Polyphenols have been shown to have antioxidant effects, which are beneficial to the heart and can protect against oxidative stress that is directly correlated with degenerative diseases, diabetes mellitus, and cancer [10]. Other evidence suggests their anti-inflammatory, antiviral, and antibacterial activity [11]. For instance, green coffee beans have been reported to contain chlorogenic and caffeic acids as the main phenols. These compounds possess antimutagenic and antioxidative effects [8]. Other studies suggest its role in neurodegenerative diseases, such as ischemic strokes and lowering blood pressure in rats [12].

A study by Masek et al. [8] on five different Ethiopian coffee brands has demonstrated a significant total phenolic content and antioxidant activities, which showed that Ethiopian coffee might be used in preventing and curing several degenerative diseases. Another study by Sentkowska et al. [5] was carried out to determine the antioxidant capacity of different green coffee extracts. It has been reported that Robusta green coffee from Laos had the highest antioxidant capacity due to the high concentration of chlorogenic acids. Duarte et al. [13] concluded that roasting is inversely proportional to the polyphenol and antioxidant activity, where light brewed coffee had the highest antioxidant capacity, while dark roasted coffee had the lowest.

While many studies were conducted in Jordan to study the total phenolic content and antioxidant capacity of various plant species [14], only a few of them were carried out on coffee. A previous study by Kandah et al. [14] analyzed total phenolic content and antioxidant activity using methyl linoleate (MeLo) assay for nine samples of green and roasted coffee beans obtained from the Jordanian market. It demonstrated that extraction time, temperature, and particle size were important variables that affected total phenolic content. Another study compared the antioxidant activity of roasted barley and roasted dates with that of two different roasted coffee samples and green coffee samples (Saudi and Colombian origins) using ABTS and DPPH assays. Results indicated that the highest antioxidant activity was observed for Saudi roasted coffee, followed by Colombian roasted coffee, roasted barley, Colombian green coffee, and roasted dates [15].

To the best of our knowledge, this is the first extensive study conducted in Jordan on 52 different coffee samples with the aim to evaluate the effect of roasting and geographical origin on the antioxidant and total phenolic content of coffee beans available in the Jordanian market, using 1,1-diphenyl-2-picrylhydrazyl (DPPH) and Folin–Ciocalteu (FC) assays, respectively, for this purpose. All samples were prepared by water extraction similar to the way normally used when preparing the beverage, to investigate the individual's intake of antioxidant phenols by consuming coffee brew. The secondary aim was to find a correlation (if any) between the antioxidant activity and total phenolic content of different types of coffee beans, with their caffeine, chlorogenic acids, and heavy metal content, the latter that were previously determined by our research group (Albals et al.) (Awwad et al.) [7,16].

2. Results
2.1. Total Polyphenols Content

The average total phenolic content, expressed as milligrams of gallic acid equivalents per gram of dry coffee extract (GAE mg/g), was determined for 52 *Coffea arabica* samples of different origins and roasting degrees, as shown in Supplementary Table S1. These samples have shown variations in total phenolic content ranging from 14.92 mg/g to 16.55 mg/g (Figure 1), reported for dark roasted coffee and medium roasted coffee, respectively. Except for green coffee beans, the variety from Kenya showed to have the lowest total phenolic content, while the variety from Colombia showed to have the highest total phenolic content (Table 1). For green coffee beans, the variety from Kenya showed to have the highest total phenolic content, which was decreasing in its content by the roasting process in varying amounts. While the variety from Colombia showed the opposite behavior upon roasting, as its content of phenols tends to increase by increasing the roasting degree.

Table 1. Comparison of different roasting degrees for coffee samples from different geographical origins according to their total phenolic and antioxidant activity (GAE mg/g and TEAC mg/g, respectively).

Geographical Origin	GAE mg/g ± SD	TEAC mg/g ± SD
Green Coffee		
Kenya	17.25 ± 0.14	1.29 ± 0.04
Ethiopia	14.55 ± 0.10	1.13 ± 0.11
Brazil	13.82 ± 0.13	0.97 ± 0.14
Colombia	16.72 ± 0.05	0.88 ± 0.08
Average	15.59 ± 1.65	1.07 ± 0.18
Light Coffee		
Kenya	9.84 ± 0.02	0.84 ± 0.05
Ethiopia	16.35 ± 0.05	0.96 ± 0.03
Brazil	17.11 ± 0.05	0.94 ± 0.07
Colombia	19.05 ± 0.13	1.07 ± 0.14
Average	15.59 ± 4.00 *	0.95 ± 0.09 *
Medium Coffee		
Kenya	12.31 ± 0.02	0.65 ± 0.12
Ethiopia	14.93 ± 0.14	1.13 ± 0.01
Brazil	14.66 ± 0.05	1.07 ± 0.07
Colombia	24.28 ± 0.11	0.08 ± 0.02
Average *	16.55 ± 5.29 **	0.73 ± 0.48 **
Dark Coffee		
Kenya	9.44 ± 0.11	0.60 ± 0.10
Ethiopia	12.24 ± 0.14	0.43 ± 0.10
Brazil	16.60 ± 0.14	0.47 ± 0.17
Colombia	21.41 ± 0.20	0.46 ± 0.15
Average	14.92 ± 5.23	0.49 ± 0.07

*, ** Results are statistically significant with p-value < 0.05, R^2 > 0.95.

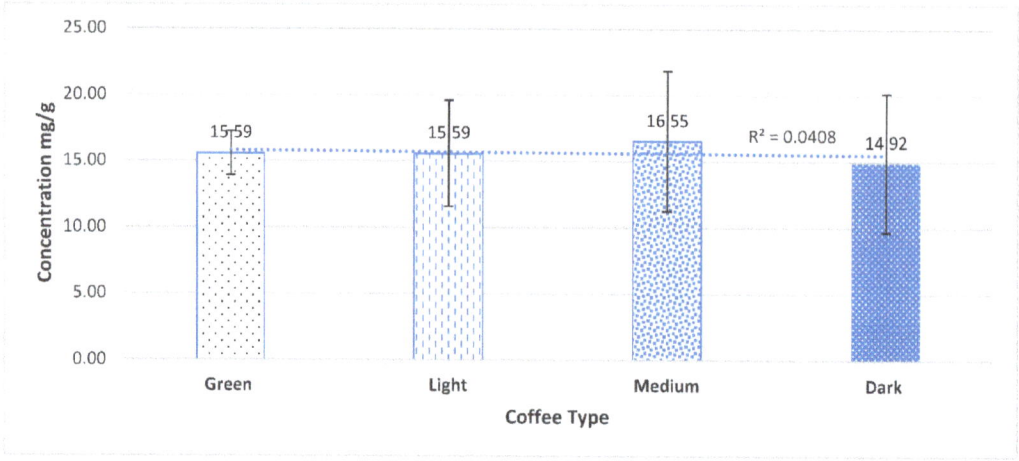

Figure 1. The average total phenolic content (mg/g GAE) for each roasting degree with (different geographical origin). No correlation was found between roasting degree and total phenolic content. (R^2 = 0.041). Results are statistically insignificant with a p-value > 0.05.

2.2. *Antioxidant Activity of Coffee Samples*

The average antioxidant activity—expressed as milligrams of Trolox equivalents antioxidant capacity per gram dry coffee extract (TEAC mg/g)—was determined for 52 *Coffea arabica* samples of different origins and roasting degrees, as shown in Supplementary Table S1. The values ranged from 0.49 mg/g to 1.07 mg/g, reported for dark roasted coffee and green coffee, respectively (Figure 2). The antioxidant activity was correlated with the roasting degree of coffee beans. Although the content of phenols showed a consistent pattern in terms of the geographical origin of coffee beans, no correlation was found for the antioxidant activity with the origin (Table 1). Only light and medium roasted coffee showed a significant correlation between the average total phenolic content and the average antioxidant content.

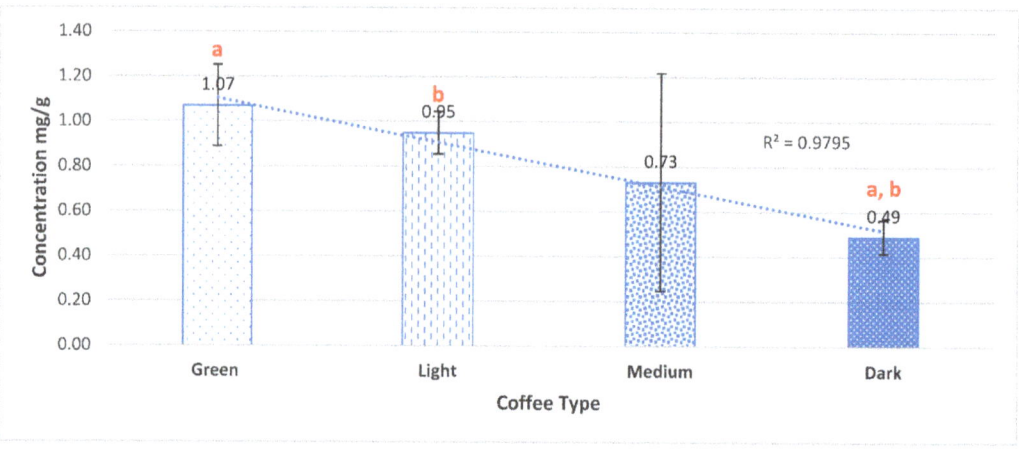

Figure 2. The average antioxidant activity (mg/g TEAC) for each roasting degree (different geographical origin) A negative correlation was found between roasting degree and TEAC mg/g (R^2 = 0.98). Bars labeled [a,b] are statistically significant with a p-value < 0.05.

2.3. Comparison of the Content of Total Phenol, Antioxidant, and Heavy Metal in Selected Coffee Samples

Figures 3 and 4 show a comparison of coffee samples of different geographical origins (regardless of their roasting degree) and roasting degrees (regardless of their origin) in terms of average GAE in mg/g, average TEAC in mg/g, and average Zn, Pb, and Cu contents in µg/g (Data for heavy metal content was taken from Albals et al. [7]).

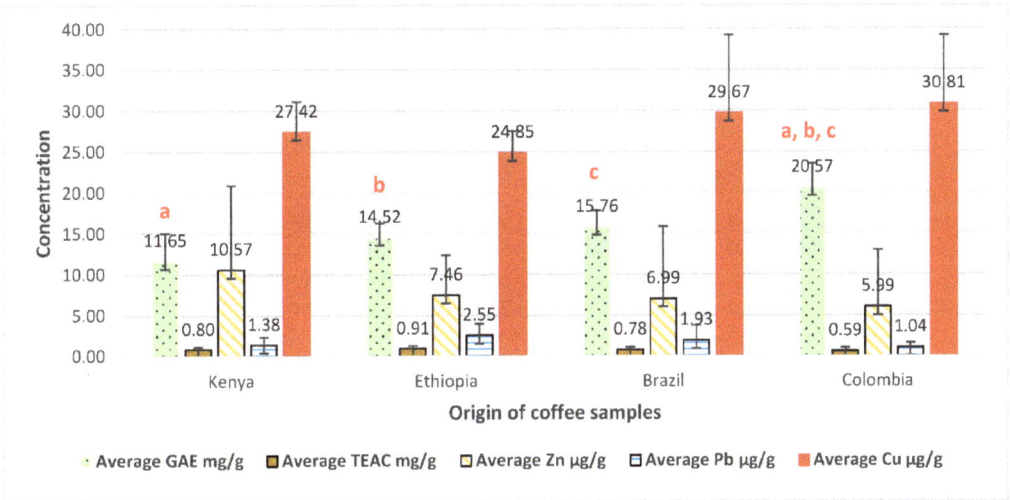

Figure 3. Comparison of coffee samples from different origins (regardless of their roasting degree) in terms of average GAE in mg/g, average TEAC in mg/g, and average Zn, Pb, and Cu contents in µg/g. (Data for heavy metal content taken from Albals et al. [7]). Bars labeled [a,b,c] are statistically significant with a p-value < 0.05.

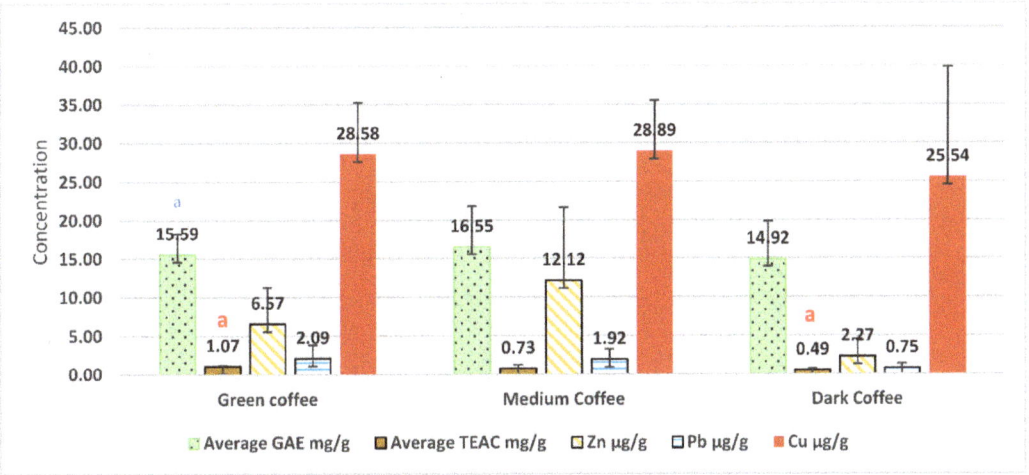

Figure 4. Comparison of coffee samples of three different roasting degrees (regardless of their origin) in terms of average GAE in mg/g, average TEAC in mg/g, and average Zn, Pb, and Cu contents in µg/g. (Data for heavy metal content was taken from Albals et al. [7]). Bars labeled [a] are statistically significant with a p-value < 0.05.

3. Discussion

The roasting degree and geographical origins are key factors affecting both the total phenolic content and the antioxidant activity of the coffee beans [17]. Therefore, this study focused on evaluating the effect of these variables on the antioxidant and phenolic compound contents in coffee samples available in the Jordanian market. In addition, this study aimed to explore the individual intake of phenols and antioxidants from coffee beverages commonly used among Jordanian citizens, known as "Turkish coffee". Since antioxidant compounds provide health benefits, coffee as a beverage is claimed to be of great interest for individuals who are trying to increase their intake of these nutrients [18]. Consequently, such findings should reveal the actual individual intake of these nutrients that would influence their health status.

The findings of the current study are in good agreement with the previous study by Król et al. [17], which showed that the highest content of total polyphenolic compounds was determined in coffee samples roasted in light and medium roasting conditions, and was better for preserving these nutrients.

A previous study by Bobková et al. [19] showed a correlation between phenolic content and antioxidant activity. On the contrary, the current findings showed that no correlation was found between phenolic content, antioxidant activity, or geographical origin. Medium roasted beans appeared to contain higher average polyphenols than the green beans, where the variety from Colombia showed to have the highest phenolic content among the roasted samples (regardless of the roasting degree), but with varied antioxidant activity.

As expected, statistical analyses of these data revealed a significant influence ($p < 0.05$. $R^2 = 0.98$) of the roasting degree on the antioxidant activity, which was decreasing with the increase in the degree of roasting. Cho et al. [20] studied the influence of roasting conditions on the antioxidant characteristics of Colombian coffee of the species *Coffea arabica* beans. They found that light-roasted coffee beans have the highest antioxidant activity, and an approximately 40–80% loss of antioxidant activity was observed after further roasting. In addition, they also detected significantly higher antioxidant activity as compared to unroasted beans, suggesting the formation of Maillard reaction products and the release of bound polyphenols from plant cells. These data suggested the formation of new phenolic compounds, other than the ones detected in the variety from Colombia, but with no effect on their antioxidant content.

Green coffee beans showed the highest content of phenols in the species from Kenya. The latter also showed the highest antioxidant activity compared to the other studied samples. In fact, different varieties of coffee samples showed a low and narrow range of phenolic content compared to the previous studies [21,22]. This could be explained by the fact that phenolic compounds are often more soluble in alcohol extracts compared to water, which was used in this study.

A previous study by Górnaś et al. [23], which investigated the contribution of phenolic acids isolated from green and roasted boiled-type coffee brews to total coffee antioxidant capacity, showed that the antioxidant effect can be poorly correlated with polyphenols' concentration when the DPPH assay method is used, which agrees with the findings of this study. The antioxidant activity changes in extracts from green, light, medium, and dark roasted coffee are negatively influenced by the intensity of the heating process and seem to be much more dependent on the roasting degree than on the geographical origin of coffee beans. Similarly, Bilge [24] conducted a study investigating the effects of geographical origin, roasting degree, particle size, and brewing method on the physicochemical and spectral properties of Arabica coffee. It showed that roasting degree and brewing method—compared with other parameters—were the most discriminating factors based on UV and fluorescence spectra of coffee brew samples, respectively. On the contrary, Muzykiewicz-Szymańska et al. [25] studied the effect of brewing process parameters on antioxidant activity in infusions of roasted and unroasted Arabica coffee beans originating from different countries. They concluded that the phenolic compound content in infusions prepared using different techniques depended on the roasting process, the bean's origin, as well as the brewing technique.

Based on the results from the study by Albals, et al. [7] which investigated the heavy metals contents for the same coffee samples that were analyzed in this study, the data showed that there was no clear correlation between the content of phenols, antioxidant compounds, and heavy metals content. Therefore, these findings suggest that roasting degree would affect the antioxidant activity, regardless of the geographical origin or heavy metal content. Nevertheless, the geographical origin had shown an impact only on the total phenolic content, with no effect on any of the other measured variables.

Results from a study—that was recently published by our research group—which determined the chlorogenic acids (CGAs) and caffeine content for the same coffee samples, have shown that the highest content of caffeine was found in the medium roasted coffee (203.63 mg/L), and the highest content of CGA was found in the green coffee (543.23 mg/L). The results demonstrated a negative correlation between the CGA levels with the degree of roasting, while it showed a positive correlation between the caffeine levels with the degree of roasting before it starts to decline in the dark roasted coffee [16].

Comparing these results with the current study, it can be concluded that the coffee samples with the highest CGA content have also shown the highest antioxidant activity, which suggests that CGAs alone rather than total phenolic content contribute to the antioxidant activity of coffee. Furthermore, the geographical origin did not seem to affect the content of either CGA or caffeine, as it did not affect the total phenolic content and antioxidant activity determined in this study [16].

This study showed the need to perform more research using different assays to investigate the relationship between the antioxidant activity with the total phenolic content.

4. Methods and Materials

4.1. Chemicals and Standards

For the determination of the total phenolic content, the FC reagent (2 N, Sigma-Aldrich, Schaffhausen, Switzerland), gallic acid (GA) (99% purity, Sigma), and anhydrous sodium carbonate (99% purity, Sigma) were used. For the determination of the antioxidant activity, DPPH (Aldrich, Darmstadt, Germany), Trolox (Sigma-Aldrich, Switzerland), and methanol (for HPLC > 99.9%, Sigma-Aldrich, St. Quentin Fallavier, France) were employed.

For this study, fifty-two samples of ground coffee beans (*Coffea arabica*), including green (11 samples), light-roasted (14 samples), medium roasted (11 samples), and dark roasted (16 samples), from different origins were purchased from different grocery stores across Jordan (Amman, Irbid) in 2019 and stored in the freezer.

4.2. Sample Preparation and Extraction

The coffee samples were freshly extracted according to the extraction procedure described by Perez Hernandez et al. [26], with few modifications. The coffee samples were extracted with hot water at 75–80 °C at a 1/100 coffee-to-solvent ratio, where 1 g of coffee was extracted with 100 mL of water. Then, ultrasonication was performed for 5 min to homogenize the solutions using an ultrasonic bath (OVAN). Afterwards, the samples were centrifuged for 15 min at $7900 \times g$ using an (MPW-260R) centrifuge system. Next, all coffee solutions were filtered with Whatman No. 2 filter paper. Finally, the coffee extracts were stored at a temperature of −20 °C until the day of analysis (May–August 2020).

4.3. Determination of Total Phenolic Content

Total phenolic content was measured using the FC method by Singleton et al. [27]. A stock solution of 10 mg/mL of extract in water was prepared for each sample. Three different concentrations were made for each extract by serial dilution; 5 mg/mL, 2.5 mg/mL and 1.25 mg/mL. An aliquot of 80 µL of each aqueous solution of extract was added to 400 µL of dilute Folin–Ciocalteu (1:10) reagent in a test tube. Then 320 µL of 7.5% sodium carbonate solution was added. The solution was covered with aluminum foil and incubated in a water bath at 45 °C for 30 min. The absorbance was recorded at 765 nm using a UV-Vis spectrophotometer against the blank solution (water only). The measurement was compared to a calibration curve prepared with GA solution at a concentration range from (0–0.12 mg/mL), and the total phenolic content was expressed as GAE mg/g, using the standard curve equation:

$$y = 7.515x + 0.0308, R^2 = 0.9927$$

where y is the absorbance at 765 nm and x is the total phenolic content in the different extracts expressed in mg/mL.

4.4. Determination of Antioxidant Capacity

Aliquots of 2 mL of each plant extract were dried in the oven overnight at a temperature of 50 °C to dry all the water. An amount of 2 mL of methanol was added to each dried sample and vortexed for 3 min to allow a homogenous stock solution of concentration 10 mg/mL. Three different concentrations were made for each extract by serial dilution; 5 mg/mL, 2.5 mg/mL and 1.25 mg/mL. A solution of 0.1 mM DPPH in methanol was prepared fresh for the assay. Then, 300 µL of the sample was added to 100 µL of DPPH solution, shaken, and incubated for 30 min in the dark at room temperature. Absorbance was monitored at 517 nm using a UV-Vis spectrophotometer. The reaction mixture containing control and reference standard (300 µL methanol and 100 µL DPPH solution) were also measured. The measurement was compared to a calibration curve prepared with Trolox solution at a concentration range from (0–50 µM). The antioxidant capacity was expressed as TEAC mg/g, using the standard curve equation:

$$y = -0.024x + 1.36, R^2 = 0.992$$

where y is the DPPH absorbance at 517 nm and x is the Trolox concentration in the different extracts expressed in µM, which is then expressed as mg TEAC/g of dry extract.

4.5. Statistical Analysis

All measurements were performed in triplicates and results were reported as mean ± standard deviation. The results were analyzed statistically using one-way analysis of variance (ANOVA) on Microsoft Excel with its data analysis add-ins. The mean values of GAE mg/g were compared with TEAC mg/g to assess the existence of statistical significance using measurements of p-value and squared correlation coefficients (R^2). The level of significance was set to 0.05 and 0.95, respectively.

Supplementary Materials: The following supporting information can be downloaded online, Table S1: The distribution of Coffea arabica samples (52) obtained from the Jordanian market according to their origin and roasting degree. Roasting temperatures for light roasted = 155–165 °C, for medium roasted = 175–185 °C, and for dark roasted = 205–215 °C. Table S2: The Average concentration (mg/g), SD, SE, Min, Max, and confidence interval for GAE in coffee beans obtained from the Jordanian market. Table S3: The Average concentration (mg/g), SD, SE, Min, Max, and confidence interval for TEAC in coffee beans obtained from the Jordanian market.

Author Contributions: L.A.: samples analysis, and writing the first manuscript. R.I.: corresponding author, methodology design, and conceptualization. S.A.: Experimental analysis, writing—review and editing. D.A.: samples collection and resources. I.A.-M: Data Analysis, editing the manuscript. All authors have read and agreed to the published version of the manuscript.

Funding: This research received no external funding.

Institutional Review Board Statement: This study did not require ethical approval.

Informed Consent Statement: Not applicable.

Data Availability Statement: Not applicable.

Acknowledgments: The authors are grateful to the Pharmacological and Diagnostic Research Center (PDRC), Faculty of Pharmacy, and Al-Ahliyya Amman University, for the full support for this research project.

Conflicts of Interest: The authors report no conflicts of interest in this work.

Sample Availability: Samples of the standard compounds are available from the authors.

References

1. Myhrvold, N. Coffee. Encyclopedia Britannica. Available online: https://www.britannica.com/topic/coffee (accessed on 19 February 2022).
2. Urgert, R.; Katan, M.B. The Cholesterol-Raising Factor from Coffee Beans. *J. R. Soc. Med.* **1996**, *89*, 618–623. [CrossRef] [PubMed]
3. Miyanari, W. *Aloha Coffee Island*; Savant Books & Publications: Singapore, 2008; ISBN 978-0-615-18348-0.
4. Tewabe, B. Determination of Caffeine Content and Antioxidant Activity of Coffee. *Am. J. Appl. Chem.* **2015**, *3*, 69. [CrossRef]
5. Sentkowska, A. Comparative Studies on the Antioxidant Properties of Different Green Coffee Extracts. *MOJ Food Process. Technol.* **2016**, *3*, 71. [CrossRef]
6. Mayer, F.; Czerny, M.; Grosch, W. Influence of Provenance and Roast Degree on the Composition of Potent Odorants in Arabica Coffees. *Eur. Food Res. Technol.* **1999**, *209*, 242–250. [CrossRef]
7. Albals, D.; Al-Momani, I.F.; Issa, R.; Yehya, A. Multi-Element Determination of Essential and Toxic Metals in Green and Roasted Coffee Beans: A Comparative Study among Different Origins Using ICP-MS. *Sci. Prog.* **2021**, *104*, 1–17. [CrossRef]
8. Masek, A.; Latos-Brozio, M.; Kałużna-Czaplińska, J.; Rosiak, A.; Chrzescijanska, E. Antioxidant Properties of Green Coffee Extract. *Forests* **2020**, *11*, 557. [CrossRef]
9. Engida, A.M.; Faika, S.; Nguyen-Thi, B.T.; Ju, Y.-H. Analysis of Major Antioxidants from Extracts of Myrmecodia Pendans by UV/Visible Spectrophotometer, Liquid Chromatography/Tandem Mass Spectrometry, and High-Performance Liquid Chromatography/UV Techniques. *J. Food Drug Anal.* **2015**, *23*, 303–309. [CrossRef]
10. Scalbert, A.; Johnson, I.T.; Saltmarsh, M. Polyphenols: Antioxidants and Beyond. *Am. J. Clin. Nutr.* **2005**, *81*, 215S–217S. [CrossRef]
11. Farah, A.; Donangelo, C. Phenolic Compounds in Coffee. *Braz. J. Plant Physiol.* **2006**, *18*, 23–36. [CrossRef]
12. Jeszka-Skowron, M.; Stanisz, E.; Peña, M.P.D. Relationship between Antioxidant Capacity, Chlorogenic Acids and Elemental Composition of Green Coffee. *LWT* **2016**, *73*, 243–250. [CrossRef]
13. Duarte, S.; Abreu, C.; Menezes, H.; Santos, M.; Gouvea, C. Effect of Processing and Roasting on the Antioxidant Activity of Coffee Brews. *Food Sci. Technol.* **2005**, *25*, 387–393. [CrossRef]
14. Kandah, M.I.; Ereifej, K.I.; Al-Azzeh, M.A. Characterization and Quantification of Phenolic Compounds in Coffee Beans and Waste. *Int. J. Adv. Sci. Eng. Technol.* **2019**, *7*, 33–41.
15. Tarawneh, M.; Al-Jaafreh, A.; Dalaeen, H.; Qaralleh, H.; Alqaraleh, M.; Khataibeh, M. Roasted Date and Barley Beans as an Alternative's Coffee Drink: Micronutrient and Caffeine Composition, Antibacterial and Antioxidant Activities. *Syst. Rev. Pharm.* **2021**, *12*, 1079–1083. [CrossRef]
16. Awwad, S.; Issa, R.; Alnsour, L.; Albals, D.; Al-Momani, I. Quantification of Caffeine and Chlorogenic Acid in Green and Roasted Coffee Samples Using HPLC-DAD and Evaluation of the Effect of Degree of Roasting on Their Levels. *Molecules* **2021**, *26*, 7502. [CrossRef] [PubMed]
17. Król, K.; Gantner, M.; Tatarak, A.; Hallmann, E. The Content of Polyphenols in Coffee Beans as Roasting, Origin and Storage Effect. *Eur. Food Res. Technol.* **2020**, *246*, 33–39. [CrossRef]
18. Castaldo, L.; Narváez, A.; Izzo, L.; Graziani, G.; Ritieni, A. In Vitro Bioaccessibility and Antioxidant Activity of Coffee Silverskin Polyphenolic Extract and Characterization of Bioactive Compounds Using UHPLC-Q-Orbitrap HRMS. *Molecules* **2020**, *25*, 2132. [CrossRef] [PubMed]
19. Bobková, A.; Hudáček, M.; Jakabová, S.; Belej, Ľ.; Capcarová, M.; Čurlej, J.; Bobko, M.; Árvay, J.; Jakab, I.; Čapla, J.; et al. The Effect of Roasting on the Total Polyphenols and Antioxidant Activity of Coffee. *J. Environ. Sci. Health B* **2020**, *55*, 495–500. [CrossRef]
20. Cho, A.R.; Park, K.W.; Kim, K.M.; Kim, S.Y.; Han, J. Influence of Roasting Conditions on the Antioxidant Characteristics of Colombian Coffee (*Coffea arabica* L.) Beans. *J. Food Biochem.* **2014**, *38*, 271–280. [CrossRef]
21. Pushpa, S. Murthy Recovery of Phenolic Antioxidants and Functional Compounds from Coffee Industry By-Products. *Food Bioprocess Technol.* **2012**, *5*, 897–903. [CrossRef]
22. Mussatto, S.I.; Ballesteros, L.F.; Martins, S.; Teixeira, J.A.C. Extraction of Antioxidant Phenolic Compounds from Spent Coffee Grounds. *Sep. Purif. Technol.* **2011**, *83*, 173–179. [CrossRef]
23. Górnaś, P.; Dwiecki, K.; Siger, A.; Tomaszewska-Gras, J.; Michalak, M.; Polewski, K. Contribution of Phenolic Acids Isolated from Green and Roasted Boiled-Type Coffee Brews to Total Coffee Antioxidant Capacity. *Eur. Food Res. Technol.* **2016**, *242*, 641–653. [CrossRef]
24. Bilge, G. Investigating the Effects of Geographical Origin, Roasting Degree, Particle Size and Brewing Method on the Physicochemical and Spectral Properties of Arabica Coffee by PCA Analysis. *J. Food Sci. Technol.* **2020**, *57*, 3345–3354. [CrossRef] [PubMed]

25. Muzykiewicz-Szymańska, A.; Nowak, A.; Wira, D.; Klimowicz, A. The Effect of Brewing Process Parameters on Antioxidant Activity and Caffeine Content in Infusions of Roasted and Unroasted Arabica Coffee Beans Originated from Different Countries. *Molecules* **2021**, *26*, 3681. [CrossRef] [PubMed]
26. Pérez-Hernández, L.; Chavez-Quiroz, K.; Medina-Juárez, L.A.; Gámez-Meza, N. Phenolic Characterization, Melanoidins, and Antioxidant Activity of Some Commercial Coffees from *Coffea arabica* and *Coffea canephora*. *J. Mex. Chem. Soc.* **2012**, *56*, 430–435.
27. Singleton, V.L.; Orthofer, R.; Lamuela-Raventós, R.M. Analysis of Total Phenols and Other Oxidation Substrates and Antioxidants by Means of Folin-Ciocalteu Reagent. *Methods Enzymol.* **1999**, *299*, 152–178.

Article

Antioxidant, Antimicrobial Activities and Fatty Acid Compositions of Wild *Berberis* spp. by Different Techniques Combined with Chemometrics (PCA and HCA)

Betül Gıdık

Department of Organic Farming Management, Bayburt University, Bayburt 69000, Turkey; betulgidik@bayburt.edu.tr

Abstract: Interest in medicinal plants and fruits has increased in recent years due to people beginning to consume natural foods. This study aims to investigate the total phenolic flavonoid content, antioxidant activity, condensed tannin content, oil content, and fatty acid compositions of five local breeds of *Berberis* spp. from Bayburt, Turkey, and their antioxidant and antimicrobial activities. The fatty acid composition of samples was performed with gas chromatography-mass spectrometry (GC-MS), and the total fatty acid content of samples was between 6.12% and 8.60%. The main fatty acids in *Berberis* spp. samples were α-linolenic acid (32.85–37.88%) and linoleic acid (30.98–34.28%) followed by oleic acid (12.85–19.56%). Two antioxidant assays produced similar results, demonstrating that extracts of wild *B. vulgaris* L. had the highest ferric reducing antioxidant power (FRAP) (621.02 μmol $FeSO_4 \cdot 7H_2O/g$) and 1,1-diphenyl-2-picrylhydrazyl radical (DPPH) (0.10 SC_{50} mg/mL) values. According to principal component analysis (PCA), four components were determined. In addition, two main groups were determined according to hierarchical cluster analysis (HCA), and wild and culture of *B. vulgaris* L. were in different subgroups. This is the first original report about the fatty acid composition and oil content of *Berberis* spp. grown in Bayburt, Turkey. The obtained results indicate that *B. integerrima* Bunge and *B. vulgaris*, which have especially remarkable fatty acid content, antioxidant, and antimicrobial activity, could be potential sources for these properties in different areas of use.

Keywords: bioactivity; medicinal plants; wild fruits; industrial crops; PCA; HCA

Citation: Gıdık, B. Antioxidant, Antimicrobial Activities and Fatty Acid Compositions of Wild *Berberis* spp. by Different Techniques Combined with Chemometrics (PCA and HCA). *Molecules* **2021**, *26*, 7448. https://doi.org/10.3390/molecules26247448

Academic Editors: Severina Pacifico and Simona Piccolella

Received: 24 October 2021
Accepted: 5 December 2021
Published: 9 December 2021

Publisher's Note: MDPI stays neutral with regard to jurisdictional claims in published maps and institutional affiliations.

Copyright: © 2021 by the author. Licensee MDPI, Basel, Switzerland. This article is an open access article distributed under the terms and conditions of the Creative Commons Attribution (CC BY) license (https://creativecommons.org/licenses/by/4.0/).

1. Introduction

The increasing world population has caused people to search for new food sources. Therefore, wild plants are gaining increasing value. Many scientific studies have been published on the nutritional content and medicinal values of wild edible fruits grown in various parts of the World [1,2]. Interest in wild plants has increased in recent years since people have recently begun consuming natural foods. For this reason, the nutrient content of wild plants should be determined and those that can be considered as a food source should be determined and their cultivation should be encouraged together with breeding studies. In fact, local people in different countries still use the fruits of wild plants to protect their health, and the pharmacological properties of these wild fruits are attributed to phenolic compounds that act as natural antioxidants [3]. When evaluated in this respect, phenolic compounds are antioxidant molecules found in all plants and whose bioactive properties are well known. Therefore, in recent years, research has focused on the identification and measurement of phenolic compounds as medicinal and food molecules in natural plant sources, especially wild plants. The protection provided by the consumption of plant products such as fruits, vegetables, and legumes is mostly related to the presence of phenolic compounds [4]. In addition, due to the increasing use of these compounds in industrial areas (development of functional food and nutraceuticals, etc.), it is very important to determine the natural sources.

The Berberidaceae family contains about 14 genera and 700 species all over the world. In addition, four wild species grow in Turkey: *B. vulgaris* L., *B. cretica* L., *B. crataegina* DC., and *B. integerrima* Bunge [5]. *Berberis* spp. fruits can be eaten raw or cooked and have been used for ornamental purposes, medical as well as food additives, especially in the form of dried fruit [6]. In addition, its rich vitamin content and highly acidic fruits are used to increase the body's resistance. Dried fruits are used as an additive in food and meals, while fresh fruits are used for making jelly, jam, syrup, sauce, fruit juice, and carbonated beverages. However, in recent years, fruits have been used as a colorant in the food industry due to their anthocyanin content [6–8].

Berberis spp. is becoming increasingly important because due to its bioactive properties such as antioxidants, antimicrobials, etc. [9]. For medical purposes, some studies have been carried out on the fruit and root of *Berberis vulgaris*. Berberine, Vitamin C, different vitamins, and salt are known to be isolated from the fruit, root, and peel of *B. vulgaris* [10]. Different parts of *Berberis* spp. plants have a wide range of phenolic compounds, vitamins, and some other metabolites, which are rich in antioxidant activity [11]. Some studies have been put forth that *Berberis* spp. contain the phenols DPPH and FRAP [12,13].

Oils are one of the essential nutrients for humans, and fatty acids form the basis of them. Fatty acids are grouped according to the availability of saturated, monounsaturated, and polyunsaturated fatty acids [14,15]. In addition, an average of 200 fatty acids are detected, and vegetable oils are known to be rich in oleic and linoleic fatty acids [16]. There are few studies published on the fatty acid composition of seeds of *Berberis* spp. [17,18].

In this study, for the first time, the oil content and fatty acid compositions of seeds of wild *Berberis vulgaris* L., *Berberis integerrima* Bunge, *Berberis crataegina* DC., the culture form of *Berberis vulgaris* L., and the hybrid of *Berberis integerrima* × *Berberis crataegina* was determined. With this, total flavonoid content, total phenolic content, and antioxidant activities were determined in order to gain an indication about the bioactive content of the samples. In addition, in vitro antibacterial activities of the samples were determined by agar well diffusion, minimum inhibition concentration (MIC), and minimum bactericidal concentration (MBC) techniques.

2. Results and Discussion

Although plants have been at the forefront of medicinal uses for thousands of years, the active ingredients of many plants are used in the production of drugs in modern technology. For this reason, it is important to determine the potential bioactive properties of different plant sources and, in addition, the bioactive active substances in detail. Especially in recent years, depending on the increase in the world population, the chemical contents of wild plants consumed by local people in different nations are a matter of curiosity, but the amount of research on this subject is limited. Therefore, in this study, some bioactive properties of different *Berberis* species—about which there was limited research before—were elucidated. For *Berberis* spp. used in the study, the total oil content of the samples and also 20 different fatty acids were screened. As a result, the oil content of *Berberis* spp. changed between 6.12% and 8.60% (Table 1). According to this, *B. crataegina* DC. had the lowest ratio and the wild *B. vulgaris* had the highest one. Similarly, the amount of *Berberis integerrima* seed oil content was found to be lower than the results reported in the literature [19]. This situation is thought to be due to the ecological differences between the collecting locations. However, when the individual fatty acid composition of *Berberis* spp. was examined, except for four fatty acids (eicosadienoic acid, arachidonic acid, tricosanoic acid, and nervonic acid), the others were detected at different rates. It is seen that the major fatty acids were α-linolenic acid (32.85 ± 3.31–37.88 ± 1.71), linoleic acid (30.98 ± 1.46–34.28 ± 1.84), and oleic acid (12.85 ± 2.88–19.56 ± 3.88). Fatty acids contain one or more covalent double bonds between carbon-carbon at various positions on the carbon chain are called unsaturated fatty acids. While oleic acid is in the group of monounsaturated fatty acids, linoleic and linolenic acids are among the polyunsaturated fatty acids. The highest amounts of α-linolenic acid, linoleic acid, and oleic acid methyl ester were determined in *B. crataegina*

DC., *B. integerrima* Bunge, and a hibrid of *B. integerrima* × *B. crataegina*, respectively. Our results were found to be higher than the studies by other researchers [20]. The difference may be related to the diversity of *Berberis* species, as well as many factors that affect the development of the plant, such as the geographical characteristics of the cultivation area where the plants grow, soil characteristics.

Table 1. Means and standard error (SE) of oil content and fatty acid compositions of *Berberis* spp.

		B. crataegina DC. (%)	*B. integerrima* × *B. crataegina* ** (%)	*B. integerrima* Bunge (%)	*B. vulgaris* * (%)	*B. vulgaris* (%)
Total Oil Content (%)		6.12 ± 1.66	7.72 ± 0.06	7.90 ± 0.11	8.57 ± 0.78	8.60 ± 0.81
Fatty acids	R.T.					
Butyric acid (C4:0)	2.89	0.12 ± 0.05	0.02 ± 0.05	0.10 ± 0.03	0.05 ± 0.01	0.04 ± 0.02
Caproic acid (C6:0)	4.73	0.21 ± 0.02	0.20 ± 0.01	0.19 ± 0.01	0.15 ± 0.04	0.21 ± 0.02
Undecanoic acid (C11:0)	14.73	0.20 ± 0.04	0.19 ± 0.02	0.14 ± 0.03	0.11 ± 0.05	0.18 ± 0.02
Lauric acid (C12:0)	17.12	0.09 ± 0.04	nd ***	0.07 ± 0.02	0.03 ± 0.01	0.05 ± 0.01
Myristic acid (C14:0)	22.67	0.22 ± 0.03	0.17 ± 0.01	0.18 ± 0.01	0.14 ± 0.04	0.21 ± 0.03
Palmitic acid (C16:0)	28.91	6.35 ± 0.73	5.75 ± 0.12	5.13 ± 0.49	5.52 ± 0.11	5.38 ± 0.25
Heptadecanoic acid (C17:0)	32.07	0.06 ± 0.01	0.68 ± 0.01	0.07 ± 0.01	0.07 ± 0.01	0.06 ± 0.01
Stearic acid (C18:0)	35.28	7.51 ± 0.69	9.20 ± 0.99	7.73 ± 0.47	7.92 ± 0.29	8.65 ± 0.45
Oleic acid (C18:1n9c)	36.61	12.85 ± 2.88	19.56 ± 3.88	14.04 ± 1.70	16.73 ± 0.99	15.49 ± 0.24
Linoleic acid (C18:2n6c)	38.97	33.26 ± 0.81	30.98 ± 1.46	34.28 ± 1.84	32.12 ± 0.33	31.60 ± 0.84
Arachidic acid (C20:0)	41.49	0.32 ± 0.11	0.44 ± 0.01	0.43 ± 0.01	0.46 ± 0.03	0.50 ± 0.07
α-Linolenic acid (C18:3n3)	41.77	37.88 ± 1.71	32.85 ± 3.31	37.09 ± 0.92	36.02 ± 0.14	36.98 ± 0.82
cis-11-Eicosenoic acid (C20:1n9)	42.72	0.17 ± 0.04	0.08 ± 0.05	0.15 ± 0.02	0.11 ± 0.02	0.14 ± 0.01
cis-11,14-Eicosadienoic acid (C20:2)	45.02	nd ***	nd ***	nd ***	nd ***	nd ***
Behenic acid (C22:0)	47.69	0.21 ± 0.03	0.27 ± 0.03	0.21 ± 0.03	0.26 ± 0.02	0.25 ± 0.01
Arachidonic acid (C20:4n6)	49.04	nd ***	nd ***	nd ***	nd ***	nd ***
Tricosanoic acid (C23:0)	50.64	nd ***	nd ***	nd ***	nd ***	nd ***
cis-5,8,11,14,17-Eicosapentaenoic acid (C20:5n3)	51.19	nd ***	0.11 ± 0.07	0.06 ± 0.02	0.06 ± 0.02	0.09 ± 0.05
Lignoceric acid (C24:0)	53.06	0.14 ± 0.10	0.13 ± 0.08	0.12 ± 0.07	0.24 ± 0.19	0.15 ± 0.11
Nervonic acid (C24:1n9)	54.09	nd ***	nd ***	nd ***	nd ***	nd ***

* Culture plant ** Hybrid *** nd: not detected.

Fatty acids that consist of a single covalent bond between carbon-carbon atoms and are generally solid at room temperature are called saturated fatty acids. Lauric acid (C12:0), palmitic acid (C16:0), myristic acid (C14:0), stearic acid (C18:0), behenic acid (C22:0), and arachidic acid (C20:0) found in vegetable oils are the most important saturated fatty acids.

Fatty acids that contain one or more covalent double bonds between carbon-carbon atoms at various positions on the carbon chain are called unsaturated fatty acids [21]. Saturated fatty acids can be synthesized in the human body, and even if no fat is consumed, these types of fatty acids can be synthesized from molecules formed by carbohydrate metabolism. In response to this, unsaturated fats are essential fatty acids that the body needs. They are liquid at room temperature and most of them are of vegetable origin [21,22]. In addition, saturated fatty acids were found at 14.38% and 16.42% total amount, while unsaturated fatty acids were found between 84.16% and 85.62. The values were changed for the saturated/unsaturated fatty acids ratio at 0.17% to 0.20%, for monounsaturated fatty acids at 13.03% to 19.64, and for polyunsaturated fatty acids at 63.94% to 71.43% (Table 2). The results for the saturated/unsaturated fatty acids ratio were similar to some studies [16,23] in the literature. The saturated fatty acid results in this study were higher than those reported by Kaya et al. [20], but they also had monounsaturated fatty acids similar to them. Similarities showed that, in general, the *Berberis* spp. seeds had the same saturated and unsaturated fatty acids ratios.

Table 2. Saturated/unsaturated fatty acids ratio, saturated, unsaturated, monounsaturated, and polyunsaturated fatty acids of *Berberis* spp.

Local Breeds	Saturated Fatty Acids (%)	Unsaturated Fatty Acids (%)	Saturated/Unsaturated Fatty Acid Ratio (%)	Monounsaturated Fatty Acids (%)	Polyunsaturated Fatty Acids (%)
B. crataegina DC.	15.84	84.16	1.88	13.03	71.13
B. integerrima × *B. crataegina* **	16.42	83.58	1.96	19.64	63.94
B. integerrima Bunge	14.38	85.62	1.68	14.19	71.43
B. vulgaris L. *	14.97	85.03	1.76	16.84	68.19
B. vulgaris L.	15.70	84.30	1.86	15.63	68.66

* Culture plant ** Hybrid.

Antimicrobials play an important role in transporting foodstuffs over long distances or extending shelf life. It is thought that antimicrobial compounds obtained from plants can be a healthy alternative in food preservation [24]. The plants that belong to the Berberidaceae were effective in many analyses employed for antioxidant and antimicrobial activity [25–27]. Our results showed that *Berberis* spp. fruit extracts have, in general, an antibacterial effect against all Gram-positive and Gram-negative bacteria at a concentration of 128 mg/mL. When we evaluated species in terms of antibacterial activity, it was observed that *B. integerrima* × *B. crataegina*, *B. integerrima* Bunge, *B. vulgaris* L. (culture), and *B. vulgaris* L. (wild) species had an in vitro inhibitory effect against all selected pathogens. However, it was observed that *B. crataegina* DC. had antibacterial effects against nine bacterial strains from target pathogens, while it did not have antibacterial effects against the remaining nine samples. In addition, when the inhibition zone diameter was evaluated, it was observed that *B. crataegina* DC. had inhibition zone diameters ranging from 10 to 17 mm, and it was observed to have a weaker inhibition effect than other samples. In addition, it has been observed that wild *B. vulgaris* L. generally has a larger inhibition zone diameter compared to other samples, and zone diameters range from 18 to 39 mm. The results of Aliakbarlu et al. [28] were similar to ours, however, Irshad et al. [29] found lower values for inhibition zone diameters of *Berberis* spp. The differences may be caused by the use of different extraction methods or solvents. On the other hand, the species can be effective at these differences. As in previous similar studies, in the results obtained in this study, *Berberis* spp. was found to be suitable for use as a food. In addition, it is thought that it will be an important source for more comprehensive studies to be conducted for *Berberis* spp., whose antimicrobial properties are determined.

Among the Gram-positive bacteria, *Bacillus cereus* ATCC 14579, *Bacillus cereus* BC 6830, *Enterococcus faecalis* ATCC 49452, *Enterococcus faecalis* NCTC 12697, *Streptococcus mutans* ATCC 35668, and *Streptococcus salivarus* ATCC 13419 strains were found to be the most susceptible strains against *Berberis* spp. fruit extracts. It was observed that these strains were sensitive to all extracts, including *B. crataegina* DC. In addition, it was observed that *Enterococcus faecium* ATCC 700211, *Staphylococus aureus* ATCC 25923, *Staphylococcus aureus* NCTC 10788, and *Staphylococcus aureus* BC 7231 strains were resistant to *B. crataegina* DC. (Table 3). The obtained results showed that *Berberis* spp. fruit samples have an antibacterial effect. In addition, it is concluded that it may be beneficial to use these fruits as food supplements for phytotherapy.

Table 3. Minimum inhibition concentration (mg/mL), minimum bactericidal concentration (mg/mL) and inhibition zone diameter (mm) of *Berberis* spp.

	Microorganisms	*B. crataegina* DC.			*B. integerrima* × *B. crataegina*			*B. integerrima* Bunge			*B. vulgaris* L. (Culture)			*B. vulgaris* L. (Wild)		
		IZD	MIC	MBC	IZD	MIC	MBC	IZD	MIC	MBC	IZDX	MIC	MBC	IZD	MIC	MBC
Gram positive	B1	13	-	-	27	16	32	29	16	32	25	32	32	27	32	32
	B2	14	-	-	30	16	32	32	8	32	29	16	32	35	8	16
	B3	17	-	-	29	16	32	33	8	32	30	16	16	39	4	8
	B4	14	-	-	28	16	32	30	16	32	34	8	16	36	8	16
	B5	-	-	-	25	32	-	19	-	-	26	32	32	30	16	32
	B6	-	-	-	20	32	-	21	32	-	23	32	-	24	32	-
	B7	-	-	-	19	32	-	22	32	-	21	32	-	27	16	32
	B8	-	-	-	21	32	-	29	16	32	24	32	-	32	16	32
	B9	13	-	-	26	32	32	23	32	32	26	32	32	26	32	32
	B10	12	-	-	23	32	-	27	16	32	24	32	32	28	16	32
Gram negative	B11	13	-	-	34	8	32	33	8	32	35	8	16	37	4	16
	B12	12	-	-	21	32	-	25	32	32	24	32	32	28	16	32
	B13	11	-	-	18	-	-	20	-	-	19	-	-	23	32	-
	B14	-	-	-	20	-	-	17	-	-	18	-	-	21	32	-
	B15	-	-	-	18	-	-	18	-	-	20	32	-	20	32	-
	B16	-	-	-	20	32	-	19	32	-	19	-	-	22	32	-
	B17	-	-	-	34	8	32	37	4	16	32	16	32	39	4	8
	B18	-	-	-	21	32	-	25	32	32	21	32	-	31	16	32

Minimum inhibition concentration (MIC) and minimum bactericidal concentration (MBC) results. Inhibition zone diameter (IZD). BEE: *Berberis* ethanolic extract; Microorganisms: B1: *Bacillus cereus* ATCC 14579; B2: *Bacillus cereus* BC 6830; B3: *Enterococcus faecalis* ATCC 49452; B4: *Enterococcus faecalis* NCTC 12697; B5: *Enterococcus faecium* ATCC 700211; B6: *Staphylococus aureus* ATCC 25923; B7: *Staphylococcus aureus* NCTC 10788; B8: *Staphylococcus aureus* BC 7231; B9: *Streptococcus mutans* ATCC 35668; B10: *Streptococcus salivarus* ATCC 13419; B11: *Acinetobacter baumannii* ATCC BA 1609; B12: *Escherichia coli* ATCC BAA 25-23; B13: *Escherichia coli* NCTC 9001; B14: *Escherichia coli* BC 1402; B15: *Pseudomonas aeruginosa* ATCC 9070; B16: *Pseudomonas aeruginosa* NCTC 12924; B17: *Salmonella Typhimurium* RSSK 95091; B19: *Yersinia enterocolitica* ATCC 27729.

Antioxidants are bodily defense mechanisms developed to prevent damage caused by the formation of reactive oxygen species (ROS). Antioxidants are substances that prevent the deterioration of the structural and functional molecules in the body, especially lipid, protein, carbohydrates, and DNA, and are effective against free radicals even at low concentrations [30]. Studies in this area of *Berberis* spp. showed that plants are valuable in terms of antioxidant content and these plants are edible. Results of the total TP, TF, CT, FRAP, and DPPH assay of *Berberis* fruits are presented in Table 4. The total phenolic content of the samples was observed to be within the range of 10.84–28.92 mg GAE/g, with the lowest and highest levels observed in *B. crataegina* DC. and *B. vulgaris* L. (wild) samples, respectively. Previous studies also revealed the total phenolic content of *Berberis* spp. fruits, and our results were lower [13,31,32] or higher [24]. TF and CT were found between the range of 0.41–2.20 mg QE/g and 1.75–6.92 CE/g, respectively. *B. crataegina* DC. has the highest value and the results were similar to those of Dimitrijević et al. [13]. FRAP and DPPH were detected in the order of 218.55–621.02 µmol $FeSO_4 \cdot 7H_2O$/g and 0.10–0.36 mg/mL; *Berberis vulgaris* L. (wild) has the highest level of FRAP and the lowest level of DPPH. Previous studies found lower FRAP and DPPH than our results [13,24] or similar results from Konc̆ic' et al. [27]. There is a difference, which, similar to other studies, shows that it can be influenced by geographical conditions and the significance of the samples' origin. Total TP, TF, CT, FRAP, and DPPH values were found to be among the values considered suitable for food use for *Berberis* spp. and especially important among wild plants.

Table 4. Means and standard error (SE), total phenolic content, total flavonoid content, condensed tannin content, and antioxidant activity *Berberis* spp. fruits.

Local Breeds	TP mg GAE/g	TF mg QE/g	CT mg CE/g	FRAP (µmol FeSO$_4$.7H$_2$O/g)	DPPH SC$_{50}$ mg/mL
B. crataegina DC.	10.84 ± 4.35	2.20 ± 0.49	6.92 ± 0.01	236.35 ± 4.64	0.36 ± 0.01
B. integerrima × *B. crataegina* **	16.10 ± 1.71	0.46 ± 0.03	2.93 ± 0.01	218.55 ± 2.46	0.20 ± 0.01
B. integerrima Bunge	17.28 ± 2.23	0.41 ± 0.01	2.00 ± 0.03	305.29 ± 25.08	0.20 ± 0.00
B. vulgaris L. *	15.37 ± 1.52	0.45 ± 0.01	1.75 ± 0.01	349.52 ± 1.49	0.14 ± 0.01
B. vulgaris L.	28.92 ± 1.94	0.56 ± 0.14	1.97 ± 0.01	621.02 ± 25.03	0.10 ± 0.01
Trolox					0.004 ± 0.00

* Culture plant ** Hybrid.

A Pearson correlation analysis was performed to determine the connection between the variables. To determine the relationship between the antioxidant activities of *Berberis* spp. and their oil content, a Pearson correlation analysis was performed. According to the Pearson correlation, a positive correlation was found between TP and FRAP ($p = 0.02$), and CT with DPPH ($p = 0.02$) at $p < 0.05$ level. In addition, a positive correlation was determined between CT and TF ($p = 0.00$) at a $p < 0.01$ level.

While a negative correlation was found between oil content and TF ($p = 0.04$) at a $p < 0.05$ level, negative correlations were found in oil content either with CT ($p = 0.01$) or DPPH ($p = 0.00$) at a $p < 0.01$ level. The correlation results showed that oil content was negatively affected by the amount of TF, CT, and DPPH. On the other hand, TP and FRAP, TF and CT, DPPH and CT, were affected positively. As revealed by the Pearson correlation analysis, although there is a negative interaction between fat ratio and TF, CT, DPPH, the obtained fat ratio and TF, CT, DPPH values show that *Berberis* spp. is suitable for consumption as food in daily life.

PCA is known as a method that reveals the variance structure of the original p variable with fewer new variables that are the linear components of the variables. According to PCA1, four components were determined for fatty acids and saturation of oil content. Eigenvalues and variance percentages of PCA1 and PCA2 analysis are provided in Table 5 and graphs are provided in Figure 1. It is indicated that variance explanation ratios over 70% were sufficient in the PCA analysis [33]. For PCA1, PC1 and PC2 explained 84.28% of the total variation, while PC1 explained 52.69% and PC2 explained 31.60% of the total variation. In PCA2, PC1 explained 51.73%, and PC2 explained 23.21%, and the total variation explained 74.94%. According to Kaiser rules, eigenvalues of greater than 1.0 are accepted as the principal component and descriptor of the variance [34]. In PCA1, eigenvalues of PC1 (11.59), PC2 (6.95), PC3 (2.29) and PC4 (1.16) were greater than 1.0. In PCA2, eigenvalues of PC1 (10.86), PC2 (4.87), PC3 (3.68) and PC4 (1.58) were greater than 1.0.

The Kaiser–Meyer–Olkin (KMO) test was used to measure sampling adequacy in principal component analysis. According to Kaiser [34], the KMO coefficient is unacceptable between 0–0.5, 0.5 is the minimum, between 0.5–0.7 is the medium, between 0.7–0.8 is good, 0.8–0.9 is considered very good, and 0.9 and above is considered perfect. According to this, PC1 was found to be related to butyric acid methyl ester, lauric acid methyl ester, stearic acid methyl ester, oleic acid methyl ester, linoleic acid methyl ester, arachidic acid methyl ester, α-linolenic acid methyl ester, cis11-eicosenoic acid methyl ester, behenic acid methyl ester, *cis*-58111417-eicosapentaenoic acid methyl ester, saturated fatty acids, unsaturated fatty acids, saturated/unsaturated fatty acid ratio, monounsaturated fatty acids, polyunsaturated fatty acids, and oil content, were indexed in PCA1 and found to be related to butyric acid methyl ester, cis58111417 eicosapentaenoic acid methyl ester, lauric acid methyl ester, cis11 eicosenoic acid methyl ester, TF, arachidic acid methyl ester, oleic acid methyl ester, behenic acid methyl ester, DPPH, CT, stearic acid methyl ester, α-linolenic acid methyl ester, linoleic acid methyl ester, palmitic acid methyl ester, and myristic acid methyl ester, which were indexed in PCA2.

Table 5. Eigenvalues and percentage of variance for investigated parameters of PCA analysis.

		PC1	PC2	PC3	PC4
PCA-1	Eigenvalue	11.59	6.95	2.30	1.16
	Variability (%)	52.69	31.60	10.43	5.28
	Cumulative (%)	52.69	84.28	94.72	100.00
PCA-2	Eigenvalue	10.86	4.87	3.68	1.58
	Variability (%)	51.73	23.21	17.54	7.52
	Cumulative (%)	51.73	74.94	92.48	100.00

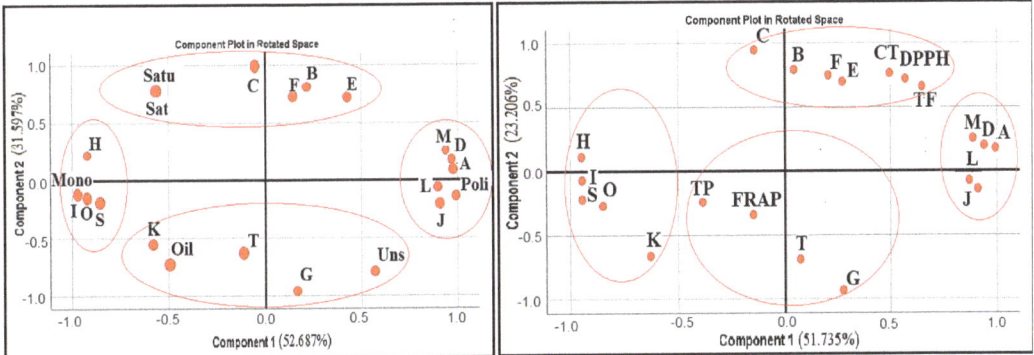

Figure 1. PCA-1 and PCA-2 score plot bases on the fatty acid composition with oil contents and the fatty acids together with antioxidant parameters. Abbreviations and labels: (**A**) Butyric acid methyl ester, (**B**) caproic acid methyl ester, (**C**) undecanoic acid methyl ester, (**D**) lauric acid methyl ester, (**E**) myristic acid methyl ester, (**F**) palmitic acid methyl ester, (**G**) heptadecanoic acid methyl Eeter, (**H**) stearic acid methyl ester, (**I**) oleic acid methyl ester, (**J**) linoleic acid methyl ester, (**K**) arachidic acid methyl ester, (**L**) α-linolenic acid methyl ester, (**M**) cis-11-eicosenoic acid methyl ester, (**O**) behenic acid methyl ester, (**S**) *cis*-58111417 eicosapentaenoic acid methyl ester, (**T**) lignoceric acid methyl ester, (**Sat**) saturated fatty acids, (**Uns**) unsaturated fatty acids, (**Satu**) saturated/unsaturated fatty acid ratio, (**Mono**) monounsaturated fatty acids, (**Poli**) polyunsaturated fatty acids, (**Oil**) oil content, (**TP**) total phenolic content, (**TF**) total flavonoid content, (**CT**) condensed tannin content, (**FRAP**) ferric reducing antioxidant power, (**DPPH**) 1,1-Diphenyl-2-picrylhydrazyl radical.

PCA1 and PCA2 graphs for fatty acid composition and fatty acid composition with antioxidant parameters are presented in Figure 1. Moreover, the score plots of the species distribution according to the main components are shown in Figure 2. The score plot shows the differences between species. Although wild and culture forms of *B. vulgaris* are the same species, there are some differences between them. Moreover, the hybrid of *B. integerrima* × *B. crataegina* is not the same as either *B. integerrima* or *B. crataegina*. All these results are supported by our determinations. According to the results obtained, the presence of both the fatty acids and TF, TP, CT, DPPH, and FRAP values among all the basic components, and that these values determined in *Berberis* spp., underline the need for further study.

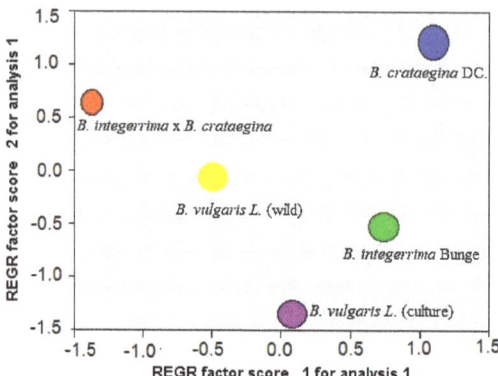

Figure 2. Distribution of the species according to the main components.

HCA is a clustering method that explores the organization of samples. Furthermore, it allows for the determination of similarities and differences within and between groups by depicting a hierarchy [35]. The results of HCA are generally presented in a dendrogram—a plot that shows the organization of samples and their relationships in tree form. There are two common approaches to resolve the grouping problem in HCA: divisive and agglomerative. According to the HCA of *Berberis* spp., two main groups were determined (Figure 3).

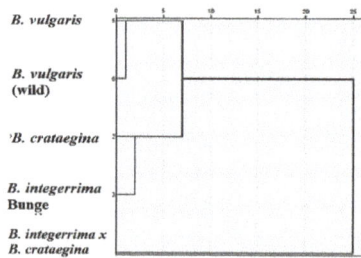

Figure 3. The HCA dendrogram according to the fatty acid compositions of *Berberis* spp.

The first one occurred in *B. integerrima* × *B. crataegina*. The second main group contained two subgroups that were separated from each other. Although culture *B. vulgaris* and wild *B. vulgaris* were in a close subgroup, they were separated from each other. These results indicate that culture and wild form plants cannot be the same. Although *Berberis* spp. are closely related plants, it has been observed that they differ in oil ratios, fatty acid compositions, antioxidant, and antimicrobial properties.

The usage areas of medicinal aromatic and wild plants are expanding day by day. There are issues in this field that have not yet been sufficiently clarified. *Berberis* spp. is among plants that grow wild in the Bayburt region of Turkey and are consumed as food by the public. No study has covered all wild *Berberis* spp. naturally grown in this region. This study, which included all wild *Berberis* spp. in the Bayburt region of Turkey, showed that it is suitable for consumption as food and can be used with different food products.

3. Materials and Methods

3.1. Plant Material

Five different local breeds of *Berberis* sp. (wild *Berberis vulgaris* L., *Berberis integerrima* Bunge, *Berberis crataegina* DC., the culture form of *Berberis vulgaris* L., and the hybrid of *Berberis integerrima* × *Berberis crataegina*) were collected from Bayburt, an eastern Black Sea city in Turkey, during September 2019. The locations were: Kop Mountain Pass, Sancaktepe and Demirozu crossroads, Sirakayalar Village, and Eski Kopuz Village Road, respectively. The altitude, latitude, and longitude data that belong to the collection location of wild *B. vulgaris* L. 2204 m, 40°03′11″ N, 40°28′42″ E, the culture form of *B. vulgaris* L. 1675 m, 40°13′22″ N, 40°03′51″ E, *B. integerrima* Bunge 1882 m, 40°06′01″ N, 40°14′03″ E, *B. crataegina* DC. 1579 m, 40°12′41″ N, 40°15′42″ E, and the hybrid of *B. integerrima* × *B. crataegina* 1592 m, 40°12′01″ N, 40°17′22″ E.

Leaves and stems were separated from fruits, and after drying for two months at room temperature and were stored in a dry environment until analyzed. The fruits and seeds of the *Berberis* sp. were photographed by a binocular microscope to support the taxonomic classification. The fruits and seeds belonging to different *Berberis* spp. are shown in Figure 4.

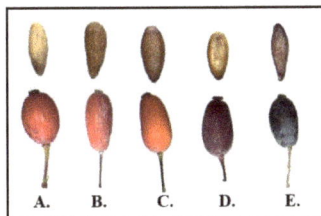

Figure 4. The fruits and seeds of *Berberis* spp. (**A**) *B. vulgaris* L., (**B**) *B. vulgaris* L. *, (**C**) *B. integerrima* Bunge, (**D**) *B. integerrima* × *B. crataegina* **, (**E**) *B. crataegina* DC., * Culture plant ** Hybrid.

3.2. Extraction of Plant Material

Ultrasonic assisted extraction of *Berberis* spp. fruit samples was performed with minor modifications from methods described by Annegowda et al. [36] and Wang et al. [37]. *Berberis* fruit samples were stored in a dry environment until use. Seedless fruit samples were ground in the grinder. Four grams of the samples was transferred to sterile Falcon tubes (15 mL), and 10 mL of ethanol was added to the tube as a solvent. This mixture was kept in an ultrasonic water bath (Kudos) for 30 min at a frequency of 35 kHz at 60 °C. Then, this reaction mixture was centrifuged for 10 min at 10,000 rpm, and the supernatants were carefully transferred to sterile Falcon tubes. The volume was made up to 10 mL with ethanol (95%). Before analysis, a portion of the supernatant was filtrated through a 0.45 μm membrane.

3.3. Determination of Total Phenolic Content (TP)

The total phenolic content in the ethanolic extract of five different *Berberis* samples was determined according to the Folin–Ciocalteu method suggested by Slinkard and Singleton [38]. Shortly, 680 μL of distilled water and 400 μL of 0.5 N Folin–Ciocalteu reagent were added to the samples in a tube. Then, a standard solution and 20 μL of extract were added into this mixture, and after 3 min, 400 μL of solution 10% Na_2CO_3 was added to the tube and the mixture was allowed to stand for 2 h with intermittent shaking. The standard concentration level was 0.03125 mg/mL^{-1} mg/mL. Finally, the absorbance was measured at 760 nm. The results were expressed as mg of gallic acid equivalent per gram of dry weight (mg GAE/g dw).

3.4. Determination of Total Flavonoid Content (TF)

The aluminum chloride colorimetric method for the determination of total flavonoids was modified from the procedure suggested by Fukumoto and Mazza [39]. In a test tube, 0.5 mL of fruit extracts, 0.1 mL of 10% $Al(NO_3)_3$, and 0.1 mL of 1M $NH_4 \cdot CH_3COO$ were combined and incubated at room temperature for 40 min. The standard concentration level was 0.03125 mg/mL^{-1} mg/mL. The absorbance of this reaction mixture was measured at 415 nm with a UV–Vis spectrophotometer. The total flavonoid content was expressed as mg of quercetin equivalents (QE) per gram of sample (mg QE/g).

3.5. Condensed Tannin Content (CT)

Condensed tannins were determined according to Julkunen-Titto's [40] method with small modifications. Twenty-five microliters of each *Berberis* sample solution was mixed with 750 µL of 4% vanillin (prepared with MeOH), then, 375 µL of concentrated HCl was added. The well-mixed solution was incubated at room temperature in the dark for 20 min. The standard concentration level was 0.03125 mg/mL^{-1} mg/mL. The absorbance was then measured at 500 nm against a blank. The results were expressed as mg of catechin equivalents (CE) per gram of dry weight (mg CE/g DW).

3.6. Antioxidant Activity

3.6.1. Ferric Reducing Antioxidant Power (FRAP) Assay

The FRAP assay was carried out according to a modification by Benzie and Strain [41]. The FRAP reagent was prepared by mixing 250 mL of acetate buffer (300 mM, pH 3.6), 25 mL of TPTZ solution (10 mM TPTZ in 40 mM HCl), and 25 mL of FeCl$_3$ (20 mM in water solution). For each *Berberis* spp. fruit sample and each standard, 50µL was added to 1.5 mL of freshly mixed FRAP reagent. The sample was vortexed, and all samples were incubated for 4 min. The standard concentration level was 31.25–1000 µM. The absorbance was measured at 593 nm against a control. The results were expressed as µmol FeSO$_4$.7H$_2$O/g.

3.6.2. 1,1-Diphenyl-2-Picrylhydrazyl Radical (DPPH) Assay

The DPPH assay was performed by using the method of Molyneux [42], with some modifications. The stock solution was prepared by dissolving 4 mg DPPH with 100 mL of 100% methanol. Shortly, for each *Berberis* spp. fruit sample, a 750 µL standard solution was added to a 750 µL DPPH methanolic solution and the mixtures were shaken vigorously and left to stand in the dark for 50 min at room temperature. Then, the absorbance was read at 517 nm. Six different concentrations were used to calculate the inhibition values of the *Berberis* spp. fruit samples. To calculate DPPH radical activity, Trolox was used as a standard.

3.7. Antibacterial Activity

The obtained ethanol extracts (as described in the title "extraction of plant material") were kept at 55 °C for 48 h and the solvents were evaporated. After the removal of the solvents, the obtained active substances were weighed (128 mg) on a precision balance and transferred to 2 mL sterile Eppendorf tubes. At the end of this period, by adding DMSO to the Eppendorf tubes, the total volume was brought to 2 mL. Extracts in these Eppendorf tubes were kept at room temperature for 24 h. After vortexing, the prepared DMSO extracts were used for in vitro antibacterial activity tests.

3.8. Microorganisms and Growth Condition

In vitro antibacterial activities of the *Berberis*, fruit samples were determined using 10 g-positive and 8 g-negative bacteria. Selected target pathogen strains were cultured for 24 h at 37 °C using Mueller Hinton Broth (MHB, Oxoid). The suspensions were adjusted to a standard turbidity of 0.5 McFarland (106 CFU/mL) and used as inoculum [43]. The microorganisms used in the study were obtained from the Department of Medical Services and Techniques, Vocational School of Health Services, Bayburt University, Turkey.

3.9. Screening for Antibacterial Activity

To determine the in vitro antibacterial activity of the *Berberis* spp. fruit samples, the agar-well diffusion (AWD) method was used. For this purpose, 8 mm diameter wells were cut into the sterile Mueller Hinton Agar (MHA) mediums using a sterile cork borer [39]. After these processes, inoculums (0.5 McFarland Turbidity Standard—10^6 CFU/mL) were seeded using sterile swabs. Next, 100 µL of DMSO extracts of the *Berberis* fruit extracts was transferred to wells and incubated at 37 °C for 24 h. Following the incubation period, the observed inhibition zones around the wells were measured with a Vernier caliper and recorded. In addition, DMSO was used as a negative control in this process. Each assay was carried out in duplicate [44].

3.10. Determination of Minimum Inhibition Concentration (MIC)

MIC values of *Berberis* spp. fruit extract were determined by the broth dilution method using 96-well round-bottom polystyrene microplates. For this purpose, first, 95 µL of sterile MHB was distributed to each well of the 96 well microtiter plates. Then, overnight-grown pathogenic microorganisms were adjusted to 0.5 McFarland turbidity, and 5 µL of inoculums were added to each well. As a result of these applications, 100 µL of inoculum + MHB medium solution was prepared in each well of a 96-well microplate. Then, 100 µL of extract (64 mg/mL of DMSO extracts) was added to all of the first wells and mixed at least three times. Afterward, 100 µL of the mixture was taken from the first well via a micropipette and transferred to the second well. This procedure was repeated successively up to the eighth well. In this manner, the starting concentration of *Berberis* fruit samples extracts was diluted at each step. After these applications, the absorbance values of the microplate were measured and recorded at a 600 nm wavelength (Thermo, Multiskan Go). The microplates were then incubated at 37 °C for 24 h. At the end of the incubation period, absorbance values were again measured and recorded. The first well, where the absorbance values increased, was considered as non-bactericidal or non-bacteriostatic concentrations, and the concentration of the upper wells was accepted as the MIC values [45,46].

3.11. Determination of Minimum Bactericidal Concentration (MBC)

After determining the MIC values of the *Berberis* spp. fruit extracts, to determine the MBC values, 10 µL suspension was removed from each well of the test microplates and transferred to the nutrient agar (NA) medium. After this process, inoculated Petri dishes were incubated at 37 °C for 24 h. At the end of this incubation period, a minimum concentration that bacterial growth was not observed was accepted as MBC. All these assays were performed twice [47].

3.12. Oil Extraction and Preparation of Methyl Esters

The *Berberis* spp. fruits were chosen randomly from the collected samples and separated seeds to detect the fatty acid composition. Ten grams of the dried and completely ground *Berberis* spp. seed samples was placed in the cartridge in the extractor section of the soxhlet apparatus and extracted with hexane for 6 h. The solvent of the resulting hexane–oil mixture was removed with the evaporator. To prevent hexane from remaining in the mixture obtained, it was kept in the oven for 1 h at 90 °C. When the process was completed, calculations were made by weighing the balloon, previously weighted, again and the oil content was defined. To determine the fatty acid composition of the obtained oils, 20 mg was taken and dissolved in 5 mL of hexane. To make methyl esters, 5 mL of methanol was prepared with 2N KOH and shook strongly. Gas chromatography was analyzed with a flame ionization detector (GC-FID) by taking the upper phase [48,49].

3.13. Statistical Analysis

Statistical analysis of this study was performed using the SPSS 25.0 software program. Descriptive statistics of TP, TF, CT, FRAP, DPPH, and oil content of *Berberis* spp. were determined, and Pearson correlation analysis was performed, in addition to calculating the standard deviations errors of the data. In this study, principal component analysis (PCA) was applied twice to determine both the main components of the fatty acid composition and fatty acids together with antioxidant parameters. In addition, hierarchical cluster analysis (HCA) was applied to determine the relationship between *Berberis* spp. In this study, principal component analysis (PCA) was applied two times to determine both the main components of the fatty acid composition and fatty acids together with antioxidant parameters. In addition, hierarchical cluster analysis (HCA) was applied to determine the relationship of *Berberis* spp.

4. Conclusions

Wild plants and medicinal aromatic plants have been used for many purposes from the past to the present. It is important to determine the fatty acid composition, antioxidant, and antimicrobial activities of these plants when consumed as food. In this study, wild *B. vulgaris* L., *B. integerrima* Bunge, *B.crataegina* DC., *B. integerrima*, the culture form of *B. vulgaris* L., and the hybrid of *B. integerrima* × *B. crataegina* were collected from Bayburt Province, Turkey. Their oil content, fatty acid compositions, antimicrobial, and antioxidant properties were detected. The oil contents, fatty acids, and phytochemical characteristics of the *Berberis* spp. fruits were affected by the variety of local breeds, species, and the collection location, considering the role of geographical conditions. The PCA and HCA results showed that the wild and cultured plants are different, and the *B. integerrima* × *B. crataegina* hybrid was diagnosed significantly different. PCA1 and PCA2 showed that fatty acids and antioxidants are principal components for *Berberis* spp. In addition, differences in fatty acids and antioxidants between hybrid and culture *Berberis* spp. support the importance of varieties. All the results obtained in this study suggest that wild *Berberis* spp. fruits collected from nature are suitable for use in the food industry.

Funding: This research received no external funding.

Institutional Review Board Statement: Not applicable.

Informed Consent Statement: Not applicable.

Data Availability Statement: Not available.

Conflicts of Interest: The author declare no conflict of interest.

Sample Availability: Samples of the compounds are not available from the authors.

References

1. Kamiloglu, O.; Ercisli, S.; Sengul, M.; Toplu, C.; Serce, S. Total phenolics and antioxidant activity of jujube (*Zizyphus jujube* Mill.) genotypes selected from Turkey. *Afr. J. Biotechnol.* **2009**, *8*, 303–307.
2. Tosun, M.; Ercisli, S.; Karlidag, H.; Şengül, M. Characterization of Red Raspberry (*Rubus idaeus* L.) Genotypes for Their Physicochemical Properties. *J. Food Sci.* **2009**, *74*, C575–C579. [CrossRef]
3. Ahmad, N.; Zuo, Y.; Anwar, F.; Abbas, A.; Shahid, M.; Hassan, A.A.; Bilal, M.; Rasheed, T. Ultrasonic-assisted extraction as a green route for hydrolysis of bound phenolics in selected wild fruits: Detection and systematic characterization using GC–MS–TIC method. *Process. Biochem.* **2021**, *111*, 79–85. [CrossRef]
4. Khoddami, A.; Wilkes, M.A.; Roberts, T.H. Techniques for Analysis of Plant Phenolic Compounds. *Molecules* **2013**, *18*, 2328–2375. [CrossRef] [PubMed]
5. Christenhusz, J.M.; Byng, J.W. The number of known plants species in the world and its annual increase. *Phytotaxa* **2016**, *261*, 201–217. [CrossRef]
6. Sharifi, A.; Niakousari, M.; Mortazavi, S.A.; Elhamirad, A.H. High-pressure CO_2 extraction of bioactive compounds of bar-berry fruit (*Berberis vulgaris*): Process optimization and compounds characterization. *J. Food Meas. Charac-Terization* **2019**, *13*, 1139–1146. [CrossRef]
7. Kafi, M.; Balandary, A.; Rashed-Mohasel, M.H.; Koochaki, A.; Molafilabi, A. *Berberis: Production and Processing*; Zaban va adab Press: Mashhad, Iran, 2002; pp. 1–209.
8. Alemardan, A.; Asadi, W.; Rezaei, M.; Tabrizi, L.; Mohammadi, S. Cultivation of Iranian seedless barberry (*Berberis inte-gerrima* 'Bidaneh'): A medicinal shrub. *Ind. Crop. Prod.* **2013**, *50*, 276–287. [CrossRef]
9. Srivastava, S.K.; Rai, V.; Srivastava, M.; Rawat, A.K.S.; Mehrotra, S. Estimation of Heavy Metals in Different *Berberis* Species and Its Market Samples. *Environ. Monit. Assess.* **2006**, *116*, 315–320. [CrossRef] [PubMed]
10. Özgen, M.; Saracoglu, O.; Geçer, E.N. Antioxidant capacity and chemical properties of selected barberry (*Berberis vulgaris* L.) fruits. *Hortic. Environ. Biotechnol.* **2012**, *53*, 447–451. [CrossRef]
11. Dashti, Z.; Shariatifar, N.; Nafchi., A.M. Study on antibacterial and antioxidant activity of *Berberis vulgaris* aqueous extracts from Iran. *Int. J. Pharma Sci. Res.* **2014**, *5*, 705–708.
12. Brighente, I.; Dias, M.; Verdi, L.; Pizzolatti, M. Antioxidant Activity and Total Phenolic Content of Some Brazilian Species. *Pharm. Biol.* **2007**, *45*, 156–161. [CrossRef]

13. Dimitrijević, M.V.; Mitić., V.D.; Ranković, G.Ž.; Miladinović, D.L. Survey of Antioxidant Properties of Barberry: A Chemical and Chemometric Approach. *Anal. Lett.* **2020**, *53*, 671–682. [CrossRef]
14. Burdge, G.C.; Calder, P. Conversion of α-linolenic acid to longer-chain polyunsaturated fatty acids in human adults. *Reprod. Nutr. Dev.* **2005**, *45*, 581–597. [CrossRef] [PubMed]
15. Mišurcová, L.; Ambrožová, J.; Samek, D. Seaweed lipids as nutraceuticals (Chapter 27). In *Advances in Food and Nutrition Research*, 1st ed.; Se-Kwon, K., Ed.; Academic Press: Burlington, NJ, USA, 2011; Volume 64, pp. 339–355. [CrossRef]
16. Kayahan, M. *Oil Chemistry*; METU Publishing: Ankara, Turkey, 2003; p. 220.
17. Sztarker, N.D.; Cattaneo, P. Chemical composition of *Berberis buxifolia* Lam. (calafate) ripe fruits. *Anales de la Asociación Química Argentina* **1976**, *64*, 281.
18. Mazzuca, M.; Kraus, W.; Balzaretti, V.T. Fatty Acids and Sterols in Seeds from Wild Species of *Berberis* in Argentine Patagonia. *J. Herb. Pharmacother.* **2003**, *3*, 31. [CrossRef] [PubMed]
19. Atefeh, T.; Mohammad, A.S.; Mohsen, B. Antioxidant activity of *Berberis integerrima* seed oil as a natural antioxidant on the oxidative stability of soybean oil. *Int. J. Food Prop.* **2017**, *20*, 2914–2925.
20. Kaya, M.; Ravikumara, P.; Ilk, S.; Mujtabaa, M.; Akyuz, L.; Labidie, J.; Salaberriae, A.M.; Cakmak, Y.S.; Erkul, S.K. Production and characterization of chitosan based edible films from *Berberis crataegina*'s fruit extract and seed oil. *Innov. Food Sci. Emerg. Technol.* **2018**, *45*, 287–297. [CrossRef]
21. Karaca, E.; Aytac, S. The factors affecting on fatty acid composition of oil crops. *J. Agric. Fac. Ondokuz Mayıs Univ.* **2007**, *22*, 123–131.
22. Nas, S.; Gokalp, Y.H.; Unsal, M. *Vegetable Oil technology*; Pamukkale University Faculty of Architecture Printing House: Denizli, Turkey, 2001; p. 322.
23. Apmd, S. Asia Pacific. *J. Clin. Nutr.* **2002**, *11*, 163–173.
24. Eroglu, A.Y.; Cakir, O.; Sagdıc, M.; Dertli, E. Bioactive Characteristics of Wild *Berberis vulgaris* and *Berberis crataegina* Fruits. *Hindawi J. Chem.* **2020**, *2020*, 8908301.
25. Freile, M.; Giannini, F.; Sortio, M.; Zamora, M.; Juárez, A.; Zacchino, S.; Enriz, D. Antifungal activity of aqueous extracts and of berberine isolated from *Berberis heterophylla*. *Acta Farm. Bonaer.* **2006**, *25*, 83.
26. Iauk, L.; Costanzo, R.; Caccamo, F.; Rapisarda, A.; Musumeci, R.; Milazzo, I.; Blandino, G. Activity of *Berberis aetnensis* root extracts on Candida strains. *Fitoterapia* **2007**, *78*, 159–161. [CrossRef]
27. Končić, M.Z.; Kremer, D.; Karlović, K.; Kosalec, I. Evaluation of antioxidant activities and phenolic content of *Berberis vulgaris* L. and *Berberis croatica* Horvat. *Food Chem. Toxicol.* **2010**, *48*, 2176–2180. [CrossRef] [PubMed]
28. Aliakbarlu, J.; Ghiasi, S.; Bazargani-Gilani, B. Effect of extraction conditions on antioxidant activity of barberry (*Berberis vulgaris* L.) fruit extracts. *Vet. Res. Forum* **2018**, *9*, 361–365.
29. Irshad, A.H.; Pervaiz, A.H.; Abrar, Y.B.; Fahelboum, I.; Awen, B.Z.S. Antibacterial Activity of *Berberis lycium* Root Extract. *Trakia J. Sci.* **2013**, *1*, 88–90.
30. Vinson, J.A. Oxidative stress in cataracts. *Pathophysiology* **2006**, *13*, 151–162. [CrossRef]
31. Yildiz, H.; Ercisli, S.; Sengül, M.; Topdas, E.F.; Beyhan, O.; Cakir, O.; Narmanlioglu, H.K.; Orhan, E. Some Physicochemical Characteristics, Bioactive Content and Antioxidant Characteristics of Non-Sprayed Barberry (*Berberis vulgaris* L.) Fruits from Turkey. *Erwerbs-Obstbau* **2014**, *56*, 123–129. [CrossRef]
32. Ersoy, N.; Kupe, M.; Sagbas, H.I.; Ercisli, S. Physicochemical Diversity among Barberry (*Berberis vulgaris* L.) Fruits from Eastern Anatolia. *Not. Bot. Horti Agrobot. Cluj-Napoca* **2018**, *46*, 336–342. [CrossRef]
33. Larrigaudiere, C.; Lentheric, I.; Puy, J.; Pinto, E. Biochemical characterization of core browning and brown heart disorder in pear by multivariate analysis. *Postharvest Biol. Technol.* **2004**, *31*, 29–39. [CrossRef]
34. Kaiser, H.F. The Application of Electronic Computers to Factor Analysis. *Educ. Psychol. Meas.* **1960**, *20*, 141–151. [CrossRef]
35. Lee, I.; Yang, J. Common clustering algorithms. In *Comprehensive Chemometrics*; Brown, S.D., Tauler, R., Walczak, B., Eds.; Elsevier: Oxford, UK, 2009; pp. 577–618.
36. Annegowda, H.V.; Mordi, M.N.; Ramanathan, S.; Hamdan, M.R.; Mansor, S.M. Effect of extraction techniques on phenolic content, antioxidant and antimicrobial activity of Bauhinia purpurea: HPLC determination of antioxidants. *Food Anal. Methods* **2012**, *5*, 226–233. [CrossRef]
37. Wang, K.; Xie, X.; Zhang, Y.; Huang, Y.; Zhou, S.; Zhang, W.; Lin, Y.; Fan, H. Combination of microwave-assisted extraction and ultrasonic-assisted dispersive liquid-liquid microextraction for separation and enrichment of pyrethroids residues in Litchi fruit prior to HPLC determination. *Food Chem.* **2018**, *240*, 1233–1242. [CrossRef]
38. Slinkard, K.; Singleton, V.L. Total phenol analyses: Automation and Comparison with Manual Methods. *Am. J. Enol. Vitic.* **1977**, *28*, 49–55.
39. Fukumoto, L.R.; Mazza, G. Assessing Antioxidant and Prooxidant Activities of Phenolic Compounds. *J. Agric. Food Chem.* **2000**, *48*, 3597–3604. [CrossRef]
40. Julkunen-Titto, R. Phenolic constituents in the leaves of northern willows: Methods for the analysis of certain phenolics. *J. Agric. Food Chem.* **1985**, *33*, 213–217. [CrossRef]
41. Benzie, I.F.F.; Strain, J.J. [2] Ferric reducing/antioxidant power assay: Direct measure of total antioxidant activity of biological fluids and modified version for simultaneous measurement of total antioxidant power and ascorbic acid concentration. *Methods Enzymol.* **1999**, *299*, 15–27. [CrossRef] [PubMed]

42. Molyneux, P. The use of the stable free radical diphenylpicrylhyrazyl (DPPH) for estimating antioxidant activity, Songklanakarin. *J. Sci. Technol.* **2004**, *26*, 211–219.
43. Sherlock, O.; Dolan, A.; Athman, R.; Power, A.; Gethin, G.; Cowman, S.; Humphreys, H. Comparison of the antimicrobial activity of Ulmo honey from Chile and Manuka honey against methicillin-resistant *Staphylococcus aureus*, *Escherichia coli* and *Pseudomonas aeruginosa*. *BMC Complement. Altern. Med.* **2010**, *10*, 47. [CrossRef]
44. Osés, S.M.; Pascual-Maté, A.; de la Fuente, D.; de Pablo, A.; Fernández-Muiño, M.A.; Sancho, M.T. Comparison of methods to determine antibacterial activity of honeys against *Staphylococcus aureus*. *NJAS-Wagening. J. Life Sci.* **2016**, *78*, 29–33. [CrossRef]
45. Zgoda, J.; Porter, J. A Convenient Microdilution Method for Screening Natural Products against Bacteria and Fungi. *Pharm. Biol.* **2001**, *39*, 221–225. [CrossRef]
46. Şahin, F.; Karaman, I.; Güllüce, M.; Öğütçü, H.; Şengül, M.; Adiguzel, A.; Öztürk, S.; Kotan, R. Evaluation of antimicrobial activities of *Satureja hortensis* L. *J. Ethnopharmacol.* **2003**, *87*, 61–65. [CrossRef]
47. Rakholiya, K.; Vaghela, P.; Rathod, T.; Chanda, S. Comparative Study of Hydroalcoholic Extracts of *Momordica charantia* L. against Foodborne Pathogens. *Indian J. Pharm. Sci.* **2014**, *76*, 148–156. [PubMed]
48. Regulation, H. Commission Regulation (EEC) No. 2568/91 of 11 July 1991 on the characteristics of olive oil and olive-residue oil and on the relevant methods of analysis. *Off. J. L.* **1991**, *248*, 1–83.
49. Paquot, C. *Standard method 2.301. Standards Methods for the Analysis of Oils, Fats and Derivatives*, 7th ed.; International Union of Pure and Applied Chemistry, Blackwell: Oxford, UK, 1992; p. 301.

Article

Effect of Spray Dryer Scale Size on the Properties of Dried Beetroot Juice

Jolanta Gawałek

Department of Dairy and Process Engineering, Poznań University of Life Sciences, 60-624 Poznań, Poland; jolanta.gawalek@up.poznan.pl

Abstract: Experiments detailing the spray drying of fruit and vegetable juices are necessary at the experimental scale in order to determine the optimum drying conditions and to select the most appropriate carriers and solution formulations for drying on the industrial scale. In this study, the spray-drying process of beetroot juice concentrate on a maltodextrin carrier was analyzed at different dryer scales: mini-laboratory (ML), semi-technical (ST), small industrial (SI), and large industrial (LI). Selected physicochemical properties of the beetroot powders that were obtained (size and microstructure of the powder particles, loose and tapped bulk density, powder flowability, moisture, water activity, violet betalain, and polyphenol content) and their drying efficiencies were determined. Spray drying with the same process parameters but at a larger scale makes it possible to obtain beetroot powders with a larger particle size, better flowability, a color that is more shifted towards red and blue, and a higher retention of violet betalain pigments and polyphenols. As the size of the spray dryer increases, the efficiency of the process expressed in powder yield also increases. To obtain a drying efficiency >90% on an industrial scale, process conditions should be selected to obtain an efficiency of a min. of 50% at the laboratory scale or 80% at the semi-technical scale. Designing the industrial process for spray dryers with a centrifugal atomization system is definitely more effective at the semi-technical scale with the same atomization system than it is at laboratory scale with a two-fluid nozzle.

Keywords: spray drying; vegetable powders; beetroot; natural colorants; violet betalain pigments; polyphenols

1. Introduction

The food market is subject to constant changes, which results, among others, from the influence of current global trends and greater consumer knowledge and awareness, as well as changes in the behaviors and lifestyles of society. Low-quality goods containing large amounts of flavor enhancers, preservatives, or artificial colorants are becoming less and less popular. Another trend, resulting from the lifestyle of a wide range of people, is the increased popularity of foods that are convenient for quick preparation, including instant foods. In this type of food, spray-dried fruit and vegetable powders are gaining popularity as ingredients in final food products. These act as carriers for flavor, aroma, and color and also have bioactive properties. One of the most important examples of these carriers is spray dried beetroot (*Beta vulgaris*) juice, which is the main ingredient in many instant soups but also in a wide range of food products, where it plays the role of a natural colorant. The development of instant meals in the global market has been characterized by very dynamic growth in recent years. It manifests itself as a huge increase in diversity, both in terms of taste and in the form of administration and packaging. As a result, the ingredients of such foods are subject to increasingly higher quality requirements so that they can meet the higher technological demands for the production of attractive, natural, and healthy instant foods. The use of a given food ingredient in the final product usually involves the precise design of its properties, and this, in turn, requires a well-designed spray drying process for that ingredient.

Spray drying is the most widely used technique for liquid food drying and encapsulation. The advantage of this method is related to its economy, flexibility, and possibility of continuous operation [1–3]. It allows the stability of the product to be increased, reducing volume of the product and extending its shelf life, which also facilitates product handling and also allows products to be stored at ambient temperature [1,3–5]. When considering the spray drying of fruit and vegetable juices, there are technological problems that are associated with the hygroscopic and thermoplastic properties of the juices (product sticking and settling on the walls of the dryer, reduction of drying efficiency and powder stability, operational problems of the dryer) [6–8]. The reason for this is the content of the organic acids and low molecular weight sugars with a low glass transition temperature that are contained in the fruit and vegetable juices [9,10]. A known way to cope with these difficulties is to add drying carriers that have high molecular weight and that can effectively lower the glass transition temperature of the dried juice [11]. The most commonly used carriers in spray drying processes are polysaccharides, proteins, and lipids [12–14]. For fruit and vegetable juice drying, polysaccharides such as hydrolyzed starches or gums are the most preferred due to their chemical similarity and plant origin but also because of their lower purchase cost compared to other carriers.

Designing the beetroot juice spray drying process and the specific properties of the resulting beetroot powder comes down to the optimal selection of the main raw material—beetroot juice concentrate, the selection of the carrier, and the selection of appropriate drying process conditions. This is often first completed on a laboratory and/or semi-technical scale, and the drying is then transferred to an industrial scale of an appropriate size, depending on the planned production batch. When designing a new industrial dryer for a specific product, laboratory tests are necessary to acquire the needed information about the behavior of the dried materials [7]. The laboratory and semi-technical scale tests performed help to design the industrial process. However, differences in product quality at different scales present some difficulties in the development and production of powder products [15]. This is largely due to the fact that spray drying involves a combination of complex physical transformations that include the mixing of the main dried component and excipients in a suitable solvent, liquid atomization, solvent evaporation (drying), gas–particle separation, and often secondary drying, which may take place in a fluidized bed dryer [16]. Scale-up studies of spray dryers (particularly in the field of pharmaceutical drying) can be found in the literature and include specific process scaling-up design procedures [15–21]. However, the limited number of literature reports on industrial scale spray drying is due to the high cost of conducting such experiments and validating theoretical models [21].

This work focuses on the determination and comparison of the physicochemical properties of dried beetroot powders that were obtained in dryers of different scales, ranging from the laboratory to large industrial scales. Determining the trends of changes in individual functional properties for this type of product (natural colorants of plant origin) will enable more the precise planning of production processes for different production batches. As mentioned above, the quality requirements for such powdered food ingredients are very high; in particular, repeatability is very important. A given product must be functionally the same regardless of whether it is produced in a large or small production batch. Additionally, it is known that the most economical drying takes place when the appropriate batch size is matched to the corresponding size of the spray dryer.

2. Results and Discussion

2.1. Powder Yield

Powder yield is a very important indicator that is related to the efficiency and economy of the production process. Designing an industrial spray drying process requires achieving a high powder yield that is at least above 90% for the production to make economic sense, and preferably, the production yield is above 97%. In other words, carriers, recipes, and drying conditions should be selected in such a way so as to minimize product losses to at least below 10%, and preferably below 3%. The determined powder yields for the conducted beetroot juice spray drying processes at different scales are presented in Figure 1. The differences in the values for individual spray dryers show statistical significance, with the larger spray-drying scale allowing for significantly higher powder yields. For both industrial-scale dryers (SI and LI), powder yields above 90% were achieved, which indicates well-chosen drying conditions for the studied beetroot juice concentrate. The powder yield values for the experimental driers on the laboratory (ML) and semi-technical (ST) scales are much lower, i.e., 54.3 and 81.6%, respectively. Other researchers who conducted research on laboratory-scale beetroot juice achieved powder yield values in the ranges of 15–18% (Poornima et al. [22]), 41.31–54.63% (Bazaria et al. [23]), and 41.33–56.55% (Teenu et al. [24]). Jedlińska et al. [25] conducted a semi-technical-scale experiment in which they spray-dried cloudy beetroot juice with dehumidified air at low inlet temperatures (90 °C and 130 °C). The temperature of 130 °C resulted in a powder yield of 76.1% and 90%, with and without maltodextrin as a carrier, respectively. However, it is noteworthy that the feed solutions in this study had a much lower concentration, i.e., 13.3 °Bx with the carrier and 8 °Bx without the carrier. As seen from the study results, the semi-technical scale definitely approximates (simulates) the industrial process better, which is also confirmed by the above-mentioned studies from other authors.

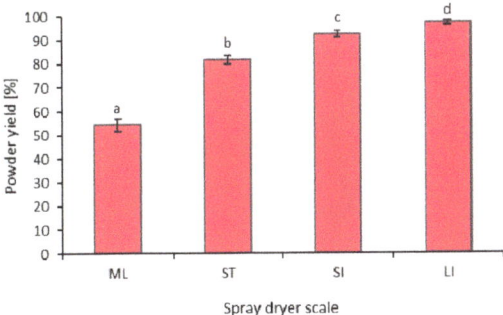

Figure 1. Powder yield of spray dried beetroot juice at different process scales (ML—mini-laboratory scale, ST—semi-technical scale, SI—small industrial scale, LI—large industrial scale; a, b, c, d—different letters show significant differences between spray dryer scale ($p \leq 0.05$)).

From this analysis of the powder yield changes, the values that must be achieved at the experimental scales to obtain satisfactory spray drying efficiency at industrial scales. It can be assumed that for the laboratory scale, it is a min. of 50%, and for the semi-technical scale, the min. is 80%. Achieving such drying effectiveness at the experimental scale will enable the transfer of the determined optimum drying conditions to the industrial scale, allowing the production effectiveness indicators of >90% to be achieved.

2.2. Microstructure of Particles

Figure 2 shows microphotographs of beetroot powder particles for different spray drying-scale sizes. The resulting particles are of various sizes and spherical shapes and have numerous folds and cracks, which are typical for crop products that have been spray-dried on maltodextrin carriers. The microstructure of the particles for the three dryers with a centrifugal liquid atomization system (ST, SI, LI) shows no significant differences, while the microstructure of the beetroot powder particles produced in the laboratory dryer (ML) is mildly different. In the first case, more cracks and wrinkles can be observed on the surface. This is related to the large loss of moisture from the particles and their subsequent cooling [25,26]. The use of the laboratory dryer (ML) resulted in much smaller but more spherical and smoother particles. This may have been caused by a different atomization system: a two-fluid nozzle, where the dried liquid is atomized by a stream of compressed air, is conducive to the formation of more small particles. The more developed evaporation surface favours a less violent exit of the contained moisture through the particle surface, which reduces surface shrinkage. Figure 3 compares the particle morphology of two different atomization systems (two-fluid nozzle and centrifugal liquid atomization) at a higher magnification (1000×).

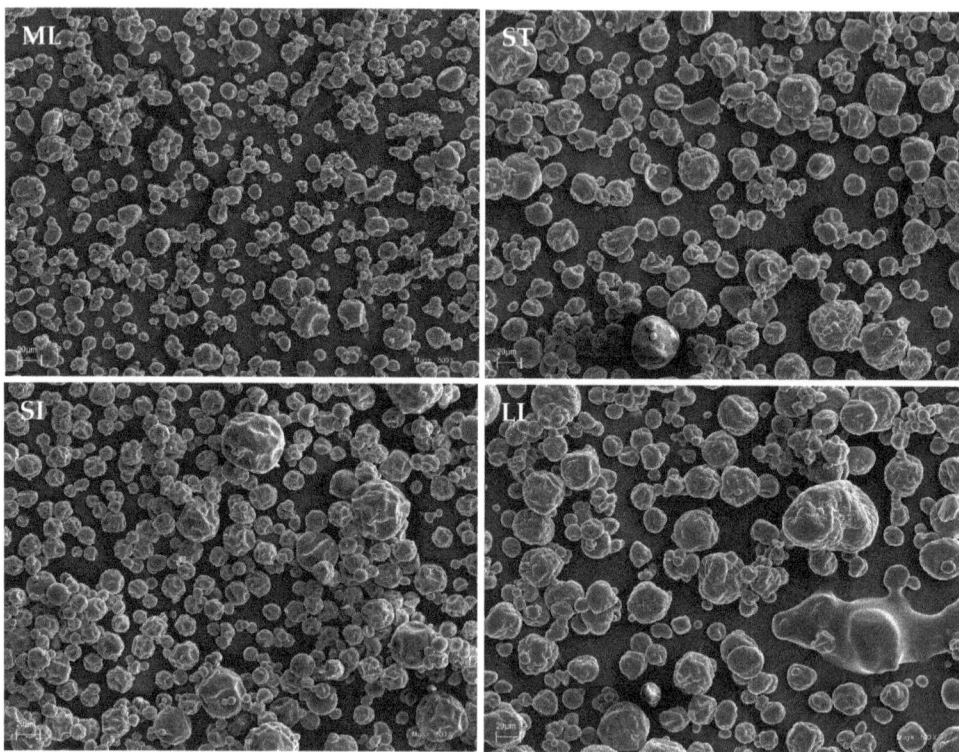

Figure 2. Microstructure of spray-dried beetroot juice particles (magnification 500×) at different process scales (ML—mini-laboratory scale, ST—semi-technical scale, SI—small industrial scale, LI—large industrial scale).

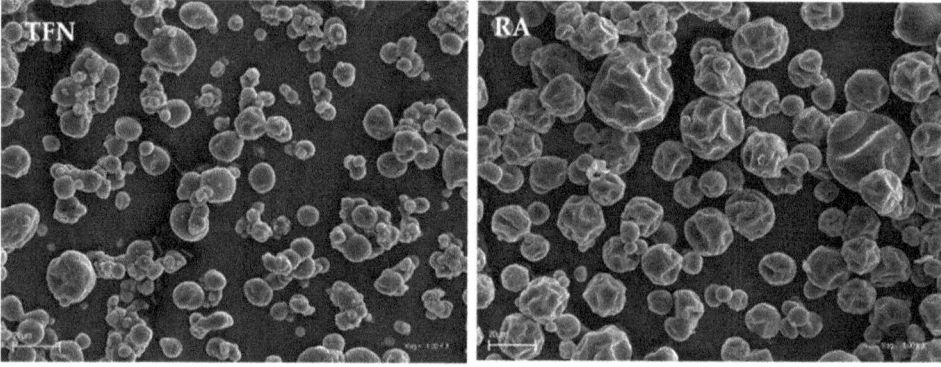

Figure 3. A comparison of particle morphology of two different liquid spray systems (magnification 1000×): two-fluid nozzle (TFN) and rotary atomizer (RA).

Jedlińska et al. [25] obtained particles with similar morphology. The researchers conducted a semi-technical-scale experiment in which they also spray-dried beet juice with a centrifugal liquid atomization system. The procedure resulted in smaller particles and fewer shrunken particles than when the ST, SI, and LI dryers were used due to the higher rotary atomizer speed, lower carrier content, and lower air inlet temperature than the ones used in our study. Janiszewska [27] also dried beetroot juice with a rotary atomizer on a semi-technical scale. The researcher used a higher content of maltodextrin as a carrier and a higher temperature (the drying conditions were closer to the parameters of our study). The microstructure of the particles was very similar to the microstructures obtained with the ST, SI, and LI spray dryers.

Singh and Hathan [26] analyzed the microstructure of the spray-dried beetroot juice particles obtained with a two-fluid nozzle laboratory dryer. The particles obtained in this study were similar in size and exhibited a similar tendency for agglomeration but had much greater shrinkage than the particles obtained with the ML dryer. The difference may have been caused the higher carrier content.

2.3. Particle Size Distribution

The particle size of the beetroot powder obtained by spray drying is a key parameter that affects several of the powders' functional properties (e.g., bulk density, flowability, compressibility, solubility, and hygroscopicity) [28]. The quantities characterizing the cumulative volumetric particle size distribution are summarized in Table 1. The determined mean values of particle size D[4,3] ranged from 11.8–43.2 μm and increased with increasing dryer size. However, the particle size of the beetroot powder obtained in the laboratory dryer (ML) was much smaller (11.8 μm) than the others (36.1–43.2 μm). Furthermore, the span of the particle size distribution for the laboratory dryer (ML) was much larger (span = 3.03) compared to the other dryers that were tested (ST, SI, LI), which achieved span values in the range of 1.93–2.19. The markedly different nature of the particle size distribution for the ML dryer may not only be due to the much smaller size but also due to the different liquid atomization system. In this case, liquid atomization is performed in a nozzle system, while in the other dryers, it is performed in a centrifugal rotary atomizer system. Janiszewska et al. [27,29] conducted a semi-technical-scale experiment with centrifugal liquid atomization. The researchers used similar amounts of maltodextrin as a carrier in beetroot juice and obtained much smaller particle sizes (9.36–12.81 μm), but they used a much higher rotary atomizer speed (39,000 rpm). Jedlińska et al. [25] used a lower rotary atomizer speed (26,500 rpm). The size of the beetroot juice particles that were spray-dried with maltodextrin was 9.1 μm, whereas the size of the particles spray-dried without maltodextrin ranged from 13.2 to 15.6 μm.

Table 1. Particle size distribution (mean ± standard deviation of five replications of D_{10}, D_{50}, D_{90}, and D[4,3], span) of spray dried beetroot powders.

Spray Dryer Scale	D_{10} (μm)	D_{50} (μm)	D_{90} (μm)	D[4,3] (μm)	Span
ML	0.58 ± 0.02 [a]	8.8 ± 0.4 [a]	27.3 ± 0.7 [a]	11.8 ± 0.4 [a]	3.03 ± 0.05 [c]
ST	0.91 ± 0.07 [ab]	32.4 ± 1.1 [b]	75.0 ± 1.9 [b]	36.1 ± 1.3 [b]	2.19 ± 0.02 [b]
SI	1.10 ± 0.03 [b]	37.6 ± 0.6 [c]	75.4 ± 0.5 [bc]	40.1 ± 0.3 [bc]	1.98 ± 0.03 [a]
LI	1.05 ± 0.19 [b]	40.9 ± 1.3 [d]	80.0 ± 1.5 [c]	43.2 ± 1.7 [c]	1.93 ± 0.03 [a]

ML—mini-laboratory, ST—semi-technical, SI—small industrial, LI—large industrial, [a], [b], [c], [d] Different letters in the same column show significant differences ($p \leq 0.05$).

2.4. Moisture Content and Water Activity

All of the beetroot powders that were obtained were characterized by low values of both moisture content (2.84–3.10%) and water activity (0.18–0.23). This indicates that adequate microbiological stability was achieved for the beetroot powders obtained in all of the tested drying scales. Among the centrifugal spray dryers (ST, SI, LI), a gentle decreasing trend in the moisture content and in the water activity of the beetroot powders can be observed with the increasing dryer size.

A comparison of the results showed that the increase in the size of the particles was correlated with a simultaneous decrease in the moisture content and water activity in spray-dried beetroot powders. This finding was in line with the typical trends observed in various studies with the same dryer but with an increase in the inlet air temperature (e.g., spray drying of orange juice by Chegini et al. [30]). In our study, the inlet temperature was the same for all of the spray-drying scales. Therefore, this decreasing trend of changes in the moisture content and water activity in beetroot powders along with the increase in the dryer scale should be associated with the decrease in the ratio of the drying air flow to the dryer volume (according to the data in Table 4). This resulted in the beetroot juice having a longer residence time in the drying chamber and a consequently smaller amount of residual moisture in the obtained beetroot powder.

Other researchers have obtained beetroot powder moisture values of 1–3.4% (Poornima et al. [22]), 3.5–6.9% (Jedlińska et al. [25]), 3.95–6.50% (Singh et al. [26]), 2.8–4.5 (Janiszewska [27]), 2.77–4.75% (Janiszewska et al. [29]), and 3.33–4.24% (Do Carmo et al. [31]), and Janiszewska [27] obtained water activity at the level of 0.44–0.75. Jedlińska et al. [25] observed much lower water activity in beetroot powders that had been spray-dried with carriers (0.107–0.148) and without carriers (0.169–0.202). Such low values may have been caused by the use of dehumidified drying air.

2.5. Bulk Density, Angle of Repose and Flowability

The values of loose and tapped bulk densities are shown in Figure 4. Loose bulk density values were obtained in the range 422–591 kg/m^3, and tapped bulk density values were obtained in the range 569–733 kg/m^3. In the case of the centrifugal spray dryer series (ST, SI, LI), the use of a larger dryer resulted in lower bulk density values, which was confirmed by statistical analysis. In the case of the mini-laboratory-scale (ML) dryer, the values that were obtained show different characteristics, in particular, a much lower value of loose bulk density was recorded in relation to the other dryer sizes. This is due to the different liquid atomization system. The pneumatic two-fluid nozzle can aerate the dryer feed solution to some extent and can create powders with encapsulated air, which consequently leads to lower bulk densities. Other researchers have obtained loose bulk densities of spray-dried beetroot juice with maltodextrin as a carrier at the laboratory and semi-technical scales of 580 kg/m^3 (Jedlińska et al. [25]), 410–615 kg/m^3 (Janiszewska et al. [29]), 309–325 kg/m^3 (Kapoor et al. [32]), and 616–698 kg/m^3 (Gawałek et al. [33]) and a tapped bulk density at values of 730 kg/m^3 (Jedlińska et al. [25]), 639–721 kg/m^3 (Singh et al. [26]), 410–615 kg/m^3 (Janiszewska et al. [29]), 352–389 kg/m^3 (Kapoor et al. [32]), and 692–938 kg/m^3 (Gawałek et al. [33]).

Comparing the differences between the loose and tapped bulk density values for each spray-drying scale, a decrease in these differences was observed with increasing dryer size, regardless of the liquid atomization system. This is directly reflected in the Hausner ratio (HR) values, which are measures of powder bulkiness. Table 2 shows the values of the HR and the angle of repose (AR), which is also a measure of powder flowability. Both of the powder flowability indices showed a statistically significant improvement in flowability when the spray dryer that was used increased in size. This dependence is related to the powder particle size (Table 1). Larger particle sizes improve the flowability of spray-dried beetroot powders. This is in line with Thomson's [34] conclusion that a lower particle size generally provides lower flowability. In Table 2, each value of the flowability index (HR, AR) was also assigned a specific degree of flowability according to Carr's classification [35]. The HR index showed a poorer flowability rating compared to the AR index. According to the HR values, beetroot powders with good or excellent flowability are only observed for industrial drying-scale-sized dryers (SI, LI), while according to the AR index, the already-tested semi-technical scale (ST) achieves good beetroot powder flowability. Similar differences in the powder flowability estimation using the loose and tapped bulk density method (HR) and the angle of repose method (AR) have been obtained by other authors: Kapoor et al. [32], Sarabandi et al. [36], and Dadi et al. [37] for spray-dried beetroot, apple, and moringa juices, respectively. Analyzing the obtained values of the flowability indices of spray-dried beetroot powders by other authors, Kapoor et al. [32] obtained beetroot powders with better flowability (HR = 1.25, AR = 33.83) at the laboratory scale with similar drying parameters. On a semi-technical scale, Gawałek et al. [33] obtained better powder flowability by comparing the Hausner coefficient (HR = 1.13) for the same drying conditions, while slightly worse flowability was obtained by using the angle of repose as a criterion (AR = 33°). However, a higher content of maltodextrin (70% d.m.) was used in this study. Jedlińska et al. [25] used a lower content of maltodextrin as a carrier (about 33% d.m.) in a semi-technical-scale experiment with dehumidified air and obtained a similar HR level (1.26) to the value found in the beetroot powder from the ST spray dryer.

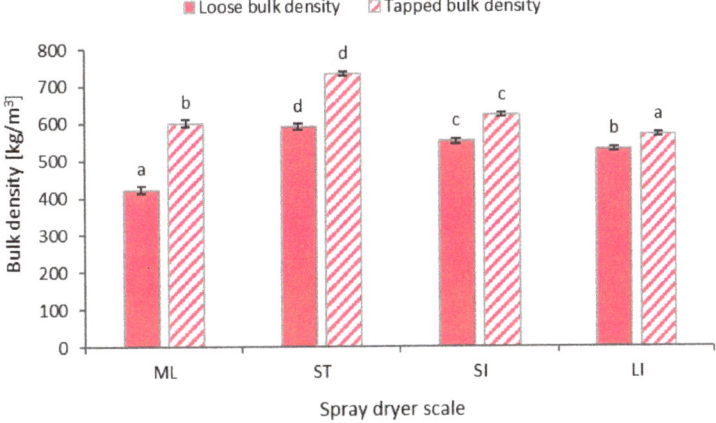

Figure 4. Loose and tapped bulk density of spray dried beetroot powder at different process scale (ML—mini-laboratory scale, ST—semi-technical scale, SI—small industrial scale, LI—large industrial scale; a, b, c, d—different letters show significant differences between spray dryer scale ($p \leq 0.05$)).

Table 2. Characteristics of spray-dried beetroot powders flowability: mean ± standard deviation of three replications of Hausner ratio (HR) and angle of repose (AR).

Spray Dryer Scale	Hausner Ratio		Angle of Repose	
	(-)	Flowability *	(°)	Flowability *
ML	1.42 ± 0.01 [d]	poor	42.2 ± 2.8 [d]	passable
ST	1.24 ± 0.01 [c]	fair	31.3 ± 1.9 [c]	good
SI	1.13 ± 0.00 [b]	good	22.6 ± 1.2 [b]	excellent
LI	1.07 ± 0.00 [a]	excellent	15.1 ± 0.9 [a]	excellent

ML—mini-laboratory, ST—semi-technical, SI—small industrial, LI—large industrial, [a, b, c, d] Different letters in the same column show significant differences ($p \leq 0.05$), * Flowability of powders according to Carr's [35] classification.

2.6. Color

When producing colorants originating from plants, a very important aspect is to obtain a proper and reproducible color scale, i.e., to obtain a color with specific attributes: brightness, hue, and saturation. The CIE Lab system color scale of aqueous solutions containing manufactured beetroot powders was tested. The results are summarized in Table 3. The brightness (L*) of the beetroot powders produced in the centrifugal spray dryers (ST, SI, LI) showed no statistical differences ($p \leq 0.05$), and only the beetroot powder from the ML dryer showed higher brightness. The larger size of the spray dryer affected the beetroot powder color changes in the red (increase in a* value) and blue (decrease in b* value) directions. The total color differences ΔE of the aqueous solutions of the beetroot powders with respect to the color of the solution prepared directly from the liquid juice concentrate showed a significant statistical decrease ($p \leq 0.05$) when the size of the dryer increased. This implies that spray drying beetroot juice on a larger scale enables color changes that can occur during drying to be minimized. It should be noted that the ΔE values for all of the beetroot powders are not large and that the differences for both industrial scales (SI and LI) are practically imperceptible to the human eye.

Table 3. Results for color parameters of 1% aqueous solutions of beetroot powders (mean ± standard deviation of 3 replications).

Spray Dryer Scale	L*	a*	b*	ΔE
ML	34.3 ± 0.3 [a]	66.8 ± 0.4 [a]	36.2 ± 0.3 [d]	5.0 ± 0.3 [d]
ST	35.6 ± 0.4 [b]	67.9 ± 0.4 [b]	35.8 ± 0.2 [c]	3.6 ± 0.2 [c]
SI	35.5 ± 0.3 [b]	68.4 ± 0.2 [b]	34.1 ± 0.2 [b]	2.2 ± 0.1 [b]
LI	36.4 ± 0.3 [b]	69.7 ± 0.4 [c]	33.4 ± 0.3 [a]	0.6 ± 0.2 [a]

ML—mini-laboratory, ST—semi-technical, SI—small industrial, LI—large industrial, [a, b, c, d] Different letters in the same column show significant differences ($p \leq 0.05$).

The above correlations correspond to changes in other parameters, which confirm smaller thermal degradation impacts for larger dryers.

2.7. Violet Betalain Content

The content of betalain, i.e., violet and yellow pigments, specifically the percentage of their retention in the spray-drying process, is a very important performance parameter when considering beetroot powders. It directly affects their coloring power when these powders are used as natural food colorants. The spray-drying process, due to thermal processes that occur during them, degrades these pigments by several to tens of percent depending on the process conditions [23,25,27,29,31,33,38]. Degradation mechanisms such as isomerization, decarboxylation, or cleavage under heat and acidic environments can be different [26,39]. Therefore, in industrial practice, spray-drying process parameters are optimized for the retention (preservation) of betalain pigments. If beetroot juices are used in industrial production as natural colorants, they are standardized to the content of violet betalain pigments, which can be calculated in terms of betanin (BT).

In the present study, the content of violet betalain pigments in the obtained beetroot powders was obtained in the range of 283.3–302.7 mg/100 g, which represents a retention rate in the drying process of 91.4–97.6%, respectively. Other researchers who have considered the spray drying of beetroot juice have achieved betalain retention values of 61–70% (Bazaria et al. [23]), 91.6% (Do Carmo et al. [31]), and 34–88.5% (Ochoa et al. [38]) at the laboratory scale, while at the semi-technical scale, values of 26.7–29.3% (Janiszewska et al. [29]), 68–76% (Janiszewska [27]), and 85.5–95.5% (Gawałek et al. [33]) have been achieved. The size of the spray dryer that was used was found to be a factor causing statistically significant differences ($p \leq 0.05$) in the content of violet betalain pigments (Figure 5A). For a range of centrifugal liquid spray dryers (ST, SI, LI), the use of a larger dryer resulted in higher levels of pigment retention. This dependence is related to the particle powder size, flowability, and powder yield. The lower stickiness of beetroot juice results in a higher yield of beetroot powder in the spray-drying process because larger particles with better flowability are formed. As a consequence, less beetroot powder remains in the drying system, and the thermal degradation of violet betalain pigments is reduced. From the above experimental results, it can be seen that by designing the drying process on a semi-technical scale and by increasing the scale, a significant reduction in the loss of violet betalain pigments can be expected. Increasing the scale from semi-technical (ST) to small industrial (SI), i.e., a sevenfold scale enlargement (7:1), resulted in a decrease in violet betalain pigments losses ranging from 9% to 6%. On the other hand, when the scale upsizing was much larger, i.e., fifty times (50:1), from semi-technical scale (ST) to industrial large scale (LI), the reduction in violet betalain pigment loss was even more significant. In this case, there was a decrease from 9% to 2%. For the laboratory dryer (ML), a higher value was achieved than would be expected due to the small scale; specifically, the expected values were similar to those achieved in the small industrial-scale (SI) dryer. This is due to the fact that the two-fluid nozzle atomization system results in milder drying thermal conditions than centrifugal atomizers despite using the same drying air inlet temperatures (T = 175 °C). The additional cool compressed air that transports and sprays the liquid raw material into the dryer chamber is responsible for this. In addition, the much higher ratio of the drying air to volume (Table 4) for the ML dryer results in the beetroot powder particles having a shorter residence time in the dryer.

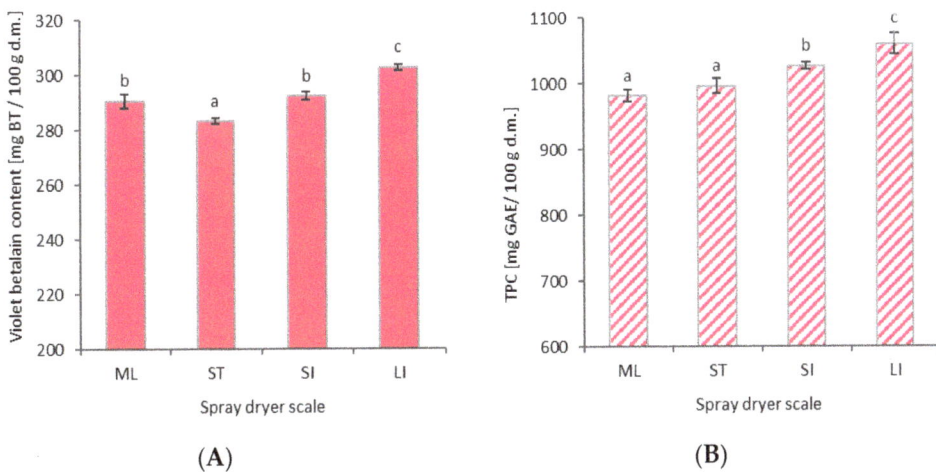

Figure 5. Violet betalain content (**A**) and total polyphenol content (TPC) (**B**) of spray-dried beetroot juice at different process scale (ML—mini-laboratory scale, ST—semi-technical scale, SI—small industrial scale, LI—large industrial scale); BT—betanin; GAE—gallic acid equivalent; d.m.—dry matter; a, b, c—different letters show significant differences between spray dryer scale ($p \leq 0.05$).

Table 4. Designations and data of spray dryers used.

Data/Parameters	Spray Dryers			
Designations	ML	ST	SI	LI
Scale	mini-laboratory spray dryer	semi-technical spray dryer	small industrial spray dryer	large industrial spray dryer
Drying chamber volume, m^3	0.013	1.4	57	179
Water evaporation capacity, kg H_2O/h	1	15	100	700
Liquid spray system	two-fluid nozzle	rotary atomizer	rotary atomizer	rotary atomizer
Spray nozzle/rotary disc diameter, mm	0.7	120	160	210
Batch size, kg	0.4	5	800	5000
Ratio of the drying air flow to the dryer volume, $m^3/m^3 \cdot h$	2692	335	87	81
Producer	Büchi Labortechnik AG, Flawil, Switzerland	Niro Atomizer, Søbork, Denmark	Combined—no name	Niro Atomizer, Søbork, Denmark
Type	B-290	FU 11 DA	-	C

2.8. Total Polyphenol Content (TPC)

Beetroot juice has bioactive properties in addition to coloring properties; hence, their retention (preservation) during spray drying was also investigated at different scales. Phenolic compounds such as phenolic acids and their derivatives (ferulic, vanillic, ellagic, caffeic, chlorogenic, p-coumaric, and sinapic acid) and flavonoids (quercetin, myricetin, kaempferol, rutin, vitexin, orientin, betagarin, betavulgarin, cochliophilin A, and dihydroisorhamnetin) can be found in beetroot juice [40–43]. Total polyphenol content (TPC) values were determined and ranged from 983–1058 mg GAE/100 g, representing retention rates in the range of 91.8–98.9%, respectively. In this case, the statistical significance of the differences between the different drying scales is much lower than that for the violet betalain pigment content, but a gentle increasing trend can be observed as the spray dryer scale increases (Figure 5B). The results obtained here may suggest that the polyphenols have a lower thermal sensitivity compared to the violet betalain pigments. TPC retention at a similar level (>90%) was also achieved during the spray drying of chokeberry juice [10,11] and bayberry juice [44]. In the case of non-optimal spray-drying conditions, lower TPC retention values can be obtained, similar to those for blueberry juice [45] and thyme extract [46]. There are also cases where the retention of TPC after the spray-drying process exceeds 100%. Such values were obtained by Zhang et al. [47] for cranberry juice (138–216%) and by Saikia et al. [48] for Khasi mandarin orange juice (417%). This effect is related to the possibility of changes in the chemical structure of phenolic compounds due to thermal treatment, which may improve the reactivity with the Folin–Ciocalteu reagent [49]. The intensity of these changes during spray drying depends on the phenolic profile of the dried material [47].

During the spray-drying of beetroot juice, betanin can be thermally decomposed into betalamic acid and cyclo-dopa-5-0-glycoside [50,51]. It is also possible to change the violet betalain pigments into yellow. Maillard reaction products may also be formed [52,53]. All of these compounds have antioxidative properties and can react with the Folin–Ciocalteu reagent to influence the TPC value. Therefore, during the spray-drying of beetroot juice, the temperature may cause a greater loss of violet pigments than the TPC values and antioxidative capacity can.

3. Materials and Methods

3.1. Materials

The research material was beetroot (*Beta vulgaris*) juice concentrate (SVZ International B.V., Breda, the Netherlands) with the following parameters: extract content 64.8 °Brix, violet betalain pigments content equal to 0.50%, density 1320 kg/m^3 at 20 °C, pH 4.0, and 2.0 g/100 g citric acid. Potato maltodextrin with a dextrose equivalent of DE 11 (PEPEES

S.A., Łomża, Poland) was used as a carrier. The same feed solution formulation was used for all of the drying processes: 34% solid content in water, 58.5% carrier content in dry mass. Assumptions were made to obtain the theoretical content of the violet betalain pigments at the level of 310 mg/kg at the moisture content of beetroot powder equal to 3%.

3.2. Spray Drying

Spray drying of the same solutions was carried out at different process scales: one laboratory-scale process, one semi-technical-scale process, and two industrial-scale processes (small and large). Table 4 summarizes the designations and data from the spray dryers that were used. A rotary atomizer system was used to spray the feed solution at the semi-technical scale (ST) and industrial scales (SI and LI). During rotary atomization, the feed is centrifugally accelerated to high velocity in the atomizer wheel before being discharged into the hot drying gas. At the laboratory scale (ML), however, atomization was performed using a two-fluid nozzle. In this system, atomization is achieved pneumatically by high-velocity compressed air making impact with the liquid feed.

In all of the drying processes, efforts were made to maintain comparable drying conditions/parameters. The principle that was adopted was to set a constant drying air inlet temperature (175 °C) and, by adjusting the flow rate of the feed solution (raw material), to maintain the outlet temperature in the range of 82–85 °C. In the case of centrifugal spray dryers (ST, SI, and LI), the same rotary atomizer speed was maintained at 15,000 rpm.

3.3. Powder Yield

Powder yield was calculated as the ratio of the dry matter content in the collected powder after each spray drying test process to the value of the dry matter content in the feed solution.

3.4. Microstructure of Particles

The beetroot powders were sputtered with gold and were examined for morphology using a scanning electron microscope SEM Zeiss Evo 40 (Carl Zeiss Microscopy Deutschland GmbH, Oberkochen, Germany; magnification: 500× and 1000×).

3.5. Particle Size Distribution

Particle size distribution was measured with a Mastersizer 2000 (Malvern Instruments Ltd., Malvern, UK) using laser diffraction. Isopropanol was used as a dispersant. Three percentiles (10th, 50th, and 90th), volume-weighted mean diameter D[4,3], and span (Equation (1)) of the volume distribution were determined.

$$\text{span} = \frac{D_{90} - D_{10}}{D_{50}}. \tag{1}$$

3.6. Moisture Content and Water Activity

The moisture content of the beetroot powder was analyzed using the oven method at 105 °C for 4 h. Water activity was measured in a Rotronic apparatus type Hygroscope DT (Rotronic AG, Bassersdorf, Switzerland) at 25 °C.

3.7. Bulk Density, Angle of Repose and Flowability

Loose bulk density (ρ_L), tapped bulk density (ρ_T), and the angle of repose (AR) were measured according to ASTM D6393 [54]. In addition to the natural angle of repose (AR), the Hausner ratio (HR) was used to evaluate the flowability of the beetroot powders. The Hausner coefficient was calculated as the ratio of tapped and loose bulk density: HR = ρ_L/ρ_T [55].

3.8. Color

Using a Jasco V630 spectrophotometer (Japan), color parameters were determined using the CIE L*a*b* system for the 1% aqueous solutions of the resulting beetroot powders.

The total color difference ΔE was also determined for all of the measurements (Equation (2)), where the solution prepared directly from the concentrate was considered as the standard.

$$\Delta E = \sqrt{(\Delta L)^2 + (\Delta a)^2 + (\Delta b)^2}. \qquad (2)$$

3.9. Violet Betalain Content

The content of the violet betalain pigments was determined using the method proposed by Nillson [56] with modifications, using a Jasco V630 spectrophotometer (JASCO International Co. Ltd., Tokyo, Japan). To 1 mL of test solution (1 g of beetroot powder + distilled water to a volume of 100 mL), 4 mL of pH 6.5 phosphate buffer was added. The absorbance was measured at 538 and 600 nm. The results were calculated in terms of betanin (BT) and were expressed as mg BT/100 g d.m. (dry matter) of beetroot powder. The retention of betalains in the spray-dried beetroot juice was calculated relative to the value measured in the feed solution before drying.

3.10. Total Polyphenol Content (TPC)

For the determination of the total polyphenol content (TPC), the Folin–Ciocalteu method, which has already been used in previous studies, was applied [11]. A sample (1 mL of solution: 0.5 g beetroot powder in 10 mL of 50% (v/v) methanol/water solution) was pipetted into a 100 mL volumetric flask and was diluted with distilled water. Determinations were performed using a Jasco V630 spectrophotometer (JASCO International Co. Ltd., Tokyo, Japan) by measuring absorbance at 765 nm. Results were expressed as mg gallic acid equivalent (GAE) per 100 g d.m. of beetroot powder. TPC retention in the dried beetroot juice was calculated relative to the value measured in the feed solution before drying.

3.11. Statistical Analysis

All of the determined physicochemical parameters were average values and were determined from measurements made after a min. of three repetitions. The statistical significance (or not) of the effect of the spray dryer-scale size on the physicochemical parameters of the beetroot powder and process efficiency was verified using one-way ANOVA analysis of variance with Tukey's HSD test at a significance level of 0.05. Statgraphics 13.1 program was used.

4. Conclusions

Conducting spray drying experiments of fruit and vegetable juices at the experimental scale is necessary to determine the optimal drying conditions and to select the optimal carriers and solution formulations for industrial scale drying. Industrial process design for centrifugal spray dryers is definitely more efficient than using a semi-technical scale with the same spray system than using a laboratory-scale dryer with a two-fluid nozzle. The laboratory-scale dryer is more convenient for research, but many properties of the beetroot powder obtained on it do not correlate with the properties of beetroot powders obtained on the industrial scale. Among other things, the particle size distribution, microstructure, bulk density, and flowability of the powder are very different. This effect is caused not only by the difference in the feed solution spraying systems, but also by the large difference in the residence time of beetroot juice in the drying chamber. The larger scale of the spray-drying process at the same drying parameters makes it possible to obtain beetroot powders with a larger particle size, better flowability, color that is more shifted towards red and blue, and the higher retention of violet betalain pigments and polyphenols. The preservation of the colouring and bioactive properties in the spray-dried beetroot powder are related to the particle powder size, flowability, and powder yield. The lower stickiness of the beetroot juice results in a higher powder yield in the spray-drying process because larger particles with better flowability are formed. As a consequence, less beetroot powder remains in the drying system, and the thermal degradation of violet betalain pigments is reduced. During the spray-drying of beetroot juice, the temperature caused a greater loss of violet pigments than polyphenols. In both cases, retention above 90% was achieved at all spray-drying scales. As the size of the spray dryer increases, the efficiency of the process expressed in powder yield also increases. To obtain a powder yield >90% on the industrial scale, the process conditions should be selected so as to obtain a powder yield of a min. of 50% at a laboratory scale or 80% at a semi-technical scale.

Funding: This publication was co-financed within the framework of the Polish Ministry of Science and Higher Education's program "Regional Initiative Excellence" in the years 2019–2022 (No. 005/RID/2018/19), with the financing amount of PLN 12,000,000.

Institutional Review Board Statement: Not applicable.

Informed Consent Statement: Not applicable.

Data Availability Statement: All data created and analyzed during the experiments were presented in this study.

Acknowledgments: The author thanks the food company Celiko (Poznań, Poland) for the opportunity to carry out the spray-drying process on an industrial scale.

Conflicts of Interest: The author declares no conflict of interest.

Sample Availability: Samples of the compounds are available from the author.

References

1. Gharsalloui, A.; Roudaut, G.; Chambin, O.; Voilley, A.; Saurel, R. Applications of spray drying in microencapsulation of food ingredients: An overview. *Int. Food Res. J.* **2007**, *40*, 1107–1121. [CrossRef]
2. Fang, Z.; Bhandari, B. Encapsulation of polyphenols—A review. *Trends Food Sci. Technol.* **2010**, *21*, 510–523. [CrossRef]
3. Shishir, M.R.I.; Chen, W. Trends of spray drying: A critical review on drying of fruit and vegetable juices. *Trends Food Sci. Technol.* **2017**, *65*, 49–67. [CrossRef]
4. Jiang, H.; Zhang, M.; Adhikari, B. Fruit and vegetable powders. In *Handbook of Food Powders: Processes and Properties*; Woodhead Publishing: Cambridge, UK, 2013; pp. 532–552.
5. Santhalakshmy, S.; Bosco, S.J.D.; Francis, S.; Sabeena, M. Effect of inlet temperature on physicochemical properties of spray-dried jamun fruit juice powder. *Powder Technol.* **2015**, *274*, 37–43. [CrossRef]
6. Bhandari, B.; Datta, N.; Crooks, R.; Howes, T.; Rigby, S. A semi-empirical approach to optimise the quantity of drying aids required to spray dry sugar-rich foods. *Dry. Technol.* **1997**, *15*, 2509–2525. [CrossRef]
7. Chegini, G.R.; Ghobadian, B. Spray dryer parameters for fruit juice drying. *World J. Agric. Res.* **2007**, *3*, 230–236.
8. Muzaffar, K.; Nayik, G.A.; Kumar, P. Production of fruit juice powders by spray drying technology. *Int. J. Adv. Res. Sci. Eng.* **2018**, *7*, 59–67.
9. Koç, B.; Kaymak-Ertekin, F. The effect of spray drying processing conditions on physical properties of spray dried malto-dextrin. In Proceedings of the 9th Baltic Conference on Food Science and Technology "Food for Consumer Well-Being" FOODBALT 2014, Jelgava, Latvia, 8–9 May 2014; Faculty of Food Technology, Latvia University of Agriculture: Jelgava, Latvia, 2014; pp. 243–247.
10. Gawałek, J.; Domian, E. Tapioca dextrin as an alternative carrier in the spray drying of fruit juices-A case study of chokeberry powder. *Foods* **2020**, *9*, 1125. [CrossRef]
11. Gawałek, J.; Domian, E.; Ryniecki, A.; Bakier, S. Effects of the spray drying conditions of chokeberry (*Aronia melanocarpa* L.) juice concentrate on the physicochemical properties of powders. *Int. J. Food Sci.* **2017**, *52*, 1933–1941. [CrossRef]
12. Bhandari, B.; Senoussi, A.; Dumoulin, E.; Lebert, A. Spray drying of concentrated fruit juices. *Dry. Technol.* **1993**, *11*, 1081–1092. [CrossRef]
13. Murugesan, R.; Orsat, V. Spray drying for the production of nutraceutical ingredients—A Review. *Food Bioprocess Technol.* **2011**, *5*, 3–14. [CrossRef]
14. Moreno, T.; De Paz, E.; Navarro, I.; Rodriguez-Rojo, S.; Matias, A.A.; Duarte, C.M.M.; Sanz-Buenhombre, M.; Cocero, M. Spray drying formulation of polyphenols-rich grape marc extract: Evaluation of operating conditions and different natural carriers. *Food Bioprocess Technol.* **2016**, *9*, 2046–2058. [CrossRef]
15. Gil, M.; Vicente, J.; Gaspar, F. Scale-up methodology for pharmaceutical spray drying. *Chem. Today* **2010**, *10*, 18–22.
16. Poozesh, S.; Bilgili, E. Scale-up of pharmaceutical spray drying using scale-up rules: A review. *Int. J. Pharm.* **2019**, *562*, 271–292. [CrossRef]
17. Masters, K. Scale-up of spray dryers. *Dry. Technol.* **1994**, *12*, 235–257. [CrossRef]
18. Schick, R.J.; Brown, K. *Spray Dryer Scale-Up: From Laboratory to Production. Spray Analysis and Research Services*; Printed in the USA; Spraying Systems Co.: Wheaton, IL, USA, 2005.

19. Thybo, P.; Hovgaard, L.; Lindeløv, J.S.; Brask, A.; Andersen, S.K. Scaling up the spray drying process from pilot to production scale using an atomized droplet size criterion. *Pharm. Res.* **2008**, *25*, 1610–1620. [CrossRef]
20. DuBose, D.; Settell, D.; Baumann, J. Efficient scale-up strategy for spray-dried dispersions. *Drug Dev. Deliv.* **2013**, *13*, 54–62.
21. Zbicinski, I. Modeling and scaling up of industrial spray dryers: A review. *J. Chem. Eng. Jpn.* **2017**, *50*, 757–767. [CrossRef]
22. Poornima, K.; Sinthiya, R. Encapsulation of beetroot extract using spray drying. *Int. J. Res. Appl. Sci. Eng. Technol.* **2017**, *5*, 346–352. [CrossRef]
23. Bazaria, B.; Kumar, P. Optimization of spray drying parameters for beetroot juice powder using response surface methodology (RSM). *J. Saudi Soc. Agric. Sci.* **2018**, *17*, 408–415. [CrossRef]
24. Teenu, M.; Kuriakose, L.; Felix, E. Optimization of spray drying parameters for beetroot juice powder on higroscopicity and powder yield using Response Surface Methodology (RSM)-Central composite design. *Best IJHAMS* **2017**, *5*, 151–156.
25. Jedlińska, A.; Barańska, A.; Witrowa-Rajchert, D.; Ostrowska-Ligęza, E.; Samborska, K. Dehumidified Air-Assisted Spray-Drying of cloudy beetroot juice at low temperature. *Appl. Sci.* **2021**, *11*, 6578. [CrossRef]
26. Singh, B.; Hathan, B.S. Process optimization of spray drying of beetroot juice. *J. Food Sci. Technol.* **2017**, *54*, 2241–2250. [CrossRef]
27. Janiszewska, E. Microencapsulated beetroot juice as a potential source of betalain. *Powder Technol.* **2014**, *264*, 190–196. [CrossRef]
28. Barbosa-Cánovas, G.; Ortega-Rivas, E.; Juliano, P.; Yan, H. *Food Powders: Physical Properties, Processing, and Functionality*; Kluwer Academic Publishers: New York, NY, USA; Plenum Publishers: New York, NY, USA, 2005.
29. Janiszewska, E.; Włodarczyk, J. Influence of spray drying conditions on beetroot pigments retention after microencapsulation process. *Acta Agrophys.* **2013**, *20*, 343–356.
30. Chegini, G.R.; Ghobadian, B. Effect of Spray-Drying conditions on physical properties of orange juice powder. *Dry. Technol.* **2005**, *23*, 657–668. [CrossRef]
31. Do Carmo, E.L. Stability of spray-dried beetroot extract using oligosaccharides and whey proteins. *Food Chem.* **2018**, *249*, 51–59. [CrossRef]
32. Kapoor, N.; Mohite, A.M.; Sharma, N.; Sharma, D. Comparative analysis of freeze dried and spray dried beet-root powder according to physico-chemical, functional and color properties. *Bull. Transilv. Univ. Bras. II For. Wood Ind. Agric. Food Eng.* **2021**, *14*, 1–202. [CrossRef]
33. Gawałek, J.; Bartczak, P. Effect of beetroot juice spray drying conditions on selected properties of produced powder. *Food Sci. Technol. Qual.* **2014**, *2*, 164–174.
34. Thomson, F.M. Storage and flow of particulate solids. In *Handbook of Powder Science and Technology*; Fayed, M.E., Otten, L., Eds.; Chapman & Hall: New York, NY, USA, 1997; pp. 389–486.
35. Carr, R.L. Evaluating Flow Properties of Solids. *Chem. Eng. J.* **1965**, *72*, 69.
36. Sarabandi, K.; Peighambardoust, S.H.; Sadeghi Mahoonak, A.R.; Samaei, S.P. Effect of different carriers on microstructure and physical characteristics of spray dried apple juice concentrate. *J. Food Sci. Technol.* **2018**, *55*, 3098–3109. [CrossRef] [PubMed]
37. Dadi, D.W.; Emire, S.A.; Hagos, A.D.; Eun, J.B. Effects of spray drying process parameters on the physical properties and digestibility of the microencapsulated product from *Moringa stenopetala* leaves extract. *Cogent Food Agric.* **2019**, *5*, 1690316. [CrossRef]
38. Ochoa-Martinez, L.A.; Garza-Juarez, S.E.; Rocha-Guzman, N.E.; Morales-Castro, J.; Gonzalez-Herrera, S.M. Functional properties, color and betalain content in Beetroot-Orange juice powder obtained by spray drying. *Res. Rev. J. Food Dairy Technol.* **2015**, *3*, 30–36.
39. Herbach, K.M.; Stinzing, F.C.; Carle, R. Impact of thermal treatment on color and pigment pattern of red beet (*Beta vulgaris* L.) preparations. *J. Food Sci.* **2004**, *69*, C491–C498. [CrossRef]
40. Koubaier, H.B.H.; Snoussi, A.; Essaidi, I.; Chaabouni, M.M.; Thonart, P.; Bouzouita, N. Betalain and phenolic compositions, antioxidant activity of tunisian red beet (*Beta vulgaris* L. *conditiva*) roots and stems extracts. *Int. J. Food Prop.* **2014**, *17*, 1934–1945. [CrossRef]
41. Deseva, I.; Stoyanova, M.; Petkova, N.; Mihaylova, D. Red beetroot juice phytochemicals bioaccessibility: An in vitro approach. *Pol. J. Food Nutr. Sci.* **2020**, *70*, 45–53. [CrossRef]
42. Płatosz, N.; Sawicki, T.; Wiczkowski, W. Profile of phenolic acids and flavonoids of red beet and its fermentation products. Does Long-Term consumption of fermented beetroot juice affect phenolics profile in human blood plasma and urine? *Pol. J. Food Nutr. Sci.* **2020**, *70*, 55–65. [CrossRef]
43. Kujala, T.S.; Vienola, M.S.; Klika, K.D. Betalain and phenolic compositions of four beetroot (*Beta vulgaris*) cultivars. *Eur. Food Res. Technol.* **2002**, *214*, 505–510. [CrossRef]
44. Fang, Z.; Bhandari, B. Effect of spray drying and storage on the stability of bayberry polyphenols. *Food Chem.* **2011**, *129*, 1139–1147. [CrossRef]
45. Darniadi, S.; Ifie, I.; Ho, P. Evaluation of total monomeric anthocyanin, total phenolic content and individual anthocyanins of foam-mat freeze-dried and spray-dried blueberry powder. *J. Food Meas. Charact.* **2019**, *13*, 1599–1606. [CrossRef]
46. Jovanović, A.A.; Lević, S.M.; Pavlović, V.B.; Marković, S.B.; Pjanović, R.V.; Dorđević, V.B.; Nedović, V.; Bugarski, B.M. Freeze vs. Spray drying for dry wild thyme (*Thymus serpyllum* L.) extract formulations: The impact of gelatin as a coating material. *Molecules* **2021**, *26*, 3933. [CrossRef] [PubMed]
47. Zhang, J.; Zhang, C.; Chen, X.; Quek, S.Y. Effect of spray drying on phenolic compounds of cranberry juice and their stability during storage. *J. Food Eng.* **2020**, *269*, 109744. [CrossRef]

48. Saikia, S.; Mahnot, N.K.; Mahanta, C.L. Effect of spray drying of four fruit juices on physicochemical, phytochemical and antioxidant properties. *J. Food Process. Preserv.* **2015**, *39*, 1656–1664. [CrossRef]
49. Robert, P.; Torres, V.; García, P.; Vergara, C.; Sáenz, C. The encapsulation of purple cactus pear (*Opuntia ficus-indica*) pulp by using polysaccharide-proteins as encapsulating agents. *LWT-Food Sci. Technol.* **2015**, *60*, 1039–1045. [CrossRef]
50. Huang, A.; Elbe, J. Kinetics of the degradation and regeneration of betanine. *J. Food Sci.* **2006**, *50*, 1115–1120. [CrossRef]
51. Miyagawa, Y.; Fujita, H.; Adachi, S. Kinetic analysis of thermal degradation of betanin at various pH values using deconvolution method. *Food Chem.* **2021**, *361*, 130165. [CrossRef]
52. Nicoli, M.C.; Anese, M.; Parpinel, M. Influence of processing on the antioxidant properties of fruit and vegetables. *Trends Food Sci. Technol.* **1999**, *10*, 94–100. [CrossRef]
53. Kidoń, M.; Czapski, J. The effect of thermal processing on betalain pigments contents and antiradical activity of red beet. *Food Sci. Technol. Qual.* **2007**, *1*, 124–131.
54. ASTM D6393-14. *Standard Test. Method for Bulk Solids Characterization by Carr Indices*; ASTM International: West Conshohocken, PA, USA, 2014. [CrossRef]
55. De Jong, J.A.; Hoffmann, A.C.; Finkers, H.J. Properly determine powder flowability to maximie plant output. *Chem. Eng. Progr.* **1999**, *95*, 25–34.
56. Nillson, T. Studies into the pigments in beetroot (*Beta vulgaris* L. ssp. *vulgaris var. rubra* L.). *Lantbrukshoegsk. Ann.* **1970**, *36*, 179–218.

Article

Multi-Targeted Molecular Docking, Pharmacokinetics, and Drug-Likeness Evaluation of Okra-Derived Ligand Abscisic Acid Targeting Signaling Proteins Involved in the Development of Diabetes

Syed Amir Ashraf [1,*], Abd Elmoneim O. Elkhalifa [1], Khalid Mehmood [2], Mohd Adnan [3], Mushtaq Ahmad Khan [4], Nagat Elzein Eltoum [1], Anuja Krishnan [5] and Mirza Sarwar Baig [5,*]

1 Department of Clinical Nutrition, College of Applied Medical Sciences, University of Hail, Hail 2440, Saudi Arabia; ao.abdalla@uoh.edu.sa (A.E.O.E.); nagacademic0509@gmail.com (N.E.E.)
2 Department of Pharmaceutics, College of Pharmacy, University of Hail, Hail 2440, Saudi Arabia; adckhalid@gmail.com
3 Department of Biology, College of Science, University of Hail, Hail 2440, Saudi Arabia; drmohdadnan@gmail.com
4 Department of Microbiology and Immunology, College of Medicine and Health Sciences, UAE University, Al Ain 15551, United Arab Emirates; mushtaq.khan@uaeu.ac.ae
5 Department of Molecular Medicine, School of Interdisciplinary Sciences & Technology, Jamia Hamdard, New Delhi 110062, India; anuja.krishnan@jamiahamdard.ac.in
* Correspondence: amirashrafy2007@gmail.com (S.A.A.); mirzasbaig@jamiahamdard.ac.in or baigmirzasarwar@gmail.com (M.S.B.); Tel.: +966-591-491-521 (S.A.A.); +966-165-358-298 or +91-999-0612-416 (M.S.B.)

Abstract: Diabetes mellitus is a global threat affecting millions of people of different age groups. In recent years, the development of naturally derived anti-diabetic agents has gained popularity. Okra is a common vegetable containing important bioactive components such as abscisic acid (ABA). ABA, a phytohormone, has been shown to elicit potent anti-diabetic effects in mouse models. Keeping its anti-diabetic potential in mind, in silico study was performed to explore its role in inhibiting proteins relevant to diabetes mellitus- 11β-hydroxysteroid dehydrogenase (11β-HSD1), aldose reductase, glucokinase, glutamine-fructose-6-phosphate amidotransferase (GFAT), peroxisome proliferator-activated receptor-gamma (PPAR-gamma), and Sirtuin family of NAD(+)-dependent protein deacetylases 6 (SIRT6). A comparative study of the ABA-protein docked complex with already known inhibitors of these proteins relevant to diabetes was compared to explore the inhibitory potential. Calculation of molecular binding energy (ΔG), inhibition constant (pKi), and prediction of pharmacokinetics and pharmacodynamics properties were performed. The molecular docking investigation of ABA with 11-HSD1, GFAT, PPAR-gamma, and SIRT6 revealed considerably low binding energy (ΔG from −8.1 to −7.3 Kcal/mol) and predicted inhibition constant (pKi from 6.01 to 5.21 µM). The ADMET study revealed that ABA is a promising drug candidate without any hazardous effect following all current drug-likeness guidelines such as Lipinski, Ghose, Veber, Egan, and Muegge.

Keywords: anti-diabetic; okra; abscisic acid; nutraceuticals; *Diabetes mellitus*; molecular docking; phytohormones

Citation: Ashraf, S.A.; Elkhalifa, A.E.O.; Mehmood, K.; Adnan, M.; Khan, M.A.; Eltoum, N.E.; Krishnan, A.; Baig, M.S. Multi-Targeted Molecular Docking, Pharmacokinetics, and Drug-Likeness Evaluation of Okra-Derived Ligand Abscisic Acid Targeting Signaling Proteins Involved in the Development of Diabetes. *Molecules* 2021, 26, 5957. https://doi.org/10.3390/molecules26195957

Academic Editors: Severina Pacifico, Simona Piccolella and Gyorgy Dorman

Received: 17 July 2021
Accepted: 28 September 2021
Published: 1 October 2021

Publisher's Note: MDPI stays neutral with regard to jurisdictional claims in published maps and institutional affiliations.

Copyright: © 2021 by the authors. Licensee MDPI, Basel, Switzerland. This article is an open access article distributed under the terms and conditions of the Creative Commons Attribution (CC BY) license (https://creativecommons.org/licenses/by/4.0/).

1. Introduction

Diabetes is one of the most prevalent epidemics, affecting almost 382 million people worldwide. According to the International Diabetes Federation (IDF) report, it is alleged that approximately 1.3 million people die from diabetes every year. IDF suggests that around 629 million people will have diabetes by 2045 worldwide [1].

Diabetes is a chronic metabolic disorder characterized by insulin resistance and pancreatic β-cell dysfunction caused by uncontrolled hyperglycemia. Altered sugar, fat, and protein metabolism in diabetes and associated complications include retinopathy, neuropathy, nephropathy, cardiovascular diseases, skin complications, and macrovascular complications [2]. This represents a major economic burden since 12% of global health expenditure is spent on the diabetic population. Type 2 diabetes mellitus (T2DM) is a complex disease characterized by high glucose plasma levels due to insufficient insulin secretion or action or both affecting people of all age groups [2,3]. Additionally, insulin resistance in target tissues and a relative deficiency of insulin secretion from pancreatic β-cells are the major metabolic issue with diabetes. In response to nutrient spillover in insulin resistance and eventual β-cell dysfunction, the general fuel homeostasis of the body gets altered. β-cell hyperplasia and hyperinsulinemia in response to insulin resistance occur in the preclinical period of the disease.

As a consequence of the failure of β-cells to compensate for insulin resistance, relative insulin deficiency progresses into diabetes [4]. Diabetes involves various cellular pathways such as insulin secretion, insulin resistance, and carbohydrate absorption. Some human proteins have been identified as key regulators in the development of diabetes like glucokinase, AMP-activated protein kinase, 11 β-hydroxysteroid dehydrogenases (11 β-HSD), insulin receptor substrate, interleukin1 beta, dipeptidyl peptidase IV, glutamine-fructose-6-phosphate amidotransferase (GFAT), peroxisome proliferator-activated receptor-gamma (PPAR-gamma), tyrosine phosphatases, tyrosine kinase insulin receptor, protein kinase B, and insulin receptor [3]. Various therapeutic interventions have been developed to treat and manage diabetes, including dietary modifications, exercise, and anti-diabetic agents. However, reports suggest that anti-diabetic agents are usually associated with severe side effects or adverse effects, and sometimes, their efficacies are controversial. Hence, attention has been shifted towards traditional and alternative medicines or food-derived products rich in anti-diabetic phytoconstituents. Bioactive components present in plants and plant-derived products such as alkaloids, flavonoids, glycosides, gum, carbohydrates, triterpenes, and verities of short-peptides are usually responsible for their therapeutic importance [5].

Okra has recently been recognized for its potential therapeutic purposes because of various important phytochemical constituents. Okra plant [*Abelmoschus esculentus* (L.) Moench] is a nutritive vegetable known by various names such as lady's finger, green ginseng, and plant Viagra [6]. Presently, okra is used for its nutritional values and nutraceutical and therapeutic properties, owing to various important bioactive compounds and their associated therapeutic properties [7].

The profile of the bioactive components present in different parts of okra has been reported, which includes polyphenolic compounds, flavanol derivatives, carotene, protein (i.e., high lysine levels), folic acid, thiamine, riboflavin, niacin, vitamin C, oxalic acid, amino acids, oligomeric catechins, and newly identified bioactive component abscisic acid [8]. ABA (MW = C15H20O4, IUPAC = (2Z,4E)-5-[(1S)-1-hydroxy-2,6,6-trimethyl-4-oxocyclohex-2-en-1-yl]-3-methylpenta-2,4-dienoic acid), shown in Figure 1, is a compound naturally present in fruits and vegetables. Its concentration varies depending on the type of vegetables ranging from 0.29 mg/kg to 0.62 mg/kg of the wet weight of vegetables and fruits, respectively. Throughout the life cycle of plants, it is involved in various physiological and developmental activities, where it regulates seed maturation, maintenance of embryo dormancy and plays a relevant role in various processes against environmental stressors [9]. In recent years, ABA was found to be associated with human diseases, and is currently being investigated for various therapeutic purposes such as diabetes, prostate cancer, Alzheimer's, and other neurodegenerative diseases, due to broad biological activity spectrum of ABA [10,11]. Mechanisms of action underlying the observed anti-diabetic effects are insulin secretion, insulin resistance, and carbohydrate absorption. The search for new therapeutic targets remains a challenge; the present work investigates the anti-diabetic potential of ABA of okra in silico approach by predicting the binding interactions between ABA with target proteins involved in the development of diabetes mellitus.

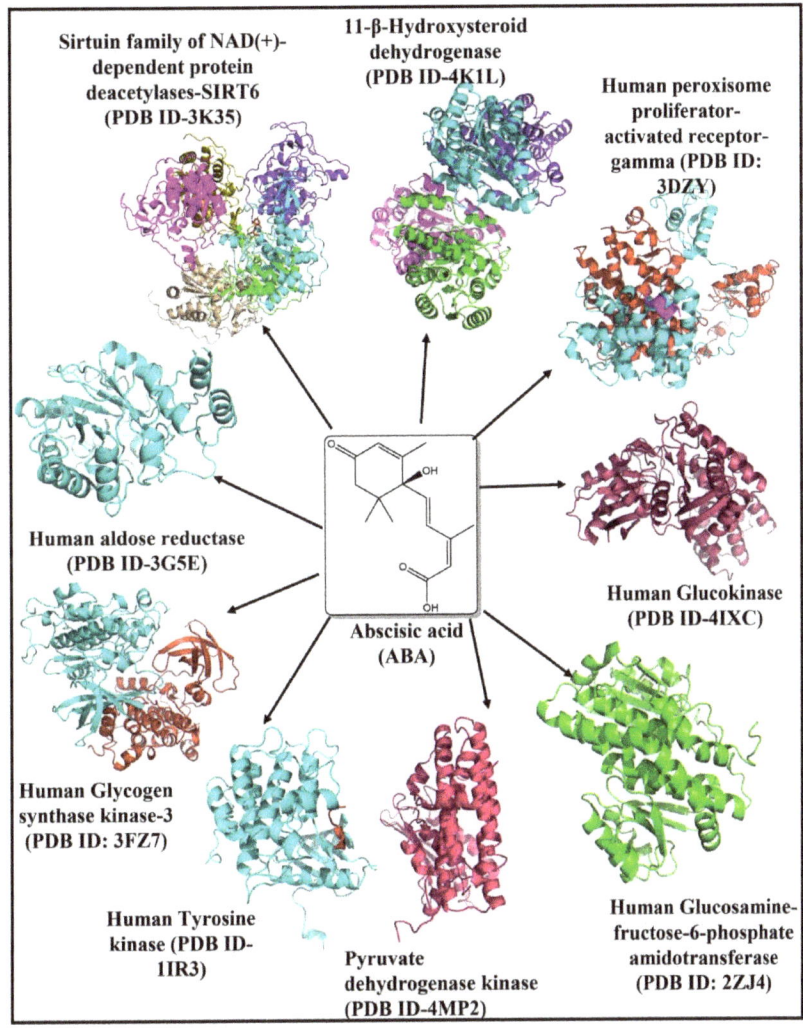

Figure 1. The chemical structure of ABA (center). The human proteins involved in the development of diabetes are 11β-Hydroxysteroid dehydrogenase [11β-HSD (PDB ID-4K1L)], aldose reductase (PDB ID-3G5E), glucokinase (PDB ID-4IXC), glycogen synthase kinase-3 (PDB ID-3F7Z), glucosamine-fructose-6-phosphate amidotransferase [GFAT (PDB ID-2ZJ4)], pyruvate dehydrogenase kinase (PDB ID-4MP2), peroxisome proliferator-activated receptor-gamma (PDB ID-3DZY), Sirtuin family of NAD(+)-dependent protein deacetylases [SIRT6 (PDB ID-3K35)] and Tyrosine kinase (PDB ID-1IR3), were individually docked with ABA.

2. Results & Discussion

Molecular docking and virtual screening are fast, economical, and reliable approaches for identifying both a potential druggable protein target as well as a novel drug (lead molecule) through rational drug designing (RDD) or computer-aided drug design (CADD). RDD or CADD is now being used to annotate and evaluate large pharmacological libraries swiftly.

This study applies molecular docking-based virtual screening to identify a promising target for T2DM. Based on literature review and available crystal structures of proteins involved in several biosynthetic pathways as a key regulator in T2DM, we selected nine human proteins 11β-hydroxysteroid dehydrogenase [11β-HSD (PDB ID-4K1L)], aldose reductase (PDB ID-3G5E), glucokinase (PDB ID-4IXC), glycogen synthase kinase-3 [GSK-3 (PDB ID-3F7Z)], glucosamine:fructose-6-phosphate amidotransferase [GFAT (PDB ID-2ZJ4)], pyruvate dehydrogenase kinase (PDB ID-4MP2), peroxisome proliferator-activated receptor-gamma (PDB ID-3DZY), sirtuin family of NAD(+)-dependent protein deacetylases [SIRT6 (PDB ID-3K35)] and Tyrosine kinase (PDB ID-1IR3), shown in Figure 1. The plausible molecular/atomic interactions of ABA with these proteins were investigated in this in silico study.

ABA was found to have a binding energy of −8.1 kcal/mol, −7.3 kcal/mol, −7.3 kcal/mol, −7.3 kcal/mol, −6.8 kcal/mol, −6.6 kcal/mol, −6.6 kcal/mol, 6.3 kcal/mol, and −6.2 kcal/mol, with 11β-HSD1, GFAT, PPAR-gamma, SIRT6, glucokinase, aldose reductase, glycogen synthase kinase-3, Pyruvate dehydrogenase kinase (PDK), and Tyrosine kinase proteins, respectively (Table 1). The binding energy (Kcal/mol) would be used to correlate and investigate the binding affinity of various ligands or inhibitors with their corresponding protein target. In general, the lower the binding energy, the greater the ligand's affinity for the receptor protein will be. As a result, the ligand with the highest affinity can be taken forward as a candidate for further research.

Table 1. AutoDock Vina results showing binding energies and inhibition constant of ABA with different proteins related to diabetes mellitus.

S. No	Protein Name (PDB ID)	Theoretical Weight (KDa)	Name of Chains	Binding Energy (ΔG) (kcal/mol)	Predicted Inhibition Constant pKi (µM)	No. of H-Bonds	H-Bond Forming Residues
1	11β-HSD1 (4K1L)	31.84	A, B, C, D	−8.1	6.01	2	TYR(A)183, SER(A)169
2.	GFAT (2ZJ4)	42.32	A	−7.3	5.21	5	CYS373, THR375, GLN421, SER422, THR425
3.	PPAR-gamma (3DZY)	51.53	A, B, C, D	−7.3	5.21	2	TYR(A)189, TYR(D)250
4.	SIRT6 (3K35)	35.15	A, B, C, D, E, F	−7.3	5.21	2	GLN(C)111, HIS(C)131
5.	Glucokinase (4IXC)	50.81	A	−6.8	5.05	3	ASN83, ARG85, GLY229
6.	Aldose reductase (3G5E)	36.18	A	−6.6	4.84	2	TRP111, CYS298
7.	Glycogen synthase kinase-3 (3F7Z)	39.88	A, B	−6.6	4.84	0	-
8.	Pyruvate dehydrogenase kinase (4MP2)	45.23	A	−6.3	4.55	1	AGR162
9.	Tyrosine kinase (1IR3)	35.03	A	−6.2	4.47	1	GLU1043

2.1. Abscisic Acid Is a Potent Inhibitor of the Human 11β-Hydroxysteroid Dehydrogenase Type 1 (11β-HSD1) Enzyme

Intracellular conversion of metabolically inert cortisone to active cortisol using NADPH as a co-factor is carried out by the 11β-HSD1 enzyme [12–15]. Cortisol increases hepatic glucose production by inducing genes involved in gluconeogenesis and glycogenolysis in the liver. Cortisol promotes pre-adipocyte differentiation into mature adipocytes, resulting in adipose tissue hyperplasia. By modulating cortisone/cortisol levels, selective inhibition of this enzyme can be a novel treatment for T2DM and hyperlipidemia [16–19]. Obesity, diabetes, wound healing, and muscular atrophy are glucocorticoid-related disorders, and inhibiting 11β-HSD1 has many therapeutic values, including T2DM and hyperlipidemia.

The X-ray crystallography investigations of the crystal structure of an inhibitor molecule (4aS,8aR)-3-(cyclohexylamino)-4a,5,6,7,8,8a-hexahydrobenzo[e][1,3,4]oxathiazine 1,1-dioxide (C13H22N2O3S) docked with human and murine 11β-HSD1 proteins revealed that its cyclohexyl-NH interacts with the key active site residue Tyr183 [17]. Furthermore, one of the sulfonyl oxygen atoms forms a hydrogen bond to the main-chain nitrogen of Ala172 of 11β-HSD1 protein. Meanwhile, the side chain Tyr177 of the human 11β-HSD1 enzyme typically forms Van der Waals interaction to the same inhibitor molecule [17].

Interestingly in our in silico analysis, we found that key residues of active site viz., TYR(A)183 and SER(A)169 of the human 11β-HSD1 enzyme interact with the two oxygen atoms of the 3-Methylpenta-2,4-dienoic acid substructure of abscisic acid [(2Z,4E)-5-[(1S)-1-hydroxy-2,6,6-trimethyl-4-oxocyclohex-2-en-1-yl]-3-methylpenta-2,4-dienoic acid)]. Nine other amino acid residues of the human 11β-HSD1 enzyme, LEU(A)126, VAL(A)168, SER(A)170, ALA(A)172, VAL(A)180, LYS(A)187, LEU(A)215, GLY(A)216, and VAL(A)231 formed Van-der-Waals interaction with ABA (Figure 2A–D). This interaction showed the lowest binding energy of −8.1 kcal/mol and the highest inhibition constant of 6.01 μM, respectively (Table 1). The pi-pi stacking interaction by LEU171, TYR177, LEU217, ILE218, and ALA223 with other pi-cation and pi-alkyl interactions may help in stabilizing the ABA bounded with the active residues of the 11β-HSD1 enzyme (Figure 2D). Based on these very similar binding patterns and docking complex analysis, we can say that ABA is the true, potent inhibitor of the human 11β-HSD1 enzyme, thus possibly help in controlling T2DM and hyperlipidemia.

2.2. Abscisic Acid Significantly Binds and Inhibits Glutamine: Fructose-6-Phosphate Amidotransferase (GFAT) Effectively

Glutamine-fructose-6-phosphate amidotransferase (GFAT) is a rate-limiting enzyme in the hexosamine biosynthetic pathway a key regulator in T2DM [20–22]. In mammals, glucose integration via the hexosamine biosynthetic pathway is regarded as a cellular nutrition sensor. This pathway is one of the strategies by which hyperglycemia induces peripheral insulin resistance [23,24]. In vitro and in vivo studies indicate the association of hyperactivity of human GFAT with insulin resistance, thus qualifying it as a promising candidate for T2DM [25].

The C-terminal 40 kDa isomerase domain of GFAT (residues Gln313–Glu680) contains the active site (near the CF helix) and is involved in converting fructose-6-phosphate (Fru6P) to glucosamine-6-phosphate (GlcN6P) utilizing ammonia (NH3) as a substrate. The X-ray diffraction pattern shows that the bound ligand AGP (2-deoxy-2-amino glucitol-6-phosphate) with GFAT (PDB ID 2ZJ4) establishes hydrogen bonds with Thr375, Ser376, Ser420, Gln421, Ser422, and Thr425 directly, and other hydrogen-bonds with Val471, His576, and Ala674 from the neighboring subunit [20].

We found a similar interaction pattern of ABA with the active site amino acid residues of GFAT (PDB ID- 2ZJ4), suggesting ABA can potentially inhibit GFAT activity. Five key amino acid residues of GFAT viz., CYS373, THR375, GLN421, SER422, and THR425 were strongly forming hydrogen bonds of bond length 2.87Å, 3.10Å, 3.08Å, 3.14Å and 2.94 Å, respectively, suggesting a strong binding to the active pocket (Figure 3A–E).

The three residues GLN421, SER422, and THR425 of the GFAT enzyme interact with two oxygen atoms of the 3-Methylpenta-2,4-dienoic acid substructure of ABA, while the remaining two residues CYS373, THR375 were forming a hydrogen bond with the oxygen atom of the 1-hydroxy-2,6,6-trimethyl-4-oxocyclohex-2-en-1-yl ring of ABA (Figure 3C). Similarly, seven residues of GFAT protein, GLY(A)374, SER(A)376, SER(A)420, GLY(A)423, SER(A)473, LYS(A)557 and SER(A)676, were forming Van der Waals interaction with ABA (Figure 3C–E). The low binding energy of −7.3 kcal/mol and inhibition constant of 5.21 μM (Table 1), and three alkyl interactions by LEU556, LYS675 and VAL677 residues with (1S)-1-hydroxy-2,6,6-trimethyl-4-oxocyclohex-2-en-1-yl ring helps in stabilizing the ABA bound with the active site. As a result, we may conclude that ABA is an effective inhibitor of the GFAT enzyme.

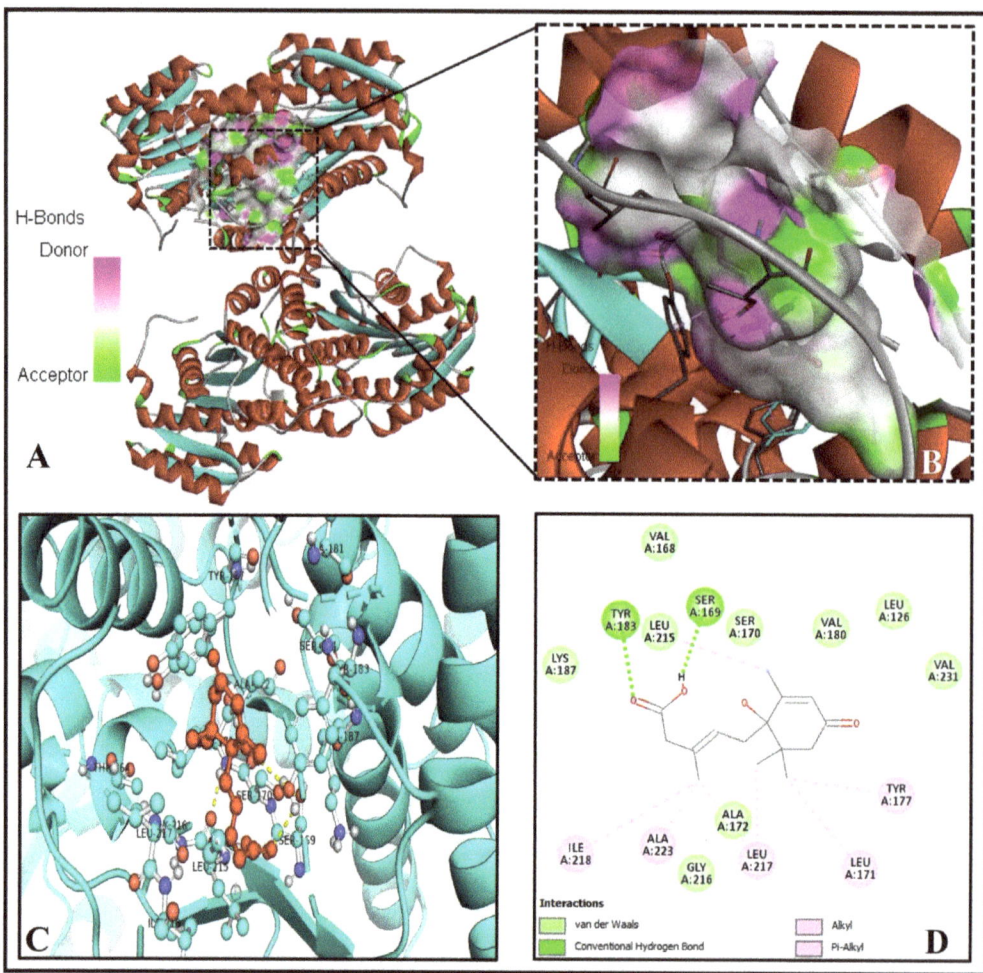

Figure 2. Significant molecular bonding of ABA with 11β-hydroxysteroid dehydrogenase1 [11β-HSD1 (PDB ID- 4K1L). (**A**) The coupling of ABA with active center residues. (**B**) The enlarged image shows the hydrogen bonds donor and acceptor amino acid residues in the junction cavity. (**C**) The 3D image shows a significant interaction of ABA (red ball and sticks) with functionally important residues (cyan ball and sticks) of human 11β-HSD1. (**D**) The 2D plot is showing the interaction of the binding pocket residues with the ABA inhibitor.

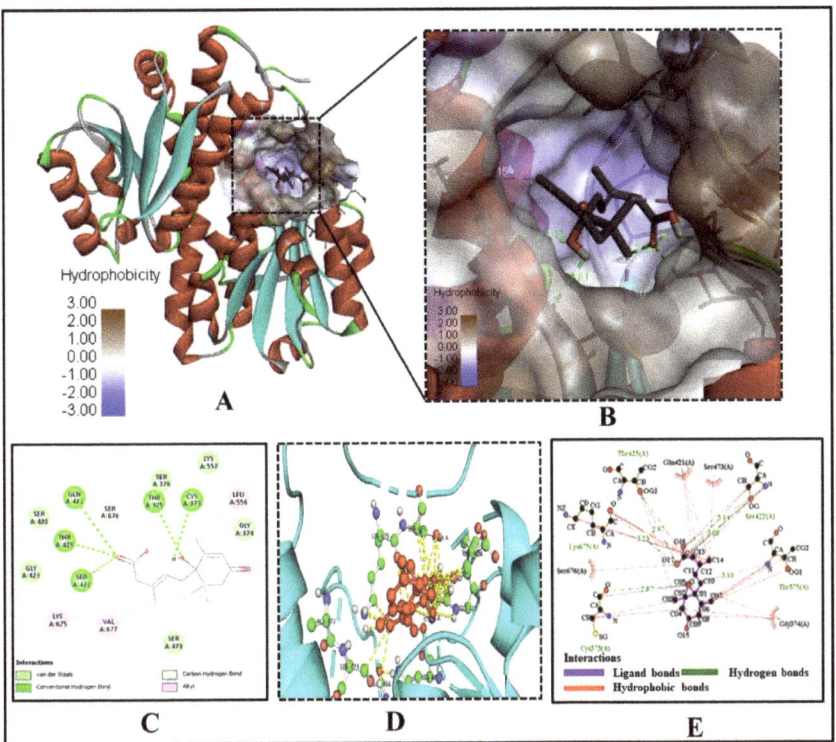

Figure 3. ABAs significant molecular binding with human glutamine-fructose-6-phosphate amidotransferase [GFAT (PDB ID- 2ZJ4)]. (**A**) ABA binding with the binding pocket. (**B**) Enlarged view showing the hydrophobicity of amino acid residues of the binding cavity (**C**) The 2D plot shows the molecular level interactions of binding pocket residues with the inhibitor ABA. (**D**) The 3D image shows strong interactions (yellow dashed lines) of ABA (red ball and sticks) with the functionally important residues (green ball and sticks) of human glutamine-fructose-6-phosphate amidotransferase. (**E**) The 2D LigPlot+ image shows the atomic level interactions of binding pocket residues with the ABA.

2.3. Binding of Abscisic Acid to the Catalytic Pocket of the Human Peroxisome Proliferator-Activated Receptor-Gamma (PPAR-Gamma)

PPAR-gamma is a major transcriptional factor (TF) that regulates adipogenesis, insulin sensitivity, and glucose homeostasis in humans [26,27]. The drug rosiglitazone, which acts as a ligand of PPAR-gamma, is an excellent insulin sensitizer, improving glucose absorption and lowering hyperglycemia and hyperinsulinemia [28–30].

The decreased ability of PPAR-gamma to bind DNA in response to rosiglitazone manifested the receptors' inability to activate transcription. The PPARs are also potential therapeutic targets that could treat atherosclerosis, inflammation, and hypertension. Studies showed that in the crystal structures of PPAR-gamma and rosiglitazone complex, binding pockets of the intact PPAR-gamma receptor interact with the rosiglitazone, especially with the Gln193, Tyr189, Leu196, Ala197, Tyr192, Glu203, Lys201, Arg202, Asp166, Lys336, Asn335, Asp337, Leu237, Phe347, Val248, Glu351, and Tyr250 residues [30].

Our investigation discovered that crucial binding pocket residues of the human PPAR-gamma, TYR(A)189 and TYR(D)250, form two hydrogen bonds with the two oxygen atoms of the 3-Methylpenta-2,4-dienoic acid substructure of ABA (Figure 4A–D).

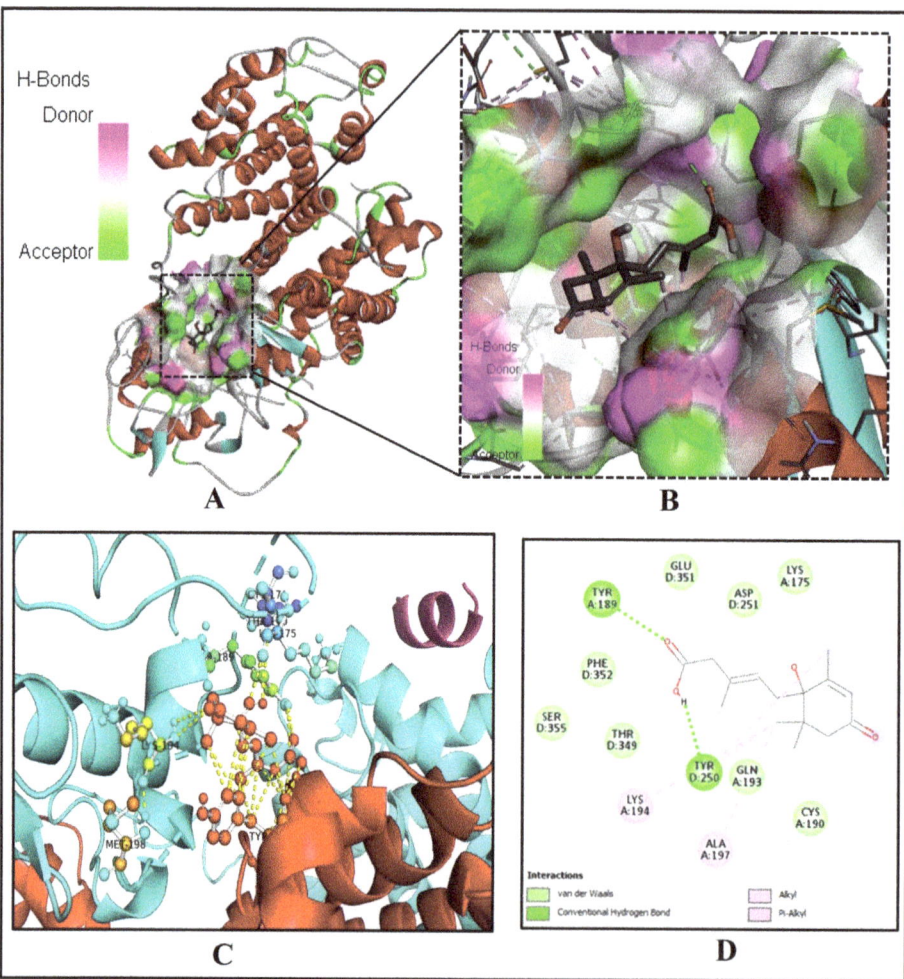

Figure 4. ABA docked with human peroxisome proliferator-activated receptor-gamma [PPAR-gamma (PDB ID: 3DZY)]. (**A**) The binding pocket residues of PPAR-gamma bind to ABA. (**B**) The binding cavity's hydrogen bond donor and acceptor amino acid residues are shown in a magnified view. (**C**) The 3D image shows significant interactions (yellow dashed lines) of ABA (red ball and sticks) with the residues of the human PPAR-gamma (green, yellow, cyan ball and sticks). (**D**) The molecular level interactions of binding-pocket residues with the ABA are depicted in the 2D plot.

The low binding energy of −7.3 kcal/mol and inhibition constant of 5.21 µM (Table 1) and two alkyl-pi-alkyl interactions by LYS194 and ALA197 with 1-hydroxy-2,6,6-trimethyl-4-oxocyclohex-2-en-1-yl ring of ABA may stabilize the ABA bounded with the active site (Figure 4D). Interestingly, eight residues of PPAR-gamma LYS(A)175, CYS(A)190, GLN(A)193, ASP(D)251, THR(D)349, GLU(D)351, PHE(D)352, and SER(D)355, were interacting via Van der Waals forces with ABA (Figure 4C,D).

Based on a similar binding and interaction pattern, we may conclude that ABA also acts as an inhibitor of PPAR-gamma, improving glucose absorption and lowering hyperglycemia.

2.4. The Binding Pattern of ABA with Human Mono-ADP-Ribosyl Transferase Sirtuin-6 (SIRT6)

SIRT6 is a prominent mammalian sirtuin (SIRT1–7) involved in various cellular processes such as glucose homeostasis maintenance, DNA repair, and longevity [31–33]. SIRT6 is NAD+- dependent deacetylases that target various acetylated proteins in mammals to regulate their cellular activity [34].

The monomeric crystal structure (2.0 Å) of human SIRT6 in complex with ADP-ribose showed that THR55, ASP61, PHE62, TRP69, ASP81, GLN111, HIS131, TRP186, GLY212, THR213, SER214, ILE217, TYR255, and VAL256 residues surround the ADP-ribose by hydrogen bonding network [31].

Our study revealed that binding pocket residues (GLN111, HIS131 of chain C) of the hexameric human SIRT6 protein make two hydrogen bonds with the one oxygen atom of the 3-Methylpenta-2,4-dienoic acid ($C_6H_8O_2$) substructure of ABA using molecular docking and pose prediction analysis (Figure 5A–D). Interestingly, four other residues of SIRT6 like ALA51, ARG63, ILE183 and LEU184 of chain C interact via Van der Waals forces with other atoms of ABA (Figure 5C,D).

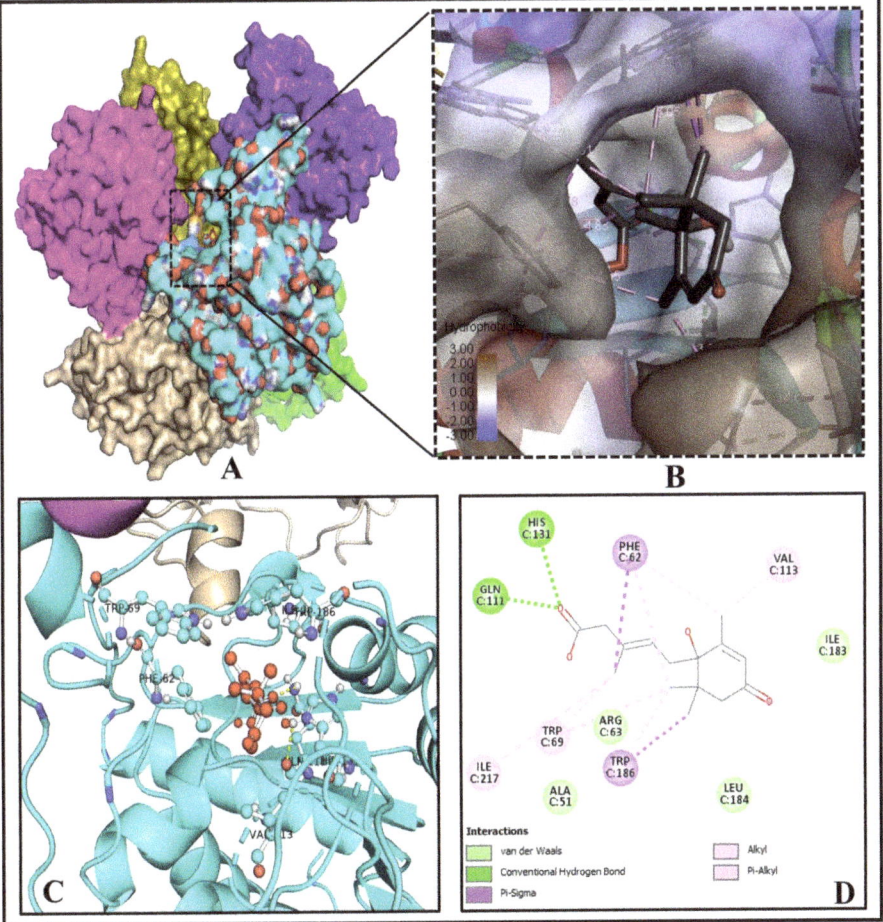

Figure 5. Molecular binding of ABA with human SIRT6 (PDB ID: 3K35). (**A**) ABA binds to the catalytic site. (**B**) Magnified view shows the hydrophobicity of amino acid residues of the binding cavity of SIRT 6 (**C**) 3D image showing interactions of ABA (red ball and sticks) with important residues of human SIRT6 (cyan ball and sticks). (**D**) 2D graph showing interactions at the molecular level pocket residues linked to the ABA.

2.5. The Binding Pattern of ABA with Glucokinase

Glucokinase is the most abundant hexokinase in the liver, and it plays a critical role in blood glucose homeostasis because it has strong control over hepatic glucose disposal and serves as the glucose sensor for insulin secretion in pancreatic β-cells [35]. Glucokinase is currently regarded as a promising target of anti-hyperglycemic medicines to control T2DM, but this protein target's mode of inhibition or activation is not fully understood.

There is a single published report on the crystal structure of human glucokinase (PDB ID- 4IXC) complexed with alpha-D-glucopyranose and (2S)-2-{[1-(3-chloropyridin-2-yl)-1H-pyrazolo[3,4-d]pyrimidin-4-yl]oxy}-N-(5-methylpyridin-2-yl)-3-(propan-2-yloxy) propanamide. These small molecules are regarded as activators of human glucokinase, but their exact mechanism of action and key amino residues involved in the interaction are not published yet.

Our docking and binding analysis exhibit a good binding pattern (binding energy = −6.8 Kcal/mol) of ABA with human glucokinase (PDB ID-4IXC) protein (Figure 6A,B), which is mediated through three hydrogen bonds forming residues ASN83, ARG85, and GLY229 and twelve via Van-der-Waals forces ASP78, GLY80, GLY81, PHE84, MET107, SER151, LYS169, ASP205, GLY227, THR228, GLY410, and SER445 (Figure 6C,D).

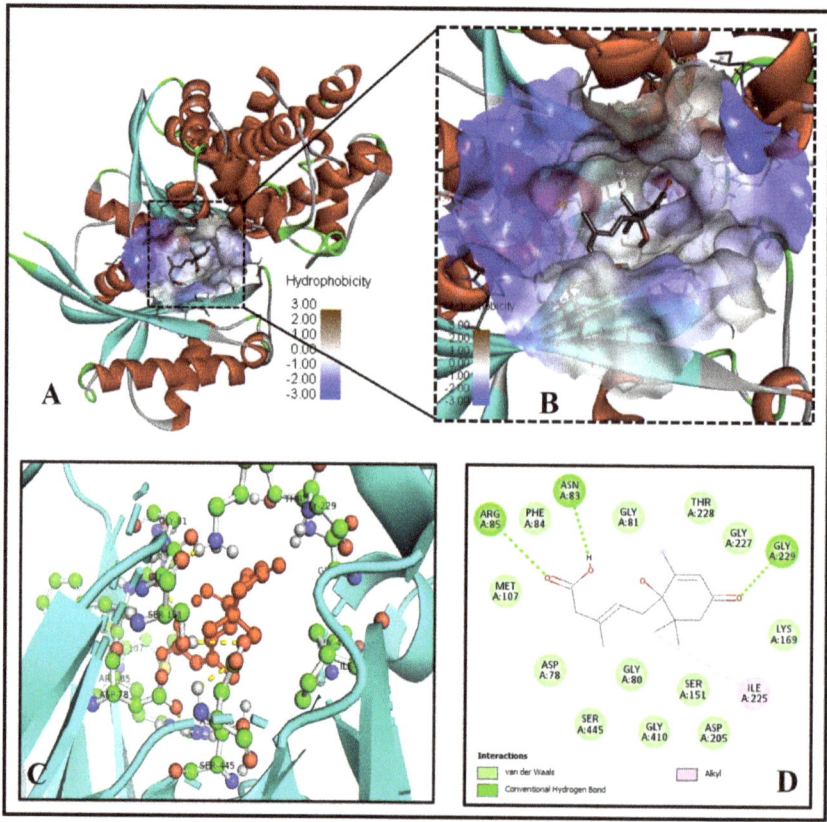

Figure 6. Human glucokinase (PDB ID- 4IXC) is docked with ABA. (**A**) ABA binds the catalytic site. (**B**) An enlarged image of the hydrophobicity of the binding cavity's amino acid residues (**C**) The 3D picture reveals substantial interactions (yellow dashed lines with bond length) of ABA (red ball and sticks) with human glucokinase residues (green ball and sticks). (**D**) The molecular level interactions of binding pocket residues with ABA are depicted in the 2D plot.

2.6. Analysis of the Molecular Binding Pattern of Abscisic Acid with Aldose Reductase

Aldose reductase is the rate-limiting enzyme in the polyol pathway. It converts excess D-glucose to D-sorbitol with NADPH as a co-factor [36]. It is crucial in the treatment of diabetic microvascular problems [37]. Aldose reductase is also involved in lipid metabolism.

Our docking and binding analysis exhibit a good binding pattern ($\Delta G = -6.6$ Kcal/mol) of ABA with aldose reductase (PDB ID- 3G5E) protein (Figure 7A–D), which is mediated through two hydrogen bonds forming residues TRP111, CYS298 and three via Van der Waals forces, GLU49, HIS110 and PHE121 (Figure 7C,D). Six alkyl-pi-alkyl interactions and one pi-pi sigma interaction by TRP20 with the 1-hydroxy-2,6,6-trimethyl-4-oxocyclohex-2-en-1-yl ring of ABA stabilize the ABA bounded with the active site of aldose reductase (Figure 7D).

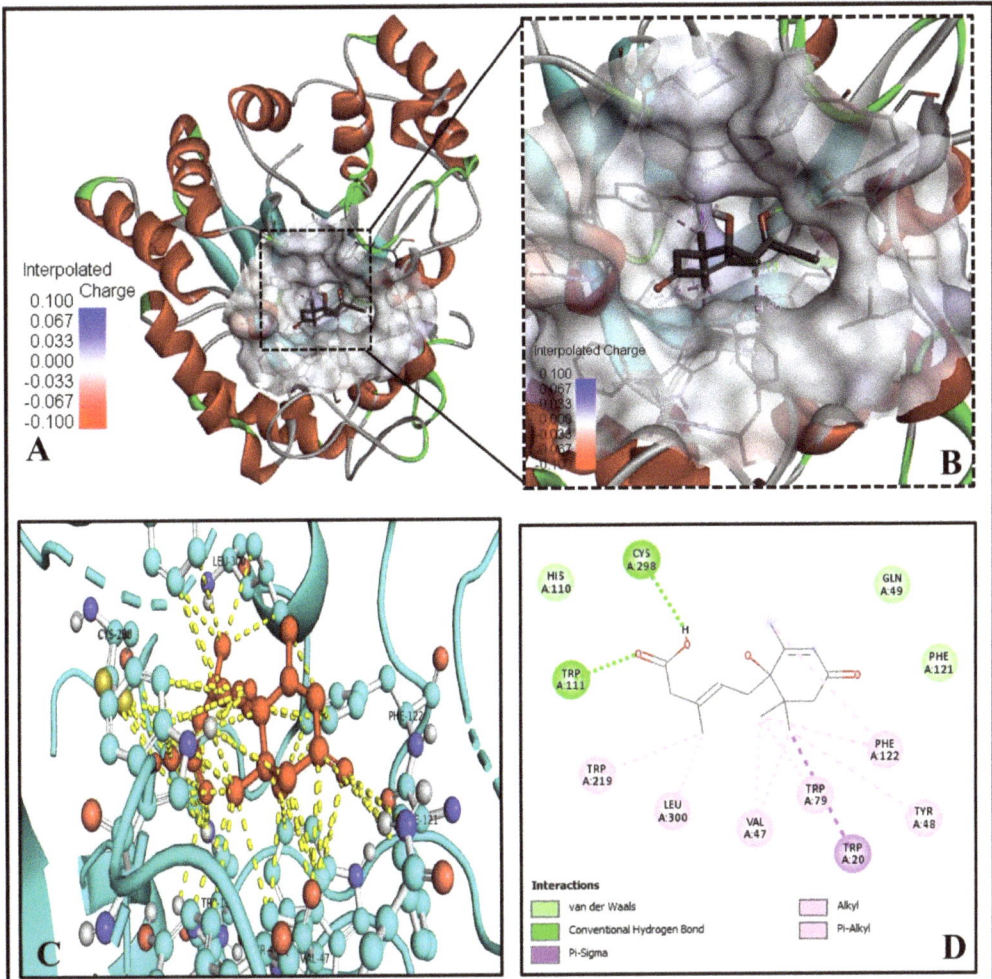

Figure 7. Molecular binding of ABA to aldose reductase (PDB ID: 3G5E). (**A**) Binding of ABA to the active site residues. (**B**) Magnified view showing the interpolated charge of amino acid residues of the binding cavity (**C**) The 3D image showing bonding (yellow dashed lines) of ABA (red ball and sticks) with residues of human aldose reductase (cyan ball and sticks). (**D**) 2D graph showing interactions at the molecular level of the binding compartment of aa residues with the ABA.

2.7. Analysis of Abscisic Acid and Glycogen Synthase Kinase-3 (GSK-3) Docked Complex

GSK-3, a unique multifunctional serine/threonine kinase, is involved in the glycolysis pathway. GSK-3 is active and capable of synthesizing glycogen in its unphosphorylated state. PKB/AKT phosphorylates GSK-3 on serine 9 in response to insulin binding [38]. As a result, it is critical to the insulin signaling pathway's transmission of regulatory and proliferative signals occurring at the cell membrane, potentially modulating blood glucose levels [38].

However, unlike previously mentioned proteins, GSK-3 (PDB ID -3F7Z) did not show a good binding pattern with ABA (binding energy = −6.6 Kcal/mol) (Figure 8A–D). There were no hydrogen bonds involved, only via Van der Waals forces, alkyl-pi-alkyl interaction, and pi-pi sigma interaction with ABA stabilizes the ABA bounded with the active site (Figure 8D).

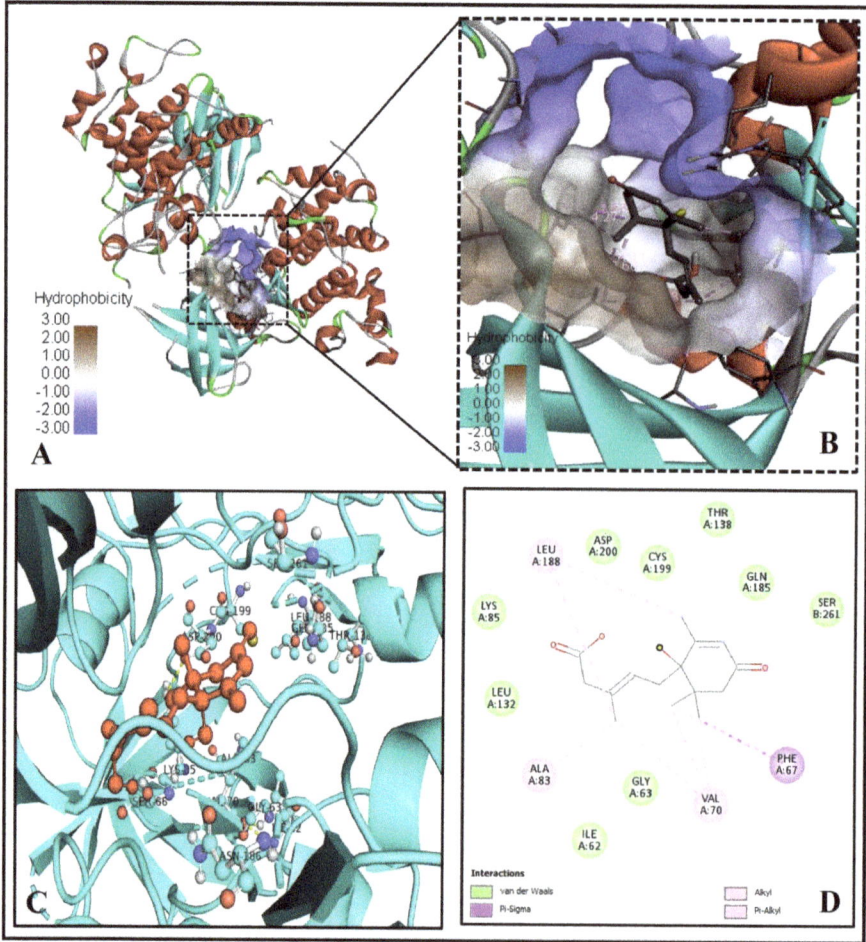

Figure 8. ABA shows mild interaction with human GSK3 (PDB ID: 3FZ7). (**A**) The ABA binds to the binding pocket near the catalytic center. (**B**) An enlarged image is showing the hydrophobicity of the binding cavity's amino acid residues. (**C**) The 3D picture reveals substantial interactions (yellow dashed lines) of ABA (red ball and sticks) with human GSK-3 residues (cyan ball and sticks). (**D**) The molecular level interactions of binding pocket residues with the ABA are depicted in the 2D plot.

2.8. Screening of Pyruvate Dehydrogenase Kinase (PKD) and Abscisic Acid Docked Complex

PKD negatively regulates the mitochondrial pyruvate dehydrogenase complex (PDC) activity by reversible phosphorylation. PDK isoforms are upregulated in obesity, diabetes, heart failure, and cancer and are potential therapeutic targets for these important human diseases [39].

Our analysis showed a poor binding pattern (binding energy = −6.3 Kcal/mol) of ABA with PKD (PDB ID- 4MP2) protein (Figure 9A–E) due to unfavorable repulsion. However, it forms hydrogen bonds, Van der Waals forces, alkyl-pi-alkyl interaction, and pi-pi sigma interaction with ABA (Figure 9C–E).

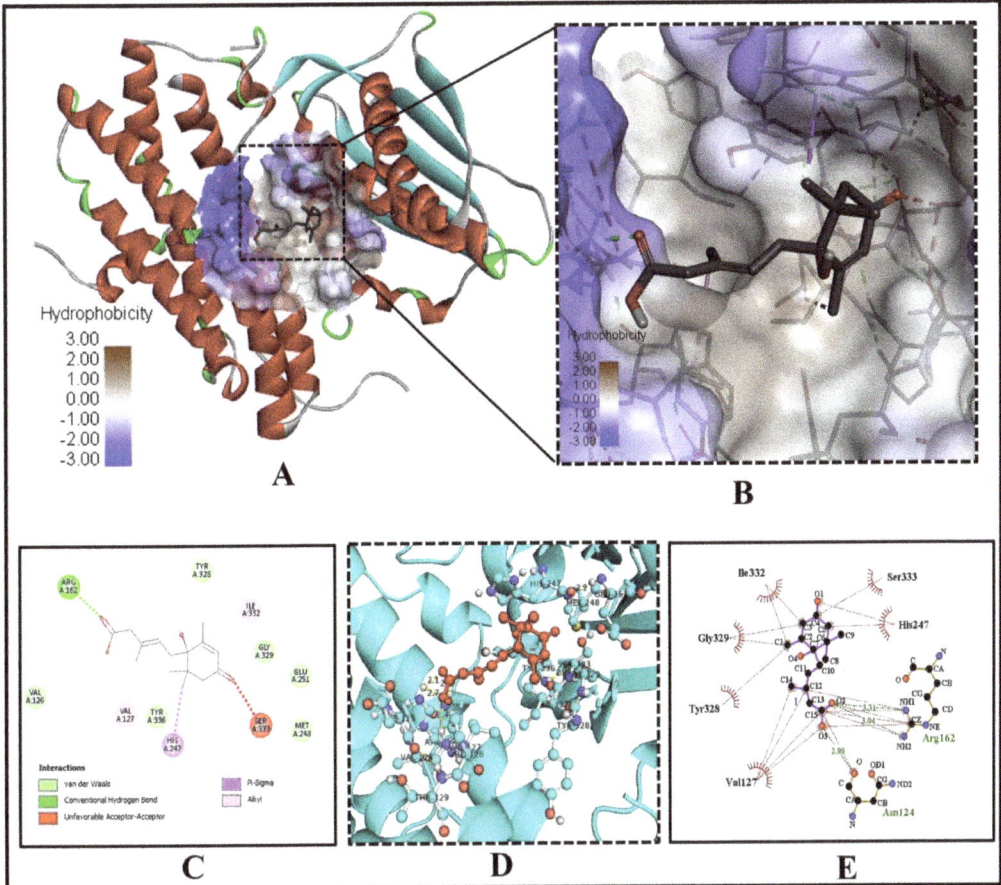

Figure 9. The docked complex of ABA with human pyruvate dehydrogenase kinase (PDB ID: 4MP2). (**A**) ABA binds with the catalytic site. (**B**) Magnified view showing the hydrophobicity of the amino acid fragments of the binding cavity (**C**) 2D graph showing the interactions at the molecular level of the bound cavity fragments with the inhibitor ABA. (**D**) The 3D image is showing interactions (with bond length) of ABA (red ball and sticks) with residues of human PKD (cyan ball and sticks). (**E**) 2D LigPlot is showing atomic level interactions of vesicle fragments of pyruvate dehydrogenase kinase bound to ABA.

2.9. Investigation of the Docked Complex of Tyrosine Kinase with Abscisic Acid

Further, ABA displayed a weak binding pattern (ΔG = −6.2 Kcal/mol) with the human tyrosine kinase (PDB ID-1IR3) protein (Figure 10A–D) due to unfavorable repulsion. However, it forms hydrogen bonds, Van der Waals forces, alkyl-pi-alkyl interaction, and pi-pi sigma interaction with ABA (Figure 10D).

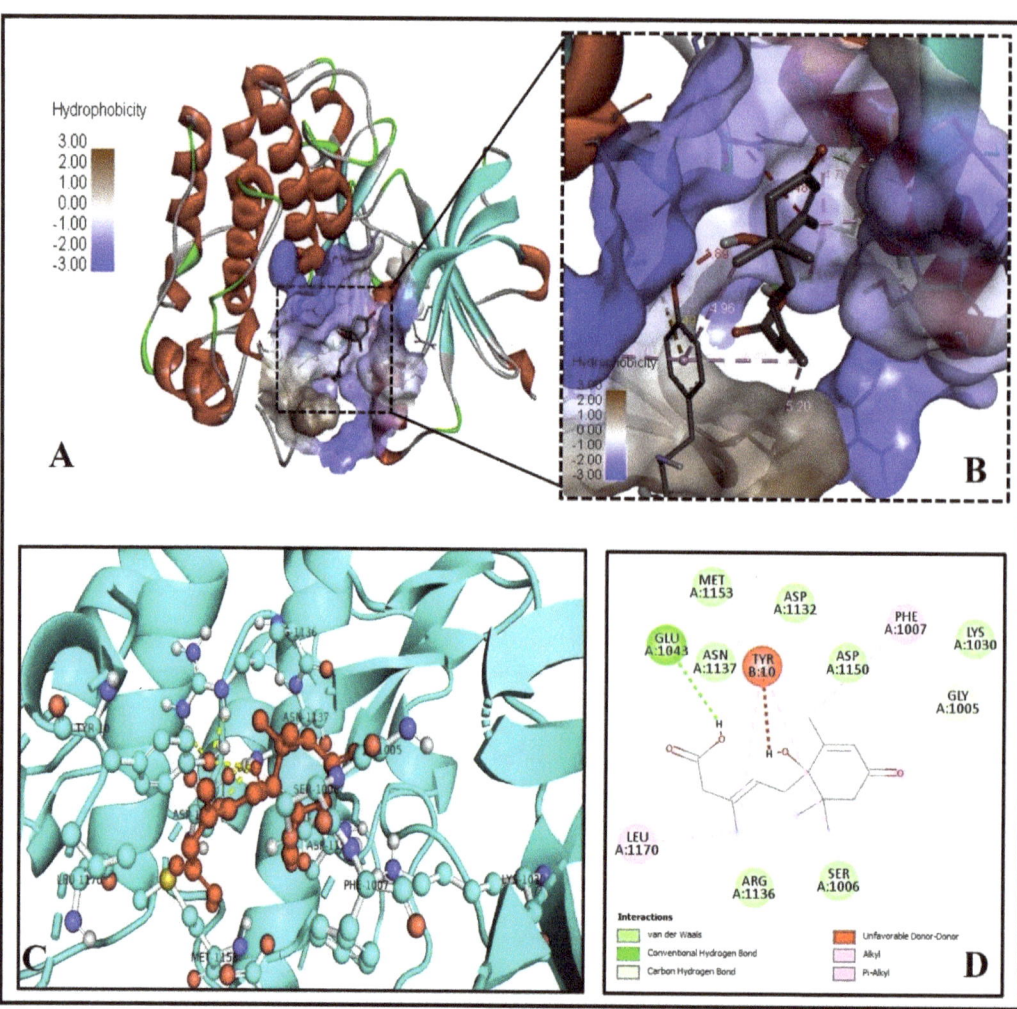

Figure 10. ABA molecular docking with human tyrosine kinase (PDB ID: 1IR3). (**A**) Binding of ABA to the active site of tyrosine kinase. (**B**) An enlarged image showing hydrophobicity of the binding cavity's amino acid residues (**C**) The 3D picture depicts substantial interactions of ABA (red ball and sticks) with human tyrosine kinase's essential residues (cyan ball and sticks) (**D**) The molecular level interactions of binding pocket residues with ABA are depicted in the 2D plot.

2.10. Computational Pharmacodynamics Screening of Abscisic Acid Ligand

The molinspiration bioactivity score (v2018.03) is calculated and presented (Table 2) for active drug-likeness towards parameters like ion channel modulators, kinase inhibitors, GPCR ligands, nuclear receptor ligands, protease inhibitors, and other enzyme inhibitors with scores for >100,000 average drug-like molecules. The score allows efficient separation of active and inactive molecules. The higher values of bioactivity score of nuclear receptor ligand, enzyme inhibitor, and Ion channel modulator of 1.06, 0.75, and 0.28, respectively, shows that ABA may act as an active inhibitor for different insulin receptor proteins.

Table 2. Prediction of bioactivity of ABA as an inhibitor against different insulin receptor proteins by molinspiration (v2018.03).

S. No	Parameters	Bioactivity Score
1	GPCR ligand	−0.01 ↓↓↓
2	Ion channel modulator	0.28 ↑
3	Kinase inhibitor	−0.61 ↓
4	Nuclear receptor ligand	1.06 ↑↑↑
5	Protease inhibitor	−0.20 ↓↓
6	Enzyme inhibitor	0.75 ↑↑

The upward arrow represents high bioactivity while downward arrow is showing low bioactivity of ABA.

2.11. In Silico Pharmacokinetics and ADMET Evaluation of Abscisic Acid

The pharmacokinetic properties and drug-likeness data are summarized in Table 3. According to the pharmacokinetic/ADMET properties, ABA showed high (96.712%) human intestinal absorption (HIA) and very low BBB permeability (−0.047 log BB). On the other hand, ABA did not affect Cytochrome P450 isomers (CYP1A2 and CYP2D6). The drug-likeness prediction was also performed using the Lipinski, Ghose, and Veber rules, as well as the bioavailability score. The Lipinski (Pfizer) filter is the pioneer to filter out any drug at the absorption or permeation level that an ideal drug has a molecular weight of less than 500 g/mol, a log P value of less than 5, and a maximum of 5 H-donor and 10 H-acceptor atoms [40]. The drug-likeness requirements are defined as follows by the Ghose filter (Amgen): The computed log P ranges from −0.4 to 5.6, the MW ranges from 160 to 480, the molar refractivity (MR) ranges from 40 to 130, and the total number of atoms ranges from 20 to 70 [41]. Veber (GSK) rule defines drug-likeness constraints as Rotatable bond count ≤ 10 and topological polar surface area (TPSA) ≤ 140 [42]. According to Martin et al., the bioavailability score was implemented to predict the probability of a compound to have at least 10% oral bioavailability in rat or measurable Caco-2 permeability [43]. AMES toxicity (non-mutagenic), hepatotoxicity, or skin sensitization was not found in the ABA. By the overall analysis of Table 3, we conclude that ABA does not violate any existing drug-likeness rules like Lipinski, Ghose, Veber, Egan and Muegge. Meanwhile, ABA has physicochemical, molecular, and ADMET properties between the upper and lower predicted values (Table 3 and Figure 11A,B).

Table 3. In silico pharmacokinetics, physico-chemical, ADMET properties, and drug-likeness of ABA.

Physicochemical Properties	Predicted Values	Absorption	Predicted Value	Distribution	Predicted Values	Metabolism	Predicted Value	Extraction and Toxicity	Predicted Value
LogP, LogS and LogD	2.342, −2.465 and 1.656	Water solubility logP	−2.253 mol/L	Volume distribution (VD) of a drug in blood plasmas	0.343 L/kg	CYP2D6 substrate and CYP3A4 substrate	No	Total drug clearance log (CLtot)	0.685 mL/min/kg
Molecular weight	264.32 g/mol	Lipd solibility LogP	1.96 Log Po/w	Plasma protein binding (PPB)	79.896%	CYP2D6 inhibitor	No	Renal organic cation transporter (OCT2) substrate	No
Number of hydrogen bond acceptors (nHA), donors (nHD), and rotatable bonds (nRot)	4, 2 and 3	Caco2 permeability	0.913 og Papp in 10−6 cm/s	The fraction unbound in blood plasmas (Fu)	9.191%	CYP3A4 inhibitor	No	AMES toxicity, hepatotoxicity, skin sensitization, hERG I & II inhibitor	No
Number of rings (nRing), rigid bonds (nRig), heteroatoms (nHet), and atoms in the biggest ring (MaxRing)	1, 10, 4 and 6	Log Kp skin permeability	−2.715 cm/s	BBB permeability	−0.047 log BB	CYP1A2 inhibitor	No	Max. tolerated dose (human)	0.304 log mg/kg/day
Formal charge (fChar)	0	Human intestinal absorption (HIA)	96.712%	CNS permeability	−2.913 log PS	CYP2C19 inhibitor	No	Oral rat acute toxicity (LD50)	1.793 mol/kg
Molecular total polar surface area (TPSA)	74.60 Å^2	P-glycoprotein substrate, P-glycoprotein I & II inhibitor	No	Bioavailability score	0.85	CYP2C9 inhibitor	No	T. Pyriformis toxicity	0.42 log ug/L

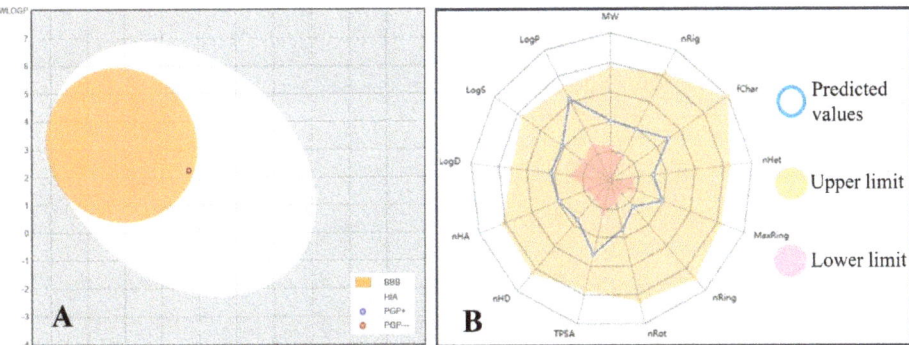

Figure 11. (**A**) A BOILED-Egg plot shows the blood-brain barrier (BBB) penetration and ABA molecules' human intestinal absorption (HIA). Here PGP- shows the P-glycoprotein substrate negative nature of ABA. (**B**) Radar graph showing upper, lower, and predicted values of various physicochemical and molecular properties of ABA (Abbreviations: MW = Molecular weight, nRig = Number of rigid bonds, fChar = Formal charge, nHet = Number of heteroatoms, MaxRing = Number of atoms in the biggest ring, nRing = Number of rings, nRot = Number of rotatable bonds, TPSA = Toplogical polar surface area, nHD = Number of hydrogen bond donors, nHA = Number of hydrogen bond acceptors, LogP = Log of the octanol/water partition coefficient, LogS = Log of the aqueous solubility, LogD = Log at physiological pH 7.4.

2.12. Boiled-Egg Plot and Radar Graph Analysis

A BOILED-Egg plot predicts the gastrointestinal absorption and brain penetration of small molecules. Here PGP+/− shows the P-glycoprotein substrate positive/negative nature of the molecule under study. The BOILED-Egg's white (white area) predicts that the molecule located in this area may be passively absorbed by the human intestinal tract (HIA). The ABA molecule (red circle) is located at the extreme periphery of BOILED-Egg's yolk (yellow area), which predicts that the ABA molecule may passively permeate through the blood-brain barrier (BBB) but have very low chances of −0.047 log BB, contrary to the chances of ABA being absorbed by the human intestinal tract (HIA) is 96.712% (Table 3 and Figure 11).

The radar graph shows that various physico-chemical and molecular characteristics of ABA like LogP (2.342), LogS (−2.465), LogD (1.656), molecular weight (264.32 g/mol), nHA (4), nHD (2), number of rotatable bonds (3), number of rings (1), rigid bonds (10), heteroatoms (4), and atoms in the biggest ring number of rings (6), formal charge (0), and topological polar surface area (74.60 Å^2) have optimum values within the upper and lower limit (Table 3 and Figure 11).

3. Materials and Methods

3.1. Retrieval and Preparation of Proteins and Ligand

The human proteins related to diabetes mellitus 11-β-hydroxysteroid dehydrogenase (PDB ID-4K1L), aldose reductase (PDB ID-3G5E), glucokinase (PDB ID-4IXC), glycogen synthase kinase-3 (PDB ID-3F7Z), glutamine:fructose-6-phosphate amidotransferase [GFAT (PDB ID-2ZJ4)], pyruvate dehydrogenase kinase (PDB ID-4MP2), peroxisome proliferator-activated receptor-gamma (PDB ID-3DZY), sirtuin family of NAD(+)-dependent protein deacetylases [SIRT6 (PDB ID-3K35)], and Tyrosine kinase (PDB ID-1IR3) were retrieved from the RCSB Protein Data Bank (Rutgers University, NJ, USA) (Figure 1) [44]. The human pyruvate dehydrogenase kinase (PDB ID: 4MP2) protein molecule showed conformational error in 12 amino acid residues (VAL373, ARG370, VAL357, MET349, PHE347, VAL261, HIS247, HIS237, LEU211, GLN82, SER29, SER27), which were rectified by protein preparation and energy minimization wizard employing Discovery Studio version 2.0. The remaining PDB IDs did not show any structural error.

The ligand molecule ABA [(PubChem ID-5280896), IUPAC name- (2Z,4E)-5-[(1S)-1-hydroxy-2,6,6-trimethyl-4-oxocyclohex-2-en-1-yl]-3-methylpenta-2,4-dienoic acid)] in 2D (SDF) format was downloaded from PubChem (https://pubchem.ncbi.nlm.nih.gov/compound/5280896) (accessed on 15 April 2021) (Figure 1). The ligand molecule was converted into 3D format (.mol2 and .pdb) employing the Chem3D 16.0 module of ChemOffice 2016 software suit.

Before docking, the protein structures downloaded from PDB were analyzed in PyMol software (The PyMOL Molecular Graphics System, Version 2.0 Schrödinger, LLC, San Diego, CA, USA), and the already docked ligands or nucleic acid or heteroatoms or water molecules were removed from the X-ray crystallographic protein-ligand complexes. Then the pure proteins as a receptor were prepared in Swiss-Pdb viewer (version 4.1.0) by optimizing bonded atoms, angles, torsions, non-bonded atoms, and improper atoms of the protein backbone and side chains.

3.2. Molecular Docking

The docking calculations were performed using the AutoDock Vina version 4.2 (ADT4.2) software suite [45]. The receptor proteins were solvated with water, and only polar hydrogens were added. The receptor grid boxes (in X, Y, Z dimension) were prepared in the ADT4.2, and the pdbqt files of proteins were generated. Similarly, the ligand was prepared with default parameters, and only Gasteiger charges were added. Flexible Ligand docking was performed applying the Lamarckian Genetic Algorithm with an exhaustiveness value of eight. The contributions of intramolecular hydrogen bonds, hydrophobic, ionic, and Van der Waals interactions between docked protein and ligand complexes were used to determine the free energy (ΔG) specifying affinity scoring of the binding. The docking poses were narrowed down using the force field's free binding energy computation. After the docked protein-ligand complexes were created, the binding sites were analyzed to construct a 2D representation of the ligand interaction for each complex.

3.3. Post-Docking Protein-Ligand Interaction Analysis

The visualization and analysis of protein-ligand complexes were performed using PyMOL software (The PyMOL Molecular Graphics System, Version 2.0 Schrödinger, LLC, San Diego, CA, USA). The receptor's active sites and interactions with the ligand or drug were determined using the PDBe [46] and PDBsum [47] servers. The protein-ligand complexes were further visualized in the Discovery Studio client v21.1.0.20298, Dassault Systemes Biovia Corp, to show the 2D diagram of ligand-receptor interaction. LigPlot+ and Maestro12.4 (Schrodinger-2020-2) were applied to visualize ligands' exact atomic level interaction with their corresponding receptor atoms.

3.4. Calculation of Inhibition Constant

Moreover, it is concluded that ABA may act as a competitive inhibitor that should compete with the known substrates to the active centers of the protein targets relevant to diabetes mellitus. Therefore, it induces a competitive type of inhibition, that inhibitors could bind to only the free enzyme and formed reversible enzyme-inhibitor (E-I) complexes. An enzyme-inhibitor complex's inhibition constant (Ki) is traditionally calculated by the basic equations of enzyme kinetics of the Lineweaver–Burk assay extrapolated on 2D plots. If there is inconsistency in the Lineweaver–Burk plots, non-linear regression of the Michaelis–Menten equation is used to validate the related constants obtained.

Sophisticated arithmetic and analytical in silico algorithms have been proposed to compute the inhibition constant (Ki) parameter, since the Ki principally depends on the binding (or association) constant (Kb) and dissociation constant (Kd) of an enzyme-inhibitor complex, which occurs in opposite directions (ln Kb = −ln Kd).

$$\Delta G = (R \times T) \ln Ki$$

Therefore, Ki is computationally calculated using the following formula:

$$Ki = \exp(\Delta G/(R \times T))$$

The binding energy ΔG is in kcal/mol, the universal gas constant R = 1.987 kcal/K/mol, at room temperature (25 °C) T = 273 + 25 = 298 K. Ki is having a unit of mM.

3.5. Screening of Ligand Abscisic Acid for Pharmacodynamics Properties

The molinspiration (https://www.molinspiration.com/cgi-bin/properties), (accessed on 29 May 2021), an online screening server based on sophisticated Bayesian statistics, was implemented to analyze the pharmacodynamics properties of ABA. It compares representative ligands' structures and determines physico-chemical properties of a particular molecule for being active molecules. There is no need to know about the target's 3D structure or binding mode. The trained model makes it possible to screen large libraries of hundreds of thousands of molecules in less than an hour to identify molecules with the highest chance of becoming active drugs, pesticides, irritants, or toxic substances. The larger the value of the bioactivity score is, the higher the probability that the particular molecule will be active.

3.6. Screening of Ligand Abscisic Acid for Pharmacokinetics and Drug-Likeness

The main causes of drug development failure are undesirable pharmacokinetics and toxicity of candidate molecules. Absorption, distribution, metabolism, excretion, and toxicity (ADMET) of chemicals have long been recognized as important considerations in the early stages of CADD.

The pkCSM [48], SwissADME [49] and ADMETLAB2.0 [50] are free web tools to evaluate Pharmacokinetics, drug-likeness, and medicinal chemistry of small molecules based on very extensive experimental data sets. The SMILES format of the molecule was entered, and 2D structure files were generated in SwissADME, pkCSM, and ADMETLAB2.0. The pkCSM is an authentic source (collaboratively developed by Instituto Rene Rachou Fiocruz Minas, The University of Melbourne and University of Cambridge) to predict small-molecule pharmacokinetics using graph-based signatures. Several parameters are analyzed to check the ADMET properties of a small molecule or inhibitor.

SwissADME [51] of Swiss Institute of Bioinformatics used to evaluate the pharmacokinetics and drug-likeness ADMET behaviors of compounds [52] employing support vector machine (SVM) algorithm [53] with well-characterized large datasets of known inhibitors/non-inhibitors as well as substrates/non-substrates.

ADMETlab 2.0 has a greater capacity to assist medicinal chemists in accelerating the drug research and development process. It allows users to calculate and predict 17 physicochemical parameters, 13 medicinal chemistry measures, 23 ADMET endpoints, 27 toxicity endpoints, and eight toxicophore rules (751 substructures) quickly and easily, allowing them to identify interesting lead compounds for further investigation.

The major target of the study was to examine if the substance in question inhibited the cytochrome P450 (CYP) family's CYP1A2 and CYP2D6 isoforms. Pharmacokinetics parameters such as human intestinal absorption, P-glycoprotein, and the BBB and drug-likeness prediction Lipinski, Ghose, and Veber criteria, as well as the bioavailability score, are crucial in judging the molecule [41,42,54]. According to several essential criteria such as molecular weight, LogP, number of HPA, and HBD, the Lipinski, Ghose, and Veber guidelines were used to assess drug-likeness to determine whether a compound is likely to be bioactive.

According to Lipinski's "Rule of 5" most "drug-like" compounds have logP $\leq$ 5, molecular weight (MW) $\leq$ 500, number of hydrogen bond acceptors (nHA) $\leq$ 10, and number of hydrogen bond donors (nHD) $\leq$ 5 [55]. The molecule that violates more than one of these principles may have problems with bioavailability. The methodology calculates logP (octanol/water partition coefficient) as a sum of fragment-based contributions and correction factors. This approach is quite reliable, and it may be used to analyze almost any organic or organometallic compound.

Topological Polar Surface Area (TPSA) is calculated based on the methodology published by Ertl et al. (2000) depending on the fragment contributions [56]. TPSA is defined as the sum of surfaces of polar atoms (typically oxygens, nitrogens, and linked hydrogens) in a molecule. TPSA is an ideal descriptor characterizing drug absorption, including human intestinal absorption, bioavailability, Caco-2 (human epithelial colorectal adenocarcinoma cell line), monolayers permeability, and BBB penetration. These parameters are quite important in predicting drug transport qualities. The number of rotatable bonds (nRot) is a simple topological parameter that measures molecular flexibility. It is a very good descriptor of the oral bioavailability of drugs [42]. Any single non-ring bond bounded to a non-terminal heavy (i.e., non-hydrogen) atom is termed a rotatable bond. Because of their large rotational energy barrier, amide C-N bonds are not considered.

4. Conclusions

Effective medications with no cytotoxicity are needed to treat diabetes mellitus, and phytohormones such as ABA are among the best natural extract with no side effects. Molecular docking investigation of ABA with nine different protein targets relevant to diabetes mellitus revealed four potential target proteins perfectly docks with ABA. Docking analysis also revealed that based on binding energy (ΔG) and predicted inhibition constant (pKi), 11β-HSD1 (4K1L) showed best binding with ABA, followed by GFAT (2ZJ4), PPAR-gamma (3DZY), and SIRT6 (3K35), which were equal in inhibition constant. The docking and interaction pattern of ligands were fantastically able to interact with the key residues of the catalytic cavity of the enzyme or located in the very close proximity of the active sites of these proteins. Following all current drug-likeness guidelines such as Lipinski, Ghose, Veber, Egan, and Muegge, the pharmacodynamic and pharmacokinetic features with ADMET study revealed that ABA could be taken best molecule without any hazardous effect. A BOILED-Egg plot and radar graph analysis confirm that all the molecular and physico-chemical properties of ABA are within the upper and lower limit fulfilling all the criteria of an ideal drug. Thus, ABA can be considered a potential candidate for developing a potent anti-diabetic drug and a promising bioactive compound of okra for developing nutraceuticals and functional foods.

Author Contributions: Conceptualization, S.A.A., M.S.B. and A.E.O.E.; methodology, M.S.B., K.M. and N.E.E.; validation, A.K., M.A.K., M.A. and A.E.O.E.; formal analysis, M.S.B., M.A. and S.A.A.; investigation, S.A.A., M.S.B. and K.M.; data curation, K.M., A.E.O.E., N.E.E. and M.A.; writing—original draft preparation, S.A.A. and M.S.B.; writing—review and editing, A.E.O.E., A.K. and M.A.K.; visualization, K.M., M.A. and S.A.A.; supervision, A.E.O.E., A.K. and K.M.; project administration, S.A.A., A.E.O.E. and K.M. All authors have read and agreed to the published version of the manuscript.

Funding: This research has been funded by Scientific Research Deanship at University of Hai'l, Saudi Arabia through project number RG-191333.

Institutional Review Board Statement: Not applicable.

Informed Consent Statement: Not applicable.

Data Availability Statement: All data generated or analyzed during this study are included in this article.

Conflicts of Interest: The authors declare no conflict of interest. The funders had no role in the design of the study, in the collection, analyses, or interpretation of data, in the writing of the manuscript, or in the decision to publish the results.

Sample Availability: Samples of the compounds are not available from the authors.

References

1. Ashraf, S.A.; Elkhalifa, A.E.O.; Siddiqui, A.J.; Patel, M.; Awadelkareem, A.M.; Snoussi, M.; Ashraf, M.S.; Adnan, M.; Hadi, S. Cordycepin for Health and Wellbeing: A Potent Bioactive Metabolite of an Entomopathogenic Cordyceps Medicinal Fungus and Its Nutraceutical and Therapeutic Potential. *Molecules* **2020**, *25*, 2735. [CrossRef]
2. Unnikrishnan, R.; Misra, A. Infections and diabetes: Risks and mitigation with reference to India. *Diabetes Metab. Syndr.* **2020**, *14*, 1889–1894. [CrossRef]
3. Damián-Medina, K.; Salinas-Moreno, Y.; Milenkovic, D.; Figueroa-Yáñez, L.; Marino-Marmolejo, E.; Higuera-Ciapara, I.; Vallejo-Cardona, A.; Lugo-Cervantes, E. In silico analysis of antidiabetic potential of phenolic compounds from blue corn (*Zea mays* L.) and black bean (*Phaseolus vulgaris* L.). *Heliyon* **2020**, *6*, e03632. [CrossRef]
4. Hameed, I.; Masoodi, S.R.; Mir, S.A.; Nabi, M.; Ghazanfar, K.; Ganai, B.A. Type 2 diabetes mellitus: From a metabolic disorder to an inflammatory condition. *World J. Diabetes* **2015**, *6*, 598–612. [CrossRef]
5. Elkhalifa, A.E.O.; Al-Shammari, E.; Adnan, M.; Alcantara, J.C.; Mehmood, K.; Eltoum, N.E.; Awadelkareem, A.M.; Khan, M.A.; Ashraf, S.A. Development and Characterization of Novel Biopolymer Derived from *Abelmoschus esculentus* L. Extract and Its Antidiabetic Potential. *Molecules* **2021**, *26*, 3609. [CrossRef] [PubMed]
6. Zhu, X.-M.; Xu, R.; Wang, H.; Chen, J.-Y.; Tu, Z.-C. Structural Properties, Bioactivities, and Applications of Polysaccharides from Okra [*Abelmoschus esculentus* (L.) Moench]: A Review. *J. Agric. Food Chem.* **2020**, *68*, 14091–14103. [CrossRef] [PubMed]
7. Elkhalifa, A.E.O.; Alshammari, E.; Adnan, M.; Alcantara, J.C.; Awadelkareem, A.M.; Eltoum, N.E.; Mehmood, K.; Panda, B.P.; Ashraf, S.A. Okra (*Abelmoschus Esculentus*) as a Potential Dietary Medicine with Nutraceutical Importance for Sustainable Health Applications. *Molecules* **2021**, *26*, 696. [CrossRef] [PubMed]
8. Durazzo, A.; Lucarini, M.; Novellino, E.; Souto, E.B.; Daliu, P.; Santini, A. *Abelmoschus esculentus* (L.): Bioactive Components' Beneficial Properties-Focused on Antidiabetic Role-For Sustainable Health Applications. *Molecules* **2018**, *24*, 38. [CrossRef]
9. Daliu, P.; Annunziata, G.; Tenore, G.C.; Santini, A. Abscisic acid identification in Okra, *Abelmoschus esculentus* L. (Moench): Perspective nutraceutical use for the treatment of diabetes. *Nat. Prod. Res.* **2020**, *34*, 3–9. [CrossRef] [PubMed]
10. Sanchez-Perez, A.M. Abscisic acid, a promising therapeutic molecule to prevent Alzheimer's and neurodegenerative diseases. *Neural Regen. Res.* **2020**, *15*, 1035–1036. [CrossRef] [PubMed]
11. Jung, Y.; Cackowski, F.C.; Yumoto, K.; Decker, A.M.; Wang, Y.; Hotchkin, M.; Lee, E.; Buttitta, L.; Taichman, R.S. Abscisic acid regulates dormancy of prostate cancer disseminated tumor cells in the bone marrow. *Neoplasia* **2021**, *23*, 102–111. [CrossRef] [PubMed]
12. Tomlinson, J.W.; Walker, E.A.; Bujalska, I.J.; Draper, N.; Lavery, G.G.; Cooper, M.S.; Hewison, M.; Stewart, P.M. 11beta-hydroxysteroid dehydrogenase type 1: A tissue-specific regulator of glucocorticoid response. *Endocr. Rev.* **2004**, *25*, 831–866. [CrossRef] [PubMed]
13. Morton, N.M.; Paterson, J.M.; Masuzaki, H.; Holmes, M.C.; Staels, B.; Fievet, C.; Walker, B.R.; Flier, J.S.; Mullins, J.J.; Seckl, J.R. Novel adipose tissue-mediated resistance to diet-induced visceral obesity in 11 beta-hydroxysteroid dehydrogenase type 1-deficient mice. *Diabetes* **2004**, *53*, 931–938. [CrossRef]
14. Morgan, S.A.; Sherlock, M.; Gathercole, L.L.; Lavery, G.G.; Lenaghan, C.; Bujalska, I.J.; Laber, D.; Yu, A.; Convey, G.; Mayers, R.; et al. 11beta-hydroxysteroid dehydrogenase type 1 regulates glucocorticoid-induced insulin resistance in skeletal muscle. *Diabetes* **2009**, *58*, 2506–2515. [CrossRef] [PubMed]
15. Hollis, G.; Huber, R. 11β-Hydroxysteroid dehydrogenase type 1 inhibition in type 2 diabetes mellitus. *Diabetes Obes. Metab.* **2011**, *13*, 1–6. [CrossRef]
16. Alberts, P.; Nilsson, C.; Selen, G.; Engblom, L.O.; Edling, N.H.; Norling, S.; Klingström, G.; Larsson, C.; Forsgren, M.; Ashkzari, M.; et al. Selective inhibition of 11 beta-hydroxysteroid dehydrogenase type 1 improves hepatic insulin sensitivity in hyperglycemic mice strains. *Endocrinology* **2003**, *144*, 4755–4762. [CrossRef] [PubMed]
17. Böhme, T.; Engel, C.K.; Farjot, G.; Güssregen, S.; Haack, T.; Tschank, G.; Ritter, K. 1,1-Dioxo-5,6-dihydro-[1,2,4]oxathiazines, a novel class of 11ß-HSD1 inhibitors for the treatment of diabetes. *Bioorganic Med. Chem. Lett.* **2013**, *23*, 4685–4691. [CrossRef]
18. Ryu, J.H.; Kim, S.; Lee, J.A.; Han, H.Y.; Son, H.J.; Lee, H.J.; Kim, Y.H.; Kim, J.S.; Park, H.G. Synthesis and optimization of picolinamide derivatives as a novel class of 11β-hydroxysteroid dehydrogenase type 1 (11β-HSD1) inhibitors. *Bioorg. Med. Chem. Lett.* **2015**, *25*, 1679–1683. [CrossRef] [PubMed]
19. Koike, T.; Shiraki, R.; Sasuga, D.; Hosaka, M.; Kawano, T.; Fukudome, H.; Kurosawa, K.; Moritomo, A.; Mimasu, S.; Ishii, H.; et al. Discovery and Biological Evaluation of Potent and Orally Active Human 11β-Hydroxysteroid Dehydrogenase Type 1 Inhibitors for the Treatment of Type 2 Diabetes Mellitus. *Chem. Pharm. Bull.* **2019**, *67*, 824–838. [CrossRef]
20. Nakaishi, Y.; Bando, M.; Shimizu, H.; Watanabe, K.; Goto, F.; Tsuge, H.; Kondo, K.; Komatsu, M. Structural analysis of human glutamine: Fructose-6-phosphate amidotransferase, a key regulator in type 2 diabetes. *FEBS Lett.* **2009**, *583*, 163–167. [CrossRef]
21. Shanak, S.; Saad, B.; Zaid, H. Metabolic and Epigenetic Action Mechanisms of Antidiabetic Medicinal Plants. *Evid.-Based Complementary Altern. Med. ECAM* **2019**, *2019*, 3583067. [CrossRef] [PubMed]
22. Belete, T.M. A Recent Achievement In the Discovery and Development of Novel Targets for the Treatment of Type-2 Diabetes Mellitus. *J. Exp. Pharmacol.* **2020**, *12*, 1–15. [CrossRef] [PubMed]

23. Lindsley, J.E.; Rutter, J. Nutrient sensing and metabolic decisions. *Comp. Biochem. Physiol. Part. B Biochem. Mol. Biol.* **2004**, *139*, 543–559. [CrossRef] [PubMed]
24. Zhang, H.; Jia, Y.; Cooper, J.J.; Hale, T.; Zhang, Z.; Elbein, S.C. Common variants in glutamine: Fructose-6-phosphate amidotransferase 2 (GFPT2) gene are associated with type 2 diabetes, diabetic nephropathy, and increased GFPT2 mRNA levels. *J. Clin. Endocrinol. Metab.* **2004**, *89*, 748–755. [CrossRef]
25. Buse, M.G. Hexosamines, insulin resistance, and the complications of diabetes: Current status. *Am. J. Physiol. Endocrinol. Metab.* **2006**, *290*, E1–E8. [CrossRef]
26. Savage, D.B.; Sewter, C.P.; Klenk, E.S.; Segal, D.G.; Vidal-Puig, A.; Considine, R.V.; O'Rahilly, S. Resistin/Fizz3 expression in relation to obesity and peroxisome proliferator-activated receptor-gamma action in humans. *Diabetes* **2001**, *50*, 2199–2202. [CrossRef]
27. Salam, N.K.; Huang, T.H.; Kota, B.P.; Kim, M.S.; Li, Y.; Hibbs, D.E. Novel PPAR-gamma agonists identified from a natural product library: A virtual screening, induced-fit docking and biological assay study. *Chem. Biol. Drug Des.* **2008**, *71*, 57–70. [CrossRef]
28. Olefsky, J.M.; Saltiel, A.R. PPAR gamma and the treatment of insulin resistance. *Trends Endocrinol. Metab. TEM* **2000**, *11*, 362–368. [CrossRef]
29. Staels, B. PPAR Agonists and the Metabolic Syndrome. *Therapies* **2007**, *62*, 319–326. [CrossRef]
30. Chandra, V.; Huang, P.; Hamuro, Y.; Raghuram, S.; Wang, Y.; Burris, T.P.; Rastinejad, F. Structure of the intact PPAR-gamma-RXR-nuclear receptor complex on DNA. *Nature* **2008**, *456*, 350–356. [CrossRef]
31. Pan, P.W.; Feldman, J.L.; Devries, M.K.; Dong, A.; Edwards, A.M.; Denu, J.M. Structure and biochemical functions of SIRT6. *J. Biol. Chem.* **2011**, *286*, 14575–14587. [CrossRef] [PubMed]
32. Kugel, S.; Mostoslavsky, R. Chromatin and beyond: The multitasking roles for SIRT6. *Trends Biochem. Sci.* **2014**, *39*, 72–81. [CrossRef] [PubMed]
33. Gertman, O.; Omer, D.; Hendler, A.; Stein, D.; Onn, L.; Khukhin, Y.; Portillo, M.; Zarivach, R.; Cohen, H.Y.; Toiber, D.; et al. Directed evolution of SIRT6 for improved deacylation and glucose homeostasis maintenance. *Sci. Rep.* **2018**, *8*, 3538. [CrossRef] [PubMed]
34. Blander, G.; Guarente, L. The Sir2 Family of Protein Deacetylases. *Annu. Rev. Biochem.* **2004**, *73*, 417–435. [CrossRef] [PubMed]
35. Agius, L. Targeting hepatic glucokinase in type 2 diabetes: Weighing the benefits and risks. *Diabetes* **2009**, *58*, 18–20. [CrossRef] [PubMed]
36. El-Kabbani, O.; Ruiz, F.; Darmanin, C.; Chung, R.P. Aldose reductase structures: Implications for mechanism and inhibition. *Cell. Mol. Life Sci. CMLS* **2004**, *61*, 750–762. [CrossRef] [PubMed]
37. Kaul, C.L.; Ramarao, P. The role of aldose reductase inhibitors in diabetic complications: Recent trends. *Methods Find. Exp. Clin. Pharmacol.* **2001**, *23*, 465–475. [CrossRef] [PubMed]
38. Doble, B.W.; Woodgett, J.R. GSK-3: Tricks of the trade for a multi-tasking kinase. *J. Cell Sci.* **2003**, *116*, 1175–1186. [CrossRef]
39. Tso, S.C.; Qi, X.; Gui, W.J.; Wu, C.Y.; Chuang, J.L.; Wernstedt-Asterholm, I.; Morlock, L.K.; Owens, K.R.; Scherer, P.E.; Williams, N.S.; et al. Structure-guided development of specific pyruvate dehydrogenase kinase inhibitors targeting the ATP-binding pocket. *J. Biol. Chem.* **2014**, *289*, 4432–4443. [CrossRef]
40. Lipinski, C.A.; Lombardo, F.; Dominy, B.W.; Feeney, P.J. Experimental and computational approaches to estimate solubility and permeability in drug discovery and development settings. *Adv. Drug Deliv. Rev.* **2012**, *64*, 4–17. [CrossRef]
41. Ghose, A.K.; Viswanadhan, V.N.; Wendoloski, J.J. A knowledge-based approach in designing combinatorial or medicinal chemistry libraries for drug discovery. 1. A qualitative and quantitative characterization of known drug databases. *J. Comb. Chem.* **1999**, *1*, 55–68. [CrossRef]
42. Veber, D.F.; Johnson, S.R.; Cheng, H.-Y.; Smith, B.R.; Ward, K.W.; Kopple, K.D. Molecular Properties That Influence the Oral Bioavailability of Drug Candidates. *J. Med. Chem.* **2002**, *45*, 2615–2623. [CrossRef]
43. Martin, Y.C. A bioavailability score. *J. Med. Chem.* **2005**, *48*, 3164–3170. [CrossRef] [PubMed]
44. Available online: http://www.rcsb.org/pdb/home (accessed on 15 April 2021).
45. Trott, O.; Olson, A.J. AutoDock Vina: Improving the speed and accuracy of docking with a new scoring function, efficient optimization, and multithreading. *J. Comput. Chem.* **2010**, *31*, 455–461. [CrossRef] [PubMed]
46. Protein Data Bank in Europe. Available online: https://www.ebi.ac.uk/pdbe (accessed on 15 April 2021).
47. PDBsum, Pictorial Database of 3D Structures in the Protein Data Bank. Available online: http://www.ebi.ac.uk/pdbsum (accessed on 17 April 2021).
48. Pharmacokinetic Properties. Available online: http://biosig.unimelb.edu.au/pkcsm/prediction (accessed on 15 April 2021).
49. SwissADME. Available online: http://www.swissadme.ch (accessed on 19 April 2021).
50. ADMET Evaluation. Available online: https://admetmesh.scbdd.com/service/evaluation/cal (accessed on 19 April 2021).
51. Zoete, V.; Daina, A.; Bovigny, C.; Michielin, O. SwissSimilarity: A Web Tool for Low to Ultra High Throughput Ligand-Based Virtual Screening. *J. Chem. Inf. Model.* **2016**, *56*, 1399–1404. [CrossRef] [PubMed]
52. Daina, A.; Michielin, O.; Zoete, V. SwissADME: A free web tool to evaluate pharmacokinetics, drug-likeness and medicinal chemistry friendliness of small molecules. *Sci. Rep.* **2017**, *7*, 42717. [CrossRef] [PubMed]
53. Cortes, C.; Vapnik, V. Support-vector networks. *Mach. Learn.* **1995**, *20*, 273–297. [CrossRef]

54. Lipinski, C.A.; Lombardo, F.; Dominy, B.W.; Feeney, P.J. Experimental and computational approaches to estimate solubility and permeability in drug discovery and development settings. *Adv. Drug Deliv. Rev.* **2001**, *46*, 3–26. [CrossRef]
55. Lipinski, C.A.; Lombardo, F.; Dominy, B.W.; Feeney, P.J. Experimental and computational approaches to estimate solubility and permeability in drug discovery and development settings. *Adv. Drug Deliv. Rev.* **1997**, *23*, 3–25. [CrossRef]
56. Ertl, P.; Rohde, B.; Selzer, P. Fast calculation of molecular polar surface area as a sum of fragment-based contributions and its application to the prediction of drug transport properties. *J. Med. Chem.* **2000**, *43*, 3714–3717. [CrossRef]

Article

Characterization and Identification of Bioactive Polyphenols in the *Trapa bispinosa* Roxb. Pericarp Extract

Yuji Iwaoka [1], Shoichi Suzuki [1], Nana Kato [1], Chisa Hayakawa [1], Satoko Kawabe [1], Natsuki Ganeko [1], Tomohiro Uemura [2] and Hideyuki Ito [1,*]

[1] Department of Nutritional Science, Faculty of Health and Welfare Sciences, Okayama Prefectural University, Okayama 719-1197, Japan; iwaoka@fhw.oka-pu.ac.jp (Y.I.); ssho5532@gmail.com (S.S.); nana-k@m.ndsu.ac.jp (N.K.); chisa5835740fastriv@gmail.com (C.H.); satoko@mw.kawasaki-m.ac.jp (S.K.); ganeko@fukuyama-u.ac.jp (N.G.)

[2] Hayashikane Sangyo Co., Ltd., 2-4-8 Yamatomachi, Shimonoseki 750-8608, Japan; tmuemura@hayashikane.co.jp

* Correspondence: hito@fhw.oka-pu.ac.jp

Abstract: In this study, we present the isolation and characterization of the structure of six gallotannins (**1–6**), three ellagitannins (**7–9**), a neolignan glucoside (**10**), and three related polyphenolic compounds (gallic acid, **11** and **12**) from *Trapa bispinosa* Roxb. pericarp extract (TBE). Among the isolates, the structure of compound **10** possessing a previously unclear absolute configuration was unambiguously determined through nuclear magnetic resonance and circular dichroism analyses. The α-glucosidase activity and glycation inhibitory effects of the isolates were evaluated. Decarboxylated rugosin A (**8**) showed an α-glucosidase inhibitory activity, while hydrolyzable tannins revealed stronger antiglycation activity than that of the positive control. Furthermore, the identification and quantification of the TBE polyphenols were investigated by high-performance liquid chromatography coupled to ultraviolet detection and electrospray ionization mass spectrometry analysis, indicating the predominance of gallic acid, ellagic acid, and galloyl glucoses showing marked antiglycation properties. These findings suggest that there is a potential food industry application of polyphenols in TBE as a functional food with antidiabetic and antiglycation activities.

Keywords: *Trapa bispinosa* Roxb.; polyphenol; ellagitannin; gallotannin; α-glucosidase inhibitor; advanced glycation end products (AGEs); antiglycation effect; LC/UV/ESIMS analysis

1. Introduction

Water chestnut (*Trapa bispinosa* Roxb.)—belonging to the family Lythraceae—is a floating annual aquatic plant originally distributed in Southeast Asia, and its cultivation is widely extended to Southern Europe, Africa, and Asia. The dried pericarp has been popular as a tea in the Fukuoka and Saga prefectures in Japan [1], and it has been used as a traditional folk medicine in China, such as an antidiarrheal and antipyretic agent [2]. The fruit and leaf extracts reportedly exhibit diverse biological, such as antioxidant [3] and anticancer [4] activities. The removal effect of industrial pollutant by this plant was reported as a sorbent [5]. The phytochemical studies on this plant revealed the presence of tannins, flavonoids, and saponins, while its detailed components remain elusive. The phytochemical and biological studies of *Trapa japonica* Flerov.—a species closely related to *Trapa bispinosa* Roxb.—described the isolation of ellagitannins (including trapanin, tellimagrandin II, trapanins A and B, rugosin D, and cornusiin G) and gallotannins (including 1,2,3- and 1,2,6-tri-*O*-galloyl-β-D-glucoses) from the leaves and pericarps, and the biological properties of the polyphenols [6–9]. Based on these studies, *Trapa bispinosa* Roxb. polyphenols are believed to contribute to various biological effects.

The accumulation of advanced glycation end products (AGEs) in the various tissues of our body is a cause of Alzheimer's disease [10] and diabetes [11], suggesting that the inhibition of glycation reaction delays the chronic glycative stress-associated diseases. From a disease prevention perspective, food components and pharmaceutical ingredients that inhibit the α-glucosidase activity and glycation reaction are beneficial for our health. The *Trapa bispinosa* Roxb. pericarp extract (TBE) reportedly inhibits the α-glucosidase and glycation reactions [1,12]. However, the TBE components that contribute to these inhibitory effects remain largely unknown. In this study, with the aim of developing functional food materials for TBE that is popular as tea in Japan and effectively utilizing the pericarp which is often wasted, we isolated and characterized TBE polyphenols and evaluated the α-glucosidase and AGE-formation inhibitory effects of the isolated compounds. In addition, the TBE polyphenols showing α-glucosidase and AGE-formation inhibitory effects were identified and quantified by high-performance liquid chromatography coupled to ultraviolet detection and electrospray ionization mass spectrometry (LC/UV/ESIMS) analysis. As the results, we believe that this study might be attract the interest of the food industry due to the development and evaluation of potential application of TBE ingredients.

2. Results and Discussion

2.1. Isolation and Structural Elucidation of TBE-Derived Polyphenols and Related Compounds

A 70% aqueous acetone extract of TBE was subsequently extracted with Et_2O, EtOAc, and water-saturated *n*-BuOH. The EtOAc extract was repeatedly purified by Toyopearl HW-40 (coarse grade), MCI gel CHP20/P120, Sephadex LH-20, Mega Bond Elute C18 and preparative HPLC to obtain eight known compounds including gallic acid; six gallotannins as 2,6-di-*O*-galloyl-β-D-glucose (**1**) [13], 1,2,3-tri-*O*-galloyl-β-D-glucose (**2**) [14], 1,2,6-tri-*O*-galloyl-β-D-glucose (**3**) [14], 2,3,6-tri-*O*-galloyl-β-D-glucose (**4**) [15], 1,2,3,6-tetra-*O*-galloyl-β-D-glucose (**5**) [16], and 1,2,4,6-tetra-*O*-galloyl-β-D-glucose (**6**) [16]; and an ellagitannin as tellimagrandin II (**7**) [17]. The *n*-BuOH extract was subsequently subjected to column chromatography over Diaion HP-20 and Toyopearl HW-40 (coarse grade) to give compounds **5** and **7**. In a separate experiment, TBE aqueous solution was chromatographed over Diaion HP-20, Toyopearl HW-40 (coarse grade), MCI gel CHP20/P120, Bond Elute Plexa, and Mega Bond Elute C18 to give five known compounds including two ellagitannins as decarboxylated rugosin A (**8**) [18] and camptothin B (**9**) [7], a neolignan as (7′*S*,8′*R*)-dihydrodehydrodiconiferyl alcohol-9′-*O*-β-D-glucoside (**10**) [19], and two ellagic acid derivatives as rubuphenol (**11**) [20] and eschweilenol A (**12**) [21]. The known compounds were identified by direct comparison with authentic specimens or by comparison of spectroscopic data with those reported in the literature (Figure 1).

The structure of neolignan glucoside (**10**) was characterized as a known dihydrodehydrodiconiferyl alcohol-9′-*O*-glucoside [19] based on the 1D and 2D-nuclear magnetic resonance (NMR) analyses including ^1H-^1H correlation spectroscopy (COSY), heteronuclear single quantum correlation (HSQC), heteronuclear multiple bond correlation (HMBC), and nuclear Overhauser effect spectroscopy (NOESY) experiments (Figure S1) and electrospray ionization-mass spectrometry (ESI-MS) analysis. However, the absolute configuration in **10** was still unknown. The stereochemistry of glucose in **10** was confirmed to D-series since the released glucose by acid hydrolysis of **10** was a positive response to the reaction with glucose oxidase [22]. The coupling constant ($J_{7',8'}$ = 6.6 Hz) between the H-7′ and H-8′ protons suggested that the relative configuration of **10** at C-7′ and C-8′ was *threo* [23,24]. This relative configuration was further supported by the NOE correlations between the H-7′ and H-9′ protons, as well as those between the H-8′ and H-2′ and H-6′ protons. The absolute configuration of neolignan with dihydrobenzo[*b*]furan skeleton has been determined in agreement with the aromatic quadrant and *P*/*M* helicity rules [25,26].

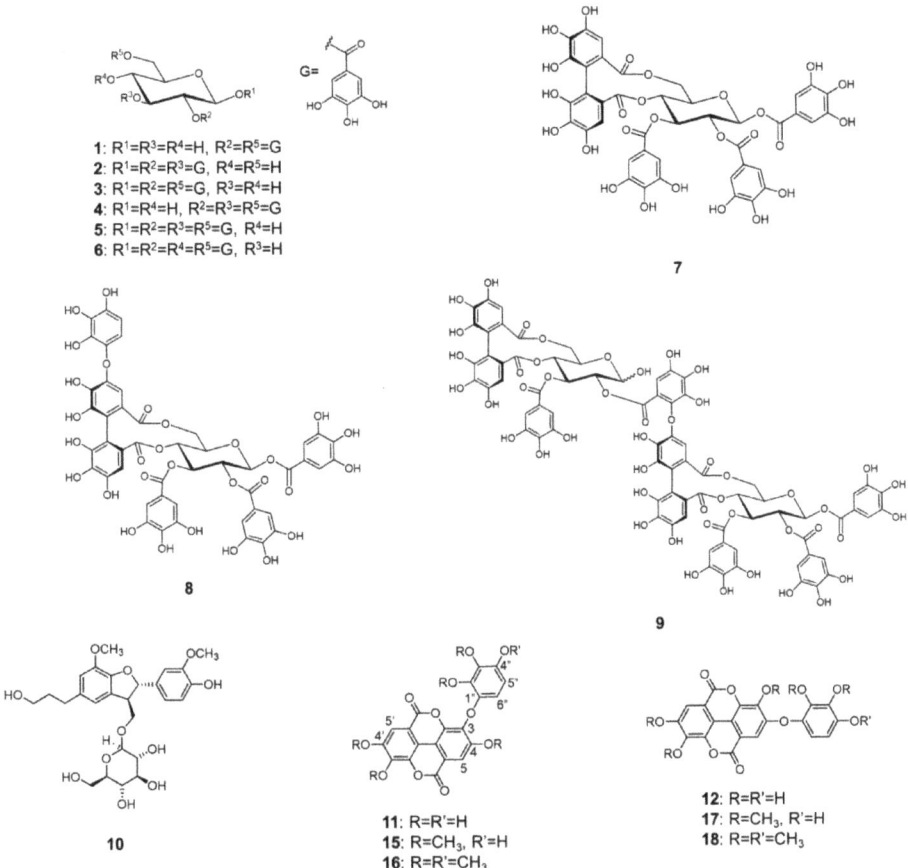

Figure 1. Chemical structures of the polyphenols isolated from TBE and the methylated compounds of **11** and **12**.

The structure of compound **13**, which was obtained by the hydrolysis of **10**, was confirmed based on ^{1}H-NMR (Figure S2), ^{13}C-NMR and atmospheric pressure ionization (APCI) MS analyses. The circular dichroism (CD) spectrum of **13** showed negative cotton at the ^1L_a (around 230 nm), indicating that the absolute configuration of C-8′ was R series based on the aromatic quadrant rule. Furthermore, the P/M helicity rule provided evidence that the 7′S, 8′R configurations of **13** by showing positive cotton at ^1L_b (around 290 nm) band in the CD spectrum. Based on these findings, the stereochemistry of **10** was determined as shown by the formula in Figure 2.

Compound **14** was also obtained as a side product from the hydrolysis of **10** (Figures 2 and S2). The product has been reported to be an intermediate of adenosine A$_1$ receptor ligand [27]. The production of **14** by acid hydrolysis of **10** was further supported the characterization of aglycone moiety of **10**.

Rubuphenol (**11**) and eschweilenol A (**12**) were identified by their ^{1}H- and ^{13}C-NMR, ^{1}H-^{1}H COSY, HSQC, HMBC, and NOESY experiments (Figure S3). The NOESY spectrum of **12** showed the correlation between H-5 and H-6″ protons, while that correlation was not observed in **11**. Compounds **11** and **12** were methylated to confirm the structures as the corresponding methylated derivatives (compounds **15–18**) based on spectral analyses (Figure S4). The ^{1}H-^{1}H COSY and NOESY spectra of compounds **15** and **16** showed the correlations between H-5 and 4-OCH$_3$ protons, while those correlations were not observed in **17** and **18**. The observations of ^{1}H-^{1}H COSY and NOE correlations between H-5′ and

4′-OCH$_3$ and, H-5″ and 4″-OCH$_3$ protons in **16** and **18** supported the positions of hydroxyl group and ether linkage of **11** and **12**. Based on these findings, the structures of **11** and **12** were established as shown by the formulas in Figure 1. In this study, compounds **1, 4, 6, 8, 11**, and **12** were firstly isolated from *Trapa* species and the absolute configuration of **10** was confirmed based on NMR and CD analyses.

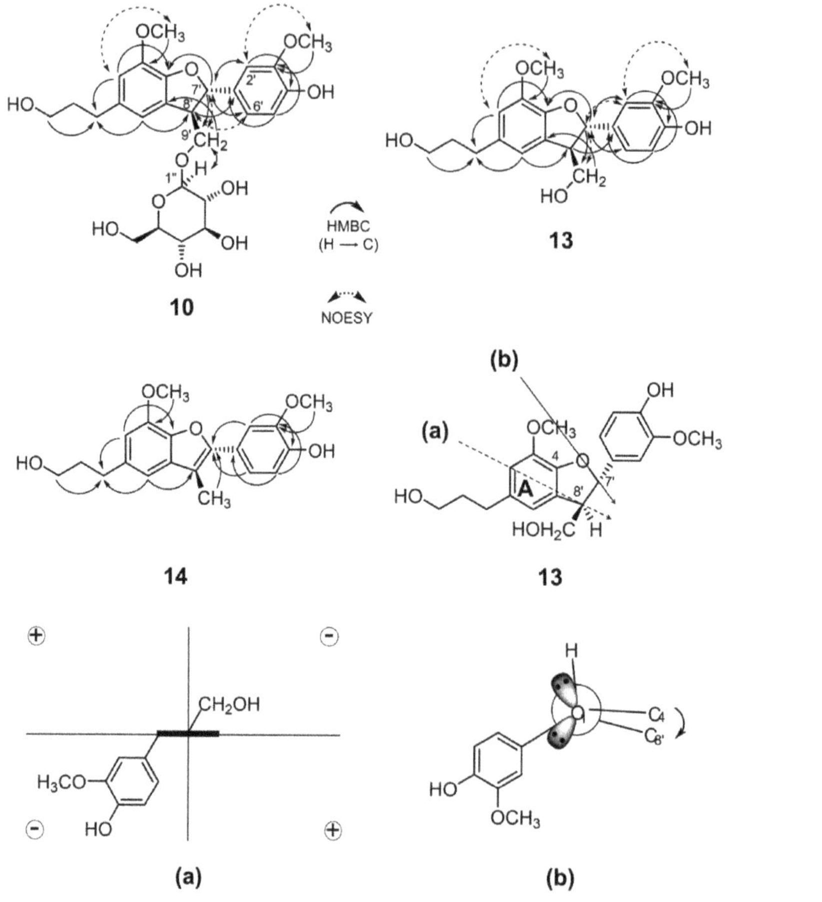

Figure 2. The key HMBC and NOESY correlations for compound **10**, compound **13** and **14** chemical structures, and projection for ^1L_a (**a**) projection in the direction of the arrow, the wedge indicates the plane of the A-ring); and ^1L_b (**b**) transition for **13**.

2.2. α-Glucosidase Inhibitory Activity of the TBE-Derived Compounds

The α-glucosidase inhibitory activities of the TBE-derived and the related polyphenols are shown in Table 1 as IC$_{50}$ (μM) values. The inhibitory activities of gallic acid and ellagic acid, which are well-known metabolites of gallotannins and ellagitannins, respectively [28], showed little effect similar to that of gallotannins. (7′S,8′R)-Dihydrodehydrodiconiferyl alcohol-9′-O-β-D-glucoside (**10**) and its hydrolysates **13** and **14**, and rubuphenol (**11**) and eschweilenol A (**12**) also showed no inhibitory activity. Among the tested compounds, 1,2,3,4,6-penta-O-galloyl-β-D-glucose and decarboxylated rugosin A (**8**) showed inhibitory activity with IC$_{50}$ values at 59.0 ± 0.4 μM and 20.7 ± 0.1 μM, respectively. However, the inhibitory activities of all tested compounds were not reached to that of acarbose as a positive inhibitor [29]. Tannin-containing plant extracts reportedly exert a strong α-glucosidase inhibitory effect [30], but little is known about the tannin contributors themselves. It has been reported that the extract of *Trapa japonica* belonging to the same genus as *T. bispinosa* and the isolated ellagitannin dimers cornusiin G and rugosin D from *T. japonica* showed the inhibitory activity on α-glucosidase comparable to that of acarbose [8]. The isolated polyphenols from *T. bispinosa* showed no effect, but the presence of cornusiin G was confirmed by HPLC analysis, which was described in Section 2.4, indicating cornusiin G and unidentified ellagitannin dimers might be contributed to inhibition on α-glucosidase. Further study is necessary to investigate the possibility that the other components besides ellagitannin dimers contribute to the activity.

Table 1. α-Glucosidase inhibitory activity of polyphenols isolated from TBE and related compounds.

Compound	Inhibitory Effects on α-Glucosidase Activity IC$_{50}$ (μM)
Gallic acid [8]	>100
Ellagic acid	>100
1,2,3-Tri-O-galloyl-β-D-glucose (**2**) [8]	>100
1,2,6-Tri-O-galloyl-β-D-glucose (**3**) [8]	>100
1,2,3,6-Tetra-O-galloyl-β-D-glucose (**5**) [8]	>100
1,2,4,6-Tetra-O-galloyl-β-D-glucose (**6**)	>100
1,2,3,4,6-Penta-O-galloyl-β-D-glucose	59.0 ± 0.4
Tellimagrandin II (**7**) [8]	>100
Decarboxylated rugosin A (**8**)	20.7 ± 0.1
Camptothin B (**9**)	>100
Compound **10**	>100
Compound **13**	>100
Compound **14**	>100
Rubuphenol (**11**)	>100
Eschweilenol A (**12**)	>100
Cornusiin G [8]	6.3 ± 0.1
Acarbose	4.0 ± 0.1

Data are expressed as the means ± SE (n = 3).

2.3. Antiglycation Effects of the TBE-Derived Compounds

The inhibitory effect of the TBE-derived polyphenols and the related compounds on AGEs, generated by the glycation reaction between human serum albumin (HSA) and glucose or fructose was evaluated. All tested compounds exhibited significantly stronger AGE-formation inhibitory activity than the positive control aminoguanidine [31] with IC$_{50}$ values in the range of 0.1 ± 0.0–14.7 ± 2.0 μM with glucose and of 0.2 ± 0.0–27.0 ± 2.6 μM with fructose (Table 2), except for (7'S, 8'R)-dihydrodehydrodiconiferyl alcohol-9'-O-β-D-glucoside (**10**) and its hydrolysates **13** and **14**. It is noteworthy that the gallotannin (**2–6**) and 1,2,3,4,6-penta-O-galloyl-β-D-glucose and ellagitannin (**7–9**) potencies were incomparably stronger than that of aminoguanidine.

AGEs are known to be generated via multiple pathways in glycation reactions [32]. Therefore, we also evaluated the AGE cross-link cleaving effects of the TBE polyphenols and the related compounds. Almost all tested compounds exhibited a stronger activity than the positive control N-phenacylthiazolium bromide (PTB) [33], except for ellagic acid, 2,6-di-O-galloyl-β-D-glucose (**1**), 1,2,3,4,6-penta-O-galloyl-β-D-glucose, **10**, **13**, and **14**. In particular, gallic acid, rubuphenol (**11**), and eschweilenol A (**12**) showed remarkable activities in this assay. Some of isolates have not been evaluated for antiglycation effects, since the isolated amount of the compounds were insufficient to test. However, these results indicated that the TBE-derived polyphenols exert AGE-formation inhibitory activity and might contribute to the antiglycative effect of TBE.

Table 2. AGE-formation inhibitory effects in HSA/glucose or fructose and AGE-derived crosslink-cleaving activities of polyphenols isolated from TBE and related compounds.

Compound	Inhibitory Effects on AGE Formation IC_{50} (μM)		Crosslink-Cleaving Activities
	Glucose	Fructose	Relative Ratio *
Gallic acid	14.7 ± 2.0	27.0 ± 2.6	720.1 ± 54.1
Ellagic acid	1.8 ± 0.1	2.5 ± 0.1	11.4 ± 0.1
2,6-Di-O-galloyl-β-D-glucose (1)	1.5 ± 0.1	3.3 ± 0.0	77.4 ± 4.0
1,2,3-Tri-O-galloyl-β-D-glucose (2)	0.4 ± 0.0	0.4 ± 0.0	190.5 ± 2.5
1,2,6-Tri-O-galloyl-β-D-glucose (3)	0.3 ± 0.0	0.4 ± 0.0	146.6 ± 12.6
2,3,6-Tri-O-galloyl-β-D-glucose (4)	0.3 ± 0.0	1.0 ± 0.0	N.T.
1,2,3,6-Tetra-O-galloyl-β-D-glucose (5)	0.3 ± 0.0	0.3 ± 0.0	209.0 ± 33.7
1,2,4,6-Tetra-O-galloyl-β-D-glucose (6)	0.3 ± 0.0	0.3 ± 0.0	159.6 ± 12.7
1,2,3,4,6-Penta-O-galloyl-β-D-glucose	0.2 ± 0.0	0.2 ± 0.0	97.9 ± 2.1
Tellimagrandin II (7)	0.2 ± 0.0	0.3 ± 0.0	230.8 ± 12.2
Decarboxylated rugosin A (8)	0.3 ± 0.0	0.3 ± 0.0	233.0 ± 5.8
Camptothin B (9)	0.1 ± 0.0	0.2 ± 0.0	180.6 ± 4.2
Compound 10	>500	>1000	16.7 ± 1.4
Compound 13	>500	>1000	17.2 ± 1.1
Compound 14	>500	>1000	0.5 ± 1.1
Rubuphenol (11)	2.4 ± 0.0	4.7 ± 0.3	514.8 ± 11.2
Eschweilenol A (12)	2.4 ± 0.0	2.9 ± 0.1	484.5 ± 12.5
Aminoguanidine	258.9 ± 6.8	801.0 ± 17.7	N.T.
N-Phenacylthiazolium bromide (PTB)	N.T.	N.T.	100

Data are expressed as the means ± SE (n = 3), N.T. means not tested, * The concentrations of the tested samples are 100 μg/mL.

2.4. LC/UV/ESIMS Analysis of TBE

The presence of phenolic compounds, such as the gallotannins and ellagitannins, has already been previously reported in the pericarp of *Trapa* species [8,34]. However, the detailed polyphenol content in the pericarp of the *Trapa* species has not yet been revealed. There are few reports on the qualitative and quantification of hydrolyzable tannins containing both gallotannins and ellagitannins by LC-MS method. Here, we could identify and quantify a total of 30 polyphenols including gallotannins and ellagitannins in TBE by LC/UV/ESIMS method with each polyphenol specimen (Figure 3). The total ion and UV at 280 nm chromatograms of TBE displayed with good separation in Figure 3A,B, respectively. Among the candidates, compounds having lactones were clearly detected at UV at 360 nm (Figure 3C) for more separation and accurate quantification. In Table 3, gallic acid (32.2 ± 0.1 mg/g) exhibiting the most potent AGE cross-link cleaving activity among the isolated polyphenols is shown as a main TBE component, suggesting that it was produced from gallotannins or ellagitannins during TBE manufacturing, as well as ellagic acid (6.9 ± 0.1 mg/g). The various gallotannins possessing significant AGE-formation inhibitory activity were contained in the range of 0.2 ± 0.1–16.8 ± 1.2 mg/g. Valoneic acid dilactone (1.8 ± 0.2 mg/g), rubuphenol (11) (4.3 ± 0.1 mg/g), and eschweilenol A (12) (0.9 ± 0.2 mg/g) were minor TBE components, implying that these polyphenols were also ascribable to TBE ellagitannins. Urolithin M5 (1.4 ± 0.4 mg/g), a well-known ellagitannin metabolite, was also found in TBE. Urolithin M5 might be produced by biosynthesis in *Trapa bispinosa*, since urolithins A and B and isourolithin A reportedly contained in the plant of the same genus, *Trapa natans* [35]. Urolithin M5 has also been isolated from *Tamarix nilotica* [36]. The presence of ellagitannin dimers, camptothin B (9) and cornusiin G, and an ellagitannin monomer, 1,2-Di-O-galloyl-4,6-hexahydroxydiphenoyl-D-glucose in TBE could also be identified by LC/UV/ESIMS analysis. Gallotannins (2–6), tellimagrandin II (7), and decarboxylated rugosin A (8), which showed strong effects of both AGE-formation inhibition and AGE-derived crosslink cleaving, were contained in TBE at high levels.

These results clearly provided the basic confirmation to the potential contribution of TBE polyphenols to antidiabetic and antiglycative effects.

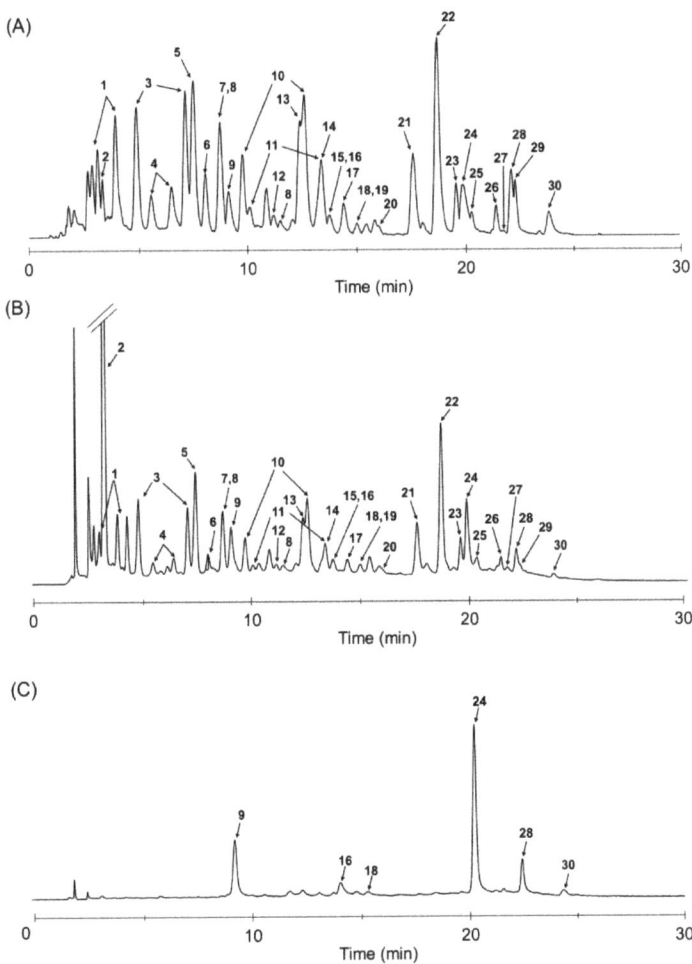

Figure 3. LC/UV/ESIMS analysis of TBE. Total ion chromatogram (**A**) and UV chromatogram at 280 nm (**B**) and 360 nm (**C**).

Table 3. TBE polyphenol content.

Peak No.	Compound	t_R (min)	MS (m/z)	Content (mg/g of Dry Weight)
1	2,3-Di-O-galloyl-β-D-glucose	3.18, 3.94	483 [M − H]⁻	15.5 ± 0.2
2	Gallic acid	3.23	339 [2M − H]⁻	32.2 ± 0.1
3	2,6-Di-O-galloyl-β-D-glucose (**1**)	4.90, 7.52	483 [M − H]⁻	N.T.
4	3,6-Di-O-galloyl-β-D-glucose	5.55, 6.58	483 [M − H]⁻	3.9 ± 0.0
5	1,6-Di-O-galloyl-β-D-glucose	7.53	483 [M − H]⁻	16.8 ± 1.2
6	Digalloyl glucose	8.07	483 [M − H]⁻	N.T.
7	1,2,3-Tri-O-galloyl-β-D-glucose (**2**)	8.75	635 [M − H]⁻	4.8 ± 0.0
8	3,4,6-Tri-O-galloyl-β-D-glucose	8.75, 11.6	635 [M − H]⁻	3.4 ± 0.1

Table 3. Cont.

Peak No.	Compound	t_R (min)	MS (m/z)	Content (mg/g of Dry Weight)
9	Brevifolincarboxylic acid	9.22	291 [M − H]$^-$	1.7 ± 0.1
10	2,3,6-Tri-O-galloyl-β-D-glucose (4)	9.86, 12.6	635 [M − H]$^-$	N.T.
11	2,4,6-Tri-O-galloyl-β-D-glucose	10.2, 13.5	635 [M − H]$^-$	0.2 ± 0.1
12	Trigalloyl glucose	11.3	635 [M − H]$^-$	N.T.
13	1,2,6-Tri-O-galloyl-β-D-glucose (3)	12.5	635 [M − H]$^-$	4.1 ± 0.5
14	1,3,6-Tri-O-galloyl-β-D-glucose	13.5	635 [M − H]$^-$	1.8 ± 0.7
15	1,2-Di-O-galloyl-4,6-hexahydroxydiphenoyl-β-D-glucose	13.8	785 [M − H]$^-$	0.9 ± 0.0
16	Valoneic acid dilactone	13.9	469 [M − H]$^-$	1.8 ± 0.2
17	Trigalloyl glucose	14.5	635 [M − H]$^-$	N.T.
18	Urolithin M5	15.1	275 [M − H]$^-$	1.4 ± 0.4
19	1,4,6-Tri-O-galloyl-β-D-glucose	15.1	635 [M − H]$^-$	0.6 ± 0.0
20	Camptothin B (9)	16.2	860 [M − 2H]$^{2-}$	N.T.
21	Tellimagrandin II (7)	17.7	937 [M − H]$^-$	5.7 ± 0.0
22	1,2,3,6-Tetra-O-galloyl-β-D-glucose (5)	18.8	787 [M − H]$^-$	13.3 ± 0.0
23	1,2,4,6-Tetra-O-galloyl-β-D-glucose (6)	19.7	787 [M − H]$^-$	1.5 ± 0.0
24	Ellagic acid	19.9	301 [M − H]$^-$	6.9 ± 0.1
25	Decarboxylated rugosin A (8)	20.3	1061 [M − H]$^-$	2.4 ± 0.0
26	1,2,3,4,6-Penta-O-galloyl-β-D-glucose	21.5	939 [M − H]$^-$	0.7 ± 0.0
27	Cornusiin G	21.7	861 [M − 2H]$^{2-}$	0.3 ± 0.0
28	Rubuphenol (11)	22.2	425 [M − H]$^-$	4.3 ± 0.1
29	Compound 10	22.1	521 [M − H]$^-$	1.3 ± 0.1
30	Eschweilenol A (12)	24.0	425 [M − H]$^-$	0.9 ± 0.2

Data are expressed as the means ± SE (n = 3), N.T. means not tested, t_R: retention time.

3. Materials and Methods

3.1. Chemicals

The TBE was prepared as follows: the *Trapa bispinosa* pericarp cultivated in Thailand was dried, sterilized, and crushed at ambient conditions, followed by extraction with hot water (approximately six times the weight of the water chestnut pericarp). Dextrin was added to the extracted liquid so that the ratio of chestnut pericarp water extract to dextrin would be 67:33 using the dry weight. TBE was obtained after spray drying the extract and its moisture content was less than 10%. Ellagic acid, *N*-phenacylthiazolium bromide (PTB), and aminoguanidine hydrochloride were obtained from Wako Pure Chemical Industries (Osaka, Japan). 1-Phenyl-1,2-propanedione (PPD) and rat intestinal acetone powder were purchased from Sigma Aldrich (St Louis, MO, USA). Trimethylsilyldiazomethane (TMS-CHN$_2$) was purchased from Tokyo Chemical Industry (Tokyo, Japan). Each polyphenol specimen was used compounds isolated from natural sources held in our library: gallotannins [37–44], 1,2-di-O-galloyl-4,6-hexahydroxydiphenoyl-D-glucose [45], cornusiin G [8], brevifolincarboxylic acid [39], urolithin M5 [46], and valoneic acid dilactone [41].

3.2. General Experimental Procedure

Optical rotations were recorded using a Jasco DIP-1000 polarimeter (Jasco, Tokyo, Japan). UV and CD spectra were measured by using Jasco V-530 spectrophotometer (Jasco, Tokyo, Japan) and Jasco J-710 spectropolarimeter (Jasco, Tokyo, Japan), respectively. ^{1}H-NMR (600 MHz) and ^{13}C-NMR (151 MHz) spectra including ^{1}H-^{1}H COSY, NOESY, HSQC, and HMBC were recorded on a Varian NMR system (Varian, Palo Alto, CA, USA) and chemical shifts are given in ppm (ppm) values relative to acetone-d_6 (2.04 ppm for ^{1}H and 29.8 ppm for ^{13}C), CD$_3$OD (3.35 ppm for ^{1}H and 49.0 ppm for ^{13}C), and CDCl$_3$ (7.26 ppm for ^{1}H and 77.0 ppm for ^{13}C). ESI or APCI mass spectra were performed on a Bruker MicrOTOF II instrument (Bruker, Billerica, MA, USA) using direct sample injection. Reversed-phase

HPLC was conducted on InertSustain C18 column (150 mm × 4.6 mm i.d., 5 μm, GL Sciences, Tokyo, Japan) at 40 °C with the mobile phase consisted of $CH_3CN:H_2O:HCOOH$ (5:90:5) (solvent A) and $CH_3CN:H_2O:HCOOH$ (45:50:5) (solvent B). The flow rate was 1.0 mL/min, and a linear gradient was programmed as follows: 0–15 min (solvent B: 10–30%), 15–20 min (solvent B: 30–50%), 20–30 min (solvent B: 10%) and the absorbance was monitored at 280 and 360 nm. Normal-phase HPLC was conducted on YMC-Pack SIL column (250 mm × 4.6 mm i.d., 5 μm, YMC, Kyoto, Japan) with n-hexane:MeOH:THF:HCOOH (55:33:11:1) containing oxalic acid (450 mg/L) by isocratic elution. The flow rate was 1.5 mL/min and the absorbance was monitored at 280 nm. Preparative HPLC was carried out under the same conditions as reversed-phase HPLC condition by using the mobile phase consisted of $CH_3CN:H_2O:HCOOH$ (10:85:5) (condition 1) or $MeOH:H_2O:HCOOH$ (10:85:5) (condition 2). Column chromatography was carried out by Diaion HP-20 (Mitsubishi Chemical, Tokyo, Japan), Toyopearl HW-40 (coarse grade) (Tosoh, Tokyo, Japan), MCI gel CHP20/P120 (Mitsubishi Chemical, Tokyo, Japan), Sephadex LH-20 (GE Healthcare, Chicago, IL, USA), Mega Bond Elut C18 (Agilent technologies, Santa Clara, CA, USA), Bond Elut Plexa (Agilent technologies, Santa Clara, CA, USA), and YMC Gel ODS-AQ-HG (YMC, Kyoto, Japan).

3.3. Extraction and Isolation

TBE (450 g) was dissolved in H_2O (700 mL) and the solution was extracted subsequently extracted with Et_2O (3 × 700 mL), EtOAc (3 × 700 mL), and water-saturated n-BuOH (3 × 700 mL), to give Et_2O (5.2 g), EtOAc (16.7 g), n-BuOH extracts (14.3 g), and H_2O soluble portion (163.5 g). A part of EtOAc extract (5.0 g) was chromatographed over Toyopearl HW-40 (coarse grade) (40 cm × 2.2 i.d. cm) with 40%, 50%, 60%, and 70% aqueous MeOH—MeOH:H_2O:acetone (7:2:1)—MeOH:H_2O:acetone (7:1:2)—70% aqueous acetone in a stepwise elution mode. The 40% aqueous MeOH fraction gave gallic acid (779.9 mg), the 50% aqueous MeOH fraction gave 1,2,3-tri-*O*-galloyl-β-D-glucose (**2**) (148.1 mg), the 60% aqueous MeOH fraction gave 1,2,3,6-tetra-*O*-galloyl-β-D-glucose (**5**) (301.1 mg), and the 70% aqueous MeOH fraction gave 1,2,4,6-tetra-*O*-galloyl-β-D-glucose (**6**) (67.7 mg) and tellimagrandin II (**7**) (189.2 mg), by column chromatographic purification on each MCI gel CHP20/P120 (40 cm × 1.1 i.d. cm). The 40% aqueous MeOH fraction (200 mg) was purified by column chromatographies over MCI gel CHP20/P120 (40 cm × 1.1 i.d. cm) with aqueous MeOH, Sephadex LH-20 (40 cm × 1.1 i.d. cm) with EtOH-MeOH solvent system, Mega Bond Elut C18 cartridge column with aqueous MeOH, and preparative HPLC under the condition 1, to give 2,6-Di-*O*-galloyl-β-D-glucose (**1**) (3.0 mg). The 50% aqueous MeOH fraction (300 mg) was further purified by Mega Bond Elut C18 with 10%, 20%, 30%, 40%, 50%, 60%, and 70% aqueous MeOH–100% MeOH–70% aqueous acetone and preparative HPLC under the condition 2 to give 1,2,6-tri-*O*-galloyl-β-D-glucose (**3**) (0.8 mg) and 2,3,6-tri-*O*-galloyl-β-D-glucose (**4**) (0.9 mg). A part of *n*-BuOH extract (10 g) was chromatographed over Diaion HP-20 (40 cm × 5.0 i.d. cm) with H_2O–10, 30, and 50% aqueous MeOH–100% MeOH–70% aqueous acetone in a stepwise elution mode. The 50% aqueous MeOH eluate (1.5 g) was further chromatographed over Toyopearl HW-40 (coarse grade) (40 cm × 2.2 i.d. cm) with 50%, 60%, and 70% aqueous MeOH–MeOH:H_2O:acetone (7:2:1)–MeOH:H_2O:acetone (7:1:2)–70% aqueous acetone in stepwise elution mode. Compounds **5** (40.3 mg) and **7** (43.8 mg) were obtained from the 60% aqueous MeOH and MeOH:H_2O:acetone (7:2:1) fractions, respectively. TBE (500 g) was dissolved in H_2O (10 L) and, the solution was subjected to Diaion HP-20 (80 cm × 5.0 i.d. cm) and eluted with H_2O increasing amounts of MeOH (0–10–30–50–100% MeOH) and 70% aqueous acetone. A part (7.75 g) of the 50% aqueous MeOH eluate (23.0 g) was chromatographed over Toyopearl HW-40 (coarse grade) (40 cm × 2.2 i.d. cm) with 50, 60, and 70% aqueous MeOH–MeOH:H_2O:acetone (7:2:1)–MeOH:H_2O:acetone (7:1:2)–70% aqueous acetone as eluent. The combined fraction (400 mg) consisted of 70% aqueous MeOH and MeOH:H_2O:acetone (7:1:2) fractions was purified by column chromatography over MCI gel CHP20/P120 (40 cm × 1.1 i.d. cm) with 30%, 40%, and 50% aqueous MeOH–100% MeOH–70% aqueous acetone. Subsequently, the 40% MeOH eluate (23.2 mg) was purified by Bond Elut Plexa with 30%, 40%, 50%, 60% aqueous MeOH and MeOH to give decarboxylated rugosin A (**8**) (8.4 mg) from the 60% aqueous MeOH eluate. The MeOH:H$_2$O:acetone (7:2:1) eluate (120 mg) was purified by Mega Bond Elut C18 and YMC Gel ODS-AQ-HG (21 cm × 1.1 i.d. cm) with H$_2$O-MeOH solvent system to give camptothin B (**9**) (7.3 mg). The 100% MeOH eluate (8.0 g) obtained by Diaion HP-20 separation was chromatographed over Toyopearl HW-40 (coarse grade) (43 cm × 2.2 i.d. cm) with H$_2$O increasing amounts of MeOH (50–60–100% MeOH) followed by 70% aqueous acetone. (7′S, 8′R)-Dihydrodehydrodiconiferyl alcohol-9′-O-β-D-glucoside (**10**) (347.2 mg) was obtained from the 50% aqueous MeOH eluate. The fraction eluted with 100% MeOH and 70% aqueous acetone was dissolved in hot MeOH and the resulting solution was stored at 4 °C to remove ellagic acid as a precipitate. The collected supernatant (2.6 g) was chromatographed over Toyopearl HW-40 (coarse grade) (44 cm × 2.2 i.d. cm) with 70% aqueous MeOH–MeOH:H$_2$O:acetone (7:2:1)–MeOH:H$_2$O:acetone (7:1:2)–70% aqueous acetone. The MeOH:H$_2$O:acetone (7:2:1)-MeOH:H$_2$O:acetone (7:1:2) eluate was purified by MCI gel CHP20/P120 (24 cm × 1.1 i.d. cm) and YMC Gel ODS-AQ-HG (27 cm × 1.1 i.d. cm) with MeOH-H$_2$O solvent system to afford rubuphenol (**11**) (33.1 mg) and eschweilenol A (**12**) (5.5 mg).

2,6-Di-O-galloyl-β-D-glucose (**1**): pale yellow amorphous powder; ^{1}H-NMR [600 MHz, acetone-d_6-D$_2$O (9:1)] δ 7.11–7.14 (4H in total each, s, galloyl-H), 5.35 (1H, d, J = 3.6 Hz, Glc α-1), 4.88 (1H, t, J = 8.4 Hz, Glc β-2), 4.83 (1H, d, J = 7.8 Hz, Glc β-1), 4.74 (1H, dd, J = 3.6, 10.2 Hz, Glc α-2), 4.56 (1H, dd, J = 1.8, 11.4 Hz, Glc β-6), 4.51 (1H, dd, J = 1.8, 11.4 Hz, Glc α-6), 4.36 (2H, m, Glc α-6, β-6), 4.13 (1H, m, Glc α-5), 4.09 (1H, t, J = 9.6 Hz, Glc α-3), 3.74 (1H, t, J = 9 Hz, Glc β-3), 3.68 (1H, m, Glc β-5), 3.5–3.7 (Glc- and β-4, overlapped with DOH). HR-ESI-MS m/z 483.0787 [M − H]$^-$ (calcd for C$_{20}$H$_{20}$O$_{14}$-H, 483.0780).

1,2,3-Tri-O-galloyl-β-D-glucose (**2**): pale yellow amorphous powder; ^{1}H-NMR [600 MHz, acetone-d_6-D$_2$O (9:1)] δ 7.09, 7.08, 7.00 (2H, each, s, galloyl-H), 6.06 (1H, d, J = 8.4 Hz, Glc H-1), 5.61 (1H, t, J = 9.6 Hz, Glc H-3), 5.40 (1H, dd, J = 8.4, 9.6 Hz, Glc H-2), 3.97 (1H, t, J = 9.6 Hz, Glc H-4), 3.93 (1H, d, J = 10.2 Hz, Glc H-6), 3.81 (2H, m, Glc H-5, 6). HR-ESI-MS m/z 635.0890 [M − H]$^-$ (calcd for C$_{27}$H$_{24}$O$_{18}$-H, 635.0890)

1,2,6-Tri-O-galloyl-β-D-glucose (**3**): pale yellow amorphous powder; ^{1}H-NMR [600 MHz, acetone-d_6-D$_2$O (9:1)] δ 7.11, 7.07, 7.03 (2H, each, s, galloyl-H), 5.90 (1H, d, J = 9.0 Hz, Glc H-1), 5.22 (1H, t, J = 9.0 Hz, Glc H-2), 4.59 (1H, dd, J = 1.8, 12.0 Hz, Glc H-6), 4.37 (1H, dd, J = 5.4, 12.0 Hz, Glc H-6), 3.96 (1H, t, J = 9.0 Hz, Glc H-3), 3.91 (1H, m, Glc H-5), 3.71 (1H, t, J = 9.0 Hz, Glc H-4). HR-ESI-MS m/z 635.0898 [M − H]$^-$ (calcd for C$_{27}$H$_{24}$O$_{18}$-H, 635.0890).

2,3,6-Tri-O-galloyl-β-D-glucose (**4**): pale yellow amorphous powder; ^{1}H-NMR [600 MHz, acetone-d_6-D$_2$O (9:1)] δ 7.00–7.14 (6H in total each, s, galloyl-H), 5.77 (1H, t, J = 9.6 Hz, Glc α-3), 5.47 (1H, d, J = 3.6 Hz, Glc α-1), 5.43 (1H, m, Glc β-3), 5.10 (1H, dd, J = 7.8, 9.6 Hz, Glc β-2), 5.01 (1H, d, J = 7.8 Hz, Glc β-1), 4.94 (1H, dd, J = 3.6, 9.6 Hz, Glc α-2), 4.59 (1H, d, J = 11.4 Hz, Glc β-6), 4.55 (1H, dd, J = 1.8, 12 Hz Glc α-6), 4.47–4.40 (3H, m, Glc α-6, β-6, β-5), 4.29 (1H, m, Glc α-5), 3.93 (1H, t, J = 9.6 Hz, Glc α-4), 3.88 (1H, d, J = 6 Hz, Glc β-4). HR-ESI-MS m/z 635.0892 [M − H]$^-$ (calcd for C$_{27}$H$_{24}$O$_{18}$-H, 635.0890).

1,2,3,6-Tetra-O-galloyl-β-D-glucose (**5**): pale yellow amorphous powder; ^{1}H-NMR [600 MHz, acetone-d_6-D$_2$O (9:1)] δ 7.16, 7.09, 7.08, 7.01 (2H, each, s, galloyl-H), 6.14 (1H, d, J = 8.4 Hz, Glc H-1), 5.69 (1H, t, J = 9.6 Hz, Glc H-3), 5.49 (1H, dd, J = 8.4, 9.6 Hz, Glc H-2), 4.66 (1H, dd, J = 1.8, 12.6 Hz, Glc H-6), 4.50 (1H, dd, J = 5.4, 12.6 Hz, Glc H-6), 4.17 (1H, m, Glc H-5), 4.08 (1H, t, J = 9.6 Hz, Glc H-4). HR-ESI-MS m/z 787.1034 [M − H]$^-$ (calcd for C$_{34}$H$_{28}$O$_{22}$-H, 787.0999).

1,2,4,6-Tetra-O-galloyl-β-D-glucose (**6**): pale yellow amorphous powder; ^{1}H-NMR [600 MHz, acetone-d_6-D$_2$O (9:1)] δ 7.14, 7.13, 7.09, 7.06 (2H, each, s, galloyl-H), 6.04 (1H, d, J = 8.4 Hz, Glc H-1), 5.40 (1H, t, J = 9.6 Hz, Glc H-4), 5.38 (1H, dd, J = 8.4, 9.6 Hz, Glc H-2), 4.53 (1H, dd, J = 1.2, 12.6 Hz, Glc H-6), 4.36 (1H, t, J = 9.6 Hz, Glc H-3), 4.30 (1H, m, Glc H-5), 4.20 (1H, dd, J = 5.4, 12.6 Hz, Glc H-6). HR-ESI-MS m/z 787.1008 [M − H]$^-$ (calcd for C$_{34}$H$_{28}$O$_{22}$-H, 787.0999).

Tellimagrandin II (**7**): pale yellow amorphous powder; ^{1}H-NMR [600 MHz, acetone-d_6-D$_2$O (9:1)] δ 7.09, 6.99, 6.95 (2H, each, s, galloyl-H), 6.63, 6.47 [1H each, HHDP-H], 6.17 (1H, d, J = 7.8 Hz, Glc H-1), 5.81 (1H, t, J = 9.6 Hz, Glc H-3), 5.59 (1H, dd, J = 7.8, 9.6 Hz, Glc H-2), 5.32 (1H, dd, J = 6.6, 13.2 Hz, Glc H-6), 5.20 (1H, t, J = 9.6 Hz, Glc H-4), 4.53 (1H, dd, J = 6.6, 9.6 Hz, Glc H-5), 3.87 (1H, d, J = 13.2 Hz, Glc H-6). HR-ESI-MS m/z 937.0956 [M − H]$^-$ (calcd for C$_{41}$H$_{30}$O$_{26}$-H, 937.0953).

Decarboxylated rugosin A (**8**): dark brown amorphous powder; ^{1}H-NMR [600 MHz, acetone-d_6-D$_2$O (9:1)] δ 7.08, 7.00, 6.98 (2H, each, s, galloyl-H), 6.49, 6.39 (1H each, s, decarboxylated valoneoyl group -H$_a$, H$_b$), 6.43, 6.38 (1H each, d, J = 9.0 Hz, decarboxylated valoneoyl group -H$_c$, H$_d$), 6.15 (1H, d, J = 7.8 Hz, Glc H-1), 5.81 (1H, t, J = 9.6 Hz, Glc H-3), 5.59 (1H, dd, J = 7.8, 9.6 Hz, Glc H-2), 5.27 (1H, dd, J = 6.6, 13.2 Hz, Glc H-6), 5.17 (1H, t, J = 9.6 Hz, Glc H-4), 4.13 (1H, dd, J = 6.6, 9.6 Hz, Glc H-5), 3.82 (1H, d, J = 13.2 Hz, Glc H-6). HR-ESI-MS m/z 1061.1114 [M − H]$^-$ (calcd for C$_{47}$H$_{34}$O$_{29}$-H, 1061.1113).

Camptothin B (**9**): dark brown amorphous powder; ^{1}H-NMR [600 MHz, acetone-d_6-D$_2$O (9:1)] δ 7.09, 6.99, 6.97, 6.84 (4/3 H, each, s, galloyl-H), 7.08, 7.02, 6.98, 6.91 (2/3 H, each, s, galloyl-H), 7.11 (1/3H, s, valoneoyl-Hc), 7.00 (2/3H, s, valoneoyl-Hc), 6.65 (2/3H, s, HHDP-Hd), 6.64 (1/3H, s, HHDP-Hd), 6.498 (1/3H, s, HHDP-He), 6.496 (2/3H, s, HHDP-He), 6.62 (1/3H, s, valoneoyl-Ha), 6.59 (2/3H, s, valoneoyl-Ha), 6.18 (1/3H, s, valoneoyl-Hb), 6.16 (2/3H, s, valoneoyl-Hb), 6.18 (2/3H, d, J = 7.8 Hz, Glc H$_R$-1), 6.16 (1/3H, d, J = 7.8 Hz, Glc H$_R$-1), 5.80 (1/3H, t, J = 9.6 Hz, Glc H$_L$-3), 5.60 (1/3H, t, J = 9.6 Hz, Glc H$_R$-3), 5.59 (2/3H, t, J = 9.6 Hz, Glc H$_R$-3), 5.55 (1/3H, dd, J = 7.8, 9.6 Hz, Glc H$_R$-2), 5.54 (2/3H, dd, J = 7.8, 9.6 Hz, Glc H$_R$-2), 5.46 (2/3H, t, J = 9.6 Hz, Glc H$_L$-3), 5.44 (1/3H, dd, J = 6.6, 13.2 Hz, Glc H$_L$-6), 5.35 (1/3H, dd, J = 4.2 Hz, Glc H$_L$-1α), 5.27 (1/3H, dd, J = 6.6, 13.2 Hz, Glc H$_R$-6), 5.21 (1/3H, d, J = 6.6, 13.2 Hz, Glc H$_R$-6′), 5.17 (2/3H, dd, J = 6.6, 13.2 Hz, Glc H$_L$-6), 5.12 (2/3H, dd, J = 8.4, 9.6 Hz, Glc H$_L$-2), 5.09 (1/3H, t, J = 9.6 Hz, Glc H$_L$-4), 5.07 (1/3H,2/3H t, J = 9.6 Hz, Glc H$_R$-4), 5.06 (1/3H, t, J = 4.2, 9.6 Hz, Glc H$_L$-2), 5.00 (2/3H, t, J = 9.6 Hz, Glc H$_L$-4), 5.00 (2/3H, t, J = 9.6 Hz, Glc H$_L$-4), 4.61 (1/3H, dd, J = 6.6, 9.6 Hz, Glc H$_R$-5), 4.57 (1/3H, dd, J = 6.6, 9.6 Hz, Glc H$_L$-5), 4.45 (2/3H, dd, J = 6.6, 9.6 Hz, Glc H$_L$-4), 4.49 (2/3H, t, J = 8.4 Hz, Glc H$_L$-1β), 3.93 (1/3H, d, J = 13.2 Hz, Glc H$_L$-6), 3.87 (2/3H, d, J = 13.2 Hz, Glc H$_R$-6′), 3.80 (2/3H, d, J = 13.2 Hz, Glc H$_L$-6), 3.74 (1/3H, d, J = 13.2 Hz, Glc H$_R$-6). HR-ESI-MS m/z 1721.1710 [M − H]$^-$ (calcd for C$_{75}$H$_{54}$O$_{48}$-H, 1721.1712).

(7′S,8′R)-Dihydrodehydrodiconiferyl alcohol-9′-O-β-D-glucoside (**10**): pale brown amorphous powder; $[α]_D^{25}$ + 12.8° (c 0.5, MeOH); UV (MeOH) λ$_{max}$ (log ε) 225 (4.30), 282 (3.92) nm; CD (MeOH) [α] (nm) − 4.3 × 10^3 (223), +1.7 × 10^4 (240), +1.0 × 10^4 (291); ^{1}H-NMR [600 MHz, CD$_3$OD] δ 6.99 (1H, d, J = 1.8 Hz, H-2′), 6.84 (1H, dd, J = 1.8, 7.8 Hz, H-6′), 6.79 (1H, brs, H-6), 6.75 (1H, d, J = 7.8 Hz, H-5′), 6.71 (1H, brd, J = 1.2 Hz, H-2), 5.57 (1H, d, J = 6.6 Hz, H-7′), 4.35 (1H, d, J = 7.8 Hz, H-1″), 4.10 (1H, dd, J = 8.4, 9.6 Hz, H-9′a), 3.86 (1H, dd, J = 2.4, 12.0 Hz, H-6″a), 3.84 (3H, s, OCH$_3$-3), 3.83 (1H, m, H-9′b), 3.81 (3H, s, OCH$_3$-3′), 3.68 (1H, dd, J = 6.0, 12.0 Hz, H-6″a), 3.64 (1H, brdd, J = 6.6, 13.2 Hz, H-8′), 3.58 (2H, t, J = 6.6 Hz, H-9), 3.37 (1H, t, J = 9.0 Hz, H-3″), 3.31 (1H, t, J = 9.0 Hz, H-4″), 3.27 (1H, ddd, J = 2.4, 6.0, 9.0 Hz, H-5″), 3.23 (1H, dd, J = 7.8, 9.0 Hz, H-2″), 2.61 (2H, brt, J = 7.8 Hz, H-7), 1.80 (2H, m, H-8); ^{13}C-NMR [151 MHz, CD$_3$OD] δ 147.6 (C-3′), 146.00 (C-4′), 145.97 (C-4), 143.8 (C-3), 135.5 (C-1), 133.2 (C-1′), 128.3 (C-5), 118.4 (C-6′), 116.8 (C-6), 114.6 (C-5′), 112.6 (C-2), 109.4 (C-2′), 102.8 (C-1″), 87.8 (C-7′), 76.7 (C-3″), 76.6 (C-5″), 73.7 (C-2″), 70.9 (C-9′), 70.2 (C-4″), 61.3 (C-6″), 60.8 (C-9), 55.3 (C-3OCH$_3$), 55.0 (C-3′OCH$_3$), 51.5 (C-8′), 34.4 (C-8), 31.5 (C-7); HR-ESI-MS m/z 521.2041 [M − H]$^-$ (calcd for C$_{26}$H$_{34}$O$_{11}$-H, 521.2028).

Acid hydrolysis of **10**: A solution of **10** (50 mg) in 1 M HCl was heated in boiled water for 1 h. The reaction mixture was purified by Mega Bond Elut C18 with MeOH-H$_2$O (0:100–10:90–20:80–30:70–40:60–50:50–60:40) in stepwise gradient. Compounds **13** (5.4 mg) and **14** (4.3 mg) were obtained from 40% and 60% MeOH fractions, respectively. The obtained H$_2$O fraction was tested by Glucose CII Test Wako kit (Wako Pure Chemical Industries, Osaka, Japan) to determine D-series of glucose in **10** [22].

Compound **13** (aglycone of **10**): off-white amorphous powder; $[\alpha]_D^{25}$ + 14.1° (*c* 0.5, MeOH); UV (MeOH) λ_{max} (log ε) 227 (4.30), 286 (3.87) nm; CD (MeOH) [α] (nm) − 3.4 × 10³ (225), + 1.5 × 10⁴ (240), + 6.7 × 10³ (292); ¹H-NMR [600 MHz, acetone-d_6-D₂O (9:1)] δ 6.99 (1H, d, *J* = 1.8 Hz, H-2′), 6.83 (1H, dd, *J* = 1.8, 7.8 Hz, H-6′), 6.78 (1H, d, *J* = 7.8 Hz, H-5′), 6.72 (1H, brs, H-6), 6.71 (1H, brs, H-2), 5.50 (1H, d, *J* = 6.6 Hz, H-7′), 3.82 (1H, m, H-9′a), 3.80 (3H, s, OCH₃-3′), 3.78 (3H, s, OCH₃-3), 3.73 (1H, dd, *J* = 7.2, 10.8 Hz, H-9′b), 3.52 (2H, t, *J* = 6.6 Hz, H-9), 3.47 (1H, brdd, *J* = 6.6, 13.2 Hz, H-8′), 2.58 (2H, brt, *J* = 7.8 Hz, H-7), 1.76 (2H, m, H-8); ¹³C-NMR [151 MHz, acetone-d_6-D₂O (9:1)] δ 147.6 (C-3′), 146.2 (C-4, 4′), 143.9 (C-3), 135.5 (C-1), 133.6 (C-1′), 128.9 (C-5), 118.4 (C-6′), 116.7 (C-6), 114.9 (C-5′), 112.8 (C-2), 109.7 (C-2′), 87.2 (C-7′), 63.6 (C-9′), 60.7 (C-9), 55.5 (C-3OCH₃), 55.4 (C-3′OCH₃), 54.0 (C-8′), 34.7 (C-8), 31.7 (C-7); HR-APCI-MS *m/z* 359.1499 [M − H]⁻ (calcd for C₂₀H₂₄O₆-H, 359.1500).

Compound **14**: off-white amorphous powder; ¹H-NMR [600 MHz, acetone-d_6-D₂O (9:1)] δ 7.32 (1H, d, *J* = 1.8 Hz, H-2′), 7.25 (1H, dd, *J* = 1.8, 8.4 Hz, H-6′), 6.97 (1H, d, *J* = 1.2 Hz, H-6), 6.95 (1H, d, *J* = 8.4 Hz, H-5′), 6.77 (1H, brd, *J* = 1.2 Hz, H-2), 3.97 (3H, s, OCH₃-3), 3.91 (3H, s, OCH₃-3′), 3.56 (2H, t, *J* = 6.6 Hz, H-9), 2.74 (2H, brt, *J* = 7.2 Hz, H-7), 2.39 (3H, s, H-9′), 1.86 (2H, m, H-8); ¹³C-NMR [151 MHz, acetone-d_6-D₂O (9:1)] δ 151.1 (C-7′), 147.8 (C-3′), 147.0 (C-4′), 144.7 (C-3), 141.0 (C-4), 137.9 (C-1), 132.9 (C-8′), 123.0 (C-5), 120.0 (C-6′), 115.4 (C-5′), 110.6 (C-6), 110.2 (C-2′), 109.5 (C-1′), 107.6 (C-2), 60.8 (C-9), 55.50 (C-3′OCH₃), 55.46 (C-3OCH₃), 34.9 (C-8), 32.3 (C-7), 8.8 (C-9′); HR-APCI-MS *m/z* 341.1394 [M − H]⁻ (calcd for C₂₀H₂₂O₅-H, 341.1394).

Rubuphenol (**11**): off-white amorphous powder; ¹H-NMR [600 MHz, CD₃OD]: δ 7.53 (1H, s, H-5′), 7.33 (1H, s, H-5), 6.19 (1H, d, *J* = 9.0 Hz, H-6″), 6.16 (1H, d, *J* = 9.0 Hz, H-5″); ¹³C-NMR [151 MHz, CD₃OD]: δ 160.7 (C-7), 160.6 (C-7′), 153.8 (C-4′), 149.5 (C-4), 143.4 (C-2 or 2′), 143.1 (C-4″), 141.2 (C-1″), 141.0 (C-3), 137.9 (C-3′), 137.15 (C-2″), 137.09 (C-2 or 2′), 135.5 (C-3″), 115.2 (C-6′), 113.4 (C-1), 113.3 (C-1′), 113.1 (C-5′), 111.8 (C-5), 108.9 (C-6), 107.4 (C-6″), 106.3 (C-5″); HR-ESI-MS *m/z* 425.0144 [M − H]⁻ (calcd for C₂₀H₁₀O₁₁-H, 425.0150).

Eschweilenol A (**12**): off-white amorphous powder; ¹H-NMR [600 MHz, CD₃OD]: δ 7.50 (1H, s, H-5′), 7.30 (1H, s, H-5), 6.51 (1H, d, *J* = 9.0 Hz, H-6″), 6.40 (1H, d, *J* = 9.0 Hz, H-5″); ¹³C-NMR [151 MHz, CD₃OD]: δ 161.34 (C-7′), 161.27 (C-7), 150.7 (C-4), 149.9 (C-4′), 145.0 (C-4″), 142.8 (C-3), 141.0 (C-3′), 139.7 (C-2″), 137.9 (2C, C-2, 2′) (2C), 137.4 (C-1″), 136.2 (C-3″), 115.9 (C-1), 113.6 (C-1′), 112.8 (C-6″), 111.9 (C-5), 111.8 (C-5′), 109.9 (C-6′), 108.7 (C-6), 107.3 (C-5″); HR-ESI-MS *m/z* 425.0134 [M − H]⁻ (calcd for C₂₀H₁₀O₁₁-H, 425.0150).

Methylation of compounds **11** and **12**: Each solution of **11** (10 mg) or **12** (10 mg) in acetone was treated with an excess of TMS-CHN₂ in hexane at room temperature overnight. Each reaction mixture was evaporated *in vacuo* and purified by preparative TLC (Merck, North Wales, PA, USA) with toluene:acetone (3:1) to obtain compounds **15** (Rf 0.55, 1.8 mg) and **16** (Rf 0.70, 2.1 mg) from **11**, and **17** (Rf 0.60, 0.7 mg) and **18** (Rf 0.74, 1.2 mg) from **12**, respectively.

Compound **15**: off-white amorphous powder; ¹H-NMR [600 MHz, CDCl₃]: δ 7.76 (1H, s, H-5′), 7.67 (1H, s, H-5), 6.52 (1H, d, *J* = 9.0 Hz, H-5″), 6.23 (1H, d, *J* = 9.0 Hz, H-6″), 4.24, 4.032, 4.026, 4.02, 3.95 (each 3H, s, OCH₃-H); HR-APCI-MS *m/z* 497.1102 [M + H]⁺ (calcd for C₂₅H₂₀O₁₁ + H, 497.1078).

Compound **16**: off-white amorphous powder; ¹H-NMR [600 MHz, CDCl₃]: δ 7.76 (1H, s, H-5′), 7.67 (1H, s, H-5), 6.47 (1H, d, *J* = 9.0 Hz, H-5″), 6.41 (1H, d, *J* = 9.0 Hz, H-6″), 4.24, 4.03, 4.00 (each 3H, s, OCH₃-H), 3.94 (6H, s, OCH₃-H), 3.82 (3H, s, OCH₃-H); HR-APCI-MS *m/z* 511.1258 [M + H]⁺ (calcd for C₂₆H₂₂O₁₁ + H, 511.1235).

Compound **17**: off-white amorphous powder; ¹H-NMR [600 MHz, CDCl₃]: δ 7.71 (1H, s, H-5′), 7.49 (1H, s, H-5), 6.79 (1H, d, *J* = 9.0 Hz, H-6″), 6.75 (1H, d, *J* = 9.0 Hz, H-5″), 4.32, 4.21, 4.04, 3.98, 3.83 (each 3H, s, OCH₃-H); HR-APCI-MS *m/z* 497.1065 [M + H]⁺ (calcd for C₂₅H₂₀O₁₁ + H, 497.1078).

Compound **18**: off-white amorphous powder; ¹H-NMR [600 MHz, CDCl₃]: δ 7.71 (1H, s, H-5′), 7.47 (1H, s, H-5), 6.85 (1H, d, *J* = 9.0 Hz, H-6″), 6.69 (1H, d, *J* = 9.0 Hz, H-5″), 4.32,

4.20, 4.04, 3.92, 3.91, 3.82 (each 3H, s, OCH$_3$-H); HR-APCI-MS m/z 511.1245 [M + H]$^+$ (calcd for C$_{26}$H$_{22}$O$_{11}$ + H, 511.1235).

3.4. α-Glucosidase Inhibitory Activity

The α-glucosidase inhibitory activity was tested according to the previously described method by Kirino et al. [47] with a slight modification. Rat intestinal acetone powder was mixed with 0.1 M phosphate buffer (pH 7) and centrifuged at 18,500× g and 4 °C for 20 min. The resulting supernatant was collected and used as a glucosidase solution for enzymatic assay. Each sample solution (160 µL) was mixed with 250 mM maltose solution (20 µL) in 0.2 M phosphate buffer (pH 7) and then incubated at 37 °C for 3 min. After incubation, the enzymatic reaction was started by adding glucosidase solution from the rat intestine (20 µL) and the resulting reaction mixtures were further incubated at 37 °C for 15 min. After 15 min, the reaction mixtures were immediately heated at 100 °C for 5 min to stop the reaction followed by cooling on ice for 5 min. The amount of glucose in the reaction mixtures was determined using the F-Kit glucose (Roche diagnostics, Co., Tokyo, Japan) by measuring the absorbance at 340 nm. A control was carried out with 0.1 M phosphate buffer (pH 7) instead of sample solution. For the blank, the glucosidase solution was replaced with distilled water. The glucosidase inhibitory rates of tested samples were calculated as;

$$\text{Inhibitory rate (\%)} = 100 - [(A_{\text{sample}} - A_{\text{blank}})/(A_{\text{control}} - A_{\text{blank}})] \times 100 \quad (1)$$

where A_{sample}, A_{control}, and A_{blank} is the absorbance of the tested sample, control, and blank, respectively. The experimental data are represented as IC$_{50}$ (µM) values.

3.5. Inhibitory Effect on AGE-Formation

The antiglycation effects of compounds isolated from TBE and several other related compounds were evaluated based on their AGE inhibitory activities as described previously [25], with slight modifications. Briefly, the sample solution was added to a reaction mixture containing 83.3 mM phosphate buffer (pH 7.2), 2.0 M glucose, 2.0 M fructose, 8.0 mg/mL human serum albumin (HSA), and distilled water (6:1:1:1:1, v/v). As a control, the vehicle was supplemented instead of the sample solution. For each blank, glucose or fructose was replaced with distilled water, and the total volume was set to 1000 µL. After incubation of the sample mixture at 60 °C for 40 h, the solutions were diluted 8-fold with distilled water, dispensed into a black microplate in 200 µL portions, and their fluorescence intensities were measured at excitation and emission wavelengths of 370 and 465 nm, respectively, using a Power Scan HT (DS Pharma Biomedical Co. Ltd., Osaka, Japan). The inhibitory rate was calculated as;

$$\text{Inhibitory rate (\%)} = 100 - [(S - SB)/(C - CB)] \times 100 \quad (2)$$

where S is the relative intensity of the sample solution, C is the relative intensity of the control solution, and SB and CB are the intensities of the glucose or fructose-omitted blank solutions. The experimental data are represented as IC$_{50}$ (µM) values.

3.6. AGE-Derived Crosslink-Cleaving Effect

The AGE crosslink-cleaving activity of the same samples was evaluated according to the previously described method by Kato et al. [28] with a slight modification. Briefly, the 1 mg/mL of tested samples prepared with H$_2$O were mixed with 1.13 mM PPD solution in MeOH:50 mM phosphate buffer (pH 7.4) (1:1) and incubated at 37 °C for 4 h. After 4 h, the reaction was stopped with 200 µL of 2 M HCl, then the stopped reaction mixtures were centrifuged at 8200× g for 5 min. The amount of benzoic acid in the supernatant was measured by reversed-phase HPLC. The following conditions were applied: column, InertSustain C18 column (150 mm × 4.6 i.d mm., 5 µm); mobile phase, 50 mM phosphate buffer (pH 2.2):CH$_3$CN (80:20) (solvent A) and 50 mM phosphate buffer (pH 2.2):CH$_3$CN

(50:50) (solvent B), the gradient program, 0–20 min (solvent B: 0–25%), 20–25 min (solvent B: 25–100%), 25–36 min (solvent B: 100–25%); column temperature, 40 °C; detection, UV at 230 nm; flow rate, 1.0 mL/min. These cleaving effects were calculated as equivalents to benzoic acid.

3.7. TBE Polyphenol Identification and Quantification by LC/UV/ESIMS Analysis

TBE (1.0 g) was sonicated with 70% aqueous acetone (3 × 10 mL) and the resulting suspension was centrifuged at 2200× g for 5 min. The supernatant was collected and dried. The obtained acetone extract from TBE (0.68 g) was dissolved in 50% aqueous MeOH to a concentration of 5 mg/mL and the solution of TBE extract was subjected to LC/UV/ESIMS. This analysis was performed on Waters 2695 separation module (Waters, Milford, MA, USA) coupled to Shimadzu SPD-6AV UV–vis spectrophotometric detector (Shimadzu, Kyoto, Japan) and Bruker MicrOTOF II instrument equipped with ESI source. The analyte was separated by InertSustain C18 column (150 mm × 4.6 mm i.d., 5 µm) at 40 °C with the mobile phase consisted of $CH_3CN:H_2O:HCOOH$ (94.5:5:0.1) (solvent A) and $CH_3CN:H_2O:HCOOH$ (54.9:45:0.1) (solvent B). The flow rate was 1.0 mL/min (splitting flow rate at 0.2 mL/min to MS unit), and a linear gradient was programmed as follows: 0–15 min (solvent B: 10–30%), 15–20 min (solvent B: 30–50%), 20–30 min (solvent B: 10%) and UV detection was monitored at 280 nm and 360 nm. MS parameters in negative ion mode were as follows: capillary voltage, 3.5 kV; nebulizer, 0.4 bar; dry gas, 4.0 L/min; dry temperature, 180 °C. The MS spectra were recorded in the range of m/z 50–3000. The polyphenol contents were expressed as mg/g (dry weight) by the absolute calibration curve method based on UV chromatogram.

4. Conclusions

In summary, we isolated the 13 known polyphenolic compounds including gallic acid, six galloyl glucoses (**1–6**), three ellagitannins (**7–9**), one neolignan (**10**), and ellagic acid derivatives **11** and **12** from TBE. The absolute configuration of **10** was confirmed by the aromatic quadrant and P/M helicity rules based on our CD analysis. Among the isolated polyphenols, decarboxylated rugosin A (**8**) and 1,2,3,4,6-penta-O-galloyl-β-D-glucose showed α-glucosidase inhibitory activities. Gallotannins and ellagitannins showed more significant inhibitory effect on AGE formation than that of gallic acid. Furthermore, gallic acid showed most potent AGE-derived crosslink cleaving activity among the tested polyphenols. A total of 30 TBE polyphenols were comprehensively identified by LC/UV/ESIMS analysis. The contents of tannins and the related polyphenols were also analyzed using LC/UV/ESIMS, indicating that gallic acid and gallotannins showing antiglycation effects were contained the major level in TBE. Further investigations are required to develop a deeper understanding of the TBE antidiabetic and antiglycation effects as well as the safety by in vivo experiments or clinical trial. Since the pericarp of this plant has experience in food as tea, it is considered that safety is guaranteed to some extent. The results of this study suggested the TBE polyphenols are a good source for antidiabetic and antiglycation effects, which could be applied as functional foods or nutritional supplements to improve human health benefits.

Supplementary Materials: The following are available online, Figure S1: 1D and 2D-NMR spectra of (7'S, 8'R)-dihydrodehydrodiconiferyl alcohol-9'-O-β-D-glucoside (**10**); Figure S2: ^{1}H-NMR spectrum of compounds **13** and **14**; Figure S3: 1D and 2D-NMR spectra of rubuphenol (**11**) and eschweilenol A (**12**); Figure S4: 1D and 2D-NMR spectra of compounds **15–18**.

Author Contributions: Conceptualization, H.I.; Methodology, Y.I., S.K. and N.G.; Validation, Y.I., S.S. and N.K.; Formal analysis, S.S. and N.K.; Investigation, Y.I., S.S., N.K., C.H. and S.K.; Data curation, Y.I., S.S., N.K., C.H., S.K. and N.G.; Visualization, Y.I.; Resources, T.U.; Funding acquisition, T.U. and H.I.; Writing—original draft preparation, Y.I.; Writing—review and editing, T.U. and H.I.; Supervision, H.I.; Project administration, H.I. All authors have read and agreed to the published version of the manuscript.

Funding: This study was funded by Okayama Prefectural University (Grant number: 2018D15 and 2019D11 to H.I.).

Institutional Review Board Statement: Not applicable.

Informed Consent Statement: Not applicable.

Data Availability Statement: Data is contained within the article and supplementary materials.

Acknowledgments: The authors grateful to the Faculty of Pharmaceutical Sciences of the Okayama University for the UV and CD spectral measurements, just as well as for the optical rotation. The authors thank to the SC-NMR Laboratory of the Okayama University for the NMR measurements.

Conflicts of Interest: The authors declare no conflict of interest.

Sample Availability: Not available.

References

1. Takeshita, S.; Yagi, M.; Uemura, T.; Yamada, M.; Yonei, Y. Peel extract of water chestnut (*Trapa bispinosa* Roxb.) inhibits glycation, degrades α-dicarbonyl compound, and breaks advanced glycation end product crosslinks. *Glycative Stress Res.* **2015**, *2*, 72–79.
2. Adkar, P.; Dongare, A.; Ambavade, S.; Bhaskar, V.H. *Trapa bispinosa* Roxb. A review on nutritional and pharmacological aspects. *Adv. Pharmacol. Pharm. Sci.* **2014**, *2014*, 959830. [CrossRef]
3. Mann, S.; Gupta, D.; Gupta, V.; Gupta, R.K. Evaluation of nutritional, phytochemical and antioxidant potential of *Trapa bispinosa* Roxb. Fruits. *Int. J. Pharm. Pharm. Sci.* **2012**, *4*, 432–436.
4. Xia, J.; Yang, C.; Wang, Y.; Yang, Y.; Yu, J. Antioxidant and antiproliferative activities of the leaf extracts from *Trapa bispinosa* and active components. *S. Afr. J. Bot.* **2017**, *113*, 377–381. [CrossRef]
5. Saeed, M.; Nadeem, R.; Yousaf, M. Removal of industrial pollutant (Reactive Orange 122 dye) using environment-friendly sorbent Trapa bispinosa's peel and fruit. *Int. J. Environ. Sci. Technol.* **2015**, *12*, 1223–1234. [CrossRef]
6. Nonaka, G.; Matsumoto, Y.; Nishioka, I. Trapanin, a new hydrolyzable tannin from *Trapa japonica* FLEROV. *Chem. Pharm. Bull.* **1981**, *29*, 1184–1187. [CrossRef]
7. Hatano, T.; Okonogi, A.; Yazaki, K.; Okuda, T. Trapanins A and B, oligomeric hydrolyzable tannins from *Trapa japonica* FLEROV. *Chem. Pharm. Bull.* **1990**, *38*, 2707–2711. [CrossRef]
8. Kawabe, S.; Ganeko, N.; Ito, H. Ellagitannin dimers from the pericarps of *Trapa japonica*. *Jpn. J. Pharmacog.* **2017**, *71*, 53–54.
9. Ngoc, T.M.; Hung, T.M.; Thuong, P.T.; Kim, J.C.; Choi, J.S.; Bae, K.; Hattori, M.; Choi, C.S.; Lee, J.S.; Min, B.S. Antioxidative activities of galloyl glucopyranosides from the stem-bark of *Juglans mandshurica*. *Biosci. Biotechnol. Biochem.* **2008**, *72*, 2158–2163. [CrossRef]
10. Ko, S.Y.; Ko, H.A.; Chu, K.H.; Shieh, T.M.; Chi, T.C.; Chen, H.I.; Chang, W.C.; Chang, S.S. The possible mechanism of advanced glycation end products (AGEs) for Alzheimer's disease. *PLoS ONE* **2015**, *10*, e0143345. [CrossRef]
11. Luévano-Contreras, C.; Garay-Sevilla, M.E.; Wrobel, K.; Malacara, J.M.; Wrobel, K. Dietary advanced glycation end products restriction diminishes inflammation markers and oxidative stress in patients with type 2 diabetes mellitus. *J. Clin. Biochem. Nutr.* **2013**, *52*, 22–26. [CrossRef] [PubMed]
12. Takeshita, S.; Ishioka, Y.; Yagi, M.; Uemura, T.; Yamada, M.; Yonei, Y. The effects of water chestnut (*Trapa bispinosa* Roxb.) on the inhibition of glycometabolism and the improvement in postprandial blood glucose levels in humans. *Glycative Stress Res.* **2016**, *3*, 24–132.
13. Kashiwada, Y.; Nonaka, G.; Nishioka, I. Tannins and related compounds. XXIII. Rhubarb (4): Isolation and structures of new classes of gallotannins. *Chem. Pharm. Bull.* **1984**, *32*, 3461–3470. [CrossRef]
14. Okuda, T.; Hatano, T.; Yazaki, K.; Ogawa, N. Rugosin A, B, C and praecoxin A, tannin having a valoneoyl group. *Chem. Pharm. Bull.* **1982**, *30*, 4230–4233. [CrossRef]
15. Ito, H.; Yamaguchi, K.; Kim, T.H.; Khennouf, S.; Gharzouli, K.; Yoshida, T. Dimeric and trimeric hydrolyzable tannins from *Quercus coccifera* and *Quercus suber*. *J. Nat. Prod.* **2002**, *65*, 339–345. [CrossRef] [PubMed]
16. Haddock, E.A.; Gupta, R.K.; Al-Shafi, S.M.K.; Haslam, E.; Magnolato, D. The metabolism of gallic acid and hexahydroxydiphenic acid in plants. Part 1. Introduction. Naturally occurring galloyl esters. *J. Chem. Soc. Perkin Trans.* **1982**, 2515–2524. [CrossRef]
17. Okuda, T.; Yoshida, T.; Ashida, M.; Yazaki, K. Tannins of *Casuarina* and *Stachyurus* species. Part 1. Structures of pedunculagin, casuarictin, strictinin, casuarinin, casuariin, and stachyurin. *J. Chem. Soc. Perkin Trans.* **1983**, 1765–1772. [CrossRef]
18. Kato, E.; Uenishi, Y.; Inagaki, Y.; Kurokawa, M.; Kawabata, J. Isolation of rugosin A, B and related compounds as dipeptidyl peptidase-IV inhibitors from rose bud extract powder. *Biosci. Biotechnol. Biochem.* **2016**, *80*, 1–6. [CrossRef]
19. Abe, F.; Yamauchi, T. Lignans from *Trachelospermum asiaticum* (Tracheolospermum. II). *Chem. Pharm. Bull.* **1986**, *34*, 4340–4345. [CrossRef]
20. Cui, C.B.; Zhao, Q.C.; Cai, B.; Yao, X.S.; Osadsa, H. Two new and four known polyphenolics obtained as new cell-cycle inhibitors from *Rubus aleaefolius* Poir. *J. Asian Nat. Prod. Res.* **2002**, *4*, 243–252. [CrossRef]
21. Yang, S.W.; Zhou, B.N.; Wisse, J.H.; Evans, R.; van der Werff, H.; Miller, J.S.; Kingston, D.G. Three new ellagic acid derivatives from the bark of *Eschweilera coriacea* from the Suriname rainforest. *J. Nat. Prod.* **1998**, *61*, 901–906. [CrossRef] [PubMed]

22. Miwa, I.; Okuda, J.; Maeua, K.; Okuda, G. Mutarotase effect on colorimetric determination of blood glucose with β-D-glucose oxidase. *Clin. Chim. Acta.* **1972**, *37*, 538–540. [CrossRef]
23. Li, S.; Iliefski, T.; Lundquist, K.; Wallis, A.F.A. Reassignment of relative stereochemistry at C-7 and C-8 in arylcoumaran neolignans. *Phytochemistry* **1997**, *46*, 929–934. [CrossRef]
24. Yuen, M.S.M.; Xue, F.; Mak, T.C.W.; Wong, H.N.C. On the absolute structure of optically active neolignan containing a dihydrobenzo[b]furan skeleton. *Tetrahedron* **1998**, *54*, 12429–12444. [CrossRef]
25. Ito, H.; Li, P.; Koreishi, M.; Nagatomo, A.; Nishida, N.; Yoshida, T. Ellagitannin oligomers and a neolignan from pomegranate arils and their inhibitory effects on the formation of advanced glycation end products. *Food Chem.* **2014**, *152*, 323–330. [CrossRef]
26. Kim, T.; Ito, H.; Hayashi, K.; Hasegawa, T.; Machiguchi, T.; Yoshida, T. Aromatic constituents from the heartwood of *Santalum album* L. *Chem. Pharm. Bull.* **2005**, *53*, 641–644. [CrossRef]
27. Yang, Z.; Liu, H.B.; Lee, C.M.; Chang, H.M.; Wong, H.N.C. Compounds from Danshen. Part 7. Regioselective introduction of carbon-3 substituents to 5-alkyl-7-methoxy-2-phenylbenzo[b]furans: Synthesis of a novel adenosine A1 receptor ligand and its derivatives. *J. Org. Chem.* **1992**, *57*, 7248–7257. [CrossRef]
28. Kato, N.; Kawabe, S.; Ganeko, N.; Yoshimura, M.; Amakura, Y.; Ito, H. Polyphenols from flowers of *Magnolia coco* and their anti-glycation effects. *Biosci. Biotechnol. Biochem.* **2017**, *81*, 1285–1288. [CrossRef]
29. Yin, Z.; Zhang, W.; Feng, F.; Hang, Y.; Kang, W. α-Glucosidase inhibitors isolated from medicinal plants. *Food Sci. Hum. Wellness.* **2014**, *3*, 136–174. [CrossRef]
30. Yun, X.; Ken, N.; Pangzhen, Z.; Robyn, D.W.; Shuibao, S.; Hsi-Yang, T.; Zijian, L.; Zhongxiang, F. In vitro α-glucosidase and α-amylase inhibitory activities of free and bound phenolic extracts from the bran and kernel fractions of five sorghum grain genotypes. *Foods* **2020**, *9*, 1301.
31. Brownlee, M.; Vlassara, H.; Kooney, A.; Ulrich, P.; Cerami, A. Aminoguanidine prevents diabetes-induced arterial wall protein cross-linking. *Science* **1986**, *232*, 1629–1632. [CrossRef] [PubMed]
32. Yeh, W.J.; Hsia, S.M.; Lee, W.H.; Wu, C.H. Polyphenols with antiglycation activity and mechanisms of action: A review of recent findings. *J. Food Drug Anal.* **2017**, *25*, 84–92. [CrossRef] [PubMed]
33. Takabe, W.; Mitsuhashi, R.; Parengkuan, L.; Yagi, M.; Yonei, Y. Cleaving effect of melatonin on crosslinks in advanced glycation end products. *Glycative Stress Res.* **2016**, *3*, 38–43.
34. Huang, H.C.; Chao, C.L.; Liaw, C.C.; Hwang, S.Y.; Kuo, Y.H.; Chang, T.C.H.; Chen, C.J.; Kuo, Y.H. Hypoglycemic constituents isolated from *Trapa natans* L. pericarps. *J. Agric. Food Chem.* **2016**, *64*, 3794–3803. [CrossRef] [PubMed]
35. Shirataki, Y.; Yoshida, S.; Toda, S. Dibenzo-α-pyrons in fluits of *Trapa natans*. *Natural Med.* **2000**, *54*, 160.
36. Nawwar, M.A.M.; Souleman, A.M.A. 3,4,8,9,10-Pentahydroxy-dibenzo[b,d]pyran-6-one from *Tamarix nilotica*. *Phytochemistry* **1984**, *23*, 2966–2967. [CrossRef]
37. Amakura, Y.; Kawada, K.; Hatano, T.; Okuda, T. Four new hydrolyzable tannins and an acylated flavonol glycoside from *Euphorbia maculata*. *Can. J. Chem.* **1996**, *75*, 727–733. [CrossRef]
38. Amakura, Y.; Yoshida, T. Tannins and related polyphenols of euphorbiaceous Plants. XIV. Euphorbin I, a new dimeric hydrolyzable tannin from *Euphorbia watanabei*. *Chem. Pharm. Bull.* **1996**, *44*, 1293–1297. [CrossRef]
39. Yoshida, T.; Itoh, H.; Matsunaga, S.; Tanaka, R.; Okuda, T. Tannins and related polyphenols of Euphorbiaceous plants. IX. Hydrolyzable tannins with $^1C_4$ glucose core from *Phyllanthus flexuosus* Muell. Arg. *Chem. Pharm. Bull.* **1992**, *40*, 53–60. [CrossRef]
40. Hatano, T.; Ogawa, N.; Kira, R.; Yasuhira, T.; Okuda, T. Tannins of cornaceous plants. I. Cornusiins A, B and C, dimeric monomeric and trimeric hydrolyzable tannins from *Cornus officinalis*, and orientation of valoneoyl group in related tannins. *Chem. Pharm. Bull.* **1989**, *37*, 2083–2090. [CrossRef]
41. Ito, H.; Miki, K.; Yoshida, T. Elaeagnatins A-G, C-glucosidic ellagitannins from *Elaeagnus umbellate*. *Chem. Pharm. Bull.* **1999**, *47*, 536–542. [CrossRef]
42. Niemetz, R.; Gross, G.G. Gallotannin biosynthesis: Purification of β-glucogallin: 1,2,3,4,6-pentagalloyl-β-D-glucose galloyltransferase from sumac leaves. *Phytochemistry* **1998**, *2*, 327–332. [CrossRef]
43. Shimozu, Y.; Kuroda, T.; Tsuchiya, T.; Hatano, T. Structures and antibacterial properties of isorugosins H–J, oligomeric ellagitannins from liquidambar formosana with characteristic bridging groups between sugar moieties. *J. Nat. Prod.* **2017**, *80*, 2723–2733. [CrossRef]
44. Yoshida, T.; Amakura, Y.; Liu, Y.Z.; Okuda, T. Tannins and related polyphenols of euphorbiaceous Plants. XI.Three new hydrolyzable tannins and a polyphenol glucoside from *Euphorbia humifusa*. *Chem. Pharm. Bull.* **1994**, *42*, 1803–1807. [CrossRef]
45. Yagi, K.; Goto, K.; Nanjo, F. Identification of a major polyphenol and polyphenolic composition in leaves of *Camellia irrawadiensis*. *Chem. Pharm. Bull.* **2009**, *11*, 1284–1288. [CrossRef] [PubMed]
46. Ito, H.; Iguchi, A.; Hatano, T. Identification of urinary and intestinal bacterial metabolites of ellagitannin geraniin in rats. *J. Agric. Food Chem.* **2008**, *56*, 393–400. [CrossRef] [PubMed]
47. Kirino, A.; Takasuka, Y.; Nishi, A.; Kawabe, S.; Kirino, A.; Takasuka, Y.; Nishi, A.; Kawabe, S.; Yamashita, H.; Kimoto, M.; et al. Analysis and functionality of major polyphenolic components of *Plygonum cuspidatum* (Itadori). *J. Nutr. Sci. Vitaminol.* **2012**, *58*, 278–286. [CrossRef]

Article

The Zebrafish Embryo as a Model to Test Protective Effects of Food Antioxidant Compounds

Cristina Arteaga [1,2], Nuria Boix [1,3], Elisabet Teixido [1,3], Fernanda Marizande [2], Santiago Cadena [4] and Alberto Bustillos [5,*]

[1] Toxicology Unit, Pharmacology, Toxicology and Therapeutical Chemistry Department, Pharmacy School, University of Barcelona, Avda Joan XXIII s/n, 08028 Barcelona, Spain; ca.arteaga@uta.edu.ec (C.A.); nuriaboix@ub.edu (N.B.); eteixido1511@ub.edu (E.T.)
[2] Faculty of Health Sciences, Nutrition and Dietetics, Technical University of Ambato, Ambato 180207, Ecuador; mf.marizande@uta.edu.ec
[3] INSA-UB Nutrition and Food Safety Research Institute, Food and Nutrition Torribera Campus, University of Barcelona, Prat de la Riba 171, 08921 Santa Coloma de Gramenet, Spain
[4] Faculty of Applied Sciences, International SEK University, Quito 170134, Ecuador; sacadena.mbme@uisek.edu.ec
[5] Faculty of Health Sciences, Medicine, Technical University of Ambato, Ambato 180207, Ecuador
* Correspondence: aa.bustillos@uta.edu.ec or abustillosortiz@gmail.com; Tel.: +593-328-48316

Abstract: The antioxidant activity of food compounds is one of the properties generating the most interest, due to its health benefits and correlation with the prevention of chronic disease. This activity is usually measured using in vitro assays, which cannot predict in vivo effects or mechanisms of action. The objective of this study was to evaluate the in vivo protective effects of six phenolic compounds (naringenin, apigenin, rutin, oleuropein, chlorogenic acid, and curcumin) and three carotenoids (lycopene B, β-carotene, and astaxanthin) naturally present in foods using a zebrafish embryo model. The zebrafish embryo was pretreated with each of the nine antioxidant compounds and then exposed to tert-butyl hydroperoxide (tBOOH), a known inducer of oxidative stress in zebrafish. Significant differences were determined by comparing the concentration-response of the tBOOH induced lethality and dysmorphogenesis against the pretreated embryos with the antioxidant compounds. A protective effect of each compound, except β-carotene, against oxidative-stress-induced lethality was found. Furthermore, apigenin, rutin, and curcumin also showed protective effects against dysmorphogenesis. On the other hand, β-carotene exhibited increased lethality and dysmorphogenesis compared to the tBOOH treatment alone.

Keywords: oxidative stress; zebrafish embryo; antioxidant effect; polyphenols; carotenoids

1. Introduction

Reactive oxygen species (ROS) and reactive nitrogen species (RNS) are originated during cell metabolism. They are essential to a normal physiological state, but they participate in pathological processes when in excess [1]. Aerobic organisms have defenses to prevent ROS-induced oxidative damage, involving antioxidant enzymes and/or non-enzymatic mechanisms, including endogenously produced antioxidant compounds or the ingestion of antioxidants in the diet [2].

The imbalance, when the concentration of reactive species is higher than the antioxidant defenses of the organism, is called oxidative stress (OS) [3]. The consequences of OS include macrophage recruitment; the inhibition of the normal functioning of lipids and proteins; and mitochondrial, membrane, and DNA damage [4–7]. These alterations have been correlated with several pathologies, such as cancer, aging, diabetes, rheumatoid arthritis, and cardiovascular and neurodegenerative diseases, among others [8–11].

Several studies have found that an organism requires the intake of antioxidants through the diet to reduce oxidative damage [12] in physiopathological situations (due to

UV exposure, smoking, polluted air, etc.), which produce excess ROS. Various antioxidants are ingested through the diet, such as phenolic compounds, vitamins, carotenoids, and flavonoids. The term "phenolic compounds" refers to any substance with a phenol group attached to aromatic or aliphatic structures. Phenolic compounds come from plants and are among the most important secondary metabolites; their presence in the animal kingdom is due to their consumption through the diet. Among these compounds, flavonoids are the most studied and abundant; their chemical structures contain a flavonic nucleus that consists of 15 carbon atoms arranged in three rings (C6–C3–C6) [13]. Their antioxidant mechanisms include the inhibition of enzymes or chelation of trace elements involved in producing free radicals, the uptake of ROS, and the protection of endogenous antioxidant defenses [14]. The mean flavonoid intake is estimated at 23 mg/day [15,16], and the primary sources are black tea, red wine, onions, apples, and beer [17,18].

Another group of compounds that have been studied because of their antioxidant activity are the carotenoids. They are pigments whose structures comprise a series of conjugated C = C bonds (polyene), which allow them to interact with free radicals; therefore, they can act as effective antioxidants [19]. Carotenoids are broadly distributed in natural systems, and their role in preventing different diseases has been studied, mainly for the compounds present in the diet such as β-carotene, lutein, and lycopene [18,20,21].

Identifying the roles of antioxidants in diseases and disorders correlated with oxidative processes is essential for analyzing protective effects in vivo. For this reason, our laboratory has developed a zebrafish (ZF) embryo model for evaluating the protective effects of these antioxidant compounds [22]. Oxidative stress is induced using tert-butyl hydroperoxide (tBOOH). tBOOH generates butoxyl radicals through Fenton's reaction [23]. The radicals formed favor the intracellular depletion of thiol groups and glutathione reserves, producing a significant increase in lethality and dysmorphogenesis in exposed zebrafish embryos. This model allows a comparison between the concentration–effect curves of lethality and dysmorphogenesis for zebrafish embryos exposed to tBOOH and the curves of embryos pretreated with antioxidants; statistical analysis can be performed to explore the protective effect of the analyzed antioxidant compound.

The objective of the present work was to evaluate the in vivo protective effects of food compounds with antioxidant activity against oxidant-induced developmental toxicity in zebrafish embryos.

2. Results

2.1. Concentration Effect Curves for tBOOH in Zebrafish Embryos

Our research group previously developed and validated a ZF embryo oxidative stress model with which to evaluate the protective activity of antioxidant substances (22).

Zebrafish embryos are exposed to tert-butyl hydroperoxide (tBOOH) to obtain lethality and dysmorphogenesis curves. The zebrafish embryos are exposed to tBOOH 24 to 48 h post-fertilization (hpf) at different concentrations, ranging from 1 to 3.5 mM (Figure 1). The lethal concentration 50 (LC_{50}) was discovered to be 2.1 mM, whereas the effective concentration 50 for dysmorphogenesis (EC_{50}) was 1.7 mM. The curves mentioned above were used to compare zebrafish embryos previously exposed, or not, to antioxidant compounds, after which they were exposed to tBOOH.

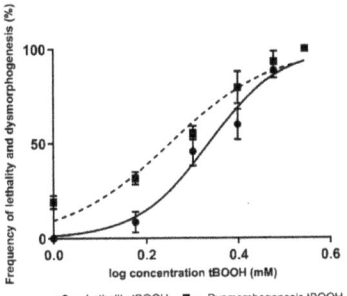

Figure 1. Concentration–response curves for lethality and dysmorphogenesis produced using tert-butyl hydroperoxide (tBOOH).

2.2. Identification of the Protective Effects of Antioxidant Compounds in Zebrafish Embryos

The previously described zebrafish model was used to evaluate the protective effects of six polyphenols and three carotenoids present in food.

Out of the six polyphenols, three were flavonoids: naringenin (20 µM), apigenin (10 µM), and rutin (10 µM). The three flavonoids produced a significant drift in the concentration–response curves for lethality. Furthermore, apigenin and rutin showed protective effects against dysmorphogenesis, whereas naringenin did not exhibit any protective effect against dysmorphogenesis (Figure 2).

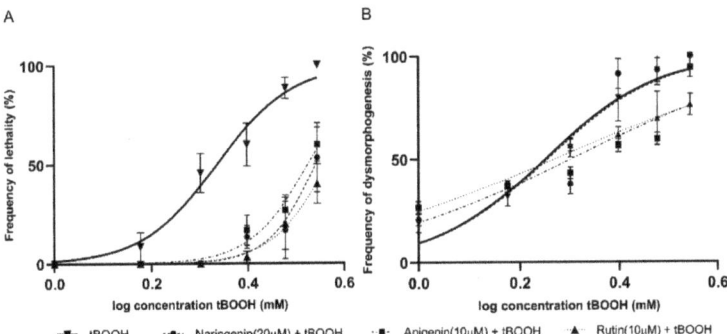

Figure 2. Concentration–response curves produced by tBOOH, in combination with different flavonoid compounds for (**A**) lethality and (**B**) dysmorphogenesis.

Oleuropein (15 µM), chlorogenic acid (20 µM), and curcumin (15 µM) were also analyzed. These polyphenols resulted in a significant drift in the concentration–response curves for lethality. Only curcumin exhibited a protective effect against dysmorphogenesis (Figure 3).

In addition, the carotenoids lycopene (20 µM), astaxanthin (20 µM), and β-carotene (25 µM) were evaluated. Lycopene and astaxanthin resulted in a significant drift in the concentration–response curves for lethality. By contrast, none of the carotenoids showed a protective effect against dysmorphogenesis (Figure 4). Furthermore, β-carotene resulted in a leftward drift in the curves for lethality and dysmorphogenesis, which could indicate a possible prooxidant effect.

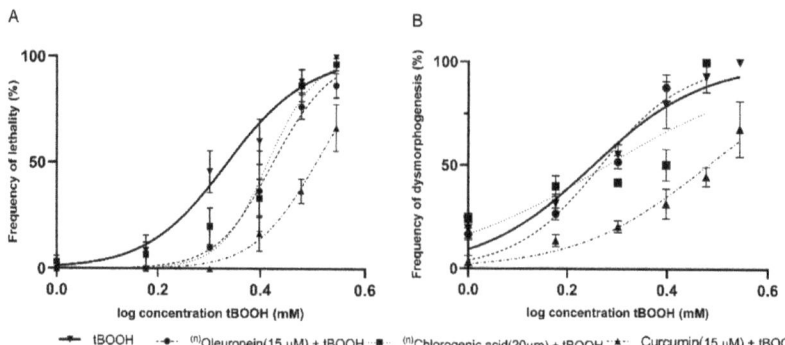

Figure 3. Concentration–response curves produced using tBOOH in combination with different polyphenolic compounds for (**A**) lethality, and (**B**) dysmorphogenesis. [n]Note: the absence of points on the graph at higher concentrations is because they already produced a 100% embryo lethality in all replicates.

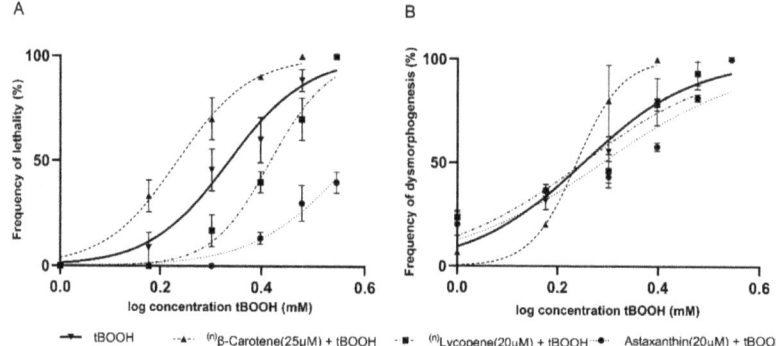

Figure 4. Concentration–response curves produced using tBOOH in combination with different carotenoid compounds for (**A**) lethality, and (**B**) dysmorphogenesis. [n]Note: The absence of points on the graph at higher concentrations is because they already produced a 100% embryo lethality in all replicates.

Table 1 presents the flavonoid and carotenoid compounds used, showing the LC_{50} and values; eight evaluated compounds exhibited protective effects against lethality after the oxidant treatment. However, for the EC_{50} values of dysmorphogenesis, only rutin, apigenin, and curcumin showed protective effects. On the other hand, β-carotene (25 μM) exhibited an increased risk of mortality and dysmorphogenesis in zebrafish after the oxidant treatment, based on a significant reduction in the LC_{50} (2.6 mM) and EC_{50} (1.5 mM) values.

Table 1. Effects of polyphenols and carotenoid compounds against an oxidant inducer (tBOOH) of developmental toxicity in zebrafish.

Compound	Lethality				Dysmorphogenesis			
	LC_{50} (mM) tBOOH (95% CI)	LC_{50} (mM) (Compound + tBOOH) (95% CI)	p-Value	Effect	EC_{50} (mM) tBOOH (95% CI)	EC_{50} (mM) (Compound + tBOOH) (95% CI)	p-Value	Effect
Naringenin (20 μM)	2.1 (2.0–2.2)	3.4 (3.3–3.7)	<0.0001	PE [1]	1.7 (1.6–1.8)	1.8 (1.5–2.0)	>0.05	S/E [2]
Apigenin (10 μM)	2.1 (2.0–2.2)	3.3 (3.2–3.4)	<0.0001	PE	1.7 (1.6–1.8)	2.0 (1.7–2.2)	0.0006	PE
Rutin (10 μM)	2.1 (2.0–2.2)	3.6 (3.5–3.9)	<0.0001	PE	1.7 (1.6–1.8)	1.8 (1.7–2.0)	<0.0008	PE

Table 1. *Cont.*

Compound	Lethality				Dysmorphogenesis			
	LC_{50} (mM) tBOOH (95% CI)	LC_{50} (mM) (Compound + tBOOH) (95% CI)	p-Value	Effect	EC_{50} (mM) tBOOH (95% CI)	EC_{50} (mM) (Compound + tBOOH) (95% CI)	p-Value	Effect
Oleuropein (15 µM)	2.1 (2.0–2.2)	2.6 (2.5–2.7)	<0.0001	PE	1.7 (1.6–1.8)	1.8 (1.7–1.9)	>0.05	S/E
Chlorogenic acid (20 µM)	2.1 (2.0–2.2)	2.5 (2.4–2.6)	0.0028	PE	1.7 (1.6–1.8)	1.8 (1.5–2.2)	>0.05	S/E
Curcumin (15 µM)	2.1 (2.0–2.2)	3.2 (3.1–3.2)	<0.0001	PE	1.7 (1.6–1.8)	3.0 (2.8–3.2)	<0.0001	PE
Lycopene (20 µM)	2.1 (2.0–2.2)	2.6 (2.5–2.6)	0.0003	PE	1.7 (1.6–1.8)	1.7 (1.5–1.9)	0.56	S/E
β-carotene (25 µM)	2.1 (2.0–2.2)	1.7 (1.6–1.7)	0.0001	IL [3]	1.7 (1.6–1.8)	1.5 (1.5–1.6)	0.01	ID [4]
Astaxanthin (20 µM)	2.1 (2.0–2.2)	3.7 (3.5–3.9)	<0.0001	PE	1.7 (1.6–1.8)	1.9 (1.7–2.1)	0.0506	S/E

[1] PE: protective effect. [2] S/E: no effect because there is no difference regarding the 95% confidence interval. [3] IL: increased lethality. [4] ID: increased dysmorphogenesis. Values of LC_{50} and EC_{50} are expressed in mM.

3. Discussion

Oxygen is essential for human life; however, at the same time, it produces toxic substances, such as free radicals and reactive oxygen species (ROS); these substances are oxidizing, unstable, and reactive. Furthermore, they can react with any macromolecule and cause cell damage [24]. To counteract these oxidizing substances, the body employs antioxidant enzymes, such as superoxide dismutase and glutathione peroxidase, and antioxidant compounds derived from the diet. Therefore, the study of the antioxidant capacity of compounds has been garnering interest in the past few years. There are several in vitro techniques for determining antioxidant activity, although they have limitations from a nutritional point of view because none reproduces a physiological situation [25]. For this reason, a method that included in vivo techniques would lead to more impactful results, because oxidative stress implies mechanisms that depend on many system conditions, especially the kinetic parts of reactions. Our team used a ZF embryo model [22], which could be a valuable in vivo method, to test the protective effects of nine antioxidant compounds that have been broadly studied in vitro. We evaluated six phenolic compounds and three carotenoids. Phenolic compounds represent an important contribution to the antioxidizing potential of the human diet; of these compounds, flavonoids are the most studied and abundant. The antioxidant activity of the flavonoids apigenin, rutin, and naringenin was studied. These flavonoids are bioactive compounds mainly found in various fruits, plants and vegetables, nuts, and onions. In vitro studies have shown that these flavonoids effectively neutralize hydroxyl radicals, superoxide, hydrogen peroxide, nitric oxide radicals, DPPH, and lipid peroxidation [26–29]. Chen et al., in 2012 [30], performed a QSAR analysis using a zebrafish larva model to evaluate the ROS-scavenging capacities of fifteen flavonoids, including rutin, against UV-induced phototoxicity. In accordance with previous studies, they concluded the importance of the two hydroxyl groups and their positions, with at least two hydroxyl groups necessary for a strong biological activity [30,31]. Furthermore, it was determined that hydroxyl groups at positions C3, C5, and C7 confer better flavone stability and activity [31]. Our results showed a protective effect against tBOOH-induced lethality for the three flavonoids. Apigenin and rutin also showed protective effects against dysmorphogenesis; however, naringenin did not show any effect against dysmorphogenesis.

In addition to those of flavonoids, the antioxidant effects of oleuropein, chlorogenic acid, and curcumin were also evaluated. In vitro and in vivo studies have shown that these three phenolic compounds have important antioxidant effects [32–34]. Oleuropein is a biophenol found in olive leaves, extra virgin olive oil, and some species of the Oleaceae family [32]. Chlorogenic acids (CGAs) are esters formed between caffeic and quinic acids and represent a group of polyphenols present in the human diet [35]. Several studies have shown that drinking beverages containing CGAs such as coffee, tea, wine, and various fruit juices reduces the risk of developing different chronic diseases [36–38]. One of the reasons for this reduction is the antioxidant capacity of the CGAs, which donate hydrogen atoms to reduce free radicals and inhibit oxidation reactions [35]. Curcumin is a polyphenol that is used for coloring and seasoning in food products. Its antioxidant activity has been studied during the last few years, and one study suggests that it can protect biomembranes against peroxidative damage [39]. Using the ZF embryo model, it was observed that pretreatment with oleuropein, chlorogenic acid, or curcumin reduced the mortality-inducing effect of tBOOH-induced oxidative stress; however, a significant protective effect against dysmorphogenesis was only observed for curcumin.

Another group with antioxidant properties is the carotenoids, a ubiquitous group of isoprenoid pigments. They are quenchers of singlet oxygen and scavengers of ROS [40]. The molecular mechanisms underlying the anti- and pro-oxidant activity of carotenoids are still not fully understood. Among the most studied carotenoids are lycopene and β-carotene. These can be found abundantly in tomatoes, tomato sauce, various fruits, algae, and vegetables [18,41]. When evaluating the protective effects of these carotenes, it was found that lycopene showed antioxidant activity, with a protective effect against embryonic lethality; however, no effect against dysmorphogenesis was found. On the other hand, β-carotene increased the incidence of lethality and dysmorphogenesis in the ZF embryos compared to the effect of the oxidant alone; this is in accordance with studies that have shown that high doses of β-carotene have antioxidant effects that are followed by a prooxidant action at high oxygen tension, which may be related to its adverse effects [42]. Moreover, a study showed that β-carotene supplementation had no protective effect on the total mortality of diabetic male smokers compared with a placebo [43]. Another carotenoid evaluated was astaxanthin; it is a xanthophyll carotenoid found in algae, yeast, salmon, trout, krill, shrimp, and crayfish. It is a red, fat-soluble antioxidant pigment that has no pro-Vitamin A activity [44]. In our study, astaxanthin exhibited a protective effect against lethality, but no effect against dysmorphogenesis was found.

In conclusion, eight of the nine molecules evaluated showed antioxidant activity with protective effects against ZF embryonic lethality. Only apigenin (10 µM), rutin (10 µM), and curcumin (15 µM) additionally exhibited protective effects against dysmorphogenesis resulting from tBOOH-induced oxidative stress. By contrast, it was found that β-carotene produced the opposite effect, increasing the mortality and dysmorphogenesis rate, because it reduced the LC_{50} and EC_{50} values. The balance and timing of oxidative and antioxidative forces is key to the proper regulation and timing of embryonic development [45]. Differences in the kinetics or mechanism of action of these antioxidants could be the leading reason for the different protective capacities against dysmorphogenesis. More studies are needed to explore why only some compounds showed protective effects on morphogenesis during embryonic development. This study tried to discriminate between embryotoxic effect (lethality) and dysmorphogenic effect (teratogenicity). In some cases, malformations are likely to precede and result in death. In other instances, lethality and malformation may be due to different causes. The independence of these two manifestations would be suspected if a compound increased the separation between the lethal and dysmorphogenic concentration–response curves. All the antioxidant compounds tested in our study did not increase the teratogenic effect over the lethal dose more than twice; therefore, no increased teratogenic potential was observed for any antioxidant compound [46].

Altogether, these results indicate that this ZF embryo model is a valuable tool with which to analyze the protective effects of the antioxidant molecules that constitute food. To determine the chemical–structural reasons for which apigenin, rutin, and curcumin showed the highest protective effects in our study, further analyses are necessary; for instance, to determine quantitative structure–activity relationships (QSARs).

4. Materials and Methods

4.1. Ethical Statement

The procedures involving zebrafish larvae and embryos were authorized by the Animal Ethics Committee of the University of Barcelona, authorization number or protocol 7971 of the Department of Livestock and Fisheries of the Government of Catalonia (Procedure DAAM 7971).

4.2. Chemicals and Solution Preparation

Tert-butyl hydroperoxide (tBOOH, CAS number: 75-91-2) and the antioxidant compounds were acquired from TCI Europe. tBOOH was dissolved in 0.3X Danieau's buffer (17.4 mM NaCl; 0.23 mM KCl; 0.12 mM $MgSO_4 \cdot 7 H_2O$; 0.18 mM $Ca(NO_3)_2$; 1.5 mM HEPES (N-(2-hydroxyethyl) piperazine-N0 -(2-ethanesulfonic acid); pH 7.4).

Naringenin (20 µM) (CAS number: 67604-48-2), oleuropein (15 µM) (CAS number: 32619-42-4), rutin (10 µM) (CAS number: 207671-50-9), chlorogenic acid (20 µM) (CAS number: 327-97-9), apigenin (10 µM) (CAS number: 520-36-5), curcumin (15 µM) (CAS number: 458-37-7), lycopene (20 µM) (CAS number: 502-65-8), astaxanthin (20 µM) (CAS number: 472-61-7), and β-carotene (25 µM) (CAS number: 7235-40-7) were acquired from Sigma-Aldrich®. Antioxidants were dissolved in 100% dimethyl sulfoxide (DMSO, Sigma-Aldrich, Madrid, Spain) and subsequently diluted in 0.3× Danieau's buffer to a final DMSO concentration of 0.05% (v/v). Antioxidants were used at different concentrations, depending on the highest concentration at which no effect on lethality or embryonic development was observed (maximum tolerable concentration, MTC)

4.3. Zebrafish Maintenance and Egg Production

Adult wild-type zebrafish were housed in standardized conditions. The embryos were collected, cleaned, and selected according to their viability. The fertilized embryos were treated with water standardized according to the International Organization for Standardization (ISO) standards 7346-1 and 7346-2 (ISO, 1998; 2 mM $CaCl_2 \cdot 2H_2O$, 0.5 mM $MgSO_4 \cdot 7H_2O$, 0.75 mM $NaHCO_3$, and 0.07 mM KCl). Fertilized eggs were staged according to previous studies by Kimmel et al., 1995 [47], and selected for subsequent exposure under a dissecting stereomicroscope (Motic SMZ168, Motic China Group, LTD., Luwan, Shanghai, China). The fish embryos were kept in glass vials at a controlled temperature of 27 ± 1 °C.

4.4. Exposure of Zebrafish Embryos to Oxidative Stress (Tert-Butyl Hydroperoxide)

For the preparation of the tBOOH curve, the methodology of Boix, 2020, was followed. This methodology is based on obtaining a concentration–lethality response (LC_{50}) curve and dysmorphogenesis (EC_{50}) by exposing the ZF embryos to an oxidative stress inducer, tert-butylhydroperoxide (tBOOH). Once zebrafish embryos are obtained, they are kept in 0.3X Danieau's medium from 0 to 24 hpf. From 24 to 48 hpf, the embryos are exposed to tBOOH solutions at different concentrations, of 1, 1.5, 2, 2.5, 3, and 3.5 mM. Embryos from 3 different clutches of zebrafish in triplicate were used (Figure 5A).

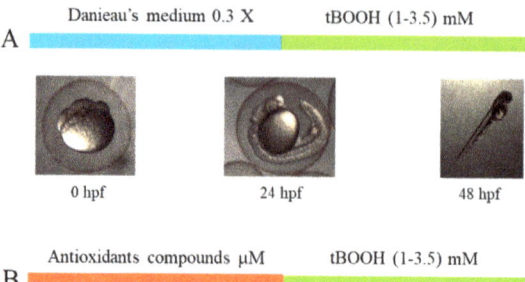

Figure 5. (**A**) Schematic overview of the process for obtaining the lethality and dysmorphogenesis curves for tBOOH. (**B**) Process for evaluating the protective effects of the antioxidant compounds.

4.5. Determination of the Protective Effects of Antioxidant Compounds

To establish whether a compound had a protective effect against oxidative stress, zebrafish embryos were first exposed to the antioxidant compound from 0 to 24 hpf. The concentrations were calculated depending on the maximum tolerable concentration assays. Then, from 24 to 48 hpf, the embryos were exposed to the stress inducer, tBOOH. Subsequently, each group of embryos was evaluated at each of the concentrations of tBOOH (Figure 5B). To determine whether there was a significant difference, the curve for exposure to tBOOH alone and the curve for pre-exposure to the antioxidant compound were compared.

Ten fertilized eggs were exposed to 2.5 mL for each substance and concentration. Three independent replications were performed, using eggs from different spawning events. The ZF embryos were pre-exposed to the antioxidant compounds for 24 h, then the antioxidant solution was removed, washing was performed using Danieau's medium, and the ZF embryos were exposed to different tBOOH concentrations. The lethality was evaluated after 48 h, and the mean of dead embryos was calculated after the appropriate assays. For the dysmorphogenesis evaluation, we followed the scoring system described by Teixidó et al. [48] to compute the embryos' dysmorphogenesis at around 48 hpf. We selected nine morphological features, described in Table 2. The frequency of abnormal embryos was calculated (defined as the embryos with a score 1 in any morphological feature) for each concentration and treated group.

Table 2. Criteria employed to evaluate dysmorphogenesis on zebrafish embryos.

Morphological Features	Morphological Abnormality	Example of Observed Characteristics.
Detachment of the tail; Tail	No tail, malformation of chorda or spinal cord. Tail necrosis, bent tail.	
Optic system; Otic system; Brain	Abnormal pigmentation, asymmetric eyes. Formation of no, one, or more than two otoliths per sacculus. Brain necrosis, hemorrhage.	
Heart	Pericardial edema, big heart, hemorrhage, abnormal chambers.	
Head-body pigmentation; Tail pigmentation	Lack of pigmentation in the tail or body.	

Table 2. *Cont.*

Morphological Features	Morphological Abnormality	Example of Observed Characteristics.
Movement	Spasms, abnormal movements, no movement at all.	The ZF embryo was touched and its movement was evaluated.
* Control	No signs of dysmorphogenesis.	

* Control of normal characteristics of a zebrafish embryo at 48 hpf.

A shift of the concentration–response curve to the right due to pre-exposure to the antioxidants indicates a protective effect against the oxidative stress inducer, because a higher concentration of inducer is required to obtain the same results as those with the exposure to tBOOH alone. Due to the pre-exposure to antioxidants, a shift to the left of the concentration–response curve implies an increase in oxidative stress.

4.6. Statistical Analysis

The concentration–response curves for mortality and dysmorphogenesis were calculated and evaluated using GraphPad 7.02 Software Inc. The extra-sum-of-squares F test was used to compare the fit of the parameters of each data group of the curve. The confidence interval was adjusted to 95%.

5. Conclusions

This study used zebrafish embryos as a model organism to test the protective capacity of six phenolic compounds and three carotenoids commonly found in foods. All the compounds, except β-carotene, showed protective effects against oxidative-stress-induced lethality. Furthermore, apigenin, rutin, and curcumin also exhibited protective effects against tBOOH-induced dysmorphogenesis. We propose that a zebrafish embryo test, as presented here, could be applied to evaluate the in vivo protective effects of novel bioactive food components with potential antioxidant capacity.

Author Contributions: Conceptualization, C.A., N.B., E.T., F.M., S.C. and A.B.; methodology, C.A., N.B., E.T. and A.B.; validation, C.A., N.B., E.T., F.M., S.C. and A.B.; formal analysis, C.A., N.B. and A.B.; investigation, C.A., N.B. and E.T.; resources, E.T. and A.B.; writing—original draft preparation, C.A. and A.B.; writing—review and editing, C.A., N.B., E.T., F.M., S.C. and A.B.; visualization, C.A., N.B., E.T. and A.B.; supervision, A.B.; project administration, E.T. and A.B.; funding acquisition, E.T. and A.B. All authors have read and agreed to the published version of the manuscript.

Funding: This research was supported by the Spanish Ministry of Economy and Competitivity (AGL2013-49083-C3-1-R).

Institutional Review Board Statement: The experimental use of zebrafish larvae and embryo in this study were authorized by the Animal Ethics Committee of the University of Barcelona, authorization number or protocol 7971 of the Department of Livestock and Fisheries of the Government of Catalonia (Procedure DAAM 7971).

Informed Consent Statement: Not applicable.

Data Availability Statement: All the data are included on this paper.

Acknowledgments: We acknowledge the Direction of Investigation and Development, DIDE, for its contribution to the "Evaluación del estrés oxidativo mediante un modelo de embrión de pez cebra y su aplicación a compuestos presentes en alimentos" project.

Conflicts of Interest: The authors declare no conflict of interest.

References

1. Tan, B.L.; Norhaizan, M.E.; Liew, W.-P.-P. Nutrients and Oxidative Stress: Friend or Foe? *Oxid. Med. Cell. Longev.* **2018**, *2018*, 9719584. [CrossRef] [PubMed]
2. Sies, H. Biochemistry of Oxidative Stress. *Angew. Chemie Int. Ed. Engl.* **1986**, *25*, 1058–1071. [CrossRef]
3. Dröge, W. Free Radicals in the Physiological Control of Cell Function. *Physiol. Rev.* **2002**, *82*, 47–95. [CrossRef] [PubMed]
4. Rendra, E.; Riabov, V.; Mossel, D.M.; Sevastyanova, T.; Harmsen, M.C.; Kzhyshkowska, J. Reactive oxygen species (ROS) in macrophage activation and function in diabetes. *Immunobiology* **2019**, *224*, 242–253. [CrossRef]
5. Lin, M.T.; Beal, M.F. Mitochondrial dysfunction and oxidative stress in neurodegenerative diseases. *Nature* **2006**, *443*, 787–795. [CrossRef] [PubMed]
6. Therond, P. Oxidative stress and damages to biomolecules (lipids, proteins, DNA). *Ann. Pharm. Fr.* **2006**, *64*, 383–389. [CrossRef]
7. Ermak, G.; Davies, K.J.A. Calcium and oxidative stress: From cell signaling to cell death. *Mol. Immunol.* **2002**, *38*, 713–721. [CrossRef]
8. Wadhwa, R.; Gupta, R.; Maurya, P.K. Oxidative Stress and Accelerated Aging in Neurodegenerative and Neuropsychiatric Disorder. *Curr. Pharm. Des.* **2018**, *24*, 4711–4725. [CrossRef]
9. Maritim, A.C.; Sanders, R.A.; Watkins, J.B. Diabetes, oxidative stress, and antioxidants: A review. *J. Biochem. Mol. Toxicol.* **2003**, *17*, 24–38. [CrossRef]
10. Sinha, N.; Dabla, P.K. Oxidative stress and antioxidants in hypertension-a current review. *Curr. Hypertens. Rev.* **2015**, *11*, 132–142. [CrossRef]
11. Klaunig, J.E. Oxidative Stress and Cancer. *Curr. Pharm. Des.* **2018**, *24*, 4771–4778. [CrossRef] [PubMed]
12. Sies, H. Oxidative stress: A concept in redox biology and medicine. *Redox Biol.* **2015**, *4*, 180–183. [CrossRef]
13. Pietta, P.-G. Flavonoids as Antioxidants. *J. Nat. Prod.* **2000**, *63*, 1035–1042. [CrossRef]
14. Halliwell, B.; Gutteridge, J.M.C. *Free Radicals in Biology and Medicine*, 5th ed.; Oxford University Press: Oxford, UK, 2015; ISBN 9780198717478.
15. Hollman, P.C.; Katan, M.B. Dietary flavonoids: Intake, health effects and bioavailability. *Food Chem. Toxicol. Int. J. Publ. Br. Ind. Biol. Res. Assoc.* **1999**, *37*, 937–942. [CrossRef]
16. Hertog, M.G.; Hollman, P.C.; Katan, M.B.; Kromhout, D. Intake of potentially anticarcinogenic flavonoids and their determinants in adults in The Netherlands. *Nutr. Cancer* **1993**, *20*, 21–29. [CrossRef]
17. Martínez, S.; González, J.; Culebras, J.; Tuñón, M. Flavonoides: Propiedades y acciones antioxidantes. *Nutr. Hosp.* **2002**, *17*, 271–278.
18. Arteaga, C.; Bustillos, A.; Gómez, J. Migración de neutrófilos en larvas de pez cebra expuestos a extractos de sofrito de tomate. *Arch. Latinoam. Nutr.* **2021**, *70*, 216–224. [CrossRef]
19. Young, A.J.; Lowe, G.L. Carotenoids—Antioxidant properties. *Antioxidants* **2018**, *7*, 28. [CrossRef] [PubMed]
20. Xavier, A.A.O.; Pérez-Gálvez, A. Carotenoids as a Source of Antioxidants in the Diet. *Subcell. Biochem.* **2016**, *79*, 359–375. [CrossRef] [PubMed]
21. Stahl, W.; Sies, H. Antioxidant activity of carotenoids. *Mol. Asp. Med.* **2003**, *24*, 345–351. [CrossRef]
22. Boix, N.; Teixido, E.; Pique, E.; Llobet, J.M.; Gomez-Catalan, J. Modulation and Protection Effects of Antioxidant Compounds against Oxidant Induced Developmental Toxicity in Zebrafish. *Antioxidants* **2020**, *9*, 721. [CrossRef]
23. Fenton, H.J.H. Oxidation of tartaric acid in presence of iron. *J. Chem. Soc. Trans.* **1894**, *65*, 899–910. [CrossRef]
24. Phaniendra, A.; Jestadi, D.B.; Periyasamy, L. Free Radicals: Properties, Sources, Targets, and Their Implication in Various Diseases. *Indian J. Clin. Biochem.* **2015**, *30*, 11–26. [CrossRef] [PubMed]
25. Fernández-Pachón, M.S.; Villaño, D.; Troncoso, A.M.; García-Parrilla, M.C. Revisión de los métodos de evaluación de la actividad antioxidante in vitro del vino y valoración de sus efectos in vivo. *Arch. Latinoam. Nutr.* **2006**, *56*, 110–122. [PubMed]
26. Cavia-Saiz, M.; Busto, M.D.; Pilar-Izquierdo, M.C.; Ortega, N.; Perez-Mateos, M.; Muñiz, P. Antioxidant properties, radical scavenging activity and biomolecule protection capacity of flavonoid naringenin and its glycoside naringin: A comparative study. *J. Sci. Food Agric.* **2010**, *90*, 1238–1244. [CrossRef] [PubMed]
27. Patel, K.; Singh, G.K.; Patel, D.K. A review on pharmacological and analytical aspects of naringenin. *Chin. J. Integr. Med.* **2014**, *24*, 551–560. [CrossRef]
28. Rashmi, R.; Magesh, S.B.; Ramkumar, K.M.; Suryanarayanan, S.; SubbaRao, M.V. Antioxidant potential of naringenin helps to protect liver tissue from streptozotocin-induced damage. *Rep. Biochem. Mol. Biol.* **2017**, *7*, 76–84.
29. Shukla, R.; Pandey, V.; Vadnere, G.P.; Lodhi, S. Chapter 18—Role of Flavonoids in Management of Inflammatory Disorders. In *Bioactive Food as Dietary Interventions for Arthritis and Related Inflammatory Diseases*; Watson, R.R., Preedy, V.R., Eds.; Academic Press: London, UK, 2019; pp. 293–322; ISBN 978-0-12-813820-5.
30. Chen, Y.H.; Yang, Z.S.; Wen, C.C.; Chang, Y.S.; Wang, B.C.; Hsiao, C.A.; Shih, T.L. Evaluation of the structure-activity relationship of flavonoids as antioxidants and toxicants of zebrafish larvae. *Food Chem.* **2012**, *134*, 717–724. [CrossRef]
31. Cushman, M.; Zhu, H.; Geahlen, R.L.; Kraker, A.J. Synthesis and Biochemical Evaluation of a Series of Aminoflavones as Potential Inhibitors of Protein-Tyrosine Kinases p56lck, EGFr, and p60v-src. *J. Med. Chem.* **1994**, *37*, 3353–3362. [CrossRef]
32. Cioffi, G.; Pesca, M.S.; De Caprariis, P.; Braca, A.; Severino, L.; De Tommasi, N. Phenolic compounds in olive oil and olive pomace from Cilento (Campania, Italy) and their antioxidant activity. *Food Chem.* **2010**, *121*, 105–111. [CrossRef]

33. Bulotta, S.; Corradino, R.; Celano, M.; D'Agostino, M.; Maiuolo, J.; Oliverio, M.; Procopio, A.; Iannone, M.; Rotiroti, D.; Russo, D. Antiproliferative and antioxidant effects on breast cancer cells of oleuropein and its semisynthetic peracetylated derivatives. *Food Chem.* **2011**, *127*, 1609–1614. [CrossRef]
34. Han, J.; Talorete, T.P.N.; Yamada, P.; Isoda, H. Anti-proliferative and apoptotic effects of oleuropein and hydroxytyrosol on human breast cancer MCF-7 cells. *Cytotechnology* **2009**, *59*, 45–53. [CrossRef] [PubMed]
35. Liang, N.; Kitts, D.D. Role of chlorogenic acids in controlling oxidative and inflammatory stress conditions. *Nutrients* **2015**, *8*, 16. [CrossRef] [PubMed]
36. Park, S.-Y.; Freedman, N.D.; Haiman, C.A.; Le Marchand, L.; Wilkens, L.R.; Setiawan, V.W. Association of Coffee Consumption With Total and Cause-Specific Mortality Among Nonwhite Populations. *Ann. Intern. Med.* **2017**, *167*, 228–235. [CrossRef] [PubMed]
37. Tajik, N.; Tajik, M.; Mack, I.; Enck, P. The potential effects of chlorogenic acid, the main phenolic components in coffee, on health: A comprehensive review of the literature. *Eur. J. Nutr.* **2017**, *56*, 2215–2244. [CrossRef] [PubMed]
38. Poole, R.; Kennedy, O.J.; Roderick, P.; Fallowfield, J.A.; Hayes, P.C.; Parkes, J. Coffee consumption and health: Umbrella review of meta-analyses of multiple health outcomes. *BMJ* **2017**, *359*, j5024. [CrossRef] [PubMed]
39. Tanvir, E.M.; Hossen, M.S.; Hossain, M.F.; Afroz, R.; Gan, S.H.; Khalil, M.I.; Karim, N. Antioxidant Properties of Popular Turmeric (*Curcuma longa*) Varieties from Bangladesh. *J. Food Qual.* **2017**, *2017*, 8471785. [CrossRef]
40. Fiedor, J.; Burda, K. Potential role of carotenoids as antioxidants in human health and disease. *Nutrients* **2014**, *6*, 466–488. [CrossRef]
41. Eggersdorfer, M.; Wyss, A. Carotenoids in human nutrition and health. *Arch. Biochem. Biophys.* **2018**, *652*, 18–26. [CrossRef]
42. Padmanabhan, P.; Cheema, A.; Paliyath, G. *Solanaceous Fruits Including Tomato, Eggplant, and Peppers*, 1st ed.; Elsevier Ltd: London, UK, 2015.
43. Kataja-Tuomola, M.K.; Kontto, J.P.; Männistö, S.; Albanes, D.; Virtamo, J.R. Effect of alpha-tocopherol and beta-carotene supplementation on macrovascular complications and total mortality from diabetes: Results of the ATBC Study. *Ann. Med.* **2010**, *42*, 178–186. [CrossRef]
44. Ambati, R.R.; Moi, P.S.; Ravi, S.; Aswathanarayana, R.G. Astaxanthin: Sources, extraction, stability, biological activities and its commercial applications—A review. *Mar. Drugs* **2014**, *12*, 128–152. [CrossRef] [PubMed]
45. Dennery, P.A. Effects of oxidative stress on embryonic development. *Birth Defects Res. C Embryo Today* **2007**, *81*, 155–162. [CrossRef] [PubMed]
46. Selderslaghs, I.W.T.; Blust, R.; Witters, H.E. Feasibility study of the zebrafish assay as an alternative method to screen for developmental toxicity and embryotoxicity using a training set of 27 compounds. *Reprod. Toxicol.* **2012**, *33*, 142–154. [CrossRef]
47. Kimmel, C.B.; Ballard, W.W.; Kimmel, S.R.; Ullmann, B.; Schilling, T.F. Stages of embryonic development of the zebrafish. *Dev. Dyn.* **1995**, *203*, 253–310. [CrossRef] [PubMed]
48. Teixidó, E.; Piqué, E.; Gómez-Catalán, J.; Llobet, J.M. Assessment of developmental delay in the zebrafish embryo teratogenicity assay. *Toxicol. In Vitro* **2013**, *27*, 469–478. [CrossRef] [PubMed]

Review

Could Polyphenols Really Be a Good Radioprotective Strategy?

Shadab Faramarzi [1,2], Simona Piccolella [1], Lorenzo Manti [3] and Severina Pacifico [1,*]

1 Department of Environmental, Biological and Pharmaceutical Sciences and Technologies, University of Campania "Luigi Vanvitelli", Via Vivaldi 43, 81100 Caserta, Italy; shadab.faramarzi@unicampania.it (S.F.); simona.piccolella@unicampania.it (S.P.)
2 Department of Plant Production and Genetics, Razi University, Kermanshah 67149-67346, Iran
3 Department of Physics E. Pancini, University of Naples "Federico II", and Istituto Nazionale di Fisica Nucleare, (INFN), Naples Section, Monte S. Angelo, Via Cinthia, 80126 Napoli, Italy; lorenzo.manti@na.infn.it
* Correspondence: severina.pacifico@unicampania.it

Abstract: Currently, radiotherapy is one of the most effective strategies to treat cancer. However, deleterious toxicity against normal cells indicate for the need to selectively protect them. Reactive oxygen and nitrogen species reinforce ionizing radiation cytotoxicity, and compounds able to scavenge these species or enhance antioxidant enzymes (e.g., superoxide dismutase, catalase, and glutathione peroxidase) should be properly investigated. Antioxidant plant-derived compounds, such as phenols and polyphenols, could represent a valuable alternative to synthetic compounds to be used as radio-protective agents. In fact, their dose-dependent antioxidant/pro-oxidant efficacy could provide a high degree of protection to normal tissues, with little or no protection to tumor cells. The present review provides an update of the current scientific knowledge of polyphenols in pure forms or in plant extracts with good evidence concerning their possible radiomodulating action. Indeed, with few exceptions, to date, the fragmentary data available mostly derive from in vitro studies, which do not find comfort in preclinical and/or clinical studies. On the contrary, when preclinical studies are reported, especially regarding the bioactivity of a plant extract, its chemical composition is not taken into account, avoiding any standardization and compromising data reproducibility.

Keywords: ionizing radiation; radioprotection; polyphenols; flavonoids; plant extracts

1. Introduction

Life on Earth has evolved in the presence of a continuous exposure to ionizing radiation (IR), whose mode of action at the biomolecular level is unique among all known mutagen and carcinogenic agents [1]. This is due to the peculiar pattern of energy deposition accompanying IR absorption at the micro- and nanometer scale [2], which is inherently nonhomogeneous, resulting in either isolated or highly clustered ionization events. As a consequence, they may generate a plethora of DNA lesions of varying severity, ranging from base damage and molecular cross-links to the most deleterious single- and double strand breaks (SSB and DSB, respectively) [3]. Indeed, cellular DNA has always been regarded as the target of choice of IR biological action because it is present in a single copy, hence any un- or mis-repaired damage can have relevant consequences, impinging on genome integrity and stability in the exposed cellular progeny. In fact, due to the ubiquitous nature of IR exposure, cellular systems have developed well-orchestrated DNA repair molecular pathways, highly specialized and differentiated to deal with the several classes or IR-induced lesions, a machinery collectively known as DNA damage response (DDR) [3]. Repair capability depends by the sheer amount of initially induced DNA damage, which is a function of the absorbed radiation dose, but also by the quality of radiation, i.e., the ionization density along radiation tracks.

Obviously, naturally occurring background radiation is not the only source of human exposure to IR [4]. At about the same time light was shed on the laws governing the process of natural radioactive decays, it became evident that IR could be artificially generated.

The impact that the discovery of X-rays by Wilhelm Conrad Roentgen in 1895 has had on many aspects of human health is still reverberating nowadays, as IR is widely used in both the diagnosis and treatment of diseases [5]. Insofar as the therapeutic use of IR is concerned, the very same DNA-damaging action by IR that classifies it as a hazard to human health is exploited by its ability to eradicate cancer cells though radiotherapy. Much is known about the way IR brings about its biological effects thanks to extensive radiobiological research that has unveiled basic mechanisms. To this aim, it is useful to classify IR as indirect and direct, based on the way energy is released in the (biological) matter. Photons, such as X-rays and γ-rays, and neutrons act indirectly, requiring a two-step action before causing any potentially biological relevant damage. In fact, photons interact with the atom shell-containing electrons thereby generating secondary fast-moving electrons, which in turn cause further ionizations, with the emission of slower electrons; neutrons interact with the nuclei of the traversed material, giving rise to charged particles such as protons and heavier nuclei [5]. Charged particles, instead, lose energy directly through Coulomb interactions, producing the above-mentioned ionizing tracks along their penetration depth [5]. Biological effects related to IR, whether caused by direct or indirect radiations, are classified into direct and indirect, too. In the first case chemical alteration of biomolecules are formed during the physico-chemical stage that temporally precedes the actual biological stage. Instead, they are indirect when they are the result of radiation products, such as free radicals generated by water radiolysis. Indirectly generated DNA damage is the only form of damage whose amount can be modulated by concomitant agents, such as antioxidant compounds. In fact, when ionizing radiation passes through water, it leads to a number of ionic and excited states that further decompose or recombine to give hydrated electrons (e^-_{aq}) and reactive species, including hydrogen radical ($H^\bullet$), hydroxyl radical ($OH^\bullet$), hydrogen peroxide (H_2O_2), oxygen (O_2), hydrogen (H_2), and hydroperoxyl radical ($HO_2^\bullet$) (Figure 1) [6].

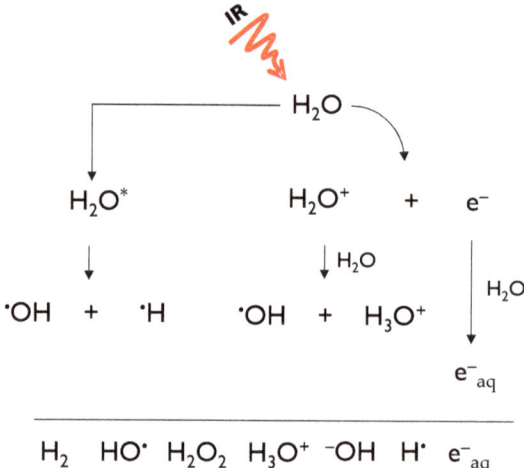

Figure 1. Hydrated electrons (e^-_{aq}), and radical and molecular species generated during water hydrolysis.

DNA DSBs are universally regarded as the most deleterious IR-induced lesion [7]. DDR can lead to cell-cycle checkpoint activation, hence cell cycle delay/arrest, in the attempt to increase time for repair. Irrespective of which mechanism the cell uses, un-/mis-repaired DSBs can lead to cell death through several pathways (e.g., mitotic failure, apoptosis), typically arising at first mitosis post-irradiation or after a few cell cycles from exposure. This is the aim of curative radiotherapy. However, failure in correct restoration of the DBS can result in rearrangements of genetic material (e.g., chromosome aberrations, micronuclei), which, if transmissible through cell division, can cause late-arising effects, leading to generalised genomic instability, and hence to an increase in the risk of malignant transformation [8].

Conventional radiotherapy by high-energy photon or electron beams is a mainstay of modern cancer treatment, with an estimated 50% of cancer patients receiving it alone or in combination with other modalities worldwide [9]. Although several improvements have been achieved in dose delivery accuracy, the mitigation of noncancerous normal tissue toxicity remains of crucial importance because of the above-mentioned secondary cancer risk affecting the unavoidably exposed normal tissues and/or organs at risk. Since photons are mainly characterized by the indirect mode of action, the amount of damage they produce during the physical stage, can be modulated during the chemical stage prior to damage fixation and before the onset of the biologically driven DDR. Therefore, modifiers/protectors can be utilized to selectively benefit normal tissues, delivering further minimal toxicity [10]. In this context, several compounds have been described, but only amifostine, the S-phospho derivative of 2-[(3-aminopropyl)amino]ethanethiol, is approved as clinical radiation protector [11]. Other thiol-containing compounds, beyond nitroxides with superoxide dismutase (SOD)-like activity, hormone analogues, antibiotics, and phytochemicals, have been investigated as radioprotectors, whereas immunomodulators, probiotics, statins are explored as mitigator agents [12]. Specialized natural compounds are playing a key role in preclinical and clinical research, thanks to their anti-oxidant and anti-inflammatory efficacy that identifies them as promising agents in the field of radioprotection and radiomitigation.

2. Radioprotection: A Valuable Approach to Counteract Radiation Exposure

Although radiotherapy is one of the most effective strategy to treat cancer, the normal-tissue response is the limiting factor for the total dose that can be safely administered to achieve tumour local control, hence reducing the chances of cure, while acute and chronic toxicities may lead to an overall poor patient's life quality. Therefore, an urgent need in protecting normal cells is actively claimed. In this context, technological improvements in IR delivery and accuracy are ongoing, whereas radiomodulating agents are considered a valuable alternative to decrease toxicity to normal tissues. This is not an emerging issue so much so that the IR research program of the National Cancer Institute classified, according to administration timing, agents with IR protective properties in three categories: (a) protection, (b) mitigation, and (c) therapeutic agents [13]. Analogously, the European Commission has paid and continues to pay great attention to radiation protection, generally addressing new research findings with potential policy and/or regulatory implications [14].

Radioprotectors and radiomitigators are valuable modulator agents. Their delivery precedes or occurs simultaneously with the radiation administration and is prompt to reduce or ameliorate normal tissue toxicity. The latter could take advantage of therapeutic compounds, when an adverse effect is established, acting after irradiation as palliation or support [15].

All these compounds need to share some functional features such as the ability to interrupt or slow down the overproduction of reactive species, which can perpetuate indefinitely the IR injury affecting different cell activities and signalling pathways. Indeed, reactive oxygen and nitrogen species reinforce IR cytotoxicity. Counteracting the onset of oxidative stress conditions prevents structural and functional disruption of nucleic acids, proteins and lipids, and a series of processes (e.g., mitochondrial depolarization), which irreversibly lead to cell death [16]. IR-induced genomic instability is the main target to encompass, as mutations, gene amplification and other cytogenetic rearrangements could be also after the initial insult [17]. Cells adaptively respond to IR by activating the Nrf2-ARE antioxidant defence [18], which is constituted by enzymatic and non-enzymatic compounds, and can benefit radioprotectors with the aim to evade free radicals, remove IR-induced toxic substances, and overall to intensify the repair and recovery processes [19]. Thus, compounds able to scavenge these species or enhance antioxidant enzymes (e.g., superoxide dismutase, catalase, and glutathione peroxidase) should properly investigated. In this context, thiols, thanks to their ability to scavenge hydroxyl radical, protect DNA, which provides, mostly in hypoxic condition, harmful DNA radicals, likely responsible for radiation lethality [10]. Furthermore, thiols are observed to prevent the oxidation of membrane phospholipids, and to modulate cell recovery and stress responses. Cysteine and cysteamine are sulfhydryl amines, and other aminothiol analogues/derivatives appeared to be radioprotective, but their side effects advised against clinical use, with the exception of the aminothiol amifostine (WR-2721) [10]. Among the deeply studied species, nitroxides are also of great interest thanks to their single-electron redox cycle ability. In particular, as a pleiotropic intracellular antioxidant, tempol was observed to reduce the incidence of radiation-induced second malignancies [20]. Indeed, tempol (4-hydroxy-2,2,6,6-tetramethylpiperidine-N-oxyl) was also shown to act as a SOD mimetic, whereas the radioprotective properties of the SOD isoforms (Cu, Zn SOD, Mn SOD and extracellular SOD) are highlighted as useful agents for their $O_2^{-\bullet}$ scavenging efficacy in cytosol, mitochondrion, and extracellular space, respectively, and their-catalysed dismutation to H_2O_2 and O_2. Other categories of radioprotective agent are cytokines and growth factors, including IL-1, TNF-α, G-CSF, GM-CSF, and erythropoietin, and angiotensin-converting enzyme inhibitors (Figure 2). These latter compounds, routinely prescribed for hypertension treatment, were observed to ameliorate radiation side effects in kidney, lung and brain and to interfere with TGF-β pathway, which could contribute to radiation-induced fibrosis [10]. Among these agents, palifermin, a recombinant N-terminal truncated form of keratinocyte growth factor, was firstly approved for the treatment of oral mucositis induced by chemo- and radio-therapy [21]. Inhibitors of PUMA (p53 Up-regulated Modulator of Apoptosis) and radiation-induced apoptosis were also investigated. PUMA inhibitors (PUMAi) are designed to inhibit PUMA-dependent and radiation-induced apoptosis and to avoid or alleviate intestinal damage and apoptosis induced by inflammatory cytokines, ROS (reactive oxygen species), or cancer therapy [22].

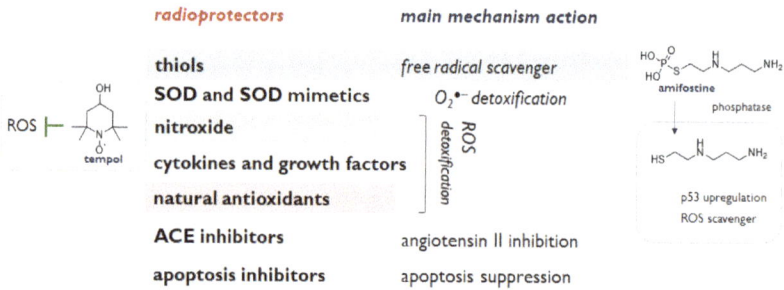

Figure 2. Radioprotectors' categories and main ascribed mechanisms of action.

Natural antioxidants also find growing interest. Vitamins (A, C, and E), L-selenomethionine, N-acetylcysteine, glutathione, and coenzyme Q-10 are suggested to be effective against radiation injury [23], while several dietary phytochemicals appeared to act as radioprotectors for normal cells, and radiosensitizers for tumour cells in a fascinating scenario. Melatonin, which is secreted by the pineal gland in the brain, from lymphocytes, the retina, and gastrointestinal system, is one of the most studied natural occurring compounds. It directly scavenges ROS species, inhibits ROS forming enzymes, and activates DNA repair enzymes [24].

The lack of harmful toxicity, together with their appreciable antioxidant and immunostimulant activities, make specialized metabolites from plants an endless reservoir of radioprotective compounds. Polyphenols, through their intrinsic antioxidant capability, are able to reduce inflammation, protecting both immune and hematopoietic systems, and preserving DNA. In particular, flavonoids, such as rutin and baicalein, the isoflavonoid genistein, and the stilbene resveratrol, represent compounds strictly related from a biosynthetic point of view and are promising radioprotective candidates. The poor bioavailability of these substance encourages new administration forms, and the phytochemical research for the discovery of new compounds from hopeful plant extracts. An update of the radioprotective natural molecules and an examination of the plant extracts enriched in these constituents are provided below.

3. Phenols and Polyphenols: Are They a Valuable Radioprotective Strategy?

The failure of synthetic compounds as effective radioprotectors allowed researchers to focus on natural substances and their radioprotective efficacy, and several botanicals, which could be less expensive than synthetic ones, have been screened for their radioprotective activity [25].

Free radical scavenging, anti-inflammation, facilitation of repair activity, regeneration of hematopoietic cells, are the main mechanisms attributable to natural radioprotectors (Figure 2). In particular, since most of the IR damage in the conventional radiotherapy setting arises from the interaction of IR-induced free radicals with biomolecules, natural substances, such as curcumin, chlorogenic acids, and different flavonoids, being able to destroy free radicals or prevent their formation, could serve as radioprotectors [26]. Thus, the use of radioprotectors for protecting normal tissue and of radiosensitizer for augmenting cancerous tissue response appeared to be innovatively maximized in a toxicity-free nutraceutical approach based on polyphenols. These natural compounds summarize the concept of an ideal protector, as, based on their dose-dependent antioxidant/pro-oxidant efficacy, could provide a high degree of protection to normal tissues, with little or no protection to tumor cells. Moreover, plant-derived polyphenols have gained a lot of attention in the long-standing quest for intrinsically low-toxic radiosensitizing drugs.

The attractive double-edged potential of pure polyphenols or polyphenol-enriched extracts provided good evidence on their possible radiomodulating action, and the polyphenol dual ability to act as both radiosensitizing and radioprotective agents would arguably hold pre-clinical significance, and, more generally, bear a significant impact on the prognosis of tumors refractory to radiation treatment.

3.1. Flavonoids: The Double-Edged Sword in Radioprotection

Over the last twenty years, the interest in radioprotective flavonoids is ongoing. These plant metabolites, commonly biosynthesized as defense compounds against UV-radiation, and other environmental stresses through chalcone precursors, are structurally characterized by a 15-carbons skeleton, consisting in two benzene rings (A and B) linked through a heterocyclic pyrone C-ring atoms. A high degree of hydroxylation, substitution, and polymerization is in flavonoid class, which consists in seven sub-classes: flavanones, dihydroflavonols, flavonols, flavones, flavandiols, anthocyanins, and catechins (Figure 3). Isoflavonoids form a well-separated flavonoid subclass, as these compounds showed a structural variant feature in which the B-aromatic ring is located at C3 carbon (Figure 3) [27]. Investigations aimed to explore structural features involved in radioprotection highlighted that some flavonoid compounds (mainly those sharing the keto group conjugated to aromatic rings) could be valid agents, because protection is related to their ability to inhibit energy transfer processes and to stabilize redox processes in irradiated cells [28].

Flavonols appeared the most valuable compounds, although glycosylation on C-3 carbon affects the reactivity, based on the saccharidic moiety identity. In fact, it was observed that flavonol glucosides decrease in reactivity when the sugar forms two intramolecular H-bonds. Furthermore, based on aglycone substitution, more phenolic functions are present, more the compound is active [28]. Among flavonol compounds, rutin ($3,3',4',5,7$-pentahydroxyflavone-3-rhamnoglucoside; Figure 4), abundant in passion flower, buckwheat, tea, and apple, is broadly investigated for its radioprotective action. Cell culture assays data highlighted its ability to protect from radiation-induced oxidative DNA damage in cells (e.g., V79).

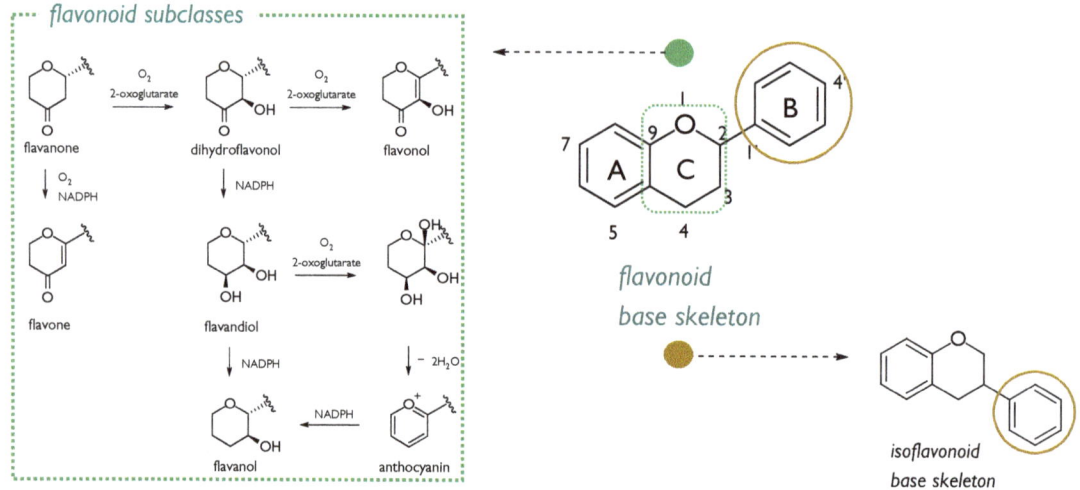

Figure 3. Flavonoid base skeleton and structural modifications of ring C which led to different flavonoid subclasses (green box). Isoflavonoids are from the 1,2-aryl shift (yellow circle).

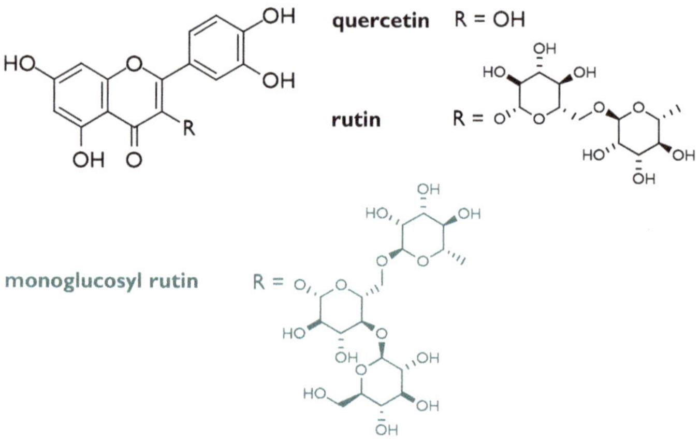

Figure 4. Chemical structures of quercetin, rutin and its monoglucosyl derivative.

Rutin daily supplementation, as well as that of its aglycone, namely quercetin, reduced the frequency of micronucleated reticulocytes in the peripheral blood of irradiated mouse. Combining podophyllotoxin and rutin in G-003 formulation, it was found a significant protection of the mice hematopoietic, gastrointestinal, and respiratory systems against lethal radiation dose [29–31]. The monoglucosyl rutin, ad hoc semi-synthetized to overcome quercetin and rutin insolubility in aqueous media, was proved to be effective towards CHO 10B2 cells, being able to increase the survival of IR-treated cells at doses greater than 2 Gy [32]. Indeed, the in vitro DNA double-strand breaks analysis, carried out on different flavonoid aglycones and glycosides, evidenced that, although quercetin derivatives reduced DNA double-strand breaks at a concentration equal to 10 µM, they low bioavailability could affect their efficacy in vivo [33].

Protection from DNA damage in γ-irradiated white blood cells [34], leucocytes [35] was also ascertained for quercetin and its enriched natural matrix propolis, so much so that further investigation in animal models was performed. In particular, the protective effect of an aqueous propolis extract against intestinal radiation damage was also evidenced in rats exposed to a γ-radiation dose of 8 Gy, able to induce intestinal mucositis [36], whereas a propolis methanolic fraction, with high content in both simple phenols and flavonoids, lowered total protein carbonyl content in UV-treated HaCat cells [37].

The radioprotective effect of flavones, such as apigenin and baicalein, was also deeply investigated (Figure 5). Apigenin, widely distributed in the leaves and stems of dietary vegetables and fruits, dose-dependently induced micronuclei in human lymphocytes treated in vitro, also suppressing adverse effects of ionizing radiation [38]. The compound appeared to exert immunostimulatory effect in vivo, thus mitigating radiation-induced hematological alterations. This outcome could be due to its ability to trigger the endogenous antioxidant status [39]. Recently, apigenin, intraperitoneally administered at a dose level equal to 15 mg/kg body weight, was found to slow down radiation-induced gastrointestinal damage in whole-body irradiated Swiss albino mice. In particular, the restoration of intestinal crypt-villus architecture appeared to occur following apigenin pre-treatment, as well as the inhibition of the radiation-induced activation of NF-κB expression in the gastrointestinal tissue [40].

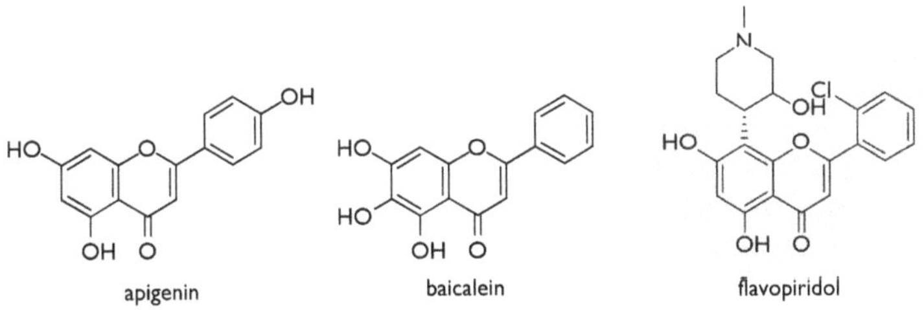

Figure 5. Chemical structures of radioprotective natural (apigenin and baicalein) and semisynthetic (flavopiridol) flavones.

Naringin, a flavanone-7-O-glycoside from Citrus species, also showed inhibitory effect towards IR-induced inflammation. The NF-κB suppression defined the alteration of pro-inflammatory factors. Moreover, naringin reinforced the intracellular defense mechanisms, through the preservation of endogenous antioxidants [41]. The oxidative stress inhibitory activity was also linked to the release of inflammatory cytokines by inducing Nrf2 activation, a common feature of other flavonoid compounds, such as naringenin and epigallocatechin-3-O-gallate (Figure 6) [42]. Moreover, regarding flavone compounds, baicalein (5,6,7-trihydroxyflavone), originally isolated from the dried roots of *Scutellaria baicalensis* and *Scutellaria lateriflora*, elicited pleiotropic activity that allowed it to protect mouse splenic lymphocytes against IR-induced cell death through its ability to suppress MKP3 and activate ERK. This is in line with mitigation of radiation-induced hematopoietic injury [43]. Recently, baicalein, administered intraperitoneally with 100 mg/kg in C57BL/6J mice, rebalanced IR-altered gut microbial composition, ameliorating intestinal structure. It down-regulated the expression of pro-apoptotic proteins (e.g., p53, caspase-3, caspase-8 and Bax), also recovering IR-induced hematopoietic dysfunction [44]. Baicalein was reported as a potent radioprotector at the concentration of 5–50 μM [45] and impacts on the NF-κB-mediated inflammatory response [46].

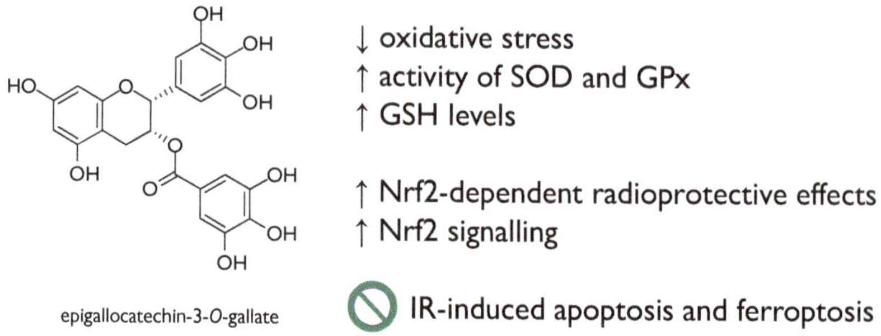

Figure 6. Epigallocatechin-3-O-gallate radioprotective mechanisms of action.

Growing evidence suggests the potential benefit from green tea flavanols. Early studies supported the hypothesis of anti-genotoxic efficacy in human lymphocytes [47], and the overall prevention against ultraviolet radiation-induced DNA damage [48].

Epigallocatechin-3-O-gallate (Figure 6), the main polyphenol in green tea, due to its antioxidant activity and the efficacy in ameliorating many oxidative stress-related diseases, is of wide interest. It promoted Nrf2-dependent radioprotective effects and Nrf2 signalling, in turn, and was found to repress IR-induced apoptosis and ferroptosis, ameliorating intestinal injury induced by total body irradiation in male C57 BL/6 J mice [49]. The radioprotection of EGCG was studied through a model of oxidative damage in ^{60}Coγ radiation mice, and data acquired evidenced the ability of the compound to enhance the activity of enzymatic antioxidants, such as superoxide dismutase and glutathione peroxidase, as well as glutathione levels [50].

Soy isoflavones mitigate vascular damage and inflammation related to lung cancer radiotherapy [51]. Genistein, a main soy isoflavone with phytoestrogen activity, enjoy dual action in radiotherapy; first, it can protect L-02 cells against radiation damage via inhibition of apoptosis, alleviation of DNA damage and chromosome aberration, down-regulation of GRP78 and up-regulation of HERP, HUS1 and hHR23A at low concentration (1.5 μM). Secondly, at high concentration (20 μM) indicate radio-sensitizing properties through the promotion of apoptosis and chromosome aberration, impairment of DNA repair, up-regulation of GRP78, and down-regulation of HUS1, SIRT1, RAD17, RAD51 and RNF8 [52]. Indeed, recently, it was observed that genistein is able to augment the radiosensitivity of hepatoblastoma cells by inducing G2/M arrest and apoptosis [53]. Administration of the genistein also demonstrated providing protection against acute radiation injury at non-toxic doses [54]. Several evidences underline genistein-induced radioprotection for the hematopoietic acute radiation injury, and the ability of the compound to act as a selective estrogen receptor β-agonist was also explored, as it is involved in its radioprotective mechanism of action [55]. The up-regulation of ER-β and FOXL-2 by genistein, with associated downregulation of TGF-β expression was implied also in reversing radiotherapy-induced premature ovarian failure [56]. Soy isoflavones are overall prone to modify clinical responses to RT, acting both with radiosensitizing and radioprotective effects. In preclinical orthotopic models of prostate cancer, renal cell carcinoma and non-small cell lung cancer, it was observed that soy isoflavones targeted signaling survival pathways radiation-upregulated, such as DNA repair and transcription factors, finally driving cancer cells to death [57].

The interest in flavonoids as radiomodulators also led to screen the properties of semisynthetic drugs, such as flavopiridol (Figure 5). This compound, also known as alvocidib, is a flavone derivative that was developed by Sanofi-Aventis, based on a flavonoid derived from the Indian indigenous plant *Dysoxylum binectariferum*. Flavopiridol, structurally based on flavonoid (2-cholorophenyl-4-one) and an alkaloid (1-methylpiperadine) moieties, is a CDK inhibitor that exhibited potent inhibition of CDK1, 2, 4, 6, 7, and 9. It was observed that the compound acts to inhibit and/or repair sublethal damage as well as the repair of DNA double-strand breakage followed by radiation therapy in malignant tumours. Indeed, it might enhance the cytotoxic effect of radiation in radioresistant tumour cells through p53 dysfunction or Bcl-2 overexpression [58].

Reducing the harmful effects of UV radiation is an important issue to pursue, as UVB (290–320 nm) could destroy the integrity of the skin causing epidermal cell apoptosis, potentially even leading to skin cancer. Thus, radioprotective compounds need to be explored and anthocyanins appeared valuable candidates. In particular, the protecting effect of cyanidin-3-*O*-glucoside against UVB-induced damage, one of the harmful factors for the benefit of human skin, to human HaCaT keratinocytes. The anthocyanin was able to decrease intracellularly reactive oxygen species, as well as phospho-p53 and phospho-ATM/ATR levels, and the expression of anti-apoptotic protein B-cell lymphoma 2 [59]. Indeed, it was also demonstrated that cyanidin-3-*O*-glucoside suppressed COX-2 expression by interaction with the MAPK and Akt signalling pathways [60]. Excess ultraviolet (UV) radiation causes numerous forms of skin damage. The encapsulation of cyanidin-3-*O*-glucoside in chitosan nanoparticles provided evidence for the efficacy of the formulation to effectively reduce UVB-induced epidermal damage through the p53-mediated apoptosis signalling pathway [61].

3.2. Other Phenols and Polyphenols with Radioprotective Efficacy

Non-flavonoid compounds were also analyzed in pure form or in mixture. Simple phenols, such as vanillin, were screened for their radioprotective activity.

The compound 4-hydroxy-3-methoxybenzaldehyde, better known as vanillin, widely used as food flavoring agent, was previously investigated as able to counteract γ-radiation-induced DNA damage in plasmid pBR322, human and mouse peripheral blood leucocytes and splenic lymphocytes. The positive action was ascribed to its radical scavenging capability, as well as to the modulation of DNA repair [62]. The antioxidant power of vanillin also includes its ability to act as lipoperoxidant. Moreover, the compound exhibits antimutagenic effects being able to inhibit X-ray and UV-induced single-strand DNA breaks, chromosomal breaks, and DNA crosslinking. Furthermore, although it was shown to also favor DNA ligation, repair and replication, its clinical use is limited by the low in vivo activity. This finding promoted the synthesis of its derivative, VND3207, which appeared to be, in a preclinical screening, radioprotective towards radiation-induced intestinal injury [63]. In this context, it was observed that radioprotection is due also, beyond the antiradical properties of the compound, to the modulation of activated p53 levels in intestinal epithelial cells. Recently, Li et al. [64] showed that treatment with VND3207 can enhance the expression of the catalytic subunit of the DNA-dependent protein kinase (DNA-PKcs) in human lymphoblastoid cells with or without γ-irradiation. Also in this case, the activity of the enzyme consisted in DNA double-strand breakage repair.

Hydroxycinnamic acids, commonly found in fruits, vegetables, and beverages, and structurally characterized by a phenylpropenoid skeleton deriving from the deamination of phenylalanine and tyrosine in plants, could be mirrored in radioprotection. In recent times, caffeic acid was found to ameliorate premature senescence of hematopoietic stem cells, due to its antioxidant capacity. In fact, senescence is mediated by ROS overproduction. On the other hand, caffeic acid can act as pro-apoptotic agent in colon cancer cells [65]. Ferulic acid was also hypothesized to be effective against accident or intentional exposures to ionizing radiation, and the repair of DNA was experimented to take place at a faster rate in ferulic acid treated mice [66].

Among hydroxycinnamoyl derivatives, chlorogenic acid prevented genomic instability induced by X-ray irradiation in non-tumorigenic human blood lymphocytes [67]. Furthermore, the treatment with this depside, at dose level equal to 200 mg/kg, one hour prior to irradiation with high doses of γ-radiation, favoured the animal survival [68].

The radioprotective ability of caffeic acid phenethyl esters, which are abundant in propolis, also was studied and it involves in prevention of oxidative and nitrosative damages induced by radiation [69]. In another study, caffeic acid phenethyl ester was found to act both as radioprotector and radiosensitizer, meaning that it can modulate the radiation response by following different mechanisms depending on the tissue type [70].

Rosmarinic acid, a depside of caffeic acid and 3,4-dihydroxyphenyl lactic acid, promoted, when administrated at a dose of 100 mg/kg, the recovery of peripheral blood cells in irradiated mice [71]. Its capability was also compared to that exerted by carnosic acid and carnosol, two aromatic diterpenes, endowed with antioxidative and antimicrobial properties, equally isolated from rosemary herb. Indeed, the radioprotective effects against γ-irradiation was in the order carnosic acid > rosmarinic acid $\geq$ carnosol [72].

Radioprotection research pays peculiar attention to curcumin, whose antioxidant and anti-inflammatory properties are well-known to target multiple signalling molecules [73].

The diferuloylmethane was demonstrated to ameliorate radiation-induced pulmonary fibrosis [74]. Its effect was through up-regulation of the cytoprotective heme oxygenase 1. Furthermore, as oxidative stress is involved in radiation pneumonopathy, the inhibition of γ-radiation-induced reactive oxygen species in murine lung primary cells was detected. The preventive curcumin outcome was also in ileum goblet cells [75], and on *Drosophila melanogaster* lifespan [76]. Inhibition of the transcription factor NF-κB is the main mode of action of curcumin and is involved also in curcumin-based radiosensitization [77].

Recent findings suggest the capability of a pre-treatment with curcumin to prevent radiotherapy-induced oxidative injury to the skin, through enhancing CAT, SOD, and GSH-Px [78]. Indeed, the low bioavailability of curcumin, due to poor absorption, rapid metabolism, and rapid systemic elimination [79], was carefully taken into account when approaches aimed at preserving its functionality were investigated. In this context, there was the formulation of nanoscale curcumin-encapsulated liposomes [80], or the design of curcumin conjugated albumin-based nanoparticles. In particular, the improvement of radioprotection of conjugated albumin-based nanoparticles was estimated in HHF-2 cells X-ray irradiated, finding that nanoparticles with curcumin at 50 μg/mL induced a 2.3-fold increase in cell viability in respect to cells which underwent only X-ray irradiation [81].

Table 1 summarizes literature data for all phenolic and polyphenolic compounds taken into account in the above discussion, ordered alphabetically, with details about the studied model (in vitro or in vivo), the used dose, and the main protective effect(s).

Table 1. Radioprotective properties of natural phenols and polyphenols herein described (BW = body weight; i.p. = intraperitonially; s.c. = single subcutaneous; i.m. = intramuscular; i.g. = intragastrical).

Compound	Studied Model	Dose	Main Protective Effects	Ref.
Apigenin	Human lymphocytes	Up to 25 μg/mL	Protection from ^{137}Cs gamma rays-induced chromosome aberrations	[38]
	Swiss albino mice	15 mg/kg BW i.p. for 7 consecutive days	Immunostimulatory effect and mitigation of radiation-induced hematological alterations	[39]
	Swiss albino mice	15 mg/kg BW i.p. for 7 days	Restoration of intestinal crypt-villus architecture Inhibition of the radiation-induced activation of NF-κB expression in the gastrointestinal tissue	[40]
Baicalein	Swiss and C57BL/6 male mice Murine T cell lymphoma cells (EL4)	10 mg/kg BW for 3 days 5–100 μM	Activation of the target molecules ERK and Nrf-2 both in vitro and in vivo	[43]
	Swiss albino mice	150 mg/kg BW	Protection from DNA damage Reduction of radiation-induced damage to mice bone marrow cells	[45]
	Human white blood cells	Up to 50 μM	Dose dependent inhibition of DNA strand breaks	
	C57BL/6 mice	5 mg/kg BW for 3 days	Protection against NFκB-mediated inflammatory response through MAPKs and the Akt pathway Up-regulation of FOXO activation, catalase and SOD activities	[46]

Table 1. Cont.

Compound	Studied Model	Dose	Main Protective Effects	Ref.
Caffeic acid	Mouse colon cancer (CT-26), human liver cancer (HepG2) and human breast cancer (MCF-7) cells lines	1–3 mM	Induction of apoptosis of colon cancer cell	[65]
	C57BL/6 mice	20 mg/kg BW 5 times (every three days) via oral gavage before and 1 after irradiation	Amelioration of ROS production and premature senescence of hematopoietic stem cells	
Caffeic acid phenethyl esters	Sprague-Dawley rats	10 µmol/kg i.p. 30 min before irradiation	Prevention of oxidative and nitrosative damages induced by radiation	[69]
Chlorogenic acid	Human blood lymphocytes	0.5, 1, 2 and 4 µg/mL	Prevention of genomic instability induced by X-ray irradiation	[67]
	Mice	100, 200 and 400 mg/kg BW 1 or 24 h before irradiation	Increase of animal survival (with 200 mg/kg dose 1 h before irradiation with high doses of γ-radiation)	[68]
Curcumin	C57BL/6 mice	standard mouse chow (5015) with 1% or 5% curcumin w/w	Amelioration of radiation-induced pulmonary fibrosis Increase of mouse survival with no impairment of tumor cell killing by radiation.	[74]
	Wistar albino rats	100 mg/kg BW orally (by intra gastric intubation)	Protection against intestinal damage	[75]
	Wistar rats	150 mg/kg BW 1 day before irradiation to 3 days after orally	Prevention of radiotherapy-induced oxidative injury to the skin, through enhancing CAT, SOD, and GSH-Px	[78]
	Human blood cells	curcumin-encapsulated liposomes (up to 1046.5 µg/mL)	Reduction in the micronuclei frequency No genotoxicity	[80]
	Human foreskin fibroblast cells (HHF-2)	conjugated albumin based nanoparticles (50 µg/mL)	Increase in cell viability	[81]
	Balb/C mice	125, 250, 500 and 1000 mg/kg via tail vain	Increase in the survival rate	
Cyanidin-3-O-glucoside (C3G)	HaCaT keratinocytes	80, 160 and 200 µM	Suppression of COX-2 expression by interaction with the MAPK and Akt signalling pathways	[60]
	Kunming mice	125 µM, 250 µM and 500 µM chitosan-C3G nanoparticles; 500 µM C3G	Reduction of UVB-induced epidermal damage through the p53-mediated apoptosis signalling pathway Higher efficiency of nanoparticles respect to C3G at the same dose	[61]
	HaCaT keratinocytes	80, 160 and 200 µM	Intracellular decrease of ROS and of the phospho-p53 and phospho-ATM/ATR levels Expression of anti-apoptotic protein B-cell lymphoma 2	[59]

Table 1. Cont.

Compound	Studied Model	Dose	Main Protective Effects	Ref.
Epigallocatechin-3-O-gallate	C57 BL/6 J mice	12.5 or 25 mg/kg BW for 5 days	Reduction of IR-induced cell death in intestinal epithelial cells through Nrf2 Repression of IR-induced apoptosis and ferroptosis Amelioration of intestinal injury induced by total body irradiation	[49]
	Kunming mice	6.25, 12.5 and 25 mg/kg BW for 30 days	Prevention of the immune system damage Enhancement of the activity of enzymatic antioxidants (e.g., SOD, GSH-Px)	[50]
	Normal fetal lung fibroblasts (MRC5) and adult skin fibroblasts (84BR) Normal human epidermal keratinocytes (NHEK)	Up to 1 mM 250 µM	Prevention against ultraviolet radiation-induced DNA damage	[48]
Ferulic acid	Swiss mice	50, 75 and 100 mg/kg BW i.p. 1 h before irradiation	Faster repair of DNA	[66]
Flavopiridol *	A172/mp53 and HeLa/bcl-2 cells (radioresistant through genetic alteration)	0.0125–0.125 µM	Radio-sensitization via inhibition of sublethal damage and DNA double-strand breakage repair	[58]
Genistein	Human embryo liver cells (L-02)	1–20 µM	Low concentration: protection against radiation damage High concentration: radiosensitizing features	[52]
	Huh-7, Hep3B and Hep G2 human HCC cells L-02 cells	0–40 µM	Low dose (5 µM): HCC cell sensitivity enhancement to X rays; no significant toxicity to L-02 cells	[53]
	CD2F1 mice	100, 200, or 400 mg/kg BW s.c.	Protection against acute radiation injury (administered 24 h before irradiation) Hypothesized indirect mechanism (e.g., cytokine release)	[54]
	CD2F1 mice	150 mg/kg BW i.m.	Selective binding to estrogen receptor β	[55]
	Sprague-Dawley rats	5 mg/kg BW i.p. for 7 days	Up-regulation of ER-β and FOXL-2 Downregulation of TGF-β expression Reversion of radiotherapy-induced premature ovarian failure	[56]
Monoglucosyl rutin *	CHO 10B2 cells	0.001–0.1%	Increase in survival of IR-treated cells at doses greater than 2 Gy	[32]
Naringin	Swiss albino mice	75 mg/kg BW for 3 days	Reversion of the liver IR-induced redox-imbalance	[41]
Quercetin	CBA mice	100 mg/kg BW i.p. for 3 days	Protection of mice white blood cells from lethal effects and DNA damage before γ-irradiation	[35]
	Human white blood cells	50 µM	Protection from DNA damage after γ-irradiation	[34]
Quercetin glycosylated derivatives	Cell-free systems	10 µM	Reduction of DNA double-strand breaks	[33]
Rosmarinic acid	Mice	100, 200 and 400 mg/kg BW (oral) for 10 days (3 before and 7 after irradiation)	Promotion of the recovery of peripheral blood cells Enhancement of 30-day survival rates	[71]

Table 1. Cont.

Compound	Studied Model	Dose	Main Protective Effects	Ref.
Rutin + podophyllo-toxin	C57BL/6 mice	2.5mg/kg BW 1 h before irradiation	ROS levels reduction Protection of cellular macromolecules Activation of antioxidant signaling pathway	[30]
	HaCat cells	0.025–0.4 µg/mL culture medium	Reduction of the cellular damage by scavenging free-radicals, cell-cycle arrest and DNA repair enhancement in the hematopoietic system	[29]
	Strain 'A' mice	2.5mg/kg BW i.m. 1 h before irradiation		
VND3207 vanillin derivative *	Human intestinal epithelial cells (HIEC)	40 µM	Promotion of intestinal repair after radiation injury by regulation of the DNA-PKcs pathway	[64]
	C57BL/6 J, NOD-SCID (Prkdcscid/scid) and BALB/c mice	100 mg/kg BW i.g. 30 min before irradiation		
	C57BL/6J mice	100 mg/kg BW (oral gavage) 30 min before/after irradiation	Intestinal repair after radiation injury by reduction of ROS-induced DNA damage and modulation of activated p53 levels in intestinal epithelial cells	[63]

* semi-synthetic compound.

4. Bioactive Plant Extracts in Radioprotection: A Still Undervalued Topic

Plant extracts possess infinite therapeutics, such as anticancer, antioxidant, antimicrobial, anti-inflammatory, and analgesic. According to the World Health Organization (WHO), about 80% of people worldwide use traditional medicine for primary healthcare needs [82]. There are nearly 20,000 medicinal plants in 91 countries which contain a wide range of substances that can be used to different therapeutic purposes. Different plant extracts rich in phenols and polyphenols and/or other specialized metabolites (e.g., carotenoids, sulphur compounds) were screened for their radioprotective effects, and potential mechanism of actions were proposed.

Indeed, although promising efficacy was suggested, literature data often lack detailed chemical composition analyses and standardization, compromising data reproducibility.

The activities in radiation protection as evaluated in extracts obtained from medicinal plants, widely used in complementary and alternative medicine, or from food plants are reported below. Alcoholic extract of the plant *Ageratum conyzoides* was able to protect mortality in mice exposed to 10 Gy of γ-radiation. Accordingly, up to a dose of 3000 mg kg^{-1} was attributed to non-toxic concentration, suggesting that the radioprotection afforded by *Ageratum conyzoides* may be in part due to the scavenging of reactive oxygen species induced by ionizing radiation [83].

The investigation on extracts from five medicinal plants including *Adhatoda vasica*, *Amaranthus paniculatus*, *Brassica compestris*, *Mentha piperita* and *Spirulina fusiformis* indicated that the antioxidant capacities of these plant extracts can be responsible for radioprotective capacities [84]. According to this study, major chemical constituents play a key role in radiation protection. These compounds were vesicine, vesicinone, betaine, vitamin C, β-carotene, and vasakin in *Adhatoda vasica*; proteins, vitamins (C and E), provitamin A, and riboflavin in *Amaranthus paniculatus*; allyl isothiocyanate, glucosinolates, indoles in *Brassica compestris*, and proteins, natural vitamins (β-carotene), and SOD in *Spirulina fusiformis*.

Several pathways of protection against ionizing radiation have been suggested in mammalian cells [85]. These mechanisms involve free radical-scavenging through the inhibition of reactive oxygen species, as well as hydrogen atom donation [86,87]. It can be concluded that phenolic compounds due to their antioxidant activities can act as free radical scavenger and, likewise radioprotectors [84]. Recently, an aqueous extract from a Southern Italian cherry cultivar, constituted by chlorogenic acids and flavonoids together with simple carbohydrates and polyols, proved to exert a radiomodulating behaviour against SH-SY5Y neuroblastoma cell line. In fact, at low doses it acted as a radioprotector agent, whereas at high doses it enhanced cytotoxic effects due to radiation exposure [88].

Olea europaea leaf is a rich source of phenols and polyphenols, whose radioprotective potential was marginally investigated in pre-UV and post-UV treatments [89]. Anticlastogenic and antiradical activities of an olive leaf extract, constituted by 24.5% in oleuropein, 1.5% in hydroxytyrosol and by almost 3% in flavone-7-glucosides, and 1% in verbascoside, was found. The effects of pure oleuropein on radiation response in nasopharyngeal carcinoma were also determined [90]. Oleanolic acid and ursolic acid, two triterpene acids from olive fruit and other dietary products, could inhibit the tumor growth and modify hematopoiesis stem cells (HSCs) after irradiation [91]. In addition, anti-tumor activity was performed by the interplay of oleanolic acid and ursolic acid, so that they might partially act as anti-cancer agents and, furthermore, decrease damages occurring on hematopoietic tissue after radiotherapy [91,92]. The radioprotection by apple polyphenols was also investigated through in vitro studies mainly aiming to clarify the polyphenols' ability to scavenge free radicals [93].

Rheum palmatum L., and its main compound emodin (6-methyl-1,3,8-trihydroxyanthraquinone), were radioprotective against γ-rays. Emodin's mechanism of action is somehow that the levels of total thiols such as glutathione and lipid peroxidation products have been decreased. Furthermore, measuring of tongue antioxidant enzymes, glutathione peroxidase, glutathione-S-transferase, γ-glutamyl transferase, and glucose-6-phosphatase, revealed the amelioration of the levels of cellular thiols and antioxidant enzymes in serum of gamma irradiated diabetic mice with emodin treatment [94].

The ethanolic extract of leaves of *Adhatoda vasica* L. Nees, a well-known plant drug in Ayurvedic and Unani medicine, showed a significant decrease in acid phosphatase level and, on the contrary, an increase in alkaline phosphatase level. Pre-treatment with Adhatoda also significantly demonstrated the prevention of radiation-induced chromosomal damage in bone marrow cells. However, it remains the necessity of mechanistic studies for its radioprotective effects as well as major ingredients in the extract thereof [95]. Leaf extract of *Adhatoda vasica* was also reported for the protective role in spleen of Swiss albino mice exposed to 6 Gy γ-radiation [96].

Radioprotective efficacy of *Amaranthus paniculatus* leaf extract have been reported [97,98]. The oral administration of *A. paniculatus* extract at 800 mg/kg body weight of Swiss albino mice for 15 consecutive days before exposure to gamma ray showed the increased endogenous spleen colonies and the spleen weight without any side effect or toxicity. The modulation of glutathione as well as lipid peroxidation were other results thereof [97].

Lamiaceae family is attributed to have radioprotector ability, with mechanisms including mainly free radicals scavenging, DNA damage protection, decreasing of lipid peroxidation and enhancing of glutathione, superoxide dismutase, catalase, and alkaline phosphatase enzymes activity levels [99]. For instance, *Mentha piperita* is a plant belonging to this family, containing eugenol, caffeic acid, rosmarinic acid, and α-tocopherol that play a key role for anti-cancer and radioprotective properties [84,100].

Remedial and pharmacological properties of aromatic plants have been reported. Amongst, essential oils are used for different aspects such as in cosmetics, fragrance, pesticides, and beverages. Since essential oils are known for their antioxidant activities and free radical scavenging, they can be also considered as radioprotectors. *Ageratum conyzoides* has shown DPPH radicals scavenging, likewise its radioprotective role with the dose of 75 mg/kg for 6–11 Gray on mice. Active agents in this plant are polyoxygenated flavonoids, triterpenes (friedelin), sterols β-sitosterol and stigmasterol, and alkaloids (lycopsamine and echinatine) [83].

Alium cepa, *Alium sativum* are two plants from Liliaceae family which enjoy antioxidant, antihypertensive and antihyperglycemic [101]. Alk(en)yl thiosulfates from onions and garlic markedly reduced damage in rat hepatoma H4IIE cells and mouse lymphoma L5178Y cells treated with 10 Gy of X-ray irradiation [102].

Recently, the radioprotective efficacy of an hydroalcoholic extract of *Pterocarpus santalinus*, a small-to-medium–sized deciduous tree belonging to the Fabaceae family [103], was verified in BALB/c mice exposed to γ-radiation. The redox homoeostasis corrupted by the radiation was ameliorated following the treatment with the extract, which probably occurred via the upregulation of Nrf2, HO-1, and GPX-1p. The UHPLC-HRMS/MS analysis of the extract highlighted its diversity in santolins, beyond other phenols and flavonoid compounds [104]. Furthermore, the *Pterocarpus santalinus* hydroalcoholic extract (PSHE) was non-toxic, and when RAW264.7 macrophages were pretreated with it, a significant inhibition of LPS-induced pro-inflammatory cytokines IL-6, and TNF-α production was observed. Polyphenols from medicinal plants, such as *Sanguisorba officinalis* and *Erigeron canadens*, were found to be able to decrease irradiation-induced oxidative stress in normal lymphocytes using ROS mechanisms, acting as a radiation modifier for normal cells [105]. An extract from *Lonicera caerulea* var. *edulis*, rich in anthocyanins, and intragastrically administered to mice once a day, prior to 5 Gy whole body ^{60}Coγ radiation, was effective to slow down the levels of malondialdehyde, while increasing superoxide dismutase and glutathione peroxidase activities and glutathione GSH content in the liver [106].

5. Conclusions

Over the last two decades, secondary metabolites in plants have been broadly considered for their therapeutic attributes as radioprotectors. Indeed, the lower toxicity and cost of natural products appear two key factors that push the research to deepen the understanding of the mechanism of action of these substances. Thus, several studies are found in literature focused on their role in counteracting IR-induced damage. Pure compounds or plant extracts showing free radical scavenging activity, anti-lipoperoxidant, and reducing properties are of predominant interest and are prone to intervene in DNA repair or in restoring chromosomal damages.

In particular, (poly)phenols could represent a valuable alternative to synthetic compounds to be used as radioprotective agents. In fact, based on their dose-dependent antioxidant/pro-oxidant efficacy, they could provide protection to normal cells, with little or no protection to tumor cells resistant to radiotherapy. Thus, they provide good evidence regarding a possible radiomodulating action.

Furthermore, considering the dietary nature of the majority of matrices investigated, another great advantage could derive from the suitability of oral administration that could be optimal during radiotherapy. However, with few exceptions, the data available to date remain fragmentary and are mostly the result of in vitro studies which, while deepening the chemical and biological knowledge of the molecules, do not find comfort in preclinical and/or clinical studies, or preclinical studies in which, especially when evaluating the bioactivity of a plant extract, its chemical composition is not taken into account.

Author Contributions: Conceptualization, S.P. (Severina Pacifico) and L.M.; investigation, S.F., S.P. (Severina Pacifico); data curation, S.P. (Severina Pacifico); writing—original draft preparation, S.F. and S.P. (Severina Pacifico); writing—review and editing, S.P. (Severina Pacifico); visualization, S.P. (Simona Piccolella); L.M.; supervision, S.P. (Severina Pacifico). All authors have read and agreed to the published version of the manuscript.

Funding: This research received no external funding.

Institutional Review Board Statement: Not applicable.

Informed Consent Statement: Not applicable.

Conflicts of Interest: The authors declare no conflict of interest.

Abbreviations

ACE	Angiotensin Converting Enzyme
Akt	Protein kinase B
Bax	Bcl-2-associated X protein
Bcl-2	B-cell lymphoma 2
CAT	Catalase
CDK	Cyclin-Dependent Kinase
COX-2	Cyclo-oxygenase-2
DDR	DNA Damage Response
DPPH	2,2-diphenyl-1-picryl hydrazyl
DSB	Double Strand Break
e^-_{aq}	Hydrated electrons
EGCG	Epigallocatechin-3-*O*-gallate
ERK	Extracellular signal-regulated kinase
ER-β	Estrogen Receptor-β
FOXL-2	Forkhead box L2 protein
G-CSF	Granulocyte-Colony Stimulating Factor
GM-CSF	Granulocyte Macrophage-Colony Stimulating Factor
GPx	Glutathione peroxidase
GRP78	78-kDa Glucose-Regulated Protein
GSH	Reduced Glutathione
HO-1	Heme Oxygenase-1
$HO_2^\bullet$	Hydroperoxyl radical
HSCs	Hematopoiesis Stem Cells
IL-1	Interleukin-1
IR	Ionizing Radiation
LPS	Lipopolysaccharide
MAPK	Mitogen-Activated Protein Kinase
MKP3	MAP Kinase Phosphatase 3
NF-κB	Nuclear factor kappa B
Nrf2-ARE	NF-E2-related factor 2-Antioxidant Responsive Element
$O_2^{-\bullet}$	Superoxide anion radical
PSHE	*Pterocarpus santalinus* hydroalcoholic extract
PUMA	p53 Up-regulated Modulator of Apoptosis
RNF8	Ring Finger Protein 8
ROS	Reactive Oxygen Species
SIRT1	Sirtuin 1
SOD	Superoxide Dismutase
SSB	Single Strand Break
TGF-β	Transforming Growth Factor-β
TNF-α	Tumor Necrosis Factor-1
UHPLC-HRMS/MS	Ultra High Performance Liquid Chromatography-High Resolution Tandem Mass Spectrometry

References

1. Belli, M.; Indovina, L. The Response of Living Organisms to Low Radiation Environment and Its Implications in Radiation Protection. *Front. Public Health* **2020**, *8*, 601711. [CrossRef]
2. Goodhead, D.T. An Assessment of the Role of Microdosimetry in Radiobiology. *Radiat. Res.* **1982**, *91*, 45. [CrossRef]
3. Jeggo, P.; Löbrich, M. Radiation-induced DNA damage responses. *Radiat. Prot. Dosim.* **2006**, *122*, 124–127. [CrossRef] [PubMed]
4. Durante, M.; Manti, L. Human response to high-background radiation environments on Earth and in space. *Adv. Space Res.* **2008**, *42*, 999–1007. [CrossRef]
5. Saha, G.B. *Physics and Radiobiology of Nuclear Medicine*; Springer Science & Business Media: Berlin, Heidelberg, Germany, 2012.
6. Varanda, E.A.; Tavares, D.C. Radioprotection: Mechanisms and Radioprotective Agents Including Honeybee Venom. *J. Venom. Anim. Toxins* **1998**, *4*, 5–21. [CrossRef]
7. Lomax, M.E.; Folkes, L.K.; O'Neill, P. Biological consequences of radiation-induced DNA damage: Relevance to radiotherapy. *Clin. Oncol.* **2013**, *25*, 578–585. [CrossRef]

8. Manti, L.; Braselmann, H.; Calabrese, M.L.; Massa, R.; Pugliese, M.; Scampoli, P.; Sicignano, G.; Grossi, G. Effects of Modulated Microwave Radiation at Cellular Telephone Frequency (1.95 GHz) on X-ray-Induced Chromosome Aberrations in Human Lymphocytes In Vitro. *Radiat. Res.* **2008**, *169*, 575–583. [CrossRef]
9. Wong, K.; Delaney, G.P.; Barton, M.B.; Information, P.E.K.F.C. Evidence-based optimal number of radiotherapy fractions for cancer: A useful tool to estimate radiotherapy demand. *Radiother. Oncol.* **2015**, *119*, 145–149. [CrossRef]
10. Johnke, R.M.; Sattler, J.A.; Allison, R.R. Radioprotective agents for radiation therapy: Future trends. *Futur. Oncol.* **2014**, *10*, 2345–2357. [CrossRef]
11. Andreassen, C.N.; Grau, C.; Lindegaard, J.C. Chemical radioprotection: A critical review of amifostine as a cytoprotector in radiotherapy. *Semin. Radiat. Oncol.* **2003**, *13*, 62–72. [CrossRef]
12. Obrador, E.; Salvador, R.; Villaescusa, J.; Soriano, J.; Estrela, J.; Montoro, A. Radioprotection and Radiomitigation: From the Bench to Clinical Practice. *Biomedicines* **2020**, *8*, 461. [CrossRef]
13. Mun, G.-I.; Kim, S.; Choi, E.; Kim, C.S.; Lee, Y.-S. Pharmacology of natural radioprotectors. *Arch. Pharmacal Res.* **2018**, *41*, 1033–1050. [CrossRef]
14. Commission Regulation (EU) No 182/2013 of 1 March 2013 (EU commission N° 182 2013). Available online: https://eur-lex.europa.eu/LexUriServ/LexUriServ.do?uri=OJ:L:2013:061:0002:0005:EN:PDF (accessed on 5 March 2013).
15. Patyar, R.R.; Patyar, S. Role of drugs in the prevention and amelioration of radiation induced toxic effects. *Eur. J. Pharmacol.* **2018**, *819*, 207–216. [CrossRef] [PubMed]
16. Sueiro-Benavides, R.A.; Leiro-Vidal, J.M.; Salas-Sánchez, A.A.; Rodríguez-González, J.A.; Ares-Pena, F.J.; López-Martín, M.E. Radiofrequency at 2.45 GHz increases toxicity, pro-inflammatory and pre-apoptotic activity caused by black carbon in the RAW 264.7 macrophage cell line. *Sci. Total Environ.* **2020**, *765*, 142681. [CrossRef] [PubMed]
17. Huang, L.; Snyder, A.R.; Morgan, W.F. Radiation-induced genomic instability and its implications for radiation carcinogenesis. *Oncogene* **2003**, *22*, 5848–5854. [CrossRef] [PubMed]
18. Mohan, S.; Gupta, D. Role of Nrf2-antioxidant in radioprotection by root extract of Inula racemosa. *Int. J. Radiat. Biol.* **2019**, *95*, 1122–1134. [CrossRef]
19. Nair, C.K.K.; Parida, D.K.; Nomura, T. Radioprotectors in Radiotherapy. *J. Radiat. Res.* **2001**, *42*, 21–37. [CrossRef] [PubMed]
20. Mitchell, J.B.; Anver, M.R.; Sowers, A.L.; Rosenberg, P.; Figueroa, M.; Thetford, A.; Krishna, M.C.; Albert, P.S.; Cook, J.A. The Antioxidant Tempol Reduces Carcinogenesis and Enhances Survival in Mice When Administered after Nonlethal Total Body Radiation. *Cancer Res.* **2012**, *72*, 4846–4855. [CrossRef]
21. Rosen, E.M.; Day, R.; Singh, V.K. New Approaches to Radiation Protection. *Front. Oncol.* **2015**, *4*, 381. [CrossRef]
22. Li, M. The role of P53 up-regulated modulator of apoptosis (PUMA) in ovarian development, cardiovascular and neurodegenerative diseases. *Apoptosis* **2021**, *26*, 235–247. [CrossRef]
23. Brown, S.L.; Kolozsvary, A.; Liu, J.; Jenrow, K.A.; Ryu, S.; Kim, J.H. Antioxidant Diet Supplementation Starting 24 Hours after Exposure Reduces Radiation Lethality. *Radiat. Res.* **2010**, *173*, 462–468. [CrossRef] [PubMed]
24. Farhood, B.; Goradel, N.H.; Mortezaee, K.; Khanlarkhani, N.; Najafi, M.; Sahebkar, A. Melatonin and cancer: From the promotion of genomic stability to use in cancer treatment. *J. Cell. Physiol.* **2019**, *234*, 5613–5627. [CrossRef] [PubMed]
25. Hazra, B.; Ghosh, S.; Kumar, A.; Pandey, B.N. The Prospective Role of Plant Products in Radiotherapy of Cancer: A Current Overview. *Front. Pharmacol.* **2012**, *2*, 94. [CrossRef] [PubMed]
26. Calvaruso, M.; Pucci, G.; Musso, R.; Bravatà, V.; Cammarata, F.P.; Russo, G.; Forte, G.I.; Minafra, L. Nutraceutical Compounds as Sensitizers for Cancer Treatment in Radiation Therapy. *Int. J. Mol. Sci.* **2019**, *20*, 5267. [CrossRef]
27. Piccolella, S.; Pacifico, S. Plant-Derived Polyphenols. *Adv. Mol. Toxicol.* **2015**, *9*, 161–214. [CrossRef]
28. Kemertelidze, E.P.; Tsitsishvili, V.G.; Alaniya, M.D.; Sagareishvili, T.G. Structure-function analysis of the radioprotective and antioxidant activity of flavonoids. *Chem. Nat. Compd.* **2000**, *36*, 54–59. [CrossRef]
29. Yashavarddhan, M.H.; Shukla, S.K.; Chaudhary, P.; Srivastava, N.N.; Joshi, J.; Suar, M.; Gupta, M.L. Targeting DNA Repair through Podophyllotoxin and Rutin Formulation in Hematopoietic Radioprotection: An In Silico, In Vitro, and In Vivo Study. *Front. Pharmacol.* **2017**, *8*, 750. [CrossRef]
30. Dutta, A.; Gupta, M.L.; Kalita, B. The combination of the active principles of Podophyllum hexandrumsupports early recovery of the gastrointestinal system via activation of Nrf2-HO-1 signaling and the hematopoietic system, leading to effective whole-body survival in lethally irradiated mice. *Free. Radic. Res.* **2015**, *49*, 317–330. [CrossRef] [PubMed]
31. Dutta, A.; Verma, S.; Sankhwar, S.; Flora, S.; Gupta, M.L. Bioavailability, antioxidant and non toxic properties of a radioprotective formulation prepared from isolated compounds of Podophyllum hexandrum: A study in mouse model. *Cell. Mol. Boil.* **2012**, *58*.
32. Sunada, S.; Fujisawa, H.; Cartwright, I.M.; Maeda, J.; Brents, C.A.; Mizuno, K.; Aizawa, Y.; Kato, T.A.; Uesaka, M. Monoglucosyl-rutin as a potential radioprotector in mammalian cells. *Mol. Med. Rep.* **2014**, *10*, 10–14. [CrossRef] [PubMed]
33. Yu, H.; Haskins, J.S.; Su, C.; Allum, A.; Haskins, A.H.; Salinas, V.A.; Sunada, S.; Inoue, T.; Aizawa, Y.; Uesaka, M.; et al. In vitro screening of radioprotective properties in the novel glucosylated flavonoids. *Int. J. Mol. Med.* **2016**, *38*, 1525–1530. [CrossRef]
34. Benković, V.; Kopjar, N.; Knežević, A.H.; Đikić, D.; Bašić, I.; Ramić, S.; Viculin, T.; Knežević, F.; Orolić, N. Evaluation of Radioprotective Effects of Propolis and Quercetin on Human White Blood Cells in Vitro. *Biol. Pharm. Bull.* **2008**, *31*, 1778–1785. [CrossRef]
35. Benkovic, V.; Knezevic, A.H.; Dikic, D.; Lisicic, D.; Orsolic, N.; Basic, I.; Kosalec, I.; Kopjar, N. Radioprotective effects of propolis and quercetin in γ-irradiated mice evaluated by the alkaline comet assay. *Phytomedicine* **2008**, *15*, 851–858. [CrossRef]

36. Khayyal, M.T.; Abdel-Naby, D.H.; El-Ghazaly, M.A. Propolis extract protects against radiation-induced intestinal mucositis through anti-apoptotic mechanisms. *Environ. Sci. Pollut. Res.* **2019**, *26*, 24672–24682. [CrossRef]
37. Karapetsas, A.; Voulgaridou, G.-P.; Konialis, M.; Tsochantaridis, I.; Kynigopoulos, S.; Lambropoulou, M.; Stavropoulou, M.-I.; Stathopoulou, K.; Aligiannis, N.; Bozidis, P.; et al. Propolis Extracts Inhibit UV-Induced Photodamage in Human Experimental In Vitro Skin Models. *Antioxidants* **2019**, *8*, 125. [CrossRef]
38. Rithidech, K.N.; Tungjai, M.; Whorton, E.B. Protective effect of apigenin on radiation-induced chromosomal damage in human lymphocytes. *Mutat. Res. Toxicol. Environ. Mutagen.* **2005**, *585*, 96–104. [CrossRef] [PubMed]
39. Prasad, R.; Thayalan, K.; Begum, N. Apigenin protects gamma-radiation induced oxidative stress, hematological changes and animal survival in whole body irradiated Swiss albino mice. *Int. J. Nutr. Pharmacol. Neurol. Dis.* **2012**, *2*, 45–52. [CrossRef]
40. Begum, N.; Prasad, N.R.; Kanimozhi, G.; Agilan, B. Apigenin prevents gamma radiation-induced gastrointestinal damages by modulating inflammatory and apoptotic signalling mediators. *Nat. Prod. Res.* **2021**, 1–5. [CrossRef]
41. Manna, K.; Khan, A.; Biswas, S.; Das, U.; Sengupta, A.; Mukherjee, D.; Chakraborty, A.; Dey, S. Naringin ameliorates radiation-induced hepatic damage through modulation of Nrf2 and NF-κB pathways. *RSC Adv.* **2016**, *6*, 23058–23073. [CrossRef]
42. Saha, S.; Buttari, B.; Panieri, E.; Profumo, E.; Saso, L. An Overview of Nrf2 Signaling Pathway and Its Role in Inflammation. *Molecules* **2020**, *25*, 5474. [CrossRef] [PubMed]
43. Patwardhan, R.; Sharma, D.; Checker, R.; Sandur, S.K. Mitigation of radiation-induced hematopoietic injury via regulation of cellular MAPK/phosphatase levels and increasing hematopoietic stem cells. *Free. Radic. Biol. Med.* **2014**, *68*, 52–64. [CrossRef] [PubMed]
44. Wang, Y.; Sano, S.; Yura, Y.; Ke, Z.; Sano, M.; Oshima, K.; Ogawa, H.; Horitani, K.; Min, K.-D.; Miura-Yura, E.; et al. Tet2-mediated clonal hematopoiesis in nonconditioned mice accelerates age-associated cardiac dysfunction. *JCI Insight* **2020**, *5*, e135204. [CrossRef]
45. Gandhi, N.M. Baicalein protects mice against radiation-induced DNA damages and genotoxicity. *Mol. Cell. Biochem.* **2013**, *379*, 277–281. [CrossRef]
46. Lee, E.K.; Kim, J.M.; Choi, J.; Jung, K.J.; Kim, D.H.; Chung, S.W.; Ha, Y.M.; Yu, B.P.; Chung, H.Y. Modulation of NF-κB and FOXOs by baicalein attenuates the radiation-induced inflammatory process in mouse kidney. *Free. Radic. Res.* **2011**, *45*, 507–517. [CrossRef]
47. Davari, H.; Haddad, F.; Moghimi, A.; Rahimi, M.F.; Ghavamnasiri, M.R. Study of Radioprotective Effect of Green Tea against Gamma Irradiation Using Micronucleus Assay on Binucleated Human Lymphocytes. *Iran. J. Basic Med. Sci.* **2012**, *15*, 1026–1031.
48. Morley, N.; Clifford, T.; Salter, L.; Campbell, S.; Gould, D.; Curnow, A. The green tea polyphenol (-)-epigallocatechin gallate and green tea can protect human cellular DNA from ultraviolet and visible radiation-induced damage. *Photodermatol. Photoimmunol. Photomed.* **2005**, *21*, 15–22. [CrossRef]
49. Xie, L.-W.; Cai, S.; Zhao, T.-S.; Li, M.; Tian, Y. Green Tea Derivative (−)-Epigallocatechin-3-Gallate (EGCG) Confers Protec-tion against Ionizing Radiation-Induced Intestinal Epithelial Cell Death Both In Vitro and In Vivo. *Free Radic. Biol. Med.* **2020**, *161*, 175–186. [CrossRef]
50. Yi, J.; Chen, C.; Liu, X.; Kang, Q.; Hao, L.; Huang, J.; Lu, J. Radioprotection of EGCG based on immunoregulatory effect and antioxidant activity against 60Coγ radiation-induced injury in mice. *Food Chem. Toxicol.* **2020**, *135*, 111051. [CrossRef] [PubMed]
51. Fountain, M.D.; McLellan, L.A.; Smith, N.L.; Loughery, B.F.; Rakowski, J.T.; Tse, H.Y.; Hillman, G.G. Isoflavone-mediated radioprotection involves regulation of early endothelial cell death and inflammatory signaling in Radiation-Induced lung injury. *Int. J. Radiat. Biol.* **2019**, *96*, 245–256. [CrossRef]
52. Song, L.; Ma, L.; Cong, F.; Shen, X.; Jing, P.; Ying, X.; Zhou, H.; Jiang, J.; Fu, Y.; Yan, H. Radioprotective effects of genistein on HL-7702 cells via the inhibition of apoptosis and DNA damage. *Cancer Lett.* **2015**, *366*, 100–111. [CrossRef]
53. Yan, H.; Jiang, J.; Du, A.; Gao, J.; Zhang, D.; Song, L. Genistein Enhances Radiosensitivity of Human Hepatocellular Carcinoma Cells by Inducing G2/M Arrest and Apoptosis. *Radiat. Res.* **2020**, *193*, 286. [CrossRef] [PubMed]
54. Landauer, M.R.; Srinivasan, V.; Seed, T.M. Genistein treatment protects mice from ionizing radiation injury. *J. Appl. Toxicol.* **2003**, *23*, 379–385. [CrossRef] [PubMed]
55. Landauer, M.R.; Harvey, A.J.; Kaytor, M.D.; Day, R.M. Mechanism and therapeutic window of a genistein nanosuspension to protect against hematopoietic-acute radiation syndrome. *J. Radiat. Res.* **2019**, *60*, 308–317. [CrossRef]
56. Haddad, Y.H.; Said, R.S.; Kamel, R.; Morsy, E.M.E.; El-Demerdash, E. Phytoestrogen genistein hinders ovarian oxidative damage and apoptotic cell death-induced by ionizing radiation: Co-operative role of ER-β, TGF-β, and FOXL. *Sci. Rep.* **2020**, *10*, 13551. [CrossRef] [PubMed]
57. Hillman, G.G. Soy Isoflavones Protect Normal Tissues While Enhancing Radiation Responses. *Semin. Radiat. Oncol.* **2018**, *29*, 62–71. [CrossRef]
58. Hara, T.; Omura-Minamisawa, M.; Kang, Y.; Cheng, C.; Inoue, T. Flavopiridol Potentiates the Cytotoxic Effects of Radiation in Radioresistant Tumor Cells in Which p53 is Mutated or Bcl-2 is Overexpressed. *Int. J. Radiat. Oncol.* **2008**, *71*, 1485–1495. [CrossRef]
59. Hu, Y.; Ma, Y.; Wu, S.; Chen, T.; He, Y.; Sun, J.; Jiao, R.; Jiang, X.; Huang, Y.; Deng, L.; et al. Protective Effect of Cyanidin-3-O-Glucoside against Ultraviolet B Radiation-Induced Cell Damage in Human HaCaT Keratinocytes. *Front. Pharmacol.* **2016**, *7*, 301. [CrossRef]

60. He, Y.; Hu, Y.; Jiang, X.; Chen, T.; Ma, Y.; Wu, S.; Sun, J.; Jiao, R.; Li, X.; Deng, L.; et al. Cyanidin-3-O-glucoside inhibits the UVB-induced ROS/COX-2 pathway in HaCaT cells. *J. Photochem. Photobiol. B Biol.* **2017**, *177*, 24–31. [CrossRef]
61. Liu, Z.; Hu, Y.; Li, X.; Mei, Z.; Wu, S.; He, Y.; Jiang, X.; Sun, J.; Xiao, J.; Deng, L.; et al. Nanoencapsulation of Cyanidin-3-O-glucoside Enhances Protection Against UVB-Induced Epidermal Damage through Regulation of p53-Mediated Apoptosis in Mice. *J. Agric. Food Chem.* **2018**, *66*, 5359–5367. [CrossRef]
62. Maurya, D.K.; Devasagayam, T.P.A. Indian medicinal herbs and ayurvedic formulations as potential radioprotectors. In *Herbal Radiomodulators: Applications in Medicine, Homeland Defence and Space*; CABI: Wallingford, UK, 2008; pp. 25–46.
63. Li, M.; Gu, M.-M.; Lang, Y.; Shi, J.; Chen, B.P.; Guan, H.; Yu, L.; Zhou, P.-K.; Shang, Z.-F. The vanillin derivative VND3207 protects intestine against radiation injury by modulating p53/NOXA signaling pathway and restoring the balance of gut microbiota. *Free Radic. Biol. Med.* **2019**, *145*, 223–236. [CrossRef]
64. Li, M.; Lang, Y.; Gu, M.-M.; Shi, J.; Chen, B.P.; Yu, L.; Zhou, P.-K.; Shang, Z.-F. Vanillin derivative VND3207 activates DNA-PKcs conferring protection against radiation-induced intestinal epithelial cells injury in vitro and in vivo. *Toxicol. Appl. Pharmacol.* **2019**, *387*, 114855. [CrossRef]
65. Sim, H.-J.; Bhattarai, G.; Lee, J.; Lee, J.-C.; Kook, S.-H. The Long-lasting Radioprotective Effect of Caffeic Acid in Mice Exposed to Total Body Irradiation by Modulating Reactive Oxygen Species Generation and Hematopoietic Stem Cell Senescence-Accompanied Long-term Residual Bone Marrow Injury. *Aging Dis.* **2019**, *10*, 1320–1327. [CrossRef] [PubMed]
66. Maurya, D.K.; Salvi, V.P.; Nair, C.K.K. Radiation protection of DNA by ferulic acid under in vitro and in vivo conditions. *Mol. Cell. Biochem.* **2005**, *280*, 209–217. [CrossRef] [PubMed]
67. Cinkilic, N.; Cetintas, S.K.; Zorlu, T.; Vatan, O.; Yilmaz, D.; Cavas, T.; Tunc, S.; Ozkan, L.; Bilaloglu, R. Radioprotection by two phenolic compounds: Chlorogenic and quinic acid, on X-ray induced DNA damage in human blood lymphocytes in vitro. *Food Chem. Toxicol.* **2013**, *53*, 359–363. [CrossRef] [PubMed]
68. Hosseinimehr, S.J.; Zakaryaee, V.; Ahmadi, A.; Akhlaghpoor, S. Radioprotective effects of chlorogenic acid against mortality induced by gamma irradiation in mice. *Methods Find. Exp. Clin. Pharmacol.* **2008**, *30*, 13–16. [CrossRef] [PubMed]
69. Taysi, S.; Demir, E.; Cinar, K.; Tarakcioglu, M. The Radioprotective Effects of Propolis and Caffeic Acid Phenethyl Ester on Radiation-Induced Oxidative/nitrosative Stress in Brain Tissue. *Free. Radic. Biol. Med.* **2016**, *100*, S111. [CrossRef]
70. Anjaly, K. Radio-Modulatory Potential of Caffeic Acid Phenethyl Ester: A Therapeutic Perspective. *Anti-Cancer Agents Med. Chem.* **2018**, *18*, 468–475. [CrossRef]
71. Xu, W.; Yang, F.; Zhang, Y.; Shen, X. Protective effects of rosmarinic acid against radiation-induced damage to the hematopoietic system in mice. *J. Radiat. Res.* **2016**, *57*, 356–362. [CrossRef]
72. Del Baño, M.J.; Castillo, J.; Benavente-García, O.; Lorente, J.; Martín-Gil, R.; Acevedo, C.; Alcaraz, M. Radioprotective−Antimutagenic Effects of Rosemary Phenolics against Chromosomal Damage Induced in Human Lymphocytes by γ-rays. *J. Agric. Food Chem.* **2006**, *54*, 2064–2068. [CrossRef]
73. Kalman, D.S.; Hewlings, S.J. The Effects of Morus alba and Acacia catechu on Quality of Life and Overall Function in Adults with Osteoarthritis of the Knee. *J. Nutr. Metab.* **2017**, *2017*, 4893104. [CrossRef]
74. Lee, J.C.; Kinniry, P.A.; Arguiri, E.; Serota, M.; Kanterakis, S.; Chatterjee, S.; Solomides, C.C.; Javvadi, P.; Koumenis, C.; Cengel, K.A.; et al. Dietary Curcumin Increases Antioxidant Defenses in Lung, Ameliorates Radiation-Induced Pulmonary Fibrosis, and Improves Survival in Mice. *Radiat. Res.* **2010**, *173*, 590–601. [CrossRef]
75. Akpolat, M.; Kanter, M.; Uzal, M.C. Protective effects of curcumin against gamma radiation-induced ileal mucosal damage. *Arch. Toxicol.* **2008**, *83*, 609–617. [CrossRef]
76. Seong, K.M.; Yu, M.; Lee, K.-S.; Park, S.; Jin, Y.W.; Min, K.-J. Curcumin Mitigates Accelerated Aging after Irradiation in Drosophila by Reducing Oxidative Stress. *BioMed Res. Int.* **2015**, *2015*, 425380. [CrossRef] [PubMed]
77. Verma, V. Relationship and interactions of curcumin with radiation therapy. *World J. Clin. Oncol.* **2016**, *7*, 275–283. [CrossRef]
78. Shabeeb, D.; Musa, A.E.; Ali, H.S.A.; Najafi, M. Curcumin Protects Against Radiotherapy-Induced Oxidative Injury to the Skin. *Drug Des. Dev. Ther.* **2020**, *14*, 3159–3163. [CrossRef]
79. Anand, P.; Kunnumakkara, A.B.; Newman, R.A.; Aggarwal, B.B. Bioavailability of Curcumin: Problems and Promises. *Mol. Pharm.* **2007**, *4*, 807–818. [CrossRef] [PubMed]
80. Nguyen, M.-H.; Pham, N.-D.; Dong, B.; Nguyen, T.-H.-N.; Bui, C.B.; Hadinoto, K. Radioprotective activity of curcumin-encapsulated liposomes against genotoxicity caused by Gamma Cobalt-60 irradiation in human blood cells. *Int. J. Radiat. Biol.* **2017**, *93*, 1267–1273. [CrossRef] [PubMed]
81. Nosrati, H.; Danafar, H.; Rezaeejam, H.; Gholipour, N.; Rahimi-Nasrabadi, M. Evaluation radioprotective effect of curcumin conjugated albumin nanoparticles. *Bioorganic Chem.* **2020**, *100*, 103891. [CrossRef]
82. Sasidharan, S.; Chen, Y.; Saravanan, D.; Sundram, K.M.; Latha, L.Y. Extraction, isolation and characterization of bioactive compounds from plants' extracts. *AJTCAM* **2011**, *8*, 1. [CrossRef] [PubMed]
83. Jagetia, G.C.; Shirwaikar, A.; Rao, S.K.; Bhilegaonkar, P.M. Evaluation of the radioprotective effect of Ageratum conyzoides Linn. extract in mice exposed to different doses of gamma radiation. *J. Pharm. Pharmacol.* **2010**, *55*, 1151–1158. [CrossRef] [PubMed]
84. Samarth, R.M.; Samarth, M. Protection against Radiation-induced Testicular Damage in Swiss Albino Mice by Mentha piperita (Linn.). *Basic Clin. Pharmacol. Toxicol.* **2009**, *104*, 329–334. [CrossRef] [PubMed]

85. Weiss, E.I.; Shaniztki, B.; Dotan, M.; Ganeshkumar, N.; Kolenbrander, P.E.; Metzger, Z. Attachment of Fusobacterium nucleatum PK1594 to mammalian cells and its coaggregation with periodontopathogenic bacteria are mediated by the same galactose-binding adhesin. *Oral Microbiol. Immunol.* **2000**, *15*, 371–377. [CrossRef] [PubMed]
86. Bump, E.A.; Brown, J. Role of glutathione in the radiation response of mammalian cells invitro and in vivo. *Pharmacol. Ther.* **1990**, *47*, 117–136. [CrossRef]
87. Murray, A.P.; Rodriguez, S.; Frontera, M.A.; Tomas, M.A.; Mulet, M.C. Antioxidant Metabolites from Limonium brasiliense (Boiss.) Kuntze. *Z. Naturforsch. C* **2004**, *59*, 477–480. [CrossRef] [PubMed]
88. Piccolella, S.; Crescente, G.; Nocera, P.; Pacifico, F.; Manti, L.; Pacifico, S. Ultrasound-assisted aqueous extraction, LC-MS/MS analysis and radiomodulating capability of autochthonous Italian sweet cherry fruits. *Food Funct.* **2018**, *9*, 1840–1849. [CrossRef] [PubMed]
89. Castillo, J.J.; Alcaraz, M.; Benavente-García, O. Antioxidant and Radioprotective Effects of Olive Leaf Extract. In *Olives and Olive Oil in Health and Disease Prevention*; Elsevier BV: Amsterdam, The Netherlands, 2010; pp. 951–958.
90. Xu, T.; Xiao, D. Oleuropein enhances radiation sensitivity of nasopharyngeal carcinoma by downregulating PDRG1 through HIF1α-repressed microRNA-519d. *J. Exp. Clin. Cancer Res.* **2017**, *36*, 3. [CrossRef] [PubMed]
91. Hsu, H.-Y.; Yang, J.-J.; Lin, C.-C. Effects of oleanolic acid and ursolic acid on inhibiting tumor growth and enhancing the recovery of hematopoietic system postirradiation in mice. *Cancer Lett.* **1997**, *111*, 7–13. [CrossRef]
92. Wang, H.; Sim, M.-K.; Loke, W.K.; Chinnathambi, A.; Alharbi, S.A.; Tang, F.R.; Sethi, G. Potential Protective Effects of Ursolic Acid against Gamma Irradiation-Induced Damage Are Mediated through the Modulation of Diverse Inflammatory Mediators. *Front. Pharmacol.* **2017**, *8*, 352. [CrossRef]
93. Chaudhary, P.; Shukla, S.; Kumar, I.P.; Namita, I.; Afrin, F.; Sharma, R.K. Radioprotective properties of apple polyphenols: An in vitro study. *Mol. Cell. Biochem.* **2006**, *288*, 37–46. [CrossRef]
94. Haggag, M.G.; Kazem, H.H. Protective Role of Emodin in Reducing the Gamma Rays Induced Hazardous Effects On The Tongue of Diabetic or Normoglycaemic Mice. *Isot. Radiat. Res.* **2013**, *45*, 359–373.
95. Kumar, M.; Samarth, R.; Kumar, M.; Selvan, S.R.; Saharan, B.; Kumar, A. Protective Effect of *Adhatoda vascia* Nees against Radiation-Induced Damage at Cellular, Biochemical and Chromosomal Levels in Swiss Albino Mice. *Evidence-Based Complement. Altern. Med.* **2007**, *4*, 343–350. [CrossRef] [PubMed]
96. Sharma, S.; Singh, M. Histological Alterations in the Spleen of Gamma-Irradiated Mice Induced by *Adhatoda vasica* Leaf Extract. *Int. J. Sci. Res.* **2016**, *5*, 1216–1219.
97. Krishna, A.; Kumar, A. Evaluation of Radioprotective Effects of Rajgira (*Amaranthus paniculatus*) Extract in Swiss Albino Mice. *J. Radiat. Res.* **2005**, *46*, 233–239. [CrossRef] [PubMed]
98. Yadav, R.K.; Bhatia, A.L.; Sisodia, R. Modulation of Radiation Induced Biochemical Changes in Testis of Swiss Albino Mice by Amaranthus Paniculatus Linn. *Asian J. Exp. Sci.* **2004**, *18*, 63–74.
99. Adamczak, A.; Ożarowski, M. Radioprotective Effects of Plants from the Lamiaceae Family. *Anti-Cancer Agents Med. Chem.* **2020**, *20*, 1–23. [CrossRef]
100. Kumar, A.; Samarth, R.M.; Yasmeen, S.; Sharma, A.; Sugahara, T.; Terado, T.; Kimura, H. Anticancer and radioprotective potentials of Mentha piperita. *BioFactors* **2004**, *22*, 87–91. [CrossRef]
101. Samarth, R.M.; Samarth, M.; Matsumoto, Y. Medicinally important aromatic plants with radioprotective activity. *Futur. Sci. OA* **2017**, *3*, FSO247. [CrossRef]
102. Chang, H.-S.; Endoh, D.; Ishida, Y.; Takahashi, H.; Ozawa, S.; Hayashi, M.; Yabuki, A.; Yamato, O. Radioprotective Effect of Alk(en)yl Thiosulfates Derived from Allium Vegetables against DNA Damage Caused by X-ray Irradiation in Cultured Cells: Antiradiation Potential of Onions and Garlic. *Sci. World J.* **2012**, *2012*, 846750. [CrossRef]
103. Bulle, S.; Reddy, V.D.; Padmavathi, P.; Maturu, P.; Varadacharyulu, N.C. Modulatory role of *Pterocarpus santalinus* against alcohol-induced liver oxidative/nitrosative damage in rats. *Biomed. Pharmacother.* **2016**, *83*, 1057–1063. [CrossRef] [PubMed]
104. Kumar, G.E.H.; Kumar, S.S.; Balaji, M.; Maurya, D.K.; Kesavulu, M. *Pterocarpus santalinus* L. extract mitigates gamma radiation-inflicted derangements in BALB/c mice by Nrf2 upregulation. *Biomed. Pharmacother.* **2021**, *141*, 111801. [CrossRef]
105. Szejk-Arendt, M.; Czubak-Prowizor, K.; Macieja, A.; Poplawski, T.; Olejnik, A.K.; Pawlaczyk-Graja, I.; Gancarz, R.; Zbikowska, H.M. Polyphenolic-polysaccharide conjugates from medicinal plants of Rosaceae/Asteraceae family protect human lymphocytes but not myeloid leukemia K562 cells against radiation-induced death. *Int. J. Biol. Macromol.* **2020**, *156*, 1445–1454. [CrossRef] [PubMed]
106. Zhao, H.; Wang, Z.; Ma, F.; Yang, X.; Cheng, C.; Yao, L. Protective Effect of Anthocyanin from *Lonicera caerulea* var. edulis on Radiation-Induced Damage in Mice. *Int. J. Mol. Sci.* **2012**, *13*, 11773–11782. [CrossRef] [PubMed]

Article

Effect of Encapsulated Beet Extracts (*Beta vulgaris*) Added to Yogurt on the Physicochemical Characteristics and Antioxidant Activity

Martha A. Flores-Mancha [1], Martha G. Ruíz-Gutiérrez [2], Rogelio Sánchez-Vega [1], Eduardo Santellano-Estrada [1] and América Chávez-Martínez [1,*]

[1] Departamento de Tecnología de Productos de Origen Animal, Facultad de Zootecnia y Ecología, Universidad Autónoma de Chihuahua, Periférico Francisco R. Almada km 1, Chihuahua 31000, CI, Mexico; 99azu.flores@gmail.com (M.A.F.-M.); rsanchezv@uach.mx (R.S.-V.); esantellano@uach.mx (E.S.-E.)

[2] Departamento de Investigación y Posgrado, Facultad de Ciencias Químicas, Universidad Autónoma de Chihuahua, Circuito Universitario s/n Campus Universitario 2, Chihuahua 31125, CI, Mexico; mruizg@uach.mx

* Correspondence: amchavez@uach.mx; Tel.: +52-614-239-8948

Abstract: Beet has been used as an ingredient for functional foods due to its high antioxidant activity, thanks to the betalains it contains. The effects of the addition of beet extract (liquid and lyophilized) on the physicochemical characteristics, color, antioxidant activity (AA), total betalains (TB), total polyphenols (TP), and total protein concentration (TPC) were evaluated on stirred yogurt. The treatments (T1-yogurt natural, T2-yogurt added with beet juice, T3-added extract of beet encapsulated with maltodextrin, and T4-yogurt added with extract of beet encapsulated with inulin) exhibited results with significant differences ($p < 0.05$). The highest TB content was observed in T2 (209.49 ± 14.91), followed by T3 (18.65 ± 1.01) and later T4 (12.96 ± 0.55). The highest AA was observed on T2 after 14 days (ABTS$^{\cdot}$ 0.819 mM TE/100 g and DPPH$^{\cdot}$ 0.343 mM TE/100 g), and the lowest was found on T1 at day 14 (ABTS$^{\cdot}$ 0.526 mM TE/100 g and DPPH$^{\cdot}$ 0.094 mM TE/100 g). A high content of TP was observed (7.13 to 9.79 mg GAE/g). The TPC varied between 11.38 to 12.56 μg/mL. The addition of beet extract significantly increased AA in yogurt, betalains being the main compounds responsible for that bioactivity.

Keywords: yogurt; betalains; encapsulation; lyophilization; antioxidant activity; polyphenols

1. Introduction

Dairy derivatives are the main segment among functional foods, representing around 74% of all functional products [1]. Yogurt is a functional dairy food that is obtained from the fermentation of milk with lactic acid bacteria (LAB); it is consumed around the world and is traditionally recognized as a healthy food due to its own characteristics [2], which can be improved by incorporating natural ingredients such as fruit extracts [3], tea [4], olive leaf extract [5], paprika juice [6], coriander and cumin seeds [7], etc. It has been reported that the high oxidative stability of yogurt is due to the peptides released during the fermentation carried out by LAB [8]. According to the latest studies, the use of various additives and fruit, in yogurt production, has a significant influence its quality [9,10]. Thus, the addition of natural food materials increases the nutritive value of yogurt. The fortification of yogurt using different bioactive compounds has been studied by different authors [4–7,11,12]; encapsulation by extrusion can significantly extend the stability of natural β-carotene with potential use as a functional ingredient in yogurt [13]. The addition of hibiscus flower essential oil can also be used to increase the antioxidant capacity of yogurt due to anthocyanin content [12]. Green, white and black tea can be used successfully to enhance the antioxidant properties of yogurt and provide sustained antioxidants during

storage [4]. On the other hand, the addition of carotenoids in a fermented-maize product, similar to yogurt, can induce a cholesterol lowering effect [14].

Beet (*Beta vulgaris* sp.) represent the main commercial source of betalains [15–17], in addition to containing polyphenols, both being powerful antioxidants. In recent decades, the functional properties that these compounds confer on health have been studied, among which their antioxidant (AA), antidiabetic, anti-inflammatory and anticancer activity stand out [18–25]. Beet extract (betanin) is an approved color under the code 73.40 by the US Food and Drug Administration [17,26] and by the European Union designated with the number E162 [22,27]. Betalains and polyphenols are unstable in the presence of light [28], high temperatures [29], alkaline pH [30,31], enzymatic activity [32], and presence of oxygen and/or metals [33,34]. Due to their low stability, their use in food has been restricted [3,35–42]. However, it has been shown that the encapsulation of betalains and polyphenols in different edible matrices (maltodextrin and inulin) can increase the stability of these compounds and therefore retain their antioxidant and anti-radical activities [21,43–49]. Therefore, the objective of this work was to evaluate the effect of adding beet extract (*Beta vulgaris* sp.) encapsulated with different carrier agents on the physicochemical characteristics, color, betalain and polyphenol content, and antioxidant capacity of yogurt during shelf life. The treatments were as follows: T1-natural yogurt as control, T2-yogurt added with beet juice, T3-added extract of beet encapsulated with maltodextrin, and T4-yogurt added with extract of beet encapsulated with inulin.

2. Results and Discussion

2.1. Physicochemical Analysis

The treatments added with encapsulated beet extract (T3 and T4) presented higher protein content (4.88 ± 0.22 and 4.84 ± 0.48, respectively), showing a significant difference ($p < 0.05$), compared to T1 (3.87 ± 0.31) and with T2 (3.88 ± 0.36) ($p < 0.05$). Regarding the percentage of fat, treatments T3 (5.49 ± 0.81) and T4 (5.36 ± 0.32) had a higher content in contrast to T1 (5.09 ± 0.45) and T2 (5.02 ± 0.86) (Table 1), although in this determination, there was no significant difference ($p > 0.05$) between the treatments. Other studies indicate that yogurts added with rubas (*Ullucus tuberosus*) concentrate, curuba (*Passiflora mollissima* Bailey) extract, chocolate, and oats also showed a significant increase in protein and fat content [50–53]. The moisture percentage showed a significant difference ($p < 0.05$) in the four treatments, this behavior could be attributed to the fact that the beet extracts were in different forms (juice or powders), since in treatments 3 and 4 added with powders, the moisture content was lower (78.10 ± 0.17 and 78.60 ± 0.33) than in T1 (78.94 ± 0.15), likewise, the moisture percentage of T2 was higher (79.92 ± 0.12) than T1. Regarding the percentage of ash, there was no significant difference ($p > 0.05$) between the treatments.

Table 1. Physicochemical analysis of yogurt added with beet extracts (*Beta vulgaris* sp.).

Treatment	% Fat	% Protein	% Moisture	% Ashes
T1	5.09 ± 0.45 [a]	3.87 ± 0.31 [b]	78.94 ± 0.15 [b]	0.18 ± 0.01 [a]
T2	5.02 ± 0.86 [a]	3.88 ± 0.36 [b]	79.92 ± 0.12 [a]	0.18 ± 0.01 [a]
T3	5.49 ± 0.81 [a]	4.88 ± 0.22 [a]	78.10 ± 0.17 [d]	0.18 ± 0.02 [a]
T4	5.36 ± 0.32 [a]	4.84 ± 0.48 [a]	78.60 ± 0.33 [c]	0.19 ± 0.02 [a]

T1 = Natural yogurt (Y), T2 = Y added with beet juice, T3 = Y added with encapsulated extract with Maltodextrin and T4 = Y added with encapsulated extract with Inulin. [a,b,c,d] Values with different superscript between rows indicate significant statistical difference between treatments ($p < 0.05$).

2.2. pH and Syneresis (SYN)

All treatments presented a decrease in pH over time ($p < 0.05$). These values were 4.13 ± 0.052 on day 1, 4.04 ± 0.030 on day 7, and 3.95 ± 0.010 on day 14 (Figure 1), and that decrease in pH could be attributed to the microbial activity of the lactic acid bacteria present in yogurt [52,54]. The degree of SYN varied from 18.01 to 35.32%, presenting significant differences ($p < 0.05$) between treatments and days of storage (Figure 2).

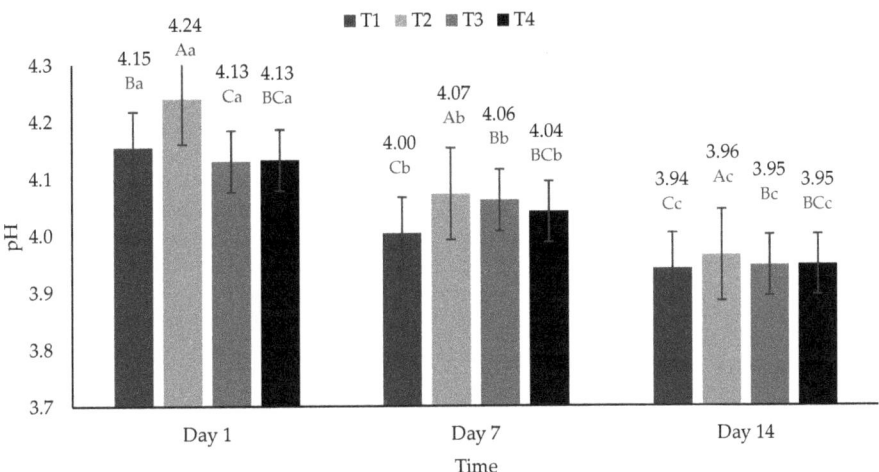

Figure 1. Hydrogen potential (pH) of yogurt added with beet extracts. T1 = Natural yogurt (Y), T2 = Y added with beet juice, T3 = Y added with extract encapsulated with Maltodextrin, and T4 = Y added with extract encapsulated with Inulin. ^{A,B,C} Values with different superscript indicate significant statistical difference between treatments ($p < 0.05$). ^{a,b,c} Values with a different superscript indicate a statistically significant difference over time ($p < 0.05$).

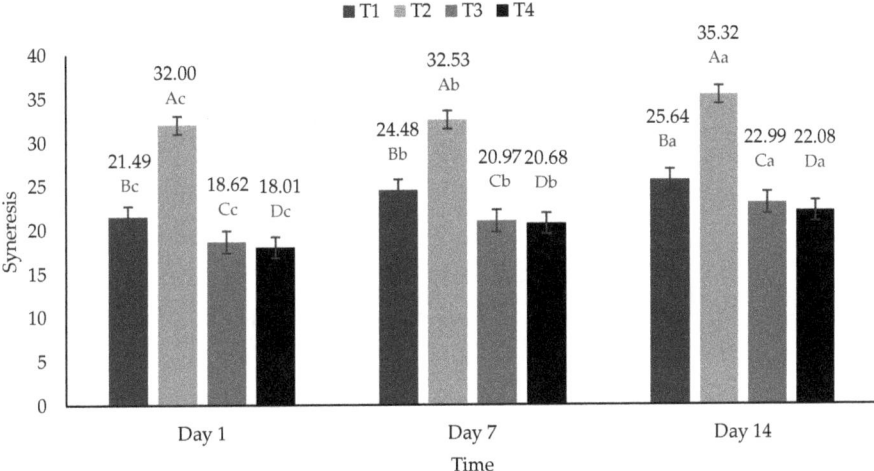

Figure 2. Percentage of syneresis of yogurt added with beet extracts. T1 = Natural yogurt (Y), T2 = Y added with beet juice, T3 = Y added with extract encapsulated with Maltodextrin, and T4 = Y added with extract encapsulated with Inulin. ^{A,B,C,D} Values with different superscript indicate significant statistical difference between treatments ($p < 0.05$). ^{a,b,c} Values with different superscript indicate statistical difference significant over time ($p < 0.05$).

These differences between treatments could be attributed to the fact that the components of the powders, specifically maltodextrin and inulin, favored water retention, because they contribute to the mesh effect in the three-dimensional network of the gel formed in the yogurt [55]. In addition, syneresis increased in all treatments during storage, possibly due to the loss of stability of the yogurt components [56]. This behavior is also due to the decrease in pH during storage, since it can have a contraction effect in the casein micelle matrix causing greater elimination of whey [57].

2.3. Color

The values of the color parameters L*, a* and b* can be observed in Table 2. The values of L* decreased significantly for T1, T3, and T4, although this value increased in T2 from 43.70 at day 1 to 50.88 at day 14 ($p < 0.05$). In relation to the parameter a* (Table 2), this was reduced in T3 after storage, and this trend was reported [58] in microcapsules of beets using M as a vehicle (day 0 to a* = 8.39, day 7 to a* = 1.71); however, the a* values for T1, T2, and T4 showed an increase over time ($p < 0.05$). In general, the b* values also showed an increase over time ($p < 0.05$). The degradation of BC leads to the formation of compounds with a yellow color and, this is reflected in the increase in the b* parameter [38]. Color saturation (Chroma) and hue (Hue angle) indicated differences ($p < 0.05$) between treatments and over time; however, treatments added with beet powder (T3 and T4) did not show differences ($p > 0.05$). The color change (ΔE) between treatments and days was also different ($p < 0.05$).

Table 2. Color values L*, a*, b*, C*, h* and ΔE of yogurt added with liquid extract and lyophilized of beet (*Beta vulgaris* sp.).

Parameter	Day	T1	T2	T3	T4
L*	1	96.354 ± 2.543 Aa	43.700 ± 1.376 Db	79.330 ± 1.441 Cb	83.678 ± 0.659 Ba
	7	88.170 ± 1.294 Ab	41.780 ± 1.828 Dc	82.079 ± 2.189 Ca	80.719 ± 0.503 Bc
	14	87.528 ± 2.448 Ac	50.884 ± 1.742 Da	76.927 ± 2.739 Cc	82.225 ± 1.033 Bb
a*	1	−3.522 ± 0.124 Cb	32.763 ± 1.729 Ab	22.705 ± 0.998 Ba	21.076 ± 0.688 Bb
	7	−3.272 ± 0.232 Ca	30.701 ± 1.254 Ab	21.581 ± 0.601 Bb	22.822 ± 0.051 Ba
	14	−3.266 ± 0.106 Ca	39.563 ± 3.047 Aa	22.137 ± 0.816 Bb	22.118 ± 0.732 Ba
b*	1	8.612 ± 0.303 Ab	2.174 ± 0.209 Bb	0.617 ± 0.097 Db	1.574 ± 0.380 Ca
	7	9.533 ± 0.281 Aa	3.254 ± 0.077 Ba	1.344 ± 0.248 Da	0.696 ± 0.024 Cb
	14	9.037 ± 0.132 Aa	3.527 ± 0.414 Ba	1.182 ± 0.117 Da	1.169 ± 0.127 Ca
C*	1	9.305 ± 0.278 Cb	32.835 ± 1.735 Ab	22.713 ± 0.998 Ba	21.135 ± 0.673 Bb
	7	10.078 ± 0.314 Ca	30.873 ± 1.241 Ab	21.623 ± 0.595 Bb	22.833 ± 0.503 Ba
	14	9.609 ± 0.134 Cb	39.720 ± 3.067 Aa	22.169 ± 0.819 Bb	22.149 ± 0.732 Ba
h*	1	2.445 ± 0.134 Cc	0.066 ± 0.005 Ac	0.027 ± 0.004 Bc	0.075 ± 0.019 Ba
	7	2.913 ± 0.168 Ca	0.106 ± 0.006 Aa	0.062 ± 0.012 Ba	0.030 ± 0.002 Bc
	14	2.767 ± 0.091 Cb	0.089 ± 0.006 Ab	0.053 ± 0.005 Bb	0.053 ± 0.006 Bb
ΔE	1	-	64.268 ± 1.584 Aa	32.273 ± 1.803 Ba	28.553 ± 1.059 Ca
	7	-	57.841 ± 1.971 Ab	26.867 ± 0.739 Bb	28.539 ± 0.690 Ca
	14	-	56.635 ± 2.469 Ab	28.626 ± 1.782 Bb	27.099 ± 1.012 Cb

T1 = Natural yogurt (Y), T2 = Y added with beet juice, T3 = Y added with extract encapsulated with Maltodextrin and T4 = Y added with extract encapsulated with Inulin. L* = luminosity, a* = green (-) and red (+), b* = blue (-) and yellow (+), C* = chroma (color saturation), h* = Hue angle (hue), and ΔE = color difference. A,B,C,D Values in columns with different superscript indicate significant statistical difference between treatments ($p < 0.05$). a,b,c Values in rows with different superscript indicate significant statistical difference over time ($p < 0.05$).

2.4. Antioxidant Activity (AA)

Statistically significant differences ($p < 0.05$) were found for the AA variable between the treatments ($p < 0.05$) in ABTS (Figure 3) and DPPH (Figure 4). The highest AA was observed on T2 after 14 days (ABTS 0.819 mM TE/100 g and DPPH 0.343 mM TE/100 g), and the lowest AA was found on T1 at day 14 (ABTS 0.526 mM TE/100 g and DPPH 0.094 mM TE/100 g). Among the treatments added with encapsulated extract, the highest AA was found in T3 on day 1 (ABTS 0.667 mM TE/100 g and DPPH 0.197 mM TE/100 g). On the other hand, T4 presented higher AA on day 7 (ABTS 0.644 mM TE/100 g and DPPH 0.145 mM TE/100 g); it is important to highlight that for this treatment, the AA was higher on day 14 (ABTS 0.613 mM TE/100 g and DPPH 0.136 mM TE/100 g) even than on day 1 (ABTS 0.604 mM TE/100 g and DPPH 0.124 mM TE/100 g) (Figure 4). Likewise, a decrease on AA was reported previously, followed by a 10 day stability period and an increase in AA by day 14 [59]. This behavior could be due to compounds with high antioxidant activity that were formed or released during storage due to the adsorption of water by encapsulation or to the interaction of some components in the stored sample with oxygen or other components of the sample [60]. In addition, it has been reported that a longer storage time and a greater water activity produce greater antioxidant activity to extinguish radicals [59,61].

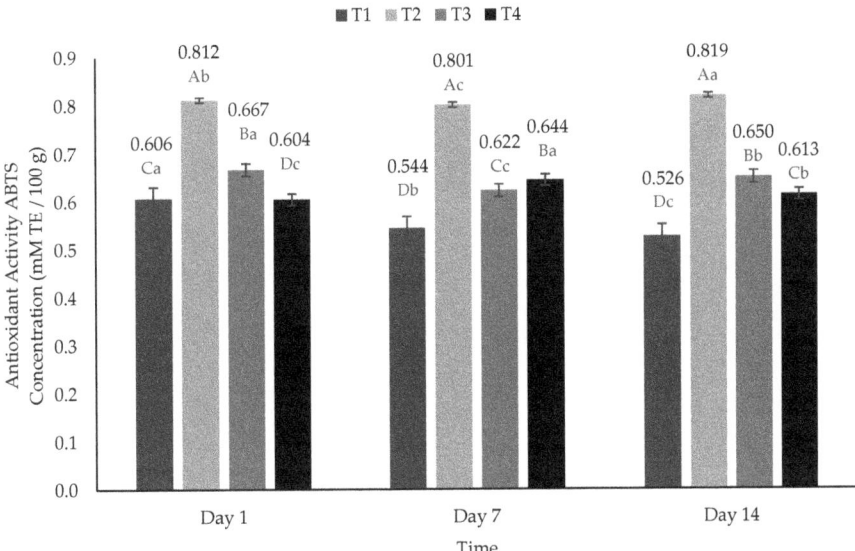

Figure 3. Antioxidant activity (ABTS· method) of yogurt added with beet extract. T1 = Natural yogurt (Y), T2 = Y added with beet juice, T3 = Y added with extract encapsulated with Maltodextrin, and T4 = Y added with extract encapsulated with Inulin. ^{A,B,C,D} Values with different superscript indicate significant statistical difference between treatments ($p < 0.05$). ^{a,b,c} Values with a different superscript indicate significant statistical difference over time ($p < 0.05$).

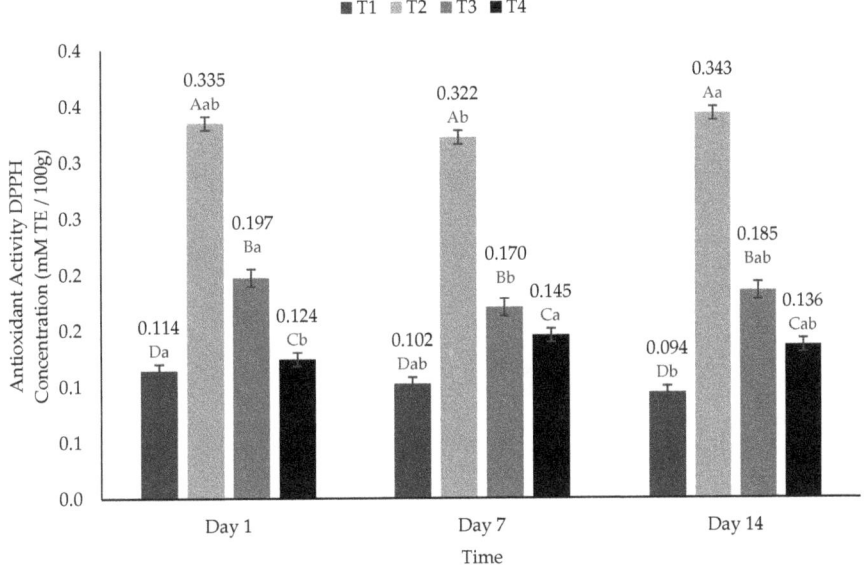

Figure 4. Antioxidant activity (DPPH· method) of yogurt added with beet extract. T1 = Natural yogurt (Y), T2 = Y added with beet juice, T3 = Y added with extract encapsulated with Maltodextrin, and T4 = Y added with extract encapsulated with Inulin. ^{A,B,C,D} Values with different superscript indicate significant statistical difference between treatments ($p < 0.05$). ^{a,b} Values with a different superscript indicate significant statistical difference over time ($p < 0.05$).

2.5. Total Polyphenols (TP)

A high content of TP was observed from 7.13 to 9.79 mg of GAE/g, (Figure 5) referring to that reported previously 4.88 mg of GAE/g [58]. In cactus pear (*Opuntia ficus* indica) powders values of 14.67 mg of GAE/g have been reported [62]. The concentration of TP showed a behavior reported previously [48], where report recoveries and polyphenol formation of more than 100% with respect to the initial time, as a consequence of the hydrolysis of the beet polyphenol conjugates during storage [48]. Storage with high water activity facilitates oxidation reactions, causing phenolic compounds to react with oxygen and produce an enriched medium that has strong radical elimination and reduction properties [59]; it is worth mentioning that these studies were carried out only on powders.

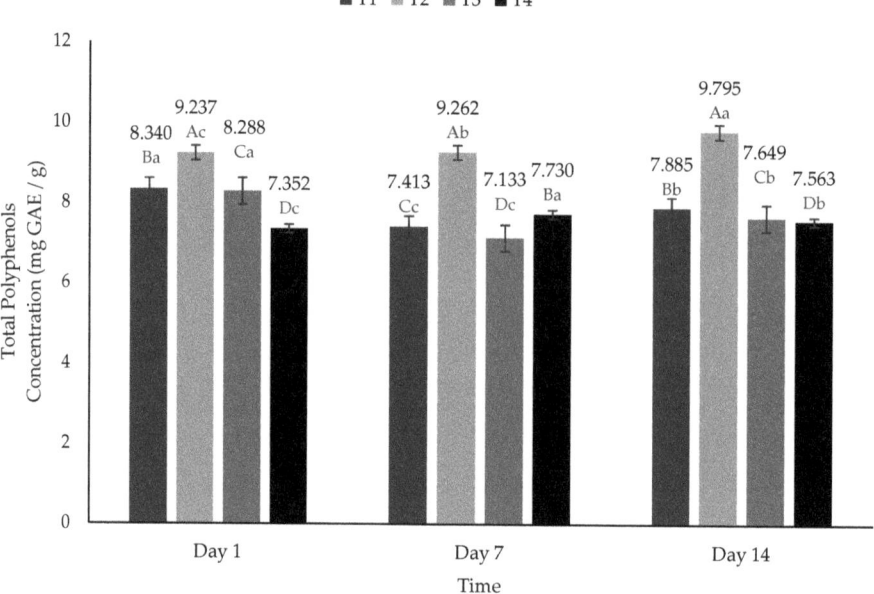

Figure 5. Total polyphenol content of yogurt added with beet extract. T1 = Natural yogurt (Y), T2 = Y added with beet juice, T3 = Y added with extract encapsulated with Maltodextrin, and T4 = Y added with extract encapsulated with Inulin. A,B,C,D Values with different superscript indicate significant statistical difference between treatments ($p < 0.05$). a,b,c Values with a different superscript indicate significant statistical difference over time ($p < 0.05$).

2.6. Total Betalains (TB)

Several investigations have been carried out on the use of encapsulating agents to prevent the degradation of betalains [35,58,63,64]; however, the most of the stability studies were conducted on powder samples. In the present study, the powder was added to a non-Newtonian fluid. The stability of the extracts was evaluated in the yogurts in terms of TB content (Figure 6), during 14 days of storage at 4 °C and in absence of light. The highest pigment content was observed in T2 after 14 days (243.20 mg/100 g); statistically significant differences ($p < 0.05$) in TB content were observed after 7 (191.65 mg/100 g) and 14 days of storage compared to its content on day 1 (209.40 mg/100 g). The TB content increased with the fermentation time, and that event could be due to acid hydrolysis and the bioconversion of condensed phenols [29,65]. In addition, during fermentation, microbial activity caused an intense release of betalains responsible for the increase in TB content [61]. Of the treatments added with encapsulated extract (T3 and T4), the highest TB content was found in T3 on day 1 (18,652 mg/100 g), and the highest TB content for T4 was on day 7 (18,024 mg/100 g). However, while T3 showed a decrease in TB on day

7 (10.24 mg/100 g) and subsequently an increase on day 14 (15.06 mg/100 g), T4 had an increase in TB on day 7 (18.02 mg/100 g) and on day 14 a decrease (15.26 mg/100 g). However, it is important to note that on day 14 the quantification of TB in T4 was higher, even than on day 1 (12.96 mg/100 g) (Figure 6).

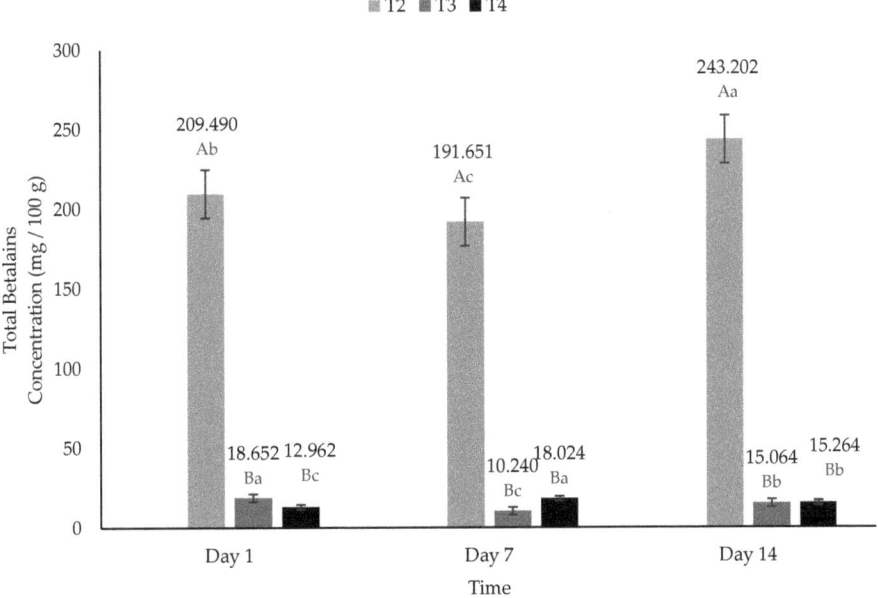

Figure 6. Total betalain content of yogurt added with beet extract. T2 = Y added with beet juice, T3 = Y added with extract encapsulated with Maltodextrin, and T4 = Y added with extract encapsulated with Inulin. [A,B] Values with different superscript indicate significant statistical difference between treatments ($p < 0.05$). [a,b,c] Values with a different superscript indicate significant statistical difference over time ($p < 0.05$).

Encapsulation with maltodextrin was effective in stabilizing betalains after 14 days of storage [66]. Unlike what was reported by Omae et al. [46], in this study, the addition of inulin did increase the stability of BC during storage, and a linear relationship was not observed from the graphs of TB content versus time. According to previous studies [28,29,67,68], the degradation of BC is the result of a hydrolysis reaction that produces cyclo-DOPA-D-glucoside (CDG) and AB; this degradation reaction is reversible, and involves a condensation of the Schiff's base of the CDG amine with the aldehyde of betalamic acid (BA) [69]. On the other hand, the degradation of betalamic pigments derived from beets was reported [46,58,59,63].

Beets have been reported to have higher levels of betacianins (BC) than betaxanthins (BX) [63,70–72]. According to the above, in this investigation, a higher content of BC was found compared to BX. The contents of BC (6.22 to 11.809 mg/100 g) and BX (4.01 to 6.85 mg/100 g) (Figures 7 and 8) in the treatments added with encapsulated beet extract (T3 and T4) in this study were lower than those reported by Janiszewska [63] and higher than those reported by Ravichandran et al. [72] (118.0 mg/100 g and 3.5 mg/100 g, respectively).

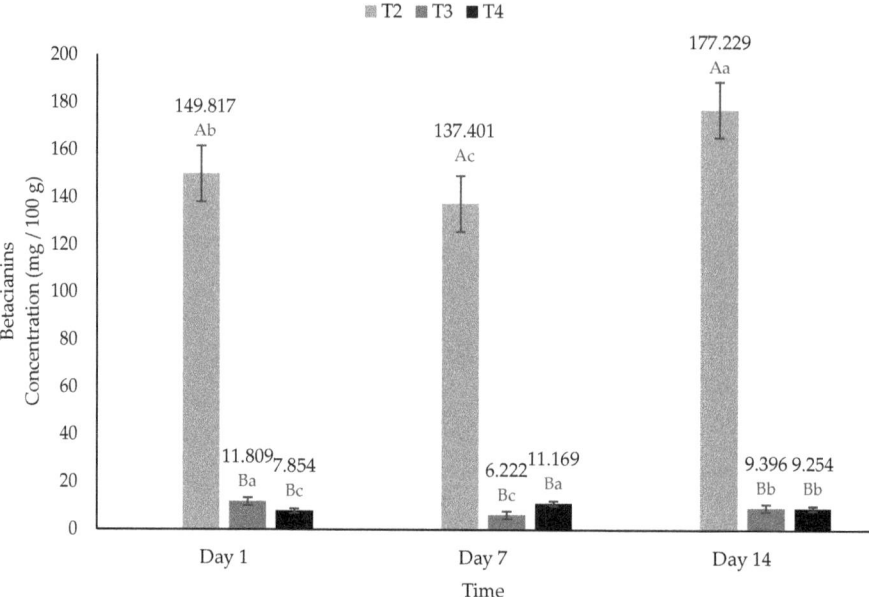

Figure 7. Betacyanins content of yogurt added with beet extract. T2 = Y added with beet juice, T3 = Y added with extract encapsulated with Maltodextrin, and T4 = Y added with extract encapsulated with Inulin. ^{A,B} Values with different superscript indicate significant statistical difference between treatments ($p < 0.05$). ^{a,b,c} Values with a different superscript indicate significant statistical difference over time ($p < 0.05$).

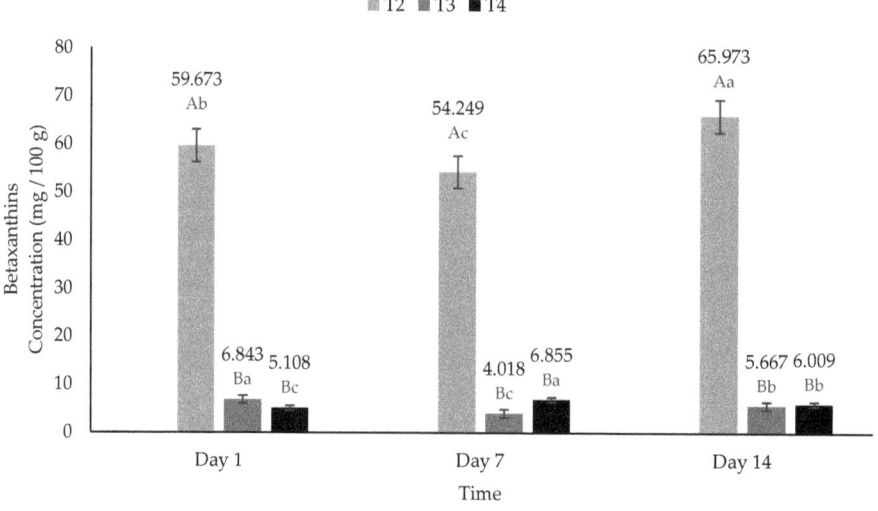

Figure 8. Betaxanthins content of yogurt added with beet extract. T2 = Y added with beet juice, T3 = Y added with extract encapsulated with Maltodextrin, and T4 = Y added with extract encapsulated with Inulin. ^{A,B} Values with different superscript indicate significant statistical difference between treatments ($p < 0.05$). ^{a,b,c} Values with a different superscript indicate significant statistical difference over time ($p < 0.05$).

2.7. Total Protein Concentration (TPC)

The TPC (Figure 9) ranged from 11.385 to 12.568 µg/mL. However, they had a statistically significant difference ($p < 0.05$) between treatments and over time. This behavior is probably due to the extracellular proteins secreted by bacteria during the fermentation time [73]. Furthermore, the increase in the total content of betalains has a positive correlation regarding the total concentration of proteins, due to the fact that they are nitrogenous compounds [74].

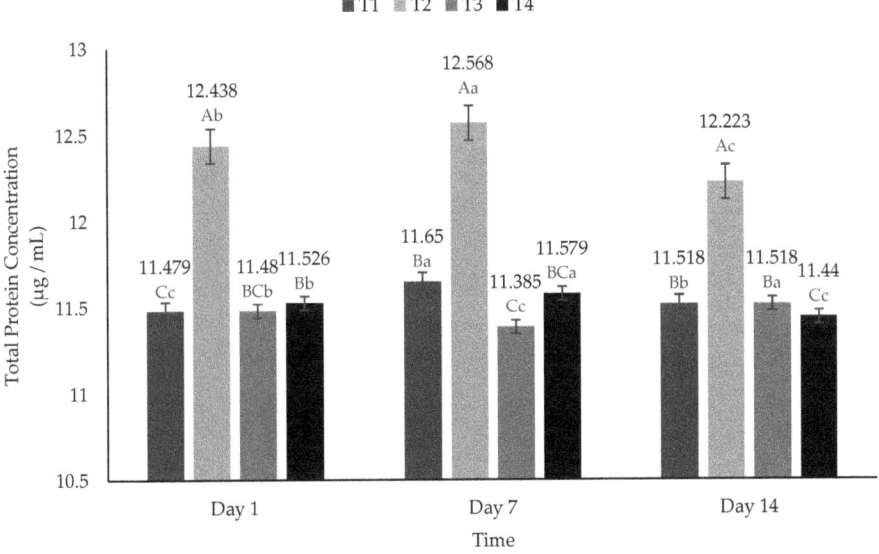

Figure 9. Total protein concentration of yogurt added with beet extract. T1 = Natural yogurt (Y), T2 = Y added with beet juice, T3 = Y added with extract encapsulated with Maltodextrin, and T4 = Y added with extract encapsulated with Inulin. [A,B,C] Values with different superscript indicate significant statistical difference between treatments ($p < 0.05$). [a,b,c] Values with a different superscript indicate significant statistical difference over time ($p < 0.05$).

2.8. Correlation between Variables

Table 3 shows the correlation coefficient (r) and level of significance between the response variables of the yogurts. A strong negative correlation ($r = -0.9410$, $p < 0.0001$) was found between L* and TB and a positive correlation ($r = 0.7054$, $p < 0.0001$) between the value of a* and TB. This indicates that the higher the TB content, the lower the luminosity and the higher the concentration of red color. These correlations have already been reported in prickly pear pulp micro particles [48,75,76] and in prickly pear pulp coloring powders [44]. Like Pitalua et al. [59], observed a strong positive correlation ($r = 0.8615$, $p < 0.0001$) between TP and TB. It is important to highlight that betalains are the main polyphenols present in beets [58]; therefore, a positive association between both variables was expected. In addition, positive correlations were obtained ($r = 0.9994$, $p < 0.0001$), ($r = 0.9960$, $p < 0.0001$) between TB and BC and between TB and BX, respectively. On the other hand, a high positive correlation ($r = 0.7946$, $p < 0.0001$) was found between TB and AA, which could be due to the presence of betalamic acid, since it acts as an antioxidant agent. However, if the acid molecule does not have hydroxyl groups, betalamic acid does not have antioxidant capacity [64]. Likewise, a positive correlation was found (ABTS$^{\cdot}$ $r = 0.8263$ and DPPH$^{\cdot}$ $r = 0.8002$, $p < 0.0001$) between AA and TP. The increase in AA during storage can occur because in these periods, new phenolic compounds capable of extinguishing radicals are released or formed [77]. A positive correlation was found between TPC and AA (ABTS$^{\cdot}$ $r = 0.8022$ and DPPH$^{\cdot}$ $r = 0.8126$, $p < 0.0001$) and TP ($r = 0.7917$, $p < 0.0001$)

and TB (r = 0.8822, $p < 0.0001$). These correlations could be to compounds with high AA that were formed or released during storage, thanks to the adsorption of water [60], and it can be observed despite the presence of betalains [64]. Storage with high water activity facilitates hydrolysis, causing phenolic compounds to react with oxygen and produce an enriched medium that has strong radical elimination and reduction properties [65,73].

Table 3. Correlation between the response variables of yogurt added with liquid extract and freeze-dried beet extracts (*Beta vulgaris* sp.).

	DPPH·	TP	TB	BC	BX	TPC	L*	a*	b*	pH	SYN
ABTS·	0.9652	0.8263	0.9281	0.9231	0.9361	0.8022	−0.9306	0.8322	−0.4192	0.3101	0.7143
p-value	<0.0001	<0.0001	<0.0001	<0.0001	<0.0001	<0.0001	<0.0001	<0.0001	<0.0001	<0.0001	<0.0001
DPPH·		0.8002	0.9395	0.9355	0.9447	0.8128	−0.9463	0.8017	−0.3518	0.2469	0.7660
p-value		<0.0001	<0.0001	<0.0001	<0.0001	<0.0001	<0.0001	<0.0001	<0.0001	<0.0001	<0.0001
TP			0.8615	0.8621	0.8552	0.7917	−0.7921	0.4929	0.0126	0.2085	0.7890
p-value			<0.0001	<0.0001	<0.0001	<0.0001	<0.0001	<0.0001	0.8217	0.0002	<0.0001
TB				0.9994	0.9960	0.8822	−0.9410	0.7054	−0.1638	0.2006	0.8773
p-value				<0.0001	<0.0001	<0.0001	<0.0001	<0.0001	0.0031	0.0003	<0.0001
BC					0.9924	0.8802	−0.9355	0.6979	−0.1525	0.1949	0.8802
p-value					<0.0001	<0.0001	<0.0001	<0.0001	0.0059	0.0004	<0.0001
BX						0.8827	−0.9502	0.7214	−0.1927	0.2147	0.8645
p-value						<0.0001	<0.0001	<0.0001	0.0005	<0.0001	<0.0001
TPC							−0.8939	0.5308	−0.0260	0.2972	0.8480
p-value							<0.0001	<0.0001	0.6416	<0.0001	<0.0001
L*								−0.7702	0.3215	−0.2102	−0.8032
p-value								<0.0001	<0.0001	0.0001	<0.0001
a*									−0.7934	0.1319	0.4245
p-value									<0.0001	0.0175	<0.0001
b*										−0.1343	0.1679
p-value										0.0156	0.0024
pH											−0.0415
p-value											0.4562

ABTS· and DPPH· = Antioxidant activity, TP = Total polyphenols, TB = Total betalains, BC = Betacyanins, BX = Betaxanthins, TPC = Total protein concentration, L* = Luminosity, a* (tendency to red), b* (tendency to green), and SYN = Syneresis. *p*-value = Level of significance of the correlation.

3. Materials and Methods

3.1. Materials

Between March and July 2018, beet, whole milk (LALA®, Gomez Palacio, Dgo., Mexico), whole cream (LALA®, Gomez Palacio, Dgo., Mexico), powdered milk (NIDO®, Ocotlan, Mex., Mexico), table sugar, and natural yogurt (LALA®, Gomez Palacio, Dgo., Mexico) were purchased in a local market in Chihuahua, Mexico. Maltodextrin (DE 10) and agave inulin were purchased from Sigma-Aldrich (St. Louis, MO, USA).

3.2. Reagents

All other reagents used were analytical grade. Reagents 2,2'-azino-bis (3-ethylbenzothiazoline-6-sulfonic acid) (ABTS), 2,2-diphenyl-1-picrylhydrazyl (DPPH), Trolox, ammonium salt, potassium persulfate, Folin–Ciocalteu, sodium carbonate, gallic acid, trichloroacetic acid, citrate, and phosphate standard reagent were analytical grade and obtained from Sigma-Aldrich (St. Louis, MO, USA). J. T. Baker provided high-performance liquid chromatography (HPLC)-grade methanol and HPLC-grade water (Mexico City, Mexico). A deionizer was used to obtain deionized water (Barnstead, Thermo Scientific, Waltham, MA, USA).

3.3. Yogurt Elaboration

To elaborate yogurts, milk was heated to 95 °C for five minutes, then cooled to 45 °C, and then, powdered milk (8%), cream (3%), sugar (3%), and the inoculum (10%) were added. Next, yogurts were incubated ((Thermo Fisher Scientific, Model 3721, USA) at 42 °C for approximately 4:30 h, until reaching a pH of 4.2. Once this acidity was reached, the yogurt was shaken and stored at 4 °C for 12 h. Subsequently, the beet extracts (liquid or lyophilized) were added to the yogurt; a shake was made, and finally, these were stored at 4 °C for 14 days.

3.4. Preparation of Beet Extracts

A household extractor (Cold Press 900W, Breville, Sydney, Australia) was used to extract the juice from the beets. The encapsulating agents (Maltodextrin-M and Inulin-I) were added to the red beet juice at room temperature at an amount of 30 g of dry matter per 100 mL, as described by Antigo et al. [58]. Next, the mixtures were homogenized (10 min) with a dispersion tool (Vortex-Ultra-Turrax IKA T18 basic) (S18N-19G, IKA Works Inc., Wilmington, NC, USA) and frozen for 48 h at 20 °C previously to freeze-dried (4 days at -85 °C and 0.035 mbar pressure) in a Labconco Niro Mobile Minor Freeze-dryer. Finally, pure beet (B), beet extract encapsulated with maltodextrin (M), and beet extract encapsulated with inulin (I) were kept at 9 ± 3 °C for further analysis. The characterization of these extracts has been reported previously [42].

3.5. Treatments

Four treatments with three repetitions were elaborated under a completely random statistical model. The description of the treatments was based on the addition of beet extracts (liquid or lyophilized and encapsulated). Control treatment was yogurt without the addition of beet extract (T1), yogurt added with 100 mL of liquid beet extract per liter (T2), yogurt added with 600 mg of beet extract encapsulated with maltodextrin per liter (T3), and yogurt added with 600 mg of beet extract encapsulated with inulin per liter (T4). Treatments were kept in refrigeration (4 °C) for 14 days, taking samples on days 1, 7, and 14 to evaluate color, pH, syneresis, betalains, and polyphenols content and antioxidant activity. On day 1, physicochemical analysis (fat, protein, moisture, and ash content) of yogurts were determined.

3.5.1. Physicochemical Analysis

The chemical composition of the yogurt samples corresponding to day 1. Moisture, ash, fat, and protein content were determined in triplicate according to the methods established by the AOAC (926.08, 935.42, 989.05, and 991.20, respectively).

3.5.2. pH and Syneresis

The pH was measured in 20 g of each sample with a digital potentiometer (Orion Versa Star) previously calibrated and syneresis was evaluated using the method described by Özturk and Óner [78]; 10 g of yogurt were taken (4 ± 1 °C), placed in Corning UHP tubes (50 mL), and centrifuged (Avanti Model J-26 XPI Beckman Coulter® centrifuge, USA) at $2800 \times g$ for 10 min, at 4 ± 1 °C. The supernatant was weighed and expressed as a percentage in relation to the initial weight of the yogurt.

3.5.3. Color

Color was measured with a Model CR-410 colorimeter (Konica Minolta®, Japan) evaluating the L*, a* and b* parameters. Where L* is an indicator of luminosity (from black to white), a* is an indicator that goes from green (negative values) to red (positive values), and b* indicates shades from blue (negative values) to yellow (positive values). The samples were placed in plastic containers with a capacity of 200 mL. The determinations were made in triplicate, and the L*, a*, and b* values were used to calculate Chroma* (C*), Hue angle (h°), and ΔE.

The Chroma value indicates the intensity or color saturation and was determined using the following equation [62]:

$$Chroma = (a^{*2} + b^{*2})^{1/2} \quad (1)$$

Hue angle refers to hues and can range from 0° (pure red), 90° (pure yellow), 180° (pure green), and 270° (pure blue). This parameter was calculated using the following equation [62]:

$$Hue° = arctan(b^* + a^*) \quad (2)$$

For the yogurt color difference (represented as ΔE^*) in each of the treatments, the average of the readings of the parameters L*, a* and b* was used using the following equation, and the control treatment was the reference of the readings [79]:

$$\Delta E = \sqrt{(L_s - L_c)^2 + (ah_s - ah_c)^2 + (bh_s - bh_c)^2} \quad (3)$$

where L_c, ah_c, and bh_c = control parameters and L_s, ah_s, and bh_s = parameters for the different treatments.

3.5.4. Yogurt Aqueous Extract

To obtain the aqueous extract, the methodology reported by Torres-Llanez et al. [80] was applied, with the modification (or elimination) of the preparation or conditioning of the sample. Briefly, 50 mL of each treatment was placed in plastic tubes (for centrifuge) and centrifuged (Avanti® J-26 XPI. Beckman Coluter®, USA) at 10,000× g for 30 min at 4 °C. The supernatant (aqueous extract) was filtered with Whatman ™ filter paper (GE Healthcare, UK) with 125mm pore diameter. Said extract was centrifuged again with the conditions mentioned above. The supernatants were filtered through a 0.22 μm pore polyethylene filters for the determination of antioxidant capacity and polyphenol content (Millipore Corp., Bedford, MA, USA). To determine betalain content, the extract was filtered on 0.45 μm pore paper (Millipore Corp., Bedford, MA, USA). All extracts were stored at −20 °C until analysis.

3.5.5. Antioxidant Activity

Antioxidant activity by the ABTS (2,2′-azino-bis (3-ethylbenzothiazoline-6-sulfonic acid)) methodology was conducted according to Thaipong et al. [81]. First, an ABTS 7.4 mM solution was elaborated (38.8 mg of crystallized ammonium salt of ABTS in 10 mL of distilled water). Next, a solution 2.6 mM of potassium persulfate was prepared (6.6 mg in 10 mL with distilled water). Then, these two solutions were then combined and let to stand at room temperature for 12 h in the dark. To make the ABTS working solution, 1 mL of ABTS free radical solution was added to 60 mL of methanol to get an absorbance of 1.1 + 0.02. After that, in a 3 mL plastic cell, 150 μL of the standard (Trolox) or sample and 2850 μL of ABTS working solution were placed and let to stand for 2 h at room temperature in the dark; then, the absorbance was measured with a UV spectrophotometer at 734 nm (UV–1800. Shimadzu, Japan). Measurements were conducted three times. Antioxidant activity was given as equivalent mM Trolox (mM TE/100 g). The resulting absorbance was then put into the regression equation (y = −1.0726x + 0.9863; r^2 = 0.9967) derived from the Trolox calibration curve.

Antioxidant activity by the DPPH (2,2-diphenyl-1-picrylhydrazyl) methodology was realized according to Thaipong et al. [81] with slight modifications. First, a 0.6 mM DPPH stock solution was elaborated (0.0240 g of DPPH in 100 mL of methanol) to get a concentration of 0.6 mM. This stock solution was kept in the dark and frozen at −20 °C until used. To obtain the working solution, 10 mL of the stock solution was mixed with methanol (45 mL) to get an absorbance of 1.1 + 0.02 and a final concentration of 0.1 mM. Later, in a 3 mL quartz cell, 150 μL of standard (Trolox) or sample and 2850 μL of the DPPH working solution were sited and let to stand in the dark at room temperature for

3 h. Finally, the absorbance was measured at 515 nm on a UV spectrophotometer (UV-1800. Shimadzu, Japan). Measurements were conducted three times. Antioxidant capacity was given as equivalent mM Trolox (mM TE/100 g). The resulting absorbance was then put into the regression equation (y = −1.3055x + 1.1077; r^2 = 0.9994) derived from the Trolox calibration curve.

The ABTS and DPPH standard curves were executed according to Thaipong et al. [81] methodology. A stock solution (31.3 mg of Trolox (6-hydroxy-2,5,7,8-tetramethylchroman-2-carboxylic acid) in 10 mL of methanol) was made and 1.20, 1.00, 0.80, 0.60, 0.40, 0.20, 0.10, 0.08, 0.05, and 0.03 mM dilutions were used to obtain the curves.

3.5.6. Polyphenols Content

The polyphenols content was determined according to Singleton and Rossi [82] with some modifications, following the Folin–Ciocalteu spectrophotometric method employing gallic acid (GA) as a standard. First, a solution of a sample extract (50 µL), distilled water (3 mL), Folin–Ciocalteu reagent (250 µL), and sodium carbonate solution at 7% (750 µL) was prepared and stirred for 10 s and let to stand at room temperature for 8 min. Next, distilled water (950 µL) was added, and the mixes were left in the dark at room temperature for 2 h. Finally, in a UV spectrophotometer (UV-1800. Shimadzu, Japan), the absorbance was taken. Triplicate measurements were taken. The absorbance measurements were linearized using the calibration curve's regression equation (y = 0.0929x − 0.0197; r^2 = 0.9991) and reported in mg gallic acid equivalent (mg GAE/g).

The standard curve for polyphenols content was obtained following Xu and Chang, 2007 [83] methodology. For this, a stock solution was made (0.5 g of gallic acid in 250 mL of distilled water) and 400, 300, 200, 150, 100, 80, 60, 40, and 20 ppm concentrations were prepared to elaborate the standard curve.

3.5.7. Extraction of Betalains

The extraction of betalains was done according to Güneşer [84]. For this, 4 mL of the aqueous extract and 4 mL of trichloroacetic acid (TCA) solution at 4% concentration were placed and mixed in Corning tubes. Then, solutions were homogenized for 3 min with a vortex (Ultra-Turrax IKA T18 basic) and centrifuged at 4032× g (Avanti® J-26 XPI. Beckman Coluter®, Indianapolis, IN, USA) for 10 min at 25 °C. Finally, using a 0.45 µm pore polyethylene filter (Millipore Corp., Bedford, MA, USA) supernatants were filtered through. Samples were stored at −20 °C for further analysis.

3.5.8. Total Betalains Content

Total betalains content was determined according to Ruíz-Gutiérrez et al. [62]. McIlvaine buffer (pH 6.5, citrate-phosphate in a 1 to 10 ratio) was used to dilute the aqueous extracts of T2. This dilution was not required for T3 and T4 to produce values at their respective absorption maxima. The following formula was used to compute TB:

$$B \text{ [mg/g]} = [(A \times DF \times MW \times V)/(\varepsilon \times L)] \qquad (4)$$

where A = value at maximum absorption (534 for BC and 480 for BX) at 600 nm, DF = dilution factor, MW = molecular weight (550 g/mol for BC and 308 g/mol for BX), V = volume of the solution (1000 mL), ε = molar extinction coefficient (60,000 Lmol^{-1} cm^{-1} for BC and 48,000 Lmol^{-1} cm^{-1} for BX), and L = length of the reading cell (1 cm).

BC and BX quantifications were done separately, and the findings were combined to get the BT content. These determinations were done in triplicate, and the results were expressed in milligrams per 100 g of powder.

3.5.9. Total Protein Concentration

The total protein concentration was evaluated following the methodology of Bradford [85]. For this, 0.1 mL of the sample and 1 mL of the Bradford reagent were mixed and let to stand in the dark for 45 min. After that, absorbance was read at 595 nm in a spectrophotometer (UV-1800 Shimadzu). The absorbance measurements were linearized using the calibration curve's regression equation (y = 0.1526x − 0.0597; r^2 = 0.9973) and reported in µ/mL of BSA.

For the calibration curve, bovine serum albumin (BSA) at different concentrations (1.20, 1.00, 0.80, 0.60, 0.40, 0.20, and 0.10 mM.) were prepared and the absorbance of these were linearized to obtain the regression equation of the calibration curve.

3.5.10. Statistical Analysis

All analyses were conducted using SAS 9.0 program (Institute Inc., Cary, NC, USA, 2006). To determine the type, strength, and significance of their linear association, an analysis of correlations between pairs of response variables was performed, using the CORR procedure. Yogurts proximal analysis were analyzed using the ANOVA procedure, through the following model:

$$Yij = \mu + Ti + \varepsilon ij. \quad (5)$$

where Yij = response variable measured in the j-th repetition of the i-th treatment, μ = general mean common to all observations, Ti = effect of the i-th treatment, and εij = random error measured in the j-th repetition of the i-th treatment, which was assumed to be identically and independently distributed in a normal way with mean μ and variance σ2. When there were differences across treatments, the Tukey's and Duncan's tests were used to perform a multiple comparison of means.

For the analysis of the physical characteristics (color, pH, and syneresis) and the bioactivities (antioxidant, betalains, and polyphenols) measured over time, an analysis was carried out with the MIXED procedure, based on the following model:

$$Yijk = \mu + Ti + Pj + \varepsilon ijk \quad (6)$$

where Yij = response variable measured in the k-th repetition of the i-th treatment evaluated in the j-th storage time, μ = general mean common to all observations, Ti = effect of the i-th treatment, Pj = effect of j-th storage time, and εijk = random error measured in the j-th repetition of the i-th treatment evaluated in the j-th storage time, which was assumed to be identically and independently distributed in a normal way with mean μ and variance σ2.

4. Conclusions

The results of this research confirm that beet extracts (liquid or lyophilized) can be added to yogurt to increase its functional properties. The addition of beet extracts significantly increased the antioxidant activity of the yogurts; where betalains and other polyphenols are the main compounds responsible for said bioactivity. Meanwhile, encapsulation of beet extract by lyophilization turned out to be an effective method to stabilize betalains. However, it was not possible to determine whether maltodextrin or inulin was better as an encapsulating agent, as the AA of the yogurt added with these varies during storage time. The mechanisms that affect antioxidant activity during fermentation are considerably varied; therefore, it is recommended to study the AA of yogurts added with liquid and/or lyophilized beet extract, with the same betalain content to obtain comparable results. The shelf life of the product could be increased to observe the behavior of betalains and polyphenols in yogurt after 14 days. In addition, it would be interesting to carry out the microbiological and economic study of this yogurt.

Author Contributions: Conceptualization, M.A.F.-M. and A.C.-M.; data curation, E.S.-E.; formal analysis, M.A.F.-M. and E.S.-E.; funding acquisition, A.C.-M.; investigation, M.A.F.-M. and R.S.-V.; methodology, M.G.R.-G.; project administration, A.C.-M.; supervision, M.G.R.-G., R.S.-V. and A.C.-M.; writing—original draft, M.A.F.-M.; writing—review and editing, M.G.R.-G., R.S.-V., E.S.-E. and A.C.-M. All authors have read and agreed to the published version of the manuscript.

Funding: This research received no external funding.

Institutional Review Board Statement: Ethical review and approval were waived for this study, due to yogurt was elaborate with pasteurized milk and it does not represent a risk to the health of the participants.

Informed Consent Statement: Informed consent was obtained from all subjects involved in the study.

Data Availability Statement: The data presented in this study are available on request from the corresponding author.

Acknowledgments: The authors acknowledge that the Universidad Autónoma de Chihuahua supported this investigation. The Science and Technology National Council of Mexico (CONACYT) provided a graduate study scholarship for Martha Azucena Flores Mancha.

Conflicts of Interest: The authors declare no conflict of interest.

Sample Availability: Samples of the compounds . . . . . . are available from the authors.

References

1. Salmerón, I. Fermented cereal beverages: From probiotic, prebiotic and synbiotic towards Nanoscience designed healthy drinks. *Lett. Appl. Microbiol.* **2017**, *65*, 114–124. [CrossRef] [PubMed]
2. Shah, N.P. Functional cultures and health benefits. *Int. Dairy J.* **2007**, *17*, 1262–1277. [CrossRef]
3. Flores-Mancha, M.A.; Ruiz-Gutiérrez, M.G.; Renteria-Monterubio, A.L.; Sanchez-Vega, R.; Juarez-Moya, J.; Santellano-Estrada, E.; Chavez-Martinez, A. Stirred yogurt added with beetrot extracts as an antioxidant source: Rheological, sensory, and physicochemical characteristics. *J. Food Process. Preserv.* **2021**, *53*, 1689–1699. [CrossRef]
4. Muniandy, P.; Shori, A.B.; Baba, A.S. Influence of green, white and black tea addition on the antioxidant activity of probiotic yogurt during refrigerated storage. *Food Packag. Shelf Life* **2016**, *8*, 1–8. [CrossRef]
5. Cho, W.Y.; Kim, D.H.; Lee, H.J.; Yeon, S.J.; Lee, C.H. Quality characteristic and antioxidant activity of yogurt containing olive leaf hot water extract. *CYTA J. Food* **2020**, *18*, 43–50. [CrossRef]
6. Hong, H.; Son, Y.-J.; Kwon, S.-H.; Kim, S. Biochemical and antioxidant activity of yogurt supplemented with paprika juice of different colors. *Food Sci. Anim. Resour.* **1390**, *40*, 613–627. [CrossRef]
7. Shori, A.B. Proteolytic activity, antioxidant, and α-Amylase inhibitory activity of yogurt enriched with coriander and cumin seeds. *LWT* **2020**, *133*, 109912. [CrossRef]
8. Sabeena-Farvin, K.H.; Baron, C.P.; Nielsen, S.; Otte, J.; Jacobsen, C. Antioxidant activity of yoghurt peptides: Part 2—Characterisation of peptide fractions. *Food Chem.* **2010**, *123*, 1090–1097. [CrossRef]
9. Cho, W.; Yeon, S.; Hong, G.; Kim, J. Antioxidant activity and quality characteristics of yogurt added Green olive powder during storage. *Korean J. Food Sci. Anim. Res.* **2017**, *37*, 865–872.
10. Yeon, S.; Hong, G.; Kim, C.; Park, W.J.; Kim, S.; Lee, C. Effects of yogurt containing fermented pepper juice on the body fat and cholesterol level in high fat and high cholesterol diet fed rat. *Korean J. Food Sci. An. Res.* **2015**, *35*, 479–485. [CrossRef]
11. Parra-Huertas, R.A. Effect of green tea (Camellia sinensis L.) on the physicochemical, microbiological, proximal and sensory characteristics of yogurt during refrigerated storage. *Aliment. Cienc. Tecnol. Aliment.* **2013**, *11*, 56–64.
12. Ramos-Arrieta, K.; Zabaleta, K.; Granados-Conde, C. Preparation of a Standardized Yogurt with the Addition of Hibiscus Sabdariffa (Jamaica Flower) with Antioxidant Functional. Property Thesis, Universidad de Cartagena, Bolívar, Colombia, 2013. Available online: https://repositorio.unicartagena.edu.co (accessed on 20 July 2021).
13. Šeregelj, V.; Pezo, L.; Šovljanski, O.; Lević, S.; Nedović, V.; Markov, S.; Tomić, A.; Čanadanović-Brunet, J.; Vulić, J.; Šaponjac, V.T.; et al. New concept of fortified yogurt formulation with encapsulated carrot waste extract. *LWT Food Sci. Technol.* **2020**, *138*, 110732. [CrossRef]
14. Gies, F.; Descalzo, A.M.; Servent, A.; Dhuique-Mayer, C. Incorporation and stability of carotenoids in a functional fermented maize yogurt-like product containing phytosterols. *LWT* **2019**, *111*, 105–110. [CrossRef]
15. Delgado-Vargas, F.; Jiménez, A.R.; Paredes-López, O. Natural pigments: Carotenoids, anthocyanins, and betalains—Characteristics, biosynthesis, processing, and stability. *Crit. Rev. Food Sci. Nutr.* **2000**, *40*, 173–289. [CrossRef] [PubMed]
16. Soriano-Santos, J.; Franco-Zavaleta, M.; Pelayo-Zaldivar, C.; Armella-Villalpando, M.; Yañez-López, M.; Guerrero-Legarreta, I. Partial characterization of the red pigment of the "Jiotilla" fruit (Escontria chiotilla [Weber] Britton & Rose). *Rev. Mex. Ing. Química* **2007**, *6*, 19–25. [CrossRef]
17. Rodriguez-Amaya, D.B. Natural food pigments and colorants. *Curr. Opin. Food Sci.* **2016**, *7*, 20–26. [CrossRef]
18. Kanner, J.; Harel, S.; Granit, R. Betalains-A New class of dietary cationized antioxidants. *J. Agric. Food Chem.* **2001**, *49*, 5178–5185. [CrossRef]
19. Kapadia, G.J.; Azuine, M.A.; Sridhar, R.; Okuda, Y.; Tsuruta, A.; Ichiishi, E.; Mukainake, T.; Takasaki, M.; Konoshima, T.; Nishino, H.; et al. Chemoprevention of DMBA-induced UV-B promoted, NOR-1-induced TPA promoted skin carcinogenesis, and DEN-induced phenobarbital promoted liver tumors in mice by extract of beetroot. *Pharmacol. Res.* **2003**, *47*, 141–148. [CrossRef]
20. Gandía-Herrero, F.; García-Carmona, F. Biosynthesis of betalains: Yellow and violet plant pigments. *Trends Plant Sci.* **2013**, *18*, 334–343. [CrossRef] [PubMed]

21. Gandía-Herrero, F.; Cabanes, J.; Escribano, J.; García-Carmona, F.; Jiménez-Atiénzar, M. Encapsulation of the most potent antioxidant betalains in edible matrixes as powders of different colors. *J. Agric. Food Chem.* **2013**, *61*, 4294–4302. [CrossRef]
22. Kapadia, G.J.; Subba, R.G. *Red Beet Biotechnology*; Neelwarne, B., Ed.; Springer: New York, NY, USA, 2013; ISBN 9781461434573.
23. Gandía-Herrero, F.; Escribano, J.; García-Carmona, F. Biological activities of plant pigments betalains biological activities of plant pigments betalains. *Crit. Rev. Food Sci. Nutr.* **2016**, *56*, 937–945. [CrossRef] [PubMed]
24. Mikołajczyk-Bator, K.; Pawlak, S. The effect of thermal treatment on antioxidant capacity and pigment contents in separated betalain fractions. *Acta Sci. Pol. Technol. Aliment.* **2016**, *15*, 257–265. [CrossRef]
25. Ciriminna, R.; Fidalgo, A.; Danzì, C.; Timpanaro, G.; Ilharco, L.M.; Pagliaro, M. Betanin: A bioeconomy insight into a valued betacyanin. *ACS Sustain. Chem. Eng.* **2018**, *6*, 2860–2865. [CrossRef]
26. Martínez, L.; Cilla, I.; Beltrán, J.A.; Roncalés, P. Comparative effect of red yeast rice (Monascus purpureus), red beet root (Beta vulgaris) and betanin (E-162) on colour and consumer acceptability of fresh pork sausages packaged in a modified atmosphere. *J. Sci. Food Agric.* **2006**, *86*, 500–508. [CrossRef]
27. Carocho, M.; Morales, P.; Ferreira, I.C.F.R. Natural food additives: Quo vadis? *Trends Food Sci. Technol.* **2015**, *45*, 284–295. [CrossRef]
28. Von Elbe, J.H.; Young-Maing, I.; Amundson, C.H. Color stability os betanin. *J. Food Sci.* **1974**, *39*, 334–337. [CrossRef]
29. Huang, A.; Von Elbe, J. Kinetics of the degradation and regeneration of betanine. *J. Food Sci.* **1985**, *50*, 1115–1120. [CrossRef]
30. Penfield, M.; Campbell, A. Fruits and vegetables. In *Experimental Food Science*; Academic Press: San Diego, CA, USA, 1990; Volume 3, pp. 294–330.
31. Reyes-Aguilar, S.L. Processing Effect on the Stability of Polyphenols in Mango Extract (*Mangifera indica* L.). Thesis, Escuela Agrícola Panamericana, Tegucigalpa, Honduras, 2014. Available online: https://bdigital.zamorano.edu/bitstream/11036/3374/1/AGI-2014-T038.pdf (accessed on 20 July 2021).
32. Slimen, I.B.; Najar, T.; Abderrabba, M. Chemical and antioxidant properties of betalains. *J. Agric. Food Chem.* **2017**, *65*, 675–689. [CrossRef] [PubMed]
33. Von Elbe, J.; Schwartz, S. Colorants. In *Food Chemistry*; Marcel Dekker: New York, NY, USA, 1996; Volume 651, p. 722.
34. Wong, Y.M.; Siow, L.F. Effects of heat, pH, antioxidant, agitation and light on betacyanin stability using red-fleshed dragon fruit (Hylocereus polyrhizus) juice and concentrate as models. *J. Food Sci. Technol.* **2015**, *52*, 3086–3092. [CrossRef] [PubMed]
35. Serris, G.S.; Biliaderis, C.G. Degradation kinetics of beetroot pigment encapsulated in polymeric matrices. *J. Sci. Food Agric.* **2001**, *700*, 691–700. [CrossRef]
36. Herbach, K.; Stintzing, F.; Carle, R. Betalain stability and degradation. *J. Food Sci.* **2006**, *71*, 41–50. [CrossRef]
37. Herbach, K.M.; Rohe, M.; Stintzing, F.C.; Carle, R. Structural and chromatic stability of purple pitaya (Hylocereus polyrhizus [Weber] Britton & Rose) betacyanins as affected by the juice matrix and selected additives. *Food Res. Int.* **2006**, *39*, 667–677. [CrossRef]
38. Herbach, K.M.; Stintzing, F.C.; Carle, R. Stability and color changes of thermally treated betanin, phyllocactin, and hylocerenin solutions. *J. Agric. Food Chem.* **2006**, *54*, 390–398. [CrossRef] [PubMed]
39. Stintzing, F.C.; Trichterborn, J.; Carle, R. Characterisation of anthocyanin-betalain mixtures for food colouring by chromatic and HPLC-DAD-MS analyses. *Food Chem.* **2006**, *94*, 296–309. [CrossRef]
40. Martins, N.; Roriz, C.; Morales, P.; Barrosa, L.; Ferreira, I. Coloring attributes of betalains: A key emphasis on stability and future applications. *Food Funct.* **2017**, *8*, 1357–1372. [CrossRef] [PubMed]
41. Flores-Mancha, M.A.; Renteria-Monterubio, A.L.; Sanchez-Vega, R.; Chavez-Martinez, A. Structure and stability of betalains. *Investig. Y Cienc.* **2019**, *44*, 318–325.
42. Flores-Mancha, M.A.; Ruíz-Gutiérrez, M.G.; Sánchez-Vega, R.; Santellano-Estrada, E.; Chávez-Martínez, A. Characterization of beet root extract (Beta vulgaris) encapsulated with maltodextrin and inulin. *Molecules* **2020**, *25*, 5498. [CrossRef]
43. Gandía-Herrero, F.; Escribano, J.; García-Carmona, F. The role of phenolic hydroxy groups in the free radical scavenging activity of betalains. *J. Nat. Prod.* **2009**, *72*, 1142–1146. [CrossRef] [PubMed]
44. Obón, J.M.; Castellar, M.R.; Alacid, M.; Fernández-lópez, J.A. Production of a red-purple food colorant from Opuntia stricta fruits by spray drying and its application in food model systems. *J. Food Eng.* **2009**, *90*, 471–479. [CrossRef]
45. López, A.; Deladino, L.; Navarro, S.; Martino, M. Encapsulation of bioactive compounds with alginates for the food industry. *Aliment. Cienc. Y Tecnol. Aliment.* **2012**, *10*, 18–27.
46. Omae, J.M.; Goto, P.A.; Rodrigues, L.M.; Santos, S.; Paraiso, C.M.; Madrona, G.S.; Bergamasco, D.C. Beetroot extract encapsulated in inulin: Storage stability and incorporation in sorbet. *Chem. Eng. Trans.* **2017**, *57*, 1843–1848. [CrossRef]
47. Diaz, Y.L.; Torres, L.S.; Serna, J.A. Effect of encapsulation in drying by atomization of biocomponents of yellow pitahaya with functional interest. *Inf. Tecnol.* **2017**, *28*, 23–34. [CrossRef]
48. Saénz, C.; Tapia, S.; Chávez, J.; Robert, P. Microencapsulation by spray drying of bioactive compounds from cactus pear (Opuntia ficusindica). *Food Chem.* **2009**, *114*, 616–622. [CrossRef]
49. Castellar, R.; Obón, J.; Alacid, M.; Fernández-López, J. Color properties and stability of betacyanins from Opuntia fruits. *J. Agric. Food Chem.* **2003**, *51*, 2772–2776. [CrossRef] [PubMed]
50. Parra-Huertas, R. Evaluacion fisicoquimica, proximal y sensorial de una bebida lactea fermentada con concentrado de rubas (Ullucus tuberosus). *Vitae* **2012**, *19*, 225–227.

51. Sanchez, N.; Sepulveda, J.; Rojano, B. Development of a milk drink with extracts of curuba (Passiflora mollissima Bailey) as a natural antioxidant. *Biotecnol. Sect. Agropecu. Agroind.* **2013**, *11*, 164–173.
52. Parra-Huertas, R. Physicochemical, sensory, proximal and microbiological characteristics of yoghurt with. *Temas Agrar.* **2014**, *19*, 146–158.
53. Parra-Huertas, R.A. Use of Rubas (Ullucus tuberosus) in the elaboration and characterization of a yogurt. *Temas Agrar.* **2015**, *20*, 91–102. [CrossRef]
54. Hernández-Rodríguez, G.; Salazar-Tijerino, M. Effect of Betalains and Total Soluble Phenols of Pitahaya (Hylocereus polyrhizus) as Antioxidants in Yogurt. Thesis, Escuela Agrícola Panamericana, Tegucigalpa, Honduras, 2017; p. 28. Available online: https://bdigital.zamorano.edu/bitstream/11036/6060/1/AGI-2017-029.pdf (accessed on 20 July 2021).
55. Macedo-Ramírez, R.C.; Vélez-Ruíz, J.F. Physicochemical and flow properties of a seated yogurt enriched with microcapsules containing omega 3 fatty acids. *Inf. Tecnol.* **2015**, *26*, 87–96. [CrossRef]
56. Diaz-Jimenez, B.; Sosa-Morales, M.; Velez-Ruiz, J. Effect of the addition of fiber and the reduction of fat on the physicochemical properties of yogurt. *Rev. Mex. Ing. Quím.* **2004**, *3*, 287–305.
57. Achanta, K.; Aryana, K.J.; Boeneke, C.A. Fat free plain set yogurts fortified with various minerals. *LWT Food Sci. Technol.* **2007**, *40*, 424–429. [CrossRef]
58. Antigo, J.L.D.; Bergamasco, R.D.C.; Madrona, G.S. Effect of ph on the stability of red beet extract (Beta vulgaris l.) microcapsules produced by spray drying or freeze drying. *Food Sci. Technol.* **2017**, *38*, 72–77. [CrossRef]
59. Pitalua, E.; Jimenez, M.; Vernon-Carter, E.J.; Beristain, C.I. Antioxidative activity of microcapsules with beetroot juice using gum Arabic as wall material. *Food Bioprod. Process.* **2010**, *88*, 253–258. [CrossRef]
60. Anese, M.; Calligaris, S.; Nicoli, M.C.; Massini, R. Influence of total solids concentration and temperature on the changes in redox potential of tomato pastes. *Int. J. Food Sci. Technol.* **2003**, *38*, 55–61. [CrossRef]
61. Sawicki, T.; Wiczkowski, W. The effects of boiling and fermentation on betalain profiles and antioxidant capacities of red beetroot products. *Food Chem.* **2018**, *259*, 292–303. [CrossRef]
62. Ruiz-Gutiérrez, M.G.; Amaya-Guerra, C.A.; Quintero-Ramos, A.; Ruiz-Anchondo, T.D.J.; Gutiérrez-Uribe, J.A.; Baez-González, J.G.; Lardizabal-Gutiérrez, D.; Campos-Venegas, K. Effect of soluble fiber on the physicochemical properties of cactus pear (Opuntia ficus indica) encapsulated using spray drying. *Food Sci. Biotechnol.* **2014**, *23*, 755–763. [CrossRef]
63. Janiszewska, E. Microencapsulated beetroot juice as a potential source of betalain. *Powder Technol.* **2014**, *264*, 190–196. [CrossRef]
64. Castro-Muñoz, R.; Barragán-Huerta, B.; Yáñez-Fernández, J. Use of gelatin-maltodextrin composite as an encapsulation support for clari fi ed juice from purple cactus pear (Opuntia stricta). *LWT Food Sci. Technol.* **2015**, *62*, 242–248. [CrossRef]
65. Ayed, L.; Hamdi, M. Manufacture of a beverage from cactus pear juice using "tea fungus" fermentation. *Ann. Microbiol.* **2015**, *65*, 2293–2299. [CrossRef]
66. Azeredo, H.; Santos, A.; Souza, A.; Mendes, K.; Andrade, M.I. Betacyanin stability during processing and storage of a microencapsulated red beetroot extract. *Am. J. Food Technol.* **2007**, *2*, 307–312. [CrossRef]
67. Attoe, E.L.; von Elbe, J.H. Oxygen involvement in betanin degradation-Oxygen uptake and influence of metal ions. *Z. Für. Lebensm. Unters. Und. Forsch.* **1984**, *179*, 232–236. [CrossRef]
68. Huang, A.; Von Elbe, J. Effect of pH on the degradation and regeneration of betanine. *J. Food Sci.* **1987**, *52*, 1689–1693. [CrossRef]
69. von Elbe, J.; Attoe, E. Oxygen involvement in betanine degradation–Measurement of active oxygen species and oxidationreduction potentials. *Food Chem.* **1985**, *16*, 49–67. [CrossRef]
70. Azeredo, H.M.C. Betalains: Properties, sources, applications, and stability-A review. *Int. J. Food Sci. Technol.* **2009**, *44*, 2365–2376. [CrossRef]
71. Gandía-Herrero, F.; Jiménez-Atiénzar, M.; Cabanes, J.; García-Carmona, F.; Escribano, J. Stabilization of the bioactive pigment of opuntia fruits through maltodextrin encapsulation. *J. Agric. Food Chem.* **2010**, *58*, 10646–10652. [CrossRef]
72. Ravichandran, K.; Palaniraj, R.; Min, N.; Thaw, M.; Gabr, A.M.M.; Ahmed, A.R.; Knorr, D.; Smetanska, I. Effects of different encapsulation agents and drying process on stability of betalains extract. *J. Food Sci. Technol.* **2014**, *51*, 2216–2221. [CrossRef]
73. Jayabalan, R.; Marimuthu, S.; Swaminathan, K. Changes in content of organic acids and tea polyphenols during kombucha tea fermentation. *Food Chem.* **2007**, *102*, 392–398. [CrossRef]
74. Khan, M.I. Stabilization of betalains: A review. *Food Chem.* **2016**, *197*, 1280–1285. [CrossRef]
75. Castillo-Garrido, I.C. Stability of Betalains in A Dry Mix for Refreshing Beverages, Based on Microencapsulated Pulp and Extract of Purple Prickly Pear (Opuntia Ficus-Indica). Repositorio Universidad de Chile. 2013. Available online: http://repositorio.uchile.cl/bitstream/handle/2250/113994/castillo_ic.pdf?sequence=1&isAllowed=y (accessed on 20 July 2021).
76. Vergara, C. Extraction and Stabilization of Betalains from Purple Prickly Pear (Opuntia Ficusindica) using Membrane Technology and Microencapsulation, as A Food Coloring. Thesis, Universidad de Chile, Santiago, Región Metropolitana, Chile, 2013. Available online: http://repositorio.uchile.cl/handle/2250/114868 (accessed on 20 July 2021).
77. Nicoli, M.; Anese, M.; Parpinel, M. Influence of processing on the antioxidant properties of fruit and vegetables. *Trends Food Sci. Technol.* **1999**, *10*, 94–100. [CrossRef]
78. Özturk, B.; Öner, M. Production and evaluation of yogurt with concentrated grape juice. *J. Food Sci.* **1999**, *64*, 530–532. [CrossRef]
79. Aportela-Palacios, A.; Sosa-Morales, M.E.; Vélez-Ruiz, J.F. Rheological and physicochemical behavior of fortified yogurt, with fiber and calcium. *J. Texture Stud.* **2005**, *36*, 333–349. [CrossRef]

80. Torres-Llanez, M.J.; Vallejo-Cordoba, B.; Díaz-Cinco, M.E.; Mazorra-Manzano, M.A.; González-Córdova, A.F. Characterization of the natural microflora of artisanal Mexican Fresco cheese. *Food Control* **2011**, *17*, 683–690. [CrossRef]
81. Thaipong, K.; Boonprakob, U.; Crosby, K.; Cisneros-Zevallos, L.; Hawkins Byrne, D. Comparison of ABTS, DPPH, FRAP, and ORAC assays for estimating antioxidant activity from guava fruit extracts. *J. Food Compos. Anal.* **2006**, *19*, 669–675. [CrossRef]
82. Singleton, V.; Rossi, J. Colorimetry of total phenolics with phosphomolybdic-phosphotungstic acid reagents. *Am. J. Enol. Vitic* **1965**, *16*, 144–158.
83. Xu, B.J.; Chang, S.K.C. A comparative study on phenolic profiles and antioxidant activities of legumes as affected by extraction solvents. *J. Food Sci.* **2007**, *72*, S159–S166. [CrossRef] [PubMed]
84. Güneşer, O. Pigment and color stability of beetroot betalains in cow milk during thermal treatment. *Food Chem.* **2016**, *196*, 220–227. [CrossRef] [PubMed]
85. Bradford, M.M. A rapid and sensitive method for the quantitation microgram quantities of protein utilizing the principle of protein-dye binding. *Anal. Biochem.* **1976**, *254*, 248–254. [CrossRef]

Review

Antiviral Potential of Plants against Noroviruses

Jolanta Sarowska [1], Dorota Wojnicz [2,*], Agnieszka Jama-Kmiecik [1], Magdalena Frej-Mądrzak [1] and Irena Choroszy-Król [1]

1. Department of Basic Sciences, Faculty of Health Sciences, Wroclaw Medical University, Chalubinskiego 4, 50-368 Wroclaw, Poland; jolanta.sarowska@umed.wroc.pl (J.S.); agnieszka.jama-kmiecik@umed.wroc.pl (A.J.-K.); magdalena.frej-madrzak@umed.wroc.pl (M.F.-M.); irena.choroszy-krol@umed.wroc.pl (I.C.-K.)
2. Department of Biology and Medical Parasitology, Faculty of Medicine, Wroclaw Medical University, Mikulicza-Radeckiego 9, 50-345 Wroclaw, Poland
* Correspondence: dorota.wojnicz@umed.wroc.pl; Tel.: +48-717-841-512

Abstract: Human noroviruses, which belong to the enterovirus family, are one of the most common etiological agents of food-borne diseases. In recent years, intensive research has been carried out regarding the antiviral activity of plant metabolites that could be used for the preservation of fresh food, because they are safer for consumption when compared to synthetic chemicals. Plant preparations with proven antimicrobial activity differ in their chemical compositions, which significantly affects their biological activity. Our review aimed to present the results of research related to the characteristics, applicability, and mechanisms of the action of various plant-based preparations and metabolites against norovirus. New strategies to combat intestinal viruses are necessary, not only to ensure food safety and reduce infections in humans but also to lower the direct health costs associated with them.

Keywords: plant secondary metabolites; antiviral activity; food; noroviruses; MNV; FCV

1. Introduction

Knowledge of food viruses is not as extensive as our understanding of bacteria or fungi, the main reason for this being the difficulties in isolating, growing, and labeling the former regarding food products. Unlike many other groups of microorganisms, food-borne viruses cannot multiply in food. However, they can apparently survive food processing and storage [1]. Food contaminated with viruses can be a source of infection in consumers. Noroviruses have been associated with many recorded major food-borne viral outbreaks worldwide, while other intestinal viruses, such as the human astrovirus (HAstV), human rotavirus (HRV), sapovirus (SaV), enterovirus (EV), or Aichi virus (AiV) have been responsible for sporadic outbreaks all over the world [2].

Human noroviruses are a major cause of epidemics and periodic acute gastroenteritis worldwide. These viruses are the most common cause of food-borne diseases in the United States and Europe, entailing the societal burden of tens of billions of dollars in estimated costs of illness [3–5]. Globally, the incidence of food-borne norovirus infections reaches 120 million cases and 35,000 deaths per year [6]. Official reports published in 2017 and 2018 list human norovirus among the most frequently reported triggers of food-borne outbreaks. These reports show that the virus was responsible for 140 outbreaks (35% of all outbreaks) in the United States, and 211 outbreaks (7.8%) in Europe [7–11]. According to the RASFF report (2019), 145 outbreaks were caused by noroviruses and other caliciviruses that were found in fish and seafood, and a further 14 outbreaks relating to non-animal products were detected in the European Union [12]. According to the CDC, norovirus was the identified etiological factor of gastrointestinal complaints in 2 outbreaks out of 4 in 2020, in 8 out of 10 in 2019, and in 5 out of 11 in 2018 [13].

The transmission of the virus to humans through the consumption of contaminated food depends on various parameters, such as virus stability, food processing methods, infectious dose, and host susceptibility [14]. It is worth noting that food ingredients can protect the virus during processing and human consumption. The infectious dose of a food-borne virus is generally low, and a small number of virus particles can cause infection. Moreover, noroviruses, as food contaminants, persist in food for a long time without loss of infectivity [2]. Many control strategies that rely on the internal and external properties of the food, e.g., pH and water activity, are ineffective against these pathogens. Heat treatment is an effective way of deactivating foodborne viruses, but it can alter the organoleptic properties (e.g., color and texture) and reduce the nutrient content (e.g., protein and vitamins) of foods [15]. Currently, consumers show an increasing demand for high-quality natural food products. One of the issues is changes to the way we eat, while another is introducing raw or mildly-heat-treated foods to everyday menus: sushi, blue beef, seafood, and insects. Shellfish, fruit, and vegetables pose a serious threat to humans because they are eaten raw [16]. These foods are prone to contamination, due to the use of fecal-contaminated water for irrigation or the lack of proper personal hygiene in the people who come into contact with food [17,18].

Noroviruses belong to a group of viruses resistant to external factors. They are not sensitive to freezing, short-term heating, ionizing radiation, organic acids, preservatives and chlorine compounds, alcohols, and other detergents. At a temperature of 60 °C, their deactivation takes place only after 30 min. In their natural environment, they can remain active for several weeks or even years [19,20]. Viral infections in which the etiological factors are viruses that contaminate food can be prevented primarily by neutralizing the source of contamination during the food sanitation processes. In the context of public health, this is a significant challenge for the food industry [21–23]. For this reason, both deactivating the virus and maintaining high standards risk lowering the food quality characteristics, presenting a challenge for food processors. Innovative non-thermal food processing technologies, including high-pressure processing (HPP), cold plasma (CP), ultraviolet (UV) light, radiation, and pulsed electric field (PEF) treatments have been tested for food-borne virus deactivation, sensory properties, and the retained nutritional value of processed foods [14].

In recent years, intensive research has been carried out on the properties of phytochemicals with antiviral activity. Unlike chemicals, these metabolites are a safe option if used as fresh food preservatives. New strategies to combat intestinal viruses are necessary, not only to ensure food safety and reduce the number of infections in humans but also to reduce the direct health costs associated with them [5].

The aim of our study is to review the results of the latest literature describing the applicability and efficacy of various metabolites of plant origin that could be used as modern and environmentally safe agents against human food-borne noroviruses.

2. Characteristics of Human Norovirus

Human Norovirus (HuNoV), formerly known as the Norwalk virus, is a non-segmented, non-enveloped RNA virus belonging to the Caliciviridae family. Caliciviruses are small viruses of 30–35 nm in size, which are visible in the microscopic image as spherical particles, devoid of envelopes and spikes [24]. Noroviruses do not multiply in vitro in cell cultures. HuNoV, as well as its surrogates that are commonly used in laboratory tests, i.e., murine norovirus (MNV) or feline calicivirus (FCV), are devoid of an envelope, contain ssRNA, and show high resistance to both antimicrobial preparations and environmental conditions [14,25].

According to the latest systematics, noroviruses are divided into seven gene groups (from GI to GVII) with 30 genotypes detected globally. GI, GII and GIV are the most common causes of human infections. Many international epidemic surveillance systems (CaliciNet and NoroNet) record the transmission of norovirus infections and provide important information about the spread of different human norovirus strains. According to

Hoa Tran et al. [26], the strains with the GII.4 genotype accounted for 70–80% of all the outbreaks reported over the past decade. The frequency of genotypes varied according to the population level and the route of transmission [27]. The GII.4 genotype is more commonly associated with dissemination via interpersonal contact, while non-GII.4 genotypes, such as GI.3, GI.6, GI.7, GII.3, GII.6 and GII.12, are most commonly transmitted by food [28]. Water transmissions occur more frequently among GI gene group strains than GII7 strains. This may be related to the fact that GI strains have higher water stability than GII strains [29]. Between 2009 and 2013, the GII.4 genotype was the cause of 2853 (72%) outbreaks in the USA, of which, 94% were GII.4 New Orleans or GII.4 Sydney [30].

Viruses do not multiply on the surfaces of raw food. Viral particles will not increase in number when introduced into raw food as their site of primary contamination. On the contrary, their numbers may drop over an extended period of storage, or change, subject to the conditions of their storage. Cold storage of raw produce, often at temperatures below 0°C, preserves the viruses present on them, leaving food still contaminated and therefore potentially infectious [31].

3. Methodology of Research Regarding the Antiviral Activity of Phytochemicals

Due to the fact that plant extracts may contain several dozen to several hundred compounds, standardization is necessary, taking into account their unique chemical profiles. In accordance with international standards, such characteristics should also include the systematic affiliation of the plant from which the oil or extract is derived and define the physicochemical properties of these phytochemicals [32].

The antiviral activity of essential oils and plant extracts is lower in food matrices, in comparison with in vitro tests. The lowest concentration of oils necessary to inhibit the growth of microorganisms in the food may be over 1000 times higher than are needed in the model conditions in in vitro studies [33].

To ensure that the range of activity of biologically active compounds in food is determined with precision, it is necessary to employ an adequately designed experimental analysis. A testing methodology of the antiviral activity of metabolites of plant origin must satisfy a number of criteria: for example, the starting titer must be determined correctly, according to the tested virus; the cytotoxicity of the plant product has no effect on the cell growth and/or cell morphology; the plant-derived phytochemicals in question show antiviral activity against the tested virus model [34].

Determining the antiviral effect of a biologically active preparation requires confirmation with appropriate tests. The use of the suspension method in the first stage of the research allows us to determine whether the active plant metabolite, being a component or one of the components of the tested preparation, exhibits antiviral activity [35]. In the next step, the test viruses are exposed to the plant product at different concentrations, contact times and temperatures, which allows the titer of the infectious virus to be determined. The virus's infectious titer is determined by assessing the presence or absence of a cytopathic effect in the cell culture. The ability of the tested plant product to deactivate the test virus is determined by decreasing its infectious titer when compared to the control mixture [36].

The virucidal activity of the tested preparation against a specific virus is confirmed if the infectious virus has decreased by at least 4 logs in the titer compared to the control mixture. This means a loss of viral infectivity of 99.99% [37].

The use of cell models in in vitro tests allows for the quick and precise determination of the antiviral activity of various preparations [38]. All results of in vitro studies on the action of plant-derived active metabolites (e.g., endpoint titration technique (EPTT), virus-induced cytopathic effect inhibition (CPE), virus yield reduction assay, MTT assay, plaque reduction assay, virus deactivation assay, virus adsorption assay, virus attachment, and virus penetration assay [39]) must also be confirmed by in vivo testing [40–42], which, at the next stage, is a necessary step in the application for registration with government food and drug control agencies and in making the preparation available to food pharmaceutical industries [43].

The results of in vitro studies regarding the activity of plant-derived metabolites require confirmation by reference tests, also conducted in vivo, to enable an application for registration with the appropriate government food control agency before the preparation is licensed for use in the food or pharmaceutical industries.

The factors significantly affecting the antimicrobial activity of plant preparations include the activity of food enzymes, water activity, pH, temperature, and the number of microbes contaminating a given food product [44,45]. Virions present in the food matrix and in foodstuffs were found to be more resistant to the antiviral activity of plant compounds than were virions present in water [46,47].

4. Mechanism of the Antiviral Action of Compounds of Plant Origin

In recent years, many laboratories around the world have engaged in research into plant extracts and their respective biological activity. Plant-derived phytochemicals exhibit various antiviral activities and employ different mechanisms of action (Figure 1) [46,48]. Individual compounds isolated from plants may show a different effect than the entire extract. Considering the fact that the effectiveness of the antimicrobial action of plant preparations is based on the mutual interaction of biologically active compounds, it is especially important to understand the structure of such molecules. Bioinformatics methods have proved to be extremely helpful in this field, making it possible to study the interactions of various low-molecular compounds with viral or cellular proteins (the so-called molecular docking). Nonetheless, wider use of plant compounds with antimicrobial activity primarily depends on determining the molecular mechanism of their action [49].

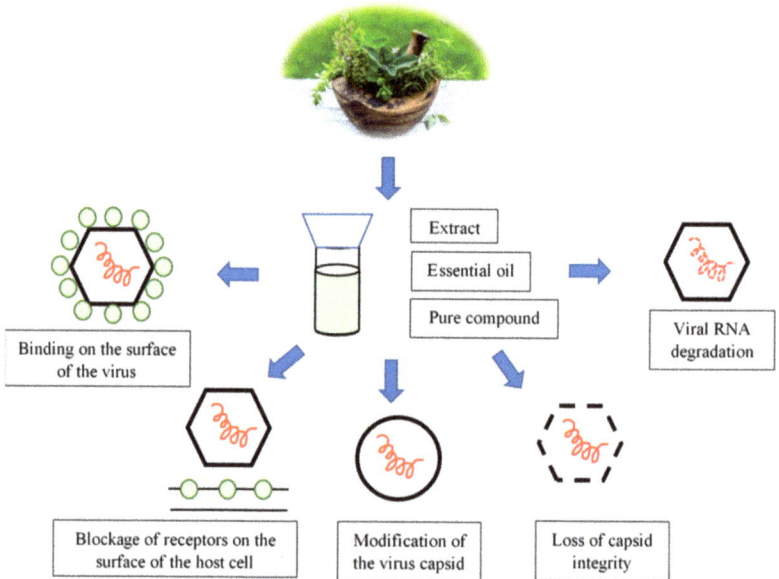

Figure 1. Representation of different possible modes of action of plant extracts, essential oils, and their constituents against noroviruses.

The biological and pharmacological activity of plant-derived secondary metabolites, such as polyphenols, terpenes, and alkaloids, has long been known and used in medicine. Plant antiviral phytochemicals can bind to particles on the surface of the virion, preventing target cell recognition and virus adsorption via the proper receptor (Figure 1). The blockage of receptors on the host cell surface is yet one more mechanism of action that is exhibited by phytochemicals. This consists of blocking the penetration of the virus into the cell or blocking the synthesis of viral nucleic acids. The activity of these compounds may also inhibit the synthesis and post-translational processing of viral proteins. It may also block the processes relating to the assembly of daughter virions, or the release of viral daughter particles from the host cell [50] (Figure 1).

The multiplication of viruses in the host cell is dependent on both cellular and viral factors. Plant metabolites exhibiting antiviral qualities and, therefore, finding uses in antiviral drugs specifically inhibit the multiplication of viruses without damaging the host cells [48]. Their most frequent target sites of action are the molecules found on the virion's surface that are responsible for the recognition, adsorption, and penetration of the virus into the cell. Nucleic acids (DNA or RNA), proteins, viral RNA replicases, and reverse transcriptase have also been recognized as attractive target sites for the action of these phytochemicals [50].

Considering the potential use of essential oils and other plant extracts to combat or deactivate food-borne viruses, their antimicrobial mechanism should be analyzed first. The available literature on this topic is still scarce, especially in the group of non-enveloped viruses, which, due to their structure, constitute a difficult objective for laboratory research. Plant antimicrobial metabolites may exhibit various mechanisms of antiviral activity, which is confirmed by the results obtained by the authors of experimental studies [51,52].

In the studies conducted by Gilling et al. [53,54], the influence and mechanisms of the antiviral activity of allspice oil, lemongrass oil, citrus oil (specifically, citral), oregano oil and its main active metabolite, carvacrol, against murine norovirus (MNV) were analyzed. As part of the research, tests were carried out on the infectivity of cell cultures, protection against RNase I, binding to receptors within the host cells, and imaging in a TEM microscope was performed [53,54]. Based on the results obtained, it was found that the effectiveness of active phytochemicals varies greatly depending on the type of virus. This is confirmed by previous observations, indicating that even small differences in the structure or genome of the virus can significantly affect its susceptibility to various antiviral agents [55,56]. In turn, the results obtained by Kovač et al. [57] indicated that the essential oils obtained from hyssop and marjoram were active against enveloped HSV viruses but did not deactivate the two non-enveloped viruses that were tested (HAdV-2 and MNV-1).

In non-enveloped viruses, the capsid protects the integrity of the viral nucleic acid. Viral RNA may remain intact, while changes in the structure of the capsid may deactivate the virus [58,59]. Modification of the virus capsid is one of the mechanisms that can lead to the inhibition of the virus adsorption process, which is associated with its deactivation. In the case of MNV, the results obtained by Gilling et al. [54] suggest that, as lemongrass oil and citral bind to the viral capsid, they most likely deactivate the virus by inducing conformational changes in the capsid proteins. The magnification of the viral particles, as seen in the TEM images, indicates that oregano oil and carvacrol affect the complete loss of the integrity of the capsid [53]. Various types of structural changes within the FCV capsid, and deformations of NoV (HuNoV GII.4) and MNV-1 particles, were also found after the application of cranberry juice and grape seed extract [55,60,61].

The blocking of the epitopes necessary for the adsorption process in viral ssRNA allowed the observation of another instance of the mechanism of action of plant-derived compounds. Thereby, the virus lost its affinity to the receptors on the surface of the host cells and was unable to infect them. In this case, the tested plant metabolites did not damage the viral RNA [54]. The exposure of FCV-F9 and MNV-1 to pomegranate juice also reduced the infectivity of the viruses studied [60].

The phytochemicals in allspice oil have been found to be virucidal against the MNV virus. They lead to the degradation of both the capsid proteins and viral RNA [54].

Our review of the literature points to the conclusion that various plant metabolites can cause a direct virucidal effect against non-enveloped virus ssRNA by degrading the capsid or viral nucleic acid. Plant-derived compounds can also bind to the surface of the virus without destroying the proteins in the capsid and, thus, interfere with its adsorption to host cells [54,62].

5. Plant Preparations as Antiviral Agents against Noroviruses

The antiviral activity of plant metabolites is the subject of many scientific studies [23,41,53–121]. The available literature includes reports of the use of various plant extracts containing essential oils and other metabolites against viruses, including noroviruses (Figure 2). The referenced publications are classified according to the various compounds of plant origin used for testing and the active metabolites they contain (Table 1). Particular attention is paid to the antiviral efficacy of the tested plant preparations and the mechanisms of their action against noroviruses. Our review presents the most interesting and promising examples of the potential use of compounds of plant origin as antiviral phytochemicals in medicine and the food industry.

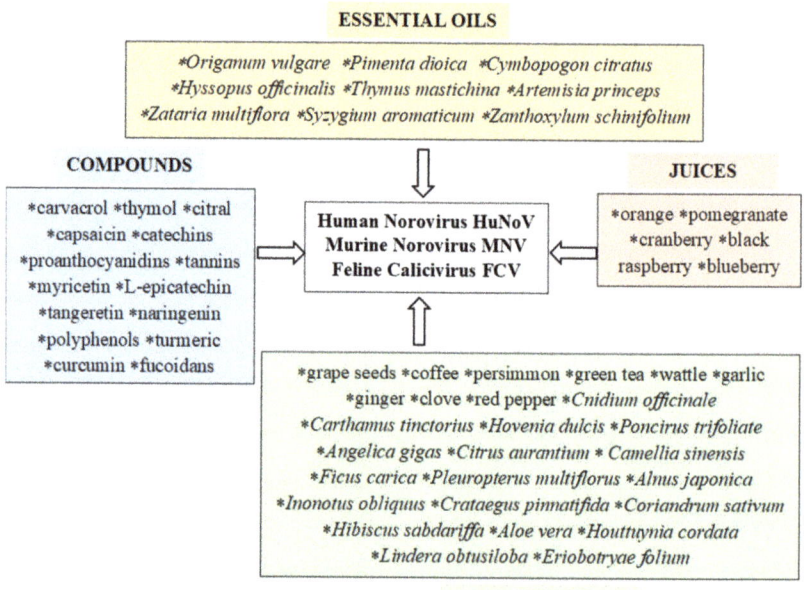

Figure 2. Antiviral activity of metabolites of plant origin against noroviruses.

Table 1. The composition of essential oils and their antiviral activity against noroviruses.

Essential Oil	Plant	Main Constituents	Group of Chemical Compounds	Content (%)	Viruses	References
Oregano	*Origanum vulgare*	Carvacrol Thymol P-cymene Gamma-terpinene Alpha-terpineol Limonene	Monoterpene Sesquiterpene lactone Related to monoterpene Monoterpene Monoterpene alcohol Monoterpene	0.3–80.8 0.96–63.7 <0.1–16.94 0.8–21.0 <0.09–12.0 0.3–0.7	MNV FCV	[53,63–67]
Marjoram	*Thymus mastichina*	Linalool 1,8-cineole Beta-pinene Alpha-pinene Alpha-terpineol Camphor Limonene	Monoterpene alcohol Monoterpene deriv. Monoterpene Monoterpene Monoterpene alcohol Monoterpene deriv. Monoterpene	24.5–73.5 9.4–55.6 0.6–5.9 0.9–4.3 0.9–3.0 0.00001–3.0 0.4–2.1	MNV-1	[57,63,68–70]

Table 1. Cont.

Essential Oil	Plant	Main Constituents	Group of Chemical Compounds	Content (%)	Viruses	References
Thyme	*Thymus vulgaris*	Thymol	Sesquiterpene lactone	27.6–100	MNV-1	[57,63,71,72]
		Trans-sabinene hydrate	Monoterpene hydrate	0.43–39.4		
		Menthol	Monoterpene alcohol	1.3–39		
		Bornyl acetate	Monoterpene	0.2–25.57		
		Limonene	Monoterpene	0.4–24.2		
		Carvacrol	Monoterpene	2.0–20.5		
		Gamma-terpinene	Monoterpene	0.6–14.9		
Zataria multiflora	*Zataria multiflora* Boiss.	Thymol	Sesquiterpene lactone	40.8	MNV FCV	[63,67,73–77]
		Carvacrol	Monoterpene	27.8		
		P-cymene	Related to monoterpene	8.4		
		Gamma-terpinene	Monoterpene	4.0		
		Beta-caryophyllene	Sesquiterpene	2.0		
		Linalol	Monoterpene alcohol	1.7		
		Alpha-terpinolene	Monoterpene	1.3		
Clove	*Syzygium aromaticum* (*Eugenia caryophyllus*)	Eugenol	Monoterpene deriv.	86.7	MNV FCV	[63,67,78,79]
		Beta-caryophyllene	Sesquiterpene	3.2		
		Allo-aromadendrene	Sesquiterpene	1.3		
		Alpha-humulene	Sesquiterpene	0.9		
Hyssop	*Hyssopus officinalis*	Linalool	Monoterpene alcohol	49.6	MNV-1	[57,63,80,81]
		1,8-cineole	Monoterpene deriv.	13.3		
		Limonene	Monoterpene	5.4–12.2		
		Beta-pinene	Monoterpene	3.0–11.1		
		Beta-caryophyllene	Sesquiterpene	1.5–2.8		
		Isopinocamphone	Bicyclic monoterpenoids	1.3–43.3		
Zanthoxylum schinifolium	*Zanthoxylum schinifolium*	Estragole	Phenylpropene	42.0	MNV-1 FCV-F9	[63,82,83]
		Oleic acid	Monounsaturated omega-9 fatty acid	20.97		
		Palmitic acid	Saturated fatty acid	19.86		
		2,4-Decadienal	Polyunsaturated fatty aldehyde	4.87		
		2-Undecenal	Aldehyde	3.81		
Allspice	*Pimenta dioica*	Eugenol	Monoterpene deriv.	45.4–83.68	MNV	[54,63,84–86]
		Beta-caryophyllene	Sesquiterpene	2.3–8.9		
		P-cymene	Related to monoterpene	1.77–1.78		
		Terpinolene	Monocyclic monoterpene	1.23–2.35		
		Alpha-cadinol	Pseudoguaianolide	1.0–5.9		
		Alpha-humulene	Sesquiterpene	0.88–5.4		
Lemongrass	*Cymbopogon citratus*	Geranial	Monoterpene aldehyde	32.7–49.9	MNV-1	[54,63,87–89]
		Neral	Monoterpene aldehyde	26.5–38.2		
		Myrcene	Monoterpene	1.7–25.3		
		Nerol	Monoterpene	0.2–12.5		
		Geraniol	Monoterpene deriv.	0.2–10.4		
		1,8-cineole	Monoterpene deriv.	0.2–2.9		
Tea tree	*Artemisia princeps* var. *orientalis*	1,8-cineole	Monoterpene deriv.	2.2–24.3	MNV-1 FCV-F9	[63,90–92]
		Borneol	Monoterpene alcohol	2.1–5.6		
		Camphor	Monoterpene deriv.	1.4–38.7		
		α-terpineol	Monoterpene alcohol	1.1–9.8		
		Beta-pinene	Monoterpene	0.6–11.7		
		Alpha-pinene	Monoterpene	0.5–9.7		
		Beta-caryophyllene	Sesquiterpene	0.4–10.6		
		Isoborneol	Monoterpene deriv.	0.1–20.9		
		Alpha-thujone	Monoterpene	0.1–16.0		

5.1. Effect of Essential Oils on Noroviruses

Essential oils (EOs) are volatile, aromatic substances that belong to secondary plant metabolites. The main components of essential oils are terpenes, including monoterpenes and sesquiterpenes (Table 1). Each oil may contain between a dozen and several dozen compounds of various concentrations and properties. The highly diverse chemical compositions of EOs support an extremely wide range of biological activity. The biological activity of oils and their ingredients has been the subject of many in vitro studies and a few in vivo tests. Table 1 lists the essential oils and their main ingredients used in research against norovirus surrogates, i.e., feline calicivirus (FCV) and murine norovirus (MNV).

Oregano essential oil (*Origanum vulgare*) successfully deactivated non-enveloped human norovirus surrogates—feline calicivirus (FCV) and murine norovirus (MNV) [53,67]. Gilling et al. [53] noted that the antiviral effect of 4% of oregano oil resulted in a statistically significant reduction in MNV within 15 min of exposure. The authors observed changes in virus particles under transmission electron microscopy (TEM) after 24 h of exposure to oregano oil. The treated virus particles were larger (40–75 nm) than the untreated virus particles (20–35 nm). Based on the results of a cell-binding assay, an RNase I protection assay, and TEM imaging, the authors drew conclusions regarding the mechanism of action of oregano essential oil on MNV and claimed that this oil is likely to disrupt the integrity of the virus capsid. Elizaquivel et al. [67] found significant reductions in both MNV and FCV at 4% of oregano essential oil. However, the reductions turned out to be temperature-dependent. The antiviral activity of oregano essential oil was recorded only at 37 °C, while no significant reduction was observed at 4 °C.

Gilling et al. [54] used allspice (*Pimenta dioica*) and lemongrass (*Cymbopogon citratus*) essential oils at concentrations of 2% and 4% to determine their antiviral efficacy against MNV. Lemongrass essential oil in both concentrations significantly reduced the viral infectivity of MNV within 6 h of exposure, while allspice oil was effective only at a concentration of 4% after 30 min of exposure. The authors also showed that the antiviral activity of allspice essential oil was both time- and concentration-dependent, while the effects of lemongrass essential oil were only time-dependent. The research of Gilling et al. [54] included an RNase I protection experiment to assess if the MNV capsid was degraded by lemongrass and allspice oils, and a cell-binding experiment to check if both tested oils inhibited the ability of MNV to bind to the RAW 264.7 cells. The test results obtained were suggestive of degradation of the viral capsid in the samples that were treated with the lemongrass and allspice oils. Nonetheless, the specific binding of MNV particles to host cells was unchanged after exposure to the tested essential oils, which means that they do not affect viral adsorption. The authors also used TEM to determine whether there were any structural changes to the virus particles after treatment with the oils. MNV particles exposed to allspice oil turned out to be slightly larger (from 25 to 75 nm) compared to untreated MNV (from 20 nm to 35 nm). Virus particles after treatment with lemongrass oil were much longer and had a size of 100–500 nm.

The effect of clove and *Zataria multiflora* essential oils on FCV and MNV at 4 °C and 37 °C was studied by Elizaquivel et al. [67]. The results obtained showed that the concentrations of 1% clove oil and 0.1% *Zataria* oil were effective against MNV and FCV at 37 °C.

Chung et al. [92] tested the antiviral effect of an essential oil obtained from the edible medicinal plant *Artemisia princeps* var. *orientalis*, which is popular in Korea. The active compounds in this essential oil, alpha-thujone (thujone), borneol, and camphor, were used in plaque tests against MNV-1 and FCV-F9, and 48% efficacy was observed for FCV-F9 and 64% for MNV-1 at 0.1% and 0.01% concentrations of essential oil. In addition, it was found that only α-thujone showed strong antiviral activity, while in the case of borneol and camphor, no inhibitory effect was observed against FCV-F9 and MNV-1. The authors point out the need for further research to elucidate the antiviral mechanisms of the action of the essential oil obtained from *Artemisia princeps* var. *orientalis* and alpha-thujone against

FCV-F9 and MNV-1, as well as the influence of temperature on the inhibition of tested noroviruses by the active phytochemicals used in the research [92].

Kovač et al. [57] investigated the ability of essential oils derived from two aromatic plants—*Hyssopus officinalis* (hyssop) and *Thymus mastichina* (marjoram)—to deactivate non-enveloped mouse norovirus (MNV-1). No significant reduction of MNV titer was observed after treatments with hyssop and marjoram at a concentration of 0.02%.

The seeds and pericarp of *Zanthoxylum schinifolium* are widely used in Korea, China, and Japan as a spice. The antiviral activity of *Z. schinifolium* essential oil (ZSE) against the foodborne viral surrogates FCV-F9 and MNV-1 was analyzed, using the cytopathic effect test [82]. In this study, RAW 264.7 or CRFK cells were exposed to ZSE at concentrations of 0.00001%, 0.0001%, and 0.001% for 72 h. Inhibition of the cytopathic effect on CRFK or RAW 264.7 cells was not detected after the incubation of FCV-F9 and MNV-1 at all tested concentrations of ZSE. These results suggested that ZSE did not deactivate viruses.

Kim et al. [89] determined the effect of lemongrass essential oil on the infectivity and replication of MNV-1. From the plaque reduction test results, this oil was found to inhibit MNV-1, both in a time-dependent and dose-dependent manner (73.09%, using a concentration of 0.02%). It has been proven that lemongrass oil, and its main component citral, deactivate the viral coat proteins necessary for viral infection and inhibit replication of the viral genome in the host cells, which was further confirmed in in vivo studies.

5.2. Effect of Plant Extracts on Noroviruses

Plant extracts, which contain innumerable ingredients, are valuable sources of new and biologically active molecules with antimicrobial properties. Reports concerning the antiviral activity of plant extracts are rather limited.

Li et al. [61] tested grape seed extract (GSE) on noroviruses—murine norovirus MNV and human norovirus NoV GII.4. MNV infectivity was detected by plaque assay, while NoV GII.4 infectivity was examined by cell-binding reverse transcription-PCR, after treatment of GSE with two solutions: 0.2 mg/mL and 2 mg/mL. The infectivity of MNV was reduced to >3-log PFU/mL. The ability of NoV GII.4 to bind to the cells of the human enterocytic Caco-2 cell line was significantly reduced by treating GSE in a dose-dependent manner. The authors also checked the effect of GSE on NoV GII.4 P particles using a saliva-binding enzyme-linked immunosorbent assay. The P domain formed the outermost surface on the NoV protein capsid, and this was needed for viral binding to carbohydrate receptors on the host cells. The binding signal (OD_{450}) of NoVs GII.4 P particles to the salivary carbohydrate coat on the ELISA plate was reduced. Based on the results obtained in the plaque assay for MNV-1, cell-binding RT-PCR for human NoV GII.4, and saliva-binding ELISA for human NoV GII.4 P particles, the authors concluded that GSE may cause the denaturation of viral capsid protein. Therefore, the morphology of NoV GII.4 before and after GSE treatment was examined by TEM. Human NoVs in the untreated control sample appeared as small spherical particles of two sizes: 18–20 nm and 30–38 nm. After treatment with GSE at 0.2 mg/mL, the viral particles clumped together. The deformation of most of the larger particles was also observed. At a dose of GSE of 2 mg/mL, the spherical particles disappeared, and a high concentration of residual protein was observed. These results provided direct evidence that GSE could effectively damage the NoV capsid protein.

The antiviral properties of GSE were described by Su and D'Souza [93]. They assessed GSE activity against the human norovirus surrogates, MNV-1 and FCV-F9, using lettuce and jalapeno peppers, which are frequently associated with foodborne outbreaks. Lettuce and jalapeno peppers were inoculated with MNV-1 and FCV-F9 at high (~7 log10 PFU/mL) or low (~5 log10 PFU/mL) titers, and were treated with 0.25, 0.5, 1 mg/mL GSE for 30 s to 5 min. At the higher titers, FCV-F9 was reduced by 2.33, 2.58, and 2.71 log10 PFU on lettuce, and 2.20, 2.74, and 3.05 log10 PFU on peppers after 1 min, respectively. Low FCV-F9 titers could not be detected after 1 min at all three GSE concentrations. The low MNV-1 titer was reduced by 0.2–0.3 log10 PFU on lettuce and 0.8 log10 PFU on paprika. High-titer MNV-1 was not reduced by GSE at all three tested concentrations.

The aim of the study by Joshi et al. [94] was to determine the antiviral activity of GSE against FCV-F9 and MNV-1 at both room temperature and 37 °C, and in complex food matrices within 24 h. Based on the results obtained, it was found that the antiviral effect of the tested extract increased proportionally to the time and dose. On the other hand, conducting tests in model food (apple juice and 2% milk) and simulated gastric conditions weakened the effect of GSE.

Oh et al. [95] determined the effect of mulberry seed extract (MSE) on FCV-F9 and MNV-1, using plaque assays. The antiviral effects of MSE at concentrations of 0.01, 0.1, and 1 mg/mL were assessed at various times during viral infection to assess the mechanism of antiviral action: cell pre-treatment, viral pre-treatment, concurrent treatment, and post-treatment. The maximum antiviral effect of MSE against MNV-1 and FCV-F9 was achieved when MSE at 1 mg/mL was added, along with viruses simultaneously added to RAW 264.7 and CRFK cells with viruses. The results obtained suggest that MAS may affect both noroviruses in the initial phase of viral replication.

The ability of persimmon, wattle, coffee, and green tea extracts to deactivate FCV and MNV was tested [96]. The results showed that the persimmon extract deactivated both viruses, inhibiting their infectivity. Both wattle and green tea extracts reduced the infectivity of FCV. Coffee extract had no suppressive effect on any virus.

Other studies have shown that green tea extract (GTE) inhibits the replication of MNV and FCV [21,52,97]. Additionally, it has been observed that GTE and catechins can deactivate these viruses by non-specific binding to their receptors, thus preventing the virus from binding to host cells [97]. MNVs were completely deactivated by GTE at 37 °C [46]. Based on subsequent research results, it was also found that the accumulation of catechin derivatives during the storage of the mature green tea extract (aged-GTE) (24 h at 25 °C) resulted in a significant increase in the antiviral activity of GTE against human GII.4 norovirus under laboratory conditions [98,99]. The results obtained by Falco et al. [99] indicate a potential use of the synergistic antiviral effect of aged-GTE, and gentle heat treatment (50 °C, 30 min) to ensure food safety, mainly in fruit juices.

Randazzo et al. [100] observed a complete inhibition of human norovirus GII.4 replication by aged-GTE at concentrations of 1 mg/mL at 37 °C, 1.75 mg/mL for 21 °C, and 2.5 mg/mL at 7 °C.

Oh et al. [101] investigated the antiviral activity of methanol extracts from medicinal plants, including spices, herbal teas, and medicinal herbs, against FCV by using a plaque reduction test. Spices: garlic, ginger, red pepper; herbal teas: rosemary, green tea; and medicinal herbs: rhizome of *Cnidium*, safflower, raisin tree, trifoliate orange, danggwi and mandarin peel were used in testing. The antiviral activity of the plant extracts was measured in a plaque reduction assay in which activity was expressed as an EC50 value. Among the investigated medicinal extracts, green tea extract showed the most effective anti-FCV activity. The EC50 value was 0.13 mg/mL. Danggwi, safflower, rosemary, orange trifoliate, and tangerine peel extracts also showed antiviral activity. The EC50 values were 0.26, 0.27, 0.34, 0.49, and 0.54 mg/mL, respectively.

Other authors used aqueous extracts obtained from cloves, fenugreek, garlic, onion, ginger, and jalapeno, which were also tested for antiviral activity, using FCV as a substitute for human norovirus. Based on the test results, it was found that the use of clove extracts (eugenol—29.5%) and ginger (1,2-propanediol—10.7%) deactivated 6.0 and 2.7 logs of the initial viral titers, respectively [102].

Seo and Choi [103] determined the activity of 29 edible Korean herbal extracts against the human norovirus surrogates, MNV and FCV. Preliminary results indicate that extracts obtained from *Camellia sinensis, Ficus carica, Pleuropterus multiflorus, Alnus japonica, Inonotus obliquus, Crataegus pinnatifida,* and *Coriandrum sativum* showed inhibitory activity against MNV and FCV, which allows their use as natural antiviral agents.

Park et al. [104] determined the antiviral activity of 5%, 10%, and 15% vinegar (6% acetic acid) against MNV-1 in edible, experimentally contaminated fresh seaweed (*Enteromorpha intestinalis*). After a 7-day storage period at 4 °C, a significant decrease in the

MNV-1 titer was observed. In other studies, capsaicin was also found to contribute to the reduction of MNV during kimchi fermentation at various temperatures [105].

Polysaccharide-rich aqueous (HWE) and alcoholic (HEE) extract of *Houttuynia cordata*, with pharmacological properties, were used by Cheng et al. [106]. Additionally, the *H. cordata* polysaccharide (HP) with a molecular weight of ~43 kDa, which consisted mainly of galacturonic acid, galactose, glucose, and xylose, was also used to determine the antiviral potential against MNV-1. HWE was shown to be the most effective in the plaque test. HP deformed and inflated virus particles. These changes made it difficult for viruses to penetrate target cells, which confirmed the antiviral properties of HP [106].

The aim of the study by Joshi et al. [107] was to determine the antiviral activity of aqueous *Hibiscus sabdariffa* extracts against FCV-F9 and MNV-1. FCV-F9 titers were reduced to undetectable levels after 15 min at concentrations of 40 and 100 mg/mL of hibiscus extract; in the case of MNV-1, a similar effect was obtained only after 24 h.

Solis-Sanchez et al. [23] examined the antiviral effect of *Lindera obtusiloba* leaf extract (LOLE) with a significant content of pinene (49.7%), phellandrene (26.2%), and limonene (17%). These compounds significantly inhibited the infectivity of MNV-1. Preincubation of viruses with LOLE at concentrations of 4, 8, or 12 mg/mL for 1 h at 25 °C reduced the infectivity of MNV-1 by 51.8%, 64.1%, and 71.2%, respectively. The results of studies concerning the antiviral activity of LOLE, as obtained by the authors, did not make it possible to establish the mechanisms of action of these phytochemicals on the viruses studied. Further experiments are needed to clarify these issues.

5.3. Effect of Bioactive Plant Compounds on Noroviruses

The antiviral activity of essential oils and plant extracts may be related to the presence of bioactive compounds.

Thyme and oregano contain significant amounts of monoterpenes, such as thymol and carvacrol. Gilling et al. [53] determined the antiviral efficacy of carvacrol, which is the main active ingredient in oregano essential oil. Depending on the geographic origin, its content can be as high as 85%. Carvacrol was tested at concentrations of 0.25% and 0.5%. Both concentrations resulted in a statistically significant reduction in MNV within 15 min, in comparison with the control sample. The authors used an RNase I protection experiment and a cell-binding experiment in the study to determine the likely mechanism of carvacrol's action on MNV. The reductions observed in cell culture infectivity for carvacrol increased with greater durations of exposure to carvacrol (e.g., from 1.28-log10 after 15 min to >4.52-log10 after 24 h of exposure to the 0.5% concentration), whereas the reductions observed in the viral RNA were initially greater. These results suggest that carvacrol partially degraded the capsid, but the virus may still be infectious. TEM images showed that all the carvacrol-treated virus particles were greatly expanded in size (100 to 900 nm). Among them were both intact particles and others completely broken into capsid components.

Carvacrol at various concentrations (0.25, 0.5, 1% for 2 h at 37 °C) was used in the MNV and FCV deactivation test at titers of about 6–7 log TCID50/mL. Carvacrol, at a concentration of 0.5%, completely deactivated both norovirus surrogates. In addition, it was also found that 0.5 or 1% carvacrol can be used in lettuce-washing water to reduce the MNV and FCV titer, which indicates the possibility of using this plant metabolite as a natural viral contamination-reducing agent in fresh vegetables [108].

Thymol was also effective in reducing the titer of norovirus surrogates in a dose-dependent manner. Thymol in concentrations of 0.5 and 1% reduced FCV titers to undetectable levels, while in the case of MNV, thymol at concentrations of 1 and 2% reduced them by 1.66 and 2.45 log TCID50/mL, respectively [109].

Antiviral activity against MNV-1 was also found using natural extracts of *Aloe vera* and *Eriobotryae folium*. Aloin and emodin are the main active metabolites of both extracts [110].

The antiviral activity of citral, one of the main active ingredients of lemongrass oil, was studied by Gilling et al. [54]. Both 2% and 4% citral concentrations significantly reduced the infectivity of the MNV cell cultures over 6 and 24 h of exposure, compared to controls. The citral-treated MNV particles were greatly enlarged to an average size of 600 nm. However, the citral-treated MNV particles appeared intact.

Catechins are an important active ingredient in green tea. The antiviral activity of four catechins—epigallocatechin (EGC), epicatechin (EC), epigallocatechin gallate (EGCG), and epicatechin gallate (EKG)—was determined by Oh et al. [101]. EGCG, which is the main component of green tea, showed the most effective activity (EC_{50}, 12 mg/mL) against FCV.

The effect of cranberry proanthocyanidins (PAC) at concentrations of 0.30, 0.60 and 1.20 mg/mL on MNV and FCV was determined [55]. At low viral titers (~5 log10 PFU/mL), FCV was undetectable after 1 h of exposure to the three tested PAC solutions, while the MNV decreased by 2.63, 2.75 and 2.95 log10 PFU/mL from 0.15, 0.30 and 0.60 mg/mL PAC, respectively. Experiments with high viral titers (~7 log10 PFU/mL) showed similar trends but with reduced effects. Su et al. [111] showed that the viral reduction within the first 10 min of PAC treatment was $\geq$50% of the total reduction. Structural changes in PAC-treated FCVs were observed under TEM.

Su et al. [112] investigated the effect of pomegranate polyphenols on the infectivity of FCV and MNV. Viruses with high (~7 log10 PFU/mL) or low (~5 log10 PFU/mL) titers were treated with pomegranate polyphenols at concentrations of 8, 16, and 32 mg/mL. FCV was undetectable after 1 h of exposure to all pomegranate polyphenols tested, using both low and high titer. MNV with low initial titers decreased by 1.30, 2.11, and 3.61 log10 PFU/mL, and at high initial titers by 1.56, 1.48, and 1.54 log10 PFU/mL, respectively, from treatment with 4.8 and 16 mg/mL of pomegranate polyphenols. Su et al. [60] described the time-dependent effect of pomegranate polyphenols at two concentrations (2 and 4 mg/mL) on the infectivity of FCV and MNV. The reduction of viral titer by pomegranate polyphenols was found to be a rapid process, with a $\geq$50% reduction in titer within the first 20 min of treatment. The FCV and MNV-1 titers were reduced by 4.02 and 0.68 log10 PFU/mL at 2 mg/mL pomegranate polyphenols. In the presence of pomegranate polyphenols at a concentration of 4 mg/mL, the FCV and MNV titers decreased by 5.09 and 1.14 log10 PFU/mL, respectively.

The antiviral activity of myricetin, L-epicatechin, tangeretin and naringenin, belonging to the flavonoids, was established by Su et al. [113]. Flavonoids at concentrations of 0.25 and 0.5 mM were used in the research. Myricetin was found to be most effective against FCV. Low-titer FCV (~5 log10 PFU/mL) decreased to undetectable levels after treatment for 2 h with myricetin, at both 0.25- and 0.5-mM concentrations. The high titer of FCV (~7 log10 PFU/mL) was reduced by 1.73 and 3.17 log10 PFU/mL with 0.25 and 0.5 mM myricetin, respectively. L-epicatechin was less effective; at 0.25 and 0.5 mM, it reduced a high-FCV titer by 0.18 and 0.72 log10 PFU/mL and a low-FCV titer by 0.33 and 1.40 log10 PFU/mL, respectively. Tangeretin and naringenin, at both concentrations tested, did not cause any significant deactivation of both high- and low-FCV titers. All flavonoids tested at 0.25 mM showed no measurable deactivation of the low-MNV titer after 2 h of incubation. Only myricetin and 0.5 mM L-epicatechin showed a negligible reduction in low-titer MNV of 0.22 log10 PFU/mL and 0.27 log10 PFU/mL, respectively. Tangeretin and naringenin at 0.5 mM showed no measurable effect on the low-MNV titer. Su et al. [113] described the effect of myricetin, L-epicatechin, tangeretin, and naringenin at concentrations of 0.25 mM on the virus adsorption and replication of FCV and MNV. Only myricetin showed a slight measurable effect on FCV adsorption to host cells. No measurable effect of all tested flavonoids on the adsorption of MNV into host cells was observed. None of the flavonoids had any effect on virus replication.

The effect of tannin-containing compounds (glucose pentagalloyl (PGG), propyl gallate (PRG), pyrogallol (PYG)) on FCV and MNV was investigated [96]. The antiviral test was performed by measuring the infectivity of the virus after treatment with tannins by the standard TCID50 method. The results obtained indicated that PGG, PRG, and PYG had a weak damping effect on FCV and MNV.

Turmeric, as an active plant component, contains 1–5% phenolic components. The antiviral properties of curcumin have been demonstrated in the example of noroviruses.

Out of 18 phytochemicals used in the study, curcumin showed the most effective neutralizing activity against MNV. The action of curcumin depends on both its concentration and the time of its incubation with pathogens. The increase in the concentration of curcumin and the extension of the incubation time resulted in an increase in the amount of neutralized MNV. The studies used curcumin concentrations at 0.25, 0.5, 0.75, 1 and 2 mg/mL. The presence of curcumin at a concentration of 2 mg/mL neutralized approximately 91% of the MNV particles. Moreover, it was found that curcumin did not inhibit viral RNA replication [34].

Another study investigating the effects of curcumin on noroviruses was based on photodynamic therapy. This method consists of the production of reactive oxygen species with the participation of light-induced photosensitizers [114]. One of these was found to be curcumin, the effect of which on FCV and MNV was assessed after initial photoactivation with an LED diode. Although antiviral activity was found against both tested viruses, it was slightly lower for MNV. These results indicate the possibility of using photoactivated curcumin as a natural additive in the food industry, to reduce food contamination with intestinal viruses [115].

Complete inhibition of virus multiplication was observed using the extract and its fraction at concentrations of 0.1–1 mg/mL [116]. Enlarged viral capsids were observed using TEM, which could interfere with the binding of the viral surface protein to host cells. Additionally, two RCS-F1-derived polyphenolic compounds were identified that inhibited replication of the tested viruses. Test results obtained by Lee et al. [116] indicate the possibility of using black raspberry seed extract in food preservation processes.

Joshi et al. [117] assessed the antiviral effect of blueberry proanthocyanidins (B-PAC) in food matrices (apple juice and 2% milk), under simulated gastrointestinal conditions, against FCV-F9 and MNV-1. Milk, which was a much more complex food matrix compared to apple juice, inhibited the antiviral activity of B-PAC.

The results obtained by Kim et al. [41] demonstrate the inhibitory effect of fucoidans obtained from three species of brown algae (*Laminaria japonica*, LJ), *Undaria pinnatifida* (UP), and *Undaria pinnatifida* sporophyll (UPS) against MNoV, FCV, and HuNoV. The use of these compounds at a concentration of 1 mg/mL showed high antiviral activity, with a mean log decrease in viral titer of 1.1 in the plaque assays. LJ showed the greatest antiviral effectiveness (54–72% inhibition at 1 mg/mL). It was observed that pre-treatment with fuconaids interfered with the attachment of the virus to the host cell receptors. It is worth noting that, according to the authors, this is the first report in which, in in vivo studies performed on mice administered with brown algae fucoidans, a 0.6 log reduction in the MNoV titer was observed, with a corresponding improvement in the survival rates of the mice in the study group compared to the animals from the control group [41].

5.4. The Effect of Juices on Noroviruses

The aim of the research conducted by Horm and D'Souza [118] was to determine the survival of human MNV-1 and FCV-F9 norovirus surrogates in orange and pomegranate juices, and a mixture of both juices, over 0.1.2, 7, 14, and 21 days in a refrigerator (4 °C). Both juices were inoculated with each virus for 21 days, then serially diluted in a cell culture medium, and plaques were tested. MNV-1 showed no titer reduction after 21 days in orange juice. A moderate reduction in titer (1.4 log) was found in the pomegranate juice. MNV-1 was completely reduced after 7 days in a mixture of orange and pomegranate juice. FCV-F9 was completely reduced after 14 days in orange and pomegranate juice. FCV-F9 was completely reduced after 1 day in a mixture of orange and pomegranate juice.

Su et al. [112] investigated the effect of pomegranate juice (PJ) on MNV-1 and FCV-F9. Viruses with high (~7 log10 PFU/mL) or low (~5 log10 PFU/mL) titers were mixed with equal volumes of PJ and incubated for 1 h at room temperature. Post-treatment viral infectivity was assessed using standard plaque tests. PJ lowered the FCV-F9 and MNV-1 titers by 2.56 and 1.32 log10 PFU/mL, respectively, for low titer, and 1.20 and 0.06 log10 PFU/mL for high titer, respectively. The same research group [60] determined the time-dependent effect of PJ on the infectivity of food-borne replacement viruses. Each virus at ~5 log10 PFU/mL was mixed with equal volumes of PJ and incubated for 0, 10, 20, 30, 45, and 60 min at room temperature. Reduction of the viral load by PJ was found to be a rapid process. Test viruses were reduced by ≥50% during the first 20 min of treatment. The titer decreased by 3.12 and 0.79 log10 PFU/mL, respectively, for FCV-F9 and MNV-1.

The effect of cranberry juice (CJ) on MNV-1 and FCV-F9 was studied by Su et al. [55]. Both viruses with high (~7 log10 PFU/mL) and low (~5 log10 PFU/mL) titers were mixed with equal volumes of CJ (pH 2.6) and incubated for 1 h at room temperature. The standardized plaque assay was used to assess viral infectivity. CJ reduced FCV-F9 at low viral load to undetectable levels in the suspension test, and MNV-1 decreased by 2.06 log10 PFU/mL. Experiments with high viral titers showed similar effects. In another time-dependent study by Su et al. [111], FCV-F9 at low viral titers was reduced by ~5 log10 PFU/mL, over 30 min when treated with CJ (pH 2.6 and pH 7.0). MNV-1 titers similarly decreased for CJ at pH 2.6 or 7.0.

Rubus coreanus is a species of black raspberry, one that is rich in polyphenols and with anti-inflammatory, antibacterial, and antiviral properties. Oh et al. [119] compared the antiviral activity of *R. coreanus* juice (black raspberry juice, BRB) and cranberry, grape, and orange juices using plaque tests. Out of all the juices tested, BRB juice was the most effective in reducing plaque formation in MNV-1 and FCV-F9. The studies attempted to determine the mechanism of action of BRB juice on viruses. The maximum antiviral effect of BRB juice on MNV-1 was observed when it was added to the cells of murine macrophage leukemic monocytes (RAW 264.7) simultaneously with the virus (co-treatment). Pretreatment of Crandell Reese Feline Kidney (CRFK) cells or FCV-F9 with BRB juice showed significant antiviral activity. On the basis of the obtained results, it can be concluded that inhibition of viral infection with BRB juice on MNV-1 and FCV-F9 probably occurs during the internalization of virions into the cell or upon the attachment of the viral surface protein to the cell receptor.

The aim of the research carried out by Joshi et al. [120] was to determine the antiviral effect of blueberry juice (BJ) and proanthocyanidins (BB-PAC) against FCV-F9 and MNV-1 (37 °C, 24 h) by reducing plaque tests. The prophylactic and therapeutic potential of commercially available juices and BB-PAC were tested in a dose- and time-dependent manner. Based on the results obtained, it was found that both BB-PAC and BJ had an influence on the processes of adsorption and replication of the intestinal viruses studied in vitro (a reduction of MNV-1 titer to undetectable levels was observed after 3 h for 1, 2, and 5 mg/mL BBPAC, and after 6 h for BJ). Determining their antiviral activity in the presence of food matrices under simulated gastric conditions is a prerequisite for the use of these preparations in therapy [120].

The antiviral activity of *Morus alba* (mulberry juice, MA) on MNV-1 and FCV-F9 was tested by cytopathic inhibition, platelet reduction, and RNA expression assays [121]. MA juice was found to be effective in reducing the infectivity of both viruses during both initial and concomitant treatment. Juice concentrations of 0.005% (equivalent to 100% natural juice) for MNV-1 and 0.25% for FCV-F9 caused a 50% decrease in viral load. 0.1% MA juice showed approximately 60% reduction in MNV-1 polymerase gene expression, confirming suppression of viral replication. It can therefore be concluded that MA juice can inhibit MNV-1 replication and the internalization of both tested viruses.

6. Practical Application of Metabolites of Plant Origin in the Food Industry

One of the most effective strategies being developed in modern methods of food preservation is the application of active packaging containing essential oils. Biologically active phytochemicals are an integral component of the packaging material [122]. The active packaging interacts with the food, limits the growth of microorganisms, and deactivates viruses. In this way, active packaging largely eliminates the risk to public health and extends the shelf life of food products [123].

In recent years, intensive work has been carried out on the use of edible films and coatings with the addition of essential oils for food preservation. The advantage of this method has been demonstrated in experimental studies of contaminated fruit, vegetable, cheese, meat, and fish, where both naturally occurring microbial contaminants and artificially introduced strains were included.

Fabra et al. [124] developed antiviral active edible membranes by adding lipids to alginate membranes. The polymer matrix prepared in this way was enriched with two natural extracts with a high phenolic compound content, green tea extract (GTE) and grape seed extract (GSE). All of these are biologically active plant metabolites and, as such, showed antiviral activity against mouse norovirus (MNV). Edible antiviral coatings benefiting from the synergistic effect of carrageenan and GTE are also an innovative strategy used to eliminate or reduce the viral contamination of berries without significantly changing their physicochemical properties [125]. In addition, it was observed that GTE solutions significantly increased their antiviral activity against MNV if left in different pH conditions for 24 h. This may be related to the formation of catechin derivatives during the storage of this preparation [52]. Additionally, it was observed that GTE solutions significantly increased their antiviral activity against MNV if left in different pH conditions for 24 h, which was associated with the formation of catechin derivatives during the storage of this preparation [98].

It was also found that the addition of aging GTE to mildly heat-treated juices increased the deactivation of MNV-1 by over 4 logs. The synergistic action of both antiviral agents reduced the infectivity of MNV-1, which confirms the hypothesis that GTE can be used as an additional control agent that improves food safety [99].

The antiviral effect of *Lindera obtusiloba* leaf extract (LOLE) on MNV-1, stemming from the synergistic action of several compounds with pinene as the key molecule, was tested on fresh lettuce, cabbage, and oysters. An hour-long incubation at 25 °C with LOLE at a concentration of 12 mg/mL resulted in a significant reduction of the viral plaques (plaque formation) of MNV-1 in lettuce (76.4%), cabbage (60.0%), and oysters (38.2%). The results of these studies suggest that LOLE can deactivate norovirus and can be used as a natural disinfectant and preservative in fresh food products [23].

Antiviral activity was also found by analyzing the effects of natural *Aloe vera* and *Eriobotryae folium* extracts. Aloin and emodin, the main active phytochemicals in the extracts of these plants, showed a preservative effect. This was confirmed, based on the results of studies in which fresh cabbage was inoculated with MNV-1 on its surface [110].

Chitosan films supplemented with green tea extract (GTE) can also be applied as active packaging materials. Chitosan is a non-toxic polysaccharide polymer that is used as an ingredient in edible packaging films, where its antimicrobial activity is used to increase the shelf life of food products. Natural plant metabolites with antimicrobial activity, e.g., essential oils and plant extracts, may be considered as possible components of edible films. It is important that all the above phytochemicals have GRAS (Generally Recognized as Safe) status. It was found that, after 24 h of incubation with the addition of 5 and 10% GTE, there was a significant reduction in the MNV-1 titer by 1.6 and 4.5 logs, respectively. Films containing 15% GTE reduced MNV-1 to undetectable levels [21].

The encapsulation of essential oils (capsules with a size of 1–1000 μm (microcapsules) or 1–100 nm (nanocapsules) offers another opportunity to preserve food using essential oils [122]. Polyethylene, carbohydrates (starch, cellulose, chitosan), proteins (casein, albumin, gelatin), fats (fatty acids, waxes, paraffin), and gums (alginates, carrageenan, acacia) are the materials most often used in this technology. Essential oils enclosed in capsules maintain greater stability, and this determines their optimal antimicrobial properties [126].

7. Summary

Noroviruses are highly resistant to environmental factors, so they can be efficiently transmitted through food, water, or surfaces contaminated with them, and pose a potential threat to public health. Antiviral metabolites of plant origin have important advantages over synthetic preservatives used as fresh food disinfectants because they are effective at safe dosages, are generally available, and use the inability of microorganisms to become resistant to plant-based viroids. As secondary metabolites of plants, essential oils, and plant extracts are part of their defense system against pathogens. Therefore, they often exhibit antimicrobial, including antiviral, activities. The activity spectrum of plant metabolites is diverse. The effectiveness of plant preparations and the possibility of their use in fighting intestinal viruses such as noroviruses is primarily dependents on the qualitative and quantitative composition of biologically active phytochemicals, and their concentration in food.

Author Contributions: Conceptualization, J.S. and D.W.; writing-original draft preparation, J.S., D.W., A.J.-K. and M.F.-M.; writing-review and editing, J.S. and D.W.; supervision, project administration, funding acquisition, J.S., D.W. and I.C.-K. All authors have read and agreed to the published version of the manuscript.

Funding: This research received no external funding.

Institutional Review Board Statement: Not applicable.

Informed Consent Statement: Not applicable.

Data Availability Statement: Not applicable.

Conflicts of Interest: The authors declare no conflict of interest.

References

1. Miranda, R.C.; Schaffner, D.W. Virus risk in the food supply chain. *Curr. Opin. Food Sci.* **2019**, *30*, 43–48. [CrossRef]
2. Sanchez, G.; Bosch, A. Survival of enteric viruses in the environment and food. *Viruses Foods* **2016**, *26*, 367–392.
3. Ahmed, S.M.; Hall, A.J.; Robinson, A.E.; Verhoef, L.; Premkumar, P.; Parashar, U.D.; Koopmans, M.; Lopman, B.A. Global prevalence of norovirus in cases of gastroenteritis: A systematic review and meta-analysis. *Lancet Infect. Dis.* **2014**, *14*, 725–730. [CrossRef]
4. Havelaar, A.H.; Kirk, M.D.; Torgerson, P.R.; Gibb, H.J.; Hald, T.; Lake, R.J.; Praet, N.; Bellinger, D.C.; de Silva, N.R.; Gargouri, N.; et al. World Health Organization global estimates and regional comparisons of the burden of foodborne disease in 2010. *PLoS Med.* **2015**, *12*, e1001923. [CrossRef] [PubMed]
5. Bartsch, S.M.; Lopman, B.A.; Ozawa, S.; Hall, A.J.; Lee, B.Y. Global economic burden of norovirus gastroenteritis. *PLoS ONE* **2016**, *11*, e0151219. [CrossRef] [PubMed]
6. World Health Organization. *WHO Estimates of the Global Burden of Foodborne Diseases*; World Health Organization: Geneva, Switzerland, 2018.
7. Pires, S.M.; Fischer-Walker, C.L.; Lanata, C.F.; Devleesschauwer, B.; Hall, A.J.; Kirk, M.D.; Duarte, A.S.R.; Black, R.E.; Angulo, F.J. Aetiology-specific estimates of the global and regional incidence and mortality of diarrhoeal diseases commonly transmitted through food. *PLoS ONE* **2015**, *10*, e0142927. [CrossRef]
8. Centers for Disease Control and Prevention. Surveillance for Foodborne Disease Outbreaks United States, 2017: Annual Report. Available online: https://www.cdc.gov/fdoss/pdf/2017_FoodBorneOutbreaks_508.pdf (accessed on 6 June 2020).
9. Neethirajan, S.; Ahmed, S.R.; Chand, R.; Buozis, J.; Nagy, E. Recent advances in biosensor development for foodborne virus detection. *Nanotheranostics* **2017**, *1*, 272–295. [CrossRef]
10. EFSA (European Food Safety Authority). The European Union summary report on trends and sources of zoonoses, zoonotic agents and food-borne outbreaks in 2017. *EFSA J.* **2018**, *16*, e5500.
11. EFSA (European Food Safety Authority). The European Union One Health 2018 Zoonoses Report. *EFSA J.* **2019**, *17*, e05926.
12. EFSA (European Food Safety Authority). The European Union One Health 2019 Zoonoses Report, European Food Safety Authority European Centre for Disease Prevention and Control. *EFSA J.* **2021**, *19*, e6406.
13. Centers for Disease Control and Prevention. Norovirus Worldwide. 2020. Available online: https://www.cdc.gov/nceh/vsp/desc/about_inspections.htm (accessed on 6 June 2020).
14. Bosch, A.; Gkogka, E.; Le Guyader, F.S.; Loisy-Hamon, F.; Lee, A.; van Lieshout, L.; Marthi, B.; Myrmel, M.; Sansom, A.; Schultz, A.C.; et al. Foodborne viruses: Detection, risk assessment, and control options in food processing. *Int. J. Food Microbiol.* **2018**, *285*, 110–128. [CrossRef]

15. Pexara, A.; Govaris, A. Foodborne viruses and innovative non-thermal food-processing technologies. *Food* **2020**, *9*, 1520. [CrossRef] [PubMed]
16. Robilotti, E.; Deresinski, S.; Pinsky, B.A. Norovirus. *Clin. Microbiol. Rev.* **2015**, *28*, 134–164. [CrossRef]
17. Callejon, R.M.; Rodriguez-Naranjo, M.I.; Ubeda, C.; Hornedo-Ortega, R.; Garcia-Parrilla, M.C.; Troncoso, A.M. Reported foodborne outbreaks due to fresh produce in the United States and European Union: Trends and causes. *Foodborne Pathog. Dis.* **2015**, *12*, 32–38. [CrossRef] [PubMed]
18. Machado-Moreira, B.; Richards, K.; Brennan, F.; Abram, F.; Burgess, C.M. Microbial contamination of fresh produce: What, where, and how? *Compr. Rev. Food Sci. Food Saf.* **2019**, *18*, 1727–1750. [CrossRef]
19. Lopman, B.A.; Reacher, M.H.; Vipond, I.B.; Sarangi, J.; Brown, D.W. Clinical manifestation of norovirus gastroenteritis in health care settings. *Clin. Infect. Dis.* **2004**, *39*, 318–324. [CrossRef] [PubMed]
20. Moore, M.D.; Goulter, R.M.; Jaykus, L.A. Human norovirus as a foodborne pathogen: Challenges and developments. *Annu. Rev. Food Sci. Technol.* **2015**, *6*, 411–433. [CrossRef] [PubMed]
21. Amankwaah, C.; Li, J.; Lee, J.; Pascall, M.A. Antimicrobial activity of chitosan-based films enriched with green tea extracts on murine norovirus, *Escherichia coli*, and *Listeria innocua*. *Int. J. Food Sci.* **2020**, *2*, 3941924. [CrossRef] [PubMed]
22. Rajiuddin, S.M.; Vigre, H.; Musavian, H.S.; Kohle, S.; Krebs, N.; Hansen, T.B.; Gantzer, C.; Schultz, A.C. Inactivation of hepatitis A virus and murine norovirus on surfaces of plastic, steel and raspberries using steam-ultrasound treatment. *Food Environ. Virol.* **2020**, *12*, 295–309. [CrossRef]
23. Solis-Sanchez, D.; Rivera-Piza, A.; Lee, S.; Kim, J.; Kim, B.; Choi, J.B.; Kim, Y.W.; Ko, G.P.; Song, M.J.; Lee, S.J. Antiviral effects of *Lindera obtusiloba* leaf extract on murine norovirus-1 (MNV-1), a human norovirus surrogate, and potential application to model Foods. *Antibiotics* **2020**, *9*, 697. [CrossRef]
24. Green, K.Y.; Ando, T.; Balayan, M.S.; Berke, T.; Clarke, I.N.; Estes, M.K. Taxonomy of the caliciviruses. *J. Infect. Dis.* **2000**, *181*, 322–330. [CrossRef] [PubMed]
25. Vinjé, J. Advances in laboratory methods for detection and typing of norovirus. *J. Clin. Microbiol.* **2015**, *53*, 373–381. [CrossRef]
26. Hoa Tran, T.N.; Trainor, E.; Nakagomi, T.; Cunliffe, N.A.; Nakagomi, O. Molecular epidemiology of noroviruses associated with acute sporadic gastroenteritis in children: Global distribution of genogroups, genotypes and GII.4 variants. *J. Clin. Virol.* **2013**, *56*, 185–193. [CrossRef] [PubMed]
27. Kroneman, A.; Verhoef, L.; Harris, J.; Vennema, H.; Duizer, E.; van Duynhoven, Y.; Gray, J.; Iturriza, M.; Böttiger, B.; Falkenhorst, G.; et al. Analysis of integrated virological and epidemiological reports of norovirus outbreaks collected within the foodborne viruses in Europe network from 1 July 2001 to 30 June 2006. *J. Clin. Microbiol.* **2008**, *46*, 2959–2965. [CrossRef] [PubMed]
28. Teunis, P.F.; Moe, C.L.; Liu, P.; Miller, S.E.; Lindesmith, L.; Baric, R.S.; Le Pendu, J.; Calderon, R.L. Norwalk virus: How infectious is it? *J. Med. Virol.* **2008**, *80*, 1468–1476. [CrossRef] [PubMed]
29. Lysen, M.; Thorhagen, M.; Brytting, M.; Hjertqvist, M.; Andersson, Y.; Hedlund, K.O. Genetic diversity among food-borne and waterborne norovirus strains causing outbreaks in Sweden. *J. Clin. Microbiol.* **2009**, *47*, 2411–2418. [CrossRef]
30. Vega, E.; Barclay, L.; Gregoricus, N.; Shirley, S.H.; Lee, D.; Vinje, J. Genotypic and epidemiologic trends of norovirus outbreaks in the United States, 2009 to 2013. *J. Clin. Microbiol.* **2014**, *52*, 147–155.
31. Hassard, F.; Sharp, J.H.; Taft, H.; LeVay, L.; Harris, J.P.; McDonald, J.E.; Tuson, K.; Wilson, J.; Jones, D.L.; Malham, S.K. Critical review on the public health impact of norovirus contamination in shellfish and the environment: A UK perspective. *Food Environ. Virol.* **2017**, *9*, 123–141. [CrossRef]
32. Bansal, A.; Chhabra, V.; Rawal, R.K.; Sharma, S. Chemometrics: A new scenario in herbal drug standardization. *J. Pharm. Anal.* **2014**, *4*, 223–233. [CrossRef]
33. Bakkali, F.; Averbeck, S.; Averbeck, D.; Idaomar, M. Biological effects of essential oils—A review. *Food Chem. Toxicol.* **2008**, *46*, 446–475. [CrossRef]
34. Yang, M.; Lee, G.; Si, J.; Lee, S.J.; You, H.J.; Ko, G. Curcumin shows antiviral properties against norovirus. *Molecules* **2016**, *21*, 1401. [CrossRef] [PubMed]
35. Drevinskas, T.; Mickiene, R.; Maruska, A.; Stankevicius, M.; Tiso, N.; Salomskas, A.; Lelesius, R.; Karpovaite, A.; Ragazinskiene, O. Confirmation of antiviral properties of medicinal plants via chemical analysis, machine learning methods and antiviral tests: Methodological approach. *Anal. Methods* **2018**, *10*, 1875–1885. [CrossRef]
36. Lee, H.Y.; Yum, J.H.; Rho, Y.K.; Oh, S.J.; Choi, H.S.; Chang, H.B.; Choi, D.H.; Leem, M.J.; Choi, E.J.; Ryu, J.M.; et al. Inhibition of HCV replicon cell growth by 2-arylbenzofuran derivatives isolated from Mori Cortex Radicis. *Planta Med.* **2007**, *73*, 1481–1485. [CrossRef] [PubMed]
37. Eggers, M.; Schwebke, I.; Suchomel, M.; Fotheringham, V.; Gebel, J.; Meyer, B.; Morace, G.; Roedger, H.J.; Roques, C.; Visa, P.; et al. The European tiered approach for virucidal efficacy testing-rationale for rapidly selecting disinfectants against emerging and re-emerging viral diseases. *Eurosurveillance* **2021**, *26*, 2000708. [CrossRef]
38. Musarra-Pizzo, M.; Pennisi, R.; Ben-Amor, I.; Mandalari, G.; Sciortino, M.T. Antiviral activity exerted by natural products against human viruses. *Viruses* **2021**, *13*, 828. [CrossRef]
39. Mukherjee, P.K. Antiviral evaluation of herbal drugs. *Qual. Control Eval. Herb. Drugs* **2019**, 599–628. [CrossRef]
40. Atanasov, A.G.; Waltenberger, B.; Pferschy-Wenzig, E.M.; Linder, T.; Wawrosch, C. Discovery and resupply of pharmacologically active plant-derived natural products: A review. *Biotechnol. Adv.* **2015**, *33*, 1582–1614. [CrossRef]

41. Kim, H.; Lim, C.Y.; Lee, D.B.; Seok, J.H.; Kim, K.H.; Chung, M.S. Inhibitory effects of *Laminaria japonica* fucoidans against noroviruses. *Viruses* **2020**, *12*, 997. [CrossRef] [PubMed]
42. Van Dycke, J.; Cuvry, A.; Knickmann, J.; Ny, A.; Rakers, S.; Taube, S.; de Witte, P.; Neyts, J.; Rocha-Pereira, J. Infection of zebrafish larvae with human norovirus and evaluation of the In Vivo efficacy of small-molecule inhibitors. *Nat. Protoc.* **2021**, *16*, 1830–1849. [CrossRef]
43. Thomford, N.E.; Senthebane, D.A.; Rowe, A.; Munro, D.; Seele, P.; Maroyi, A.; Dzobo, K. Natural products for drug discovery in the 21st century: Innovations for novel drug discovery. *Int. J. Mol. Sci.* **2018**, *19*, 1578. [CrossRef]
44. Lee, S.J.; Si, J.; Yun, H.S.; Ko, G.P. Effect of temperature and relative humidity on the survival of foodborne viruses during food storage. *Appl. Environ. Microb.* **2015**, *81*, 2075–2081. [CrossRef] [PubMed]
45. Huang, X.; Lao, Y.; Pan, Y.; Chen, Y.; Zhao, H.; Gong, L.; Xie, N.; Mo, C.H. Synergistic antimicrobial effectiveness of plant essential oil and its application in seafood preservation: A review. *Molecules* **2021**, *26*, 307.
46. Ni, Z.J.; Wang, X.; Shen, Y.; Thakur, K.; Han, J.; Zhang, J.G.; Hu, F.; Wei, Z.J. Recent updates on the chemistry, bioactivities, mode of action, and industrial applications of plant essential oils. *Trends Food Sci. Technol.* **2021**, *110*, 78–89. [CrossRef]
47. Bertrand, I.; Schijven, J.F.; Sanchez, G.; Wyn-Jones, P.; Ottoson, J.; Morin, T.; Muscillo, M.; Verani, M.; Nasser, A.; de Roda Husman, A.M.; et al. The impact of temperature on the inactivation of enteric viruses in food and water: A review. *J. Appl. Microbiol.* **2012**, *112*, 1059–1074. [CrossRef] [PubMed]
48. Ben-Shabat, S.; Yarmolinsky, L.; Porat, D.; Dahan, A. Antiviral effect of phytochemicals from medicinal plants: Applications and drug delivery strategies. *Drug Deliv. Transl. Res.* **2020**, *10*, 354–367. [CrossRef] [PubMed]
49. Andricoplo, A.D.; Ceron-Carrasco, J.P.; Mozzarelli, A. Bridging molecular docking to molecular dynamics in exploring ligand-protein recognition process: An overview. *Front. Pharmacol.* **2018**, *9*, 438.
50. Perez, R.M. Antiviral activity of compounds isolated from plants. *Pharm. Biol.* **2003**, *41*, 107–157. [CrossRef]
51. Seo, D.J.; Jeon, S.B.; Oh, H.; Lee, B.H.; Lee, S.Y.; Oh, S.H.; Jung, J.Y.; Choi, C. Comparison of the antiviral activity of flavonoids against murine norovirus and feline calicivirus. *Food Control* **2016**, *60*, 25–30. [CrossRef]
52. Randazzo, W.; Falco, I.; Aznar, R.; Sanchez, G. Effect of green tea extract on enteric viruses and its application as natural sanitizer. *Food Microbiol.* **2017**, *66*, 150–156. [CrossRef]
53. Gilling, D.H.; Kitajima, M.; Torrey, J.T.; Bright, K.R. Antiviral efficacy and mechanisms of action of oregano essential oil and its primary component carvacrol against murine norovirus. *J. Appl. Microbiol.* **2014**, *116*, 1149–1163. [CrossRef]
54. Gilling, D.H.; Kitajima, M.; Torrey, J.T.; Bright, K.R. Mechanisms of antiviral action of plant antimicrobials against murine norovirus. *Appl. Environ. Microbiol.* **2014**, *80*, 4898–4910. [CrossRef] [PubMed]
55. Su, X.; Howell, A.B.; D'Souza, D.H. The effect of cranberry juice and cranberry proanthocyanidins on the infectivity of human enteric viral surrogates. *Food Microbiol.* **2010**, *27*, 535–540. [CrossRef] [PubMed]
56. Pilau, M.R.; Alves, S.H.; Weiblen, R.; Arenhart, S.; Cueto, A.P.; Lovato, L.T. Antiviral activity of the *Lippia graveolens* (Mexican oregano) essential oil and its main compound carvacrol against human and animal viruses. *Braz. J. Microbiol.* **2011**, *42*, 1616–1624. [CrossRef] [PubMed]
57. Kovač, K.; Diez-Valcarce, M.; Raspor, P.; Hernández, M.; Rodríguez-Lázaro, D. Natural plant essential oils do not inactivate non-enveloped enteric viruses. *Food Environ. Virol.* **2012**, *4*, 209–212. [CrossRef] [PubMed]
58. Cliver, D.O. Capsid and infectivity in virus detection. *Food Environ. Virol.* **2009**, *1*, 123–128. [CrossRef] [PubMed]
59. Tubiana, T.; Boulard, Y.; Bressanelli, S. Dynamics and asymmetry in the dimer of the norovirus major capsid protein. *PLoS ONE* **2017**, *12*, e0182056. [CrossRef] [PubMed]
60. Su, X.; Sangster, M.Y.; D'Souza, D.H. Time-dependent effects of pomegranate juice and pomegranate polyphenols on foodborne virus reduction. *Foodborne Pathog. Dis.* **2011**, *8*, 1177–1183. [CrossRef] [PubMed]
61. Li, D.; Baert, L.; Zhang, D.; Xia, M.; Zhong, W.; Van Coillie, E.; Xiang, J.; Uyttendaele, M. The effect of grape seed extract on human norovirus GII.4 and murine norovirus-1 in viral suspensions, on stainless steel discs, and in lettuce wash water. *Appl. Environ. Microbiol.* **2012**, *78*, 7572–7578. [CrossRef]
62. Koch, C.; Reichling, J.; Schneele, J.; Schnitzler, P. Inhibitory effect of essential oils against herpes simplex virus type 2. *Phytomedicine* **2008**, *15*, 71–78. [CrossRef]
63. Essoil Database. Available online: http://www.nipgr.ac.in/Essoildb/ (accessed on 6 May 2021).
64. Leyva-López, N.; Gutiérrez-Grijalva, E.P.; Vazquez-Olivo, G.; Heredia, J.B. Essential oils of oregano: Biological activity beyond their antimicrobial properties. *Molecules* **2017**, *22*, 989. [CrossRef]
65. Swamy, M.K.; Akhtar, M.S.; Sinniah, U.R. Antimicrobial properties of plant essential oils against human pathogens and their mode of action: An updated review. *Evid. Based Complement. Altern. Med.* **2016**, *2016*, 3012462. [CrossRef]
66. Adam, K.; Sivropoulou, A.; Kokkini, S.; Lanaras, T.; Arsenakis, M. Antifungal activities of *Origanum vulgare* subsp. *hirtum*, *Mentha spicata*, *Lavandula angustifolia*, and *Salvia fruticosa* essential oils against human pathogenic fungi. *J. Agric. Food Chem.* **1998**, *46*, 1739–1745. [CrossRef]
67. Elizaquivel, P.; Azizkhani, M.; Aznar, R.; Sanchez, G. The effect of essential oils on norovirus surrogates. *Food Control* **2013**, *32*, 275–278. [CrossRef]
68. Cutillas, A.B.; Carrasco, A.; Martinez-Gutierrez, R.; Tomas, V.; Tudela, J. *Thymus mastichina* L. essential oils from Murcia (Spain): Composition and antioxidant, antienzymatic and antimicrobial bioactivities. *PLoS ONE* **2018**, *13*, e0190790.

69. Rodrigues, M.; Lopes, A.C.; Vaz, F.; Filipe, M.; Alves, G.; Ribeiro, M.P.; Coutinho, P.; Araujo, A.R.T.S. *Thymus mastichina*: Composition and biological properties with a focus on antimicrobial activity. *Pharmaceuticals* **2020**, *19*, 479. [CrossRef]
70. Fraternale, D.; Giamperi, L.; Ricci, D. Chemical composition and antifungal activity of essential oil obtained from In Vitro plants of *Thymus mastichina* L. *J. Essent. Oil Res.* **2003**, *15*, 278–281. [CrossRef]
71. Borugă, O.; Jianu, C.; Mişcă, C.; Goleţ, I.; Gruia, A.T.; Horhat, F.G. *Thymus vulgaris* essential oil: Chemical composition and antimicrobial activity. *J. Med. Life* **2014**, *7*, 56–60.
72. Kryvtsova, M.V.; Salamon, I.; Koscova, J.; Bucko, D.; Spivak, M. Antimicrobial, antibiofilm and biochemichal properties of *Thymus vulgaris* essential oil against clinical isolates of opportunistic infections. *Biosyst. Divers.* **2019**, *27*, 270–275. [CrossRef]
73. Mahboubi, M.; Heidarytabar, R.; Mahdizadeh, E. Antibacterial activity of *Zataria multiflora* essential oil and its main components against *Pseudomonas aeruginosa*. *Herba Pol.* **2017**, *63*, 18–24. [CrossRef]
74. Saei-Dehkordi, S.S.; Tajik, H.; Moradi, M.; Khalighi-Sigaroodi, F. Chemical composition of essential oils in *Zataria multiflora* Boiss. from different parts of Iran and their radical scavenging and antimicrobial activity. *Food Chem. Toxicol.* **2010**, *48*, 1562–1567. [CrossRef] [PubMed]
75. Eftekhar, F.; Zamani, S.; Yusefzadi, M.; Hadian, J.; Ebrahimi, S.N. Antibacterial activity of *Zataria multiflora* Boiss essential oil against extended spectrum β lactamase produced by urinary isolates of *Klebsiella pneumoniae*. *Jundishapur J. Microbiol.* **2011**, *4*, S43–S49.
76. Mahboubi, M.; Bidgoli, F.G. Antistaphylococcal activity of *Zataria multiflora* essential oil and its synergy with vancomycin. *Phytomedicine* **2010**, *17*, 548–550. [CrossRef] [PubMed]
77. Naeini, A.R.; Nazeri, M.; Shokri, H. Antifungal activity of *Zataria multiflora*, *Pelargonium graveolens* and *Cuminum cyminum* essential oils towards three species of *Malassezia* isolated from patients with pityriasis versicolor. *J. Mycol. Med.* **2011**, *21*, 87–91. [CrossRef]
78. Selles, S.M.A.; Kouidri, M.; Belhamiti, B.T.; Amrane, A.A. Chemical composition, In-Vitro antibacterial and antioxidant activities of *Syzygium aromaticum* essential oil. *J. Food Meas. Charact.* **2020**, *13*, 1–7. [CrossRef]
79. Saeed, A.; Shahwar, D. Evaluation of biological activities of the essential oil and major component of *Syzygium aromaticum*. *J. Anim. Plant Sci.* **2015**, *25*, 1095–1099.
80. Kizil, S.; Hasimi, N.; Tolan, V.; Kilinc, E.; Karatas, H. Chemical composition, antimicrobial and antioxidant activities of hyssop (*Hyssopus officinalis* L.) essential oil. *Not. Bot. Horti Agrobot. Cluj Napoca* **2010**, *38*, 99–103.
81. Mahboubi, M.; Haghi, G.; Kazempour, N. Antimicrobial activity and chemical composition of *Hyssopus officinalis* L. essential oil. *J. Biol. Act. Prod. Nat.* **2011**, *1*, 132–137.
82. Oh, M.; Chung, M.S. Effects of oils and essential oils from seeds of *Zanthoxylum schinifolium* against foodborne viral surrogates. *Evid. Based Complement. Altern. Med.* **2014**, *8*, 135797.
83. Diao, W.R.; Hu, Q.P.; Feng, S.S.; Li, W.Q.; Xu, J.G. Chemical composition and antibacterial activity of the essential oil from green huajiao (*Zanthoxylum schinifolium*) against selected foodborne pathogens. *J. Agric. Food Chem.* **2013**, *61*, 6044–6049. [CrossRef] [PubMed]
84. Dharmadasa, R.M.; Abeysinghe, D.C.; Dissanayake, D.M.N.; Fernando, N.S. Leaf essential oil composition, antioxidant activity, total phenolic content and total flavonoid content of *Pimenta dioica* (L.) Merr (Myrtaceae): A superior quality spice grown in Sri Lanka. *Univers. J. Agric. Res.* **2015**, *3*, 49–52.
85. Mérida-Reyes, M.S.; Muñoz-Wug, M.A.; Oliva-Hernández, B.E.; Gaitán-Fernández, I.C.; Simas, D.L.R.; Ribeiro da Silva, A.J.; Pérez-Sabino, J.F. Composition and antibacterial activity of the essential oil from *Pimenta dioica* (L.) Merr. from Guatemala. *Medicines* **2020**, *7*, 59. [CrossRef] [PubMed]
86. Milenkovic, A.; Stanojević, J.; Stojanović-Radić, Z.; Pejčić, M.; Cvetkovic, D.; Zvezdanović, J.B.; Stanojević, L. Chemical composition, antioxidative and antimicrobial activity of allspice (*Pimenta dioica* (L.) Merr.) essential oil and extract. *Adv. Technol.* **2020**, *9*, 27–36. [CrossRef]
87. Majewska, W.; Kozłowska, M.; Gruczyńska-Sękowska, E.; Kowalska, D.; Tarnowska, K. Lemongrass (*Cymbopogon citratus*) essential oil: Extraction, composition, bioactivity and uses for food preservation—A review. *Pol. J. Food Nutr. Sci.* **2019**, *69*, 327–341. [CrossRef]
88. Premathilake, U.G.A.T.; Wathugala, D.L.; Dharmadasa, R.M. Evaluation of chemical composition and assessment of antimicrobial activities of essential oil of lemongrass (*Cymbopogon citratus* (DC.) Stapf). *Int. J. Minor Fruits Med. Aromat. Plants* **2018**, *4*, 13–19.
89. Kim, Y.W.; You, H.J.; Lee, S.; Kim, B.; Kim, D.K.; Choi, J.B.; Kim, J.A.; Lee, H.J.; Joo, I.S.; Lee, J.S.; et al. Inactivation of norovirus by lemongrass essential oil using a norovirus surrogate system. *J. Food Prot.* **2017**, *80*, 1293–1302. [CrossRef]
90. Abad, M.J.; Bedoya, L.M.; Apaza, L.; Bermejo, P. The *Artemisia* L. genus: A review of bioactive essential oils. *Molecules* **2012**, *17*, 2542–2566. [CrossRef] [PubMed]
91. Choi, H.S. The variation of the major compounds of *Artemisia princeps* var. *orientalis* (Pampan) Hara essential oil by harvest year. *Korean J. Food Nutr.* **2015**, *28*, 533–543. [CrossRef]
92. Chung, M.S. Antiviral activities of *Artemisia princeps* var. *orientalis* essential oil and its α-thujone against norovirus surrogates. *Food Sci. Biotechnol.* **2017**, *28*, 1457–1461. [CrossRef]
93. Su, X.; D'Souza, D.H. Grape seed extract for foodborne virus reduction on produce. *Food Microbiol.* **2013**, *34*, 1–6. [CrossRef]
94. Joshi, S.S.; Su, X.; D'Souza, D.H. Antiviral effects of grape seed extract against feline calicivirus, murine norovirus, and hepatitis A virus in model food systems and under gastric conditions. *Food Microbiol.* **2015**, *52*, 1–10. [CrossRef]

95. Oh, M.; Bae, S.Y.; Chung, M.S. Mulberry (*Morus alba*) seed extract and its polyphenol compounds for control of foodborne viral surrogates. *J. Korean Soc. Appl. Biol. Chem.* **2013**, *56*, 655–660. [CrossRef]
96. Ueda, K.; Kawabata, R.; Irie, T.; Nakai, Y.; Tohya, Y.; Sakaguchi, T. Inactivation of pathogenic viruses by plant-derived tannins: Strong effects of extracts from persimmon (*Diospyros kaki*) on a broad range of viruses. *PLoS ONE* **2013**, *8*, e55343. [CrossRef]
97. Falco, I.; Randazzo, W.; Rodriguez-Diaz, J.; Gozalbo-Rovira, R.; Luque, D.; Aznar, R.; Sanchez, G. Antiviral activity of aged green tea extract in model food systems and under gastric conditions. *Int. J. Food Microbiol.* **2019**, *2*, 101–106. [CrossRef] [PubMed]
98. Falco, I.; Randazzo, W.; Gomez-Mascaraque, L.G.; Aznar, R.; Lopez-Rubio, A.; Sanchez, G. Fostering the antiviral activity of green tea extract for sanitizing purposes through controlled storage conditions. *Food Control* **2018**, *84*, 485–492. [CrossRef]
99. Falco, I.; Díaz-Reolid, A.; Randazzo, W.; Sanchez, G. Green tea extract assisted low-temperature pasteurization to inactivate enteric viruses in juices. *Int. J. Food Microbiol.* **2020**, *334*, 108809. [CrossRef]
100. Randazzo, W.; Costantini, V.; Morantz, E.K.; Vinje, J. Human intestinal enteroids to evaluate human norovirus GII.4 inactivation by aged-green tea. *Front. Microbiol.* **2020**, *18*, 1917. [CrossRef] [PubMed]
101. Oh, E.G.; Kim, K.L.; Shin, S.B.; Son, K.T.; Lee, H.J.; Kim, T.H.; Kim, Y.M.; Cho, E.J.; Kim, D.K.; Lee, E.W.; et al. Antiviral activity of green tea catechins against feline calicivirus as a surrogate for norovirus. *Food Sci. Biotechnol.* **2013**, *22*, 593–598. [CrossRef]
102. Aboubakr, H.A.; Nauertz, A.; Luong, N.T.; Agrawal, S.; El-Sohaimy, S.A.; Youssef, M.M.; Goyal, S.M. In Vitro antiviral activity of clove and ginger aqueous extracts against feline calicivirus, a surrogate for human norovirus. *J. Food Prot.* **2016**, *79*, 1001–1012. [CrossRef]
103. Seo, D.J.; Choi, C. Inhibition of murine norovirus and feline calicivirus by edible herbal extracts. *Food Environ. Virol.* **2017**, *9*, 35–44. [CrossRef]
104. Park, S.Y.; Kang, S.; Ha, S.D. Antimicrobial effects of vinegar against norovirus and *Escherichia coli* in the traditional Korean vinegared green laver (*Enteromorpha intestinalis*) salad during refrigerated storage. *Int. J. Food Microbiol.* **2016**, *5*, 208–214. [CrossRef]
105. Lee, H.M.; Kim, S.J.; Lee, J.; Park, B.; Yang, J.S.; Ha, S.D.; Choi, C.; Ha, J.H. Capsaicinoids reduce the viability of a norovirus surrogate during kimchi fermentation. *LWT* **2019**, *115*, 108460. [CrossRef]
106. Cheng, D.; Sun, L.; Zou, S.; Chen, J.; Mao, H.; Zhang, Y.; Liao, N.; Zhang, R. Antiviral effects of *Houttuynia cordata* polysaccharide extract on murine norovirus-1 (MNV-1)-a human norovirus surrogate. *Molecules* **2019**, *24*, 1835. [CrossRef]
107. Joshi, S.S.; Dice, L.; D'Souza, D.H. Aqueous extracts of hibiscus sabdariffa calyces decrease hepatitis A virus and human norovirus surrogate titers. *Food Environ. Virol.* **2015**, *7*, 366–373. [CrossRef]
108. Sanchez, C.; Aznar, R.; Sanchez, G. The effect of carvacrol on enteric viruses. *Int. J. Food Microbiol.* **2015**, *192*, 72–76. [CrossRef]
109. Sanchez, G.; Aznar, R. Evaluation of natural compounds of plant origin for inactivation of enteric viruses. *Food Environ. Virol.* **2015**, *7*, 183–187. [CrossRef] [PubMed]
110. Ng, Y.C.; Kim, Y.W.; Ryu, S.; Lee, A.; Lee, J.S.; Song, M.J. Suppression of norovirus by natural phytochemicals from *Aloe vera* and *Eriobotryae folium*. *Food Control* **2017**, *73*, 1362–1370. [CrossRef]
111. Su, X.; Howell, A.B.; D'Souza, D.H. Antiviral effects of cranberry juice and cranberry proanthocyanidins on foodborne viral surrogates—A time dependence study In Vitro. *Food Microbiol.* **2010**, *27*, 985–991. [CrossRef] [PubMed]
112. Su, X.; Sangster, M.Y.; D'Souza, D.H. In Vitro effects of pomegranate juice and pomegranate polyphenols on foodborne viral surrogates. *Foodborne Pathog. Dis.* **2010**, *7*, 1473–1479. [CrossRef] [PubMed]
113. Su, X.; D'Souza, D.H. Naturally occurring flavonoids against human norovirus surrogates. *Food Environ. Virol.* **2013**, *5*, 97–102. [CrossRef]
114. Narayanan, A.; Kehn-Hall, K.; Senina, S.; Lundberg, L.; Duyne, R.V.; Guendel, I.; Das, R.; Baer, A.; Bethel, L.; Turell, M.; et al. Curcumin inhibits rift valley fever virus replication in human cells. *J. Biol. Chem.* **2012**, *287*, 33198–33214. [CrossRef]
115. Randazzo, W.; Aznar, R.; Sanchez, G. Curcumin-mediated photodynamic inactivation of norovirus surrogates. *Food Environ. Virol.* **2016**, *8*, 244–250. [CrossRef] [PubMed]
116. Lee, J.H.; Bae, S.Y.; Oh, M.; Seok, J.H.; Kim, S.; Chung, Y.B.; Gowda, K.G.; Mun, J.Y.; Chung, M.S.; Kim, K.H. Antiviral effects of black raspberry (*Rubus coreanus*) seed extract and its polyphenolic compounds on norovirus surrogates. *Biosci. Biotechnol. Biochem.* **2016**, *80*, 1196–1204. [CrossRef] [PubMed]
117. Joshi, S.; Howell, A.B.; D'Souza, D.H. Blueberry proanthocyanidins against human norovirus surrogates in model foods and under simulated gastric conditions. *Food Microbiol.* **2017**, *63*, 263–267. [CrossRef] [PubMed]
118. Horm, K.M.; D'Souza, D.H. Survival of human norovirus surrogates in milk, orange, and pomegranate juice, and juice blends at refrigeration (4 °C). *Food Microbiol.* **2011**, *28*, 1054–1061. [CrossRef] [PubMed]
119. Oh, M.; Bae, S.Y.; Lee, J.H.; Cho, K.J.; Kim, K.H.; Chung, M.S. Antiviral effects of black raspberry (*Rubus coreanus*) juice on foodborne viral surrogates. *Foodborne Pathog. Dis.* **2012**, *9*, 915–921. [CrossRef]
120. Joshi, S.S.; Howell, A.B.; D'Souza, D.H. Reduction of enteric viruses by blueberry juice and blueberry proanthocyanidins. *Food Environ. Virol.* **2016**, *8*, 235–243. [CrossRef]
121. Lee, J.H.; Bae, S.Y.; Oh, M.; Kim, K.H.; Chung, M.S. Antiviral effects of mulberry (*Morus alba*) juice and its fractions on foodborne viral surrogates. *Foodborne Pathog. Dis.* **2014**, *11*, 224–229. [CrossRef]
122. Ribeiro-Santos, R.; Andrade, M.; de Melo, N.R.; Sanches-Silva, A. Use of essential oils in active food packaging: Recent advances and future trends. *Trends Food Sci. Technol.* **2017**, *61*, 132–140. [CrossRef]
123. Pandey, A.K.; Kumar, P.; Singh, P.; Tripathi, N.N.; Bajpai, V.K. Essential oils: Sources of antimicrobials and food preservatives. *Front. Microbiol.* **2017**, *16*, 2161. [CrossRef]

124. Fabra, M.J.; Falco, I.; Randazzo, W.; Sanchez, G.; Lopez-Rubio, A. Antiviral and antioxidant properties of active alginate edible films containing phenolic extracts. *Food Hydrocoll.* **2018**, *81*, 96–103. [CrossRef]
125. Falco, I.; Flores-Meraz, P.L.; Randazzo, W.; Sanchez, G.; Lopez-Rubio, A.; Fabra, M.J. Antiviral activity of alginate-oleic acid based coatings incorporating green tea extract on strawberries and raspberries. *Food Hydrocoll.* **2019**, *87*, 611–618. [CrossRef]
126. Ju, J.; Chen, X.; Xie, Y.; Yu, H.; Guo, Y.; Cheng, Y.; Qian, H.; Yao, W. Application of essential oil as a sustained release preparation in food packaging. *Trends Food Sci. Technol.* **2019**, *92*, 22–32. [CrossRef]

Article

Chemical Analysis, Toxicity Study, and Free-Radical Scavenging and Iron-Binding Assays Involving Coffee (*Coffea arabica*) Extracts

Nuntouchaporn Hutachok [1], Pimpisid Koonyosying [1], Tanachai Pankasemsuk [2], Pongsak Angkasith [3], Chaiwat Chumpun [3], Suthat Fucharoen [4] and Somdet Srichairatanakool [1,*]

[1] Department of Biochemistry, Faculty of Medicine, Chiang Mai University, Chiang Mai 50200, Thailand; n.hutachok@hotmail.com (N.H.); pimpisid_m@hotmail.com (P.K.)
[2] Department of Plant and Soil Sciences, Faculty of Agriculture, Chiang Mai University, Chiang Mai 50200, Thailand; tanachai.p@cmu.ac.th
[3] Royal Project Foundation, Chiang Mai 50200, Thailand; pongsak.a@cmu.ac.th (P.A.); chdeech@yahoo.com (C.C.)
[4] Thalassemia Research Center, Institute of Molecular Biosciences, Salaya Campus, Mahidol University, Nakorn Pathom 70130, Thailand; suthat.fuc@mahidol.ac.th
* Correspondence: somdet.s@cmu.ac.th; Tel.: +66-5393-5225

Abstract: We aimed to analyze the chemical compositions in Arabica coffee bean extracts, assess the relevant antioxidant and iron-chelating activities in coffee extracts and instant coffee, and evaluate the toxicity in roasted coffee. Coffee beans were extracted using boiling, drip-filtered and espresso brewing methods. Certain phenolics were investigated including trigonelline, caffeic acid and their derivatives, gallic acid, epicatechin, chlorogenic acid (CGA) and their derivatives, *p*-coumaroylquinic acid, *p*-coumaroyl glucoside, the rutin and syringic acid that exist in green and roasted coffee extracts, along with dimethoxycinnamic acid, caffeoylarbutin and cymaroside that may be present in green coffee bean extracts. Different phytochemicals were also detected in all of the coffee extracts. Roasted coffee extracts and instant coffees exhibited free-radical scavenging properties in a dose-dependent manner, for which drip coffee was observed to be the most effective ($p < 0.05$). All coffee extracts, instant coffee varieties and CGA could effectively bind ferric ion in a concentration-dependent manner resulting in an iron-bound complex. Roasted coffee extracts were neither toxic to normal mononuclear cells nor breast cancer cells. The findings indicate that phenolics, particularly CGA, could effectively contribute to the iron-chelating and free-radical scavenging properties observed in coffee brews. Thus, coffee may possess high pharmacological value and could be utilized as a health beverage.

Keywords: coffee; *Coffea arabica*; phenolic; free-radical scavenging; iron chelating; cytotoxic

1. Introduction

Coffea arabica (Arabica) and *Coffea acanephora* (Robusta) are known to be two of the most popular beverages in the world; however, Arabica coffee is more often consumed and more preferable in the global coffee market [1,2]. Coffee cherry husks, as well as green and roasted coffee beans, have all been processed to produce popular coffee beverages, of which roasted coffee beans are recognized as the most popular. Green coffee extract is made of unroasted green coffee beans. It is available as a dietary supplement and contains phenolic amides, as well as other phytochemicals [3,4]. In fact, the physical aspects, the species of the coffee bean, and the roasting and brewing processes are all important factors that influence the chemical composition of coffee beverages [5]. These chemical compositions include caffeine (CF), chlorogenic acid (CGA), caffeic acid (CA) and Maillard reaction products (e.g., melanoidins) [1,5]. Besides providing an alerting effect, coffee consumption is also associated with a range of health complications such as insomnia, tremors, nausea, polyuria, diarrhea, polyphagia, hypertension and a decrease in iron absorption. All of which have been attributed to the CF and melanoidins content in coffee-based beverages [6–9]. Beneficially, coffee intake can reduce the risks associated with type 2 diabetes mellitus, Parkinson's disease, colorectal cancer, hepatic injury, cirrhosis and hepatocellular carcinoma [10–15]. These benefits have been attributed to the actions of nitrogenous compounds, acids, esters and CGA [5,16–18]. CF (1,3,7-trimethylxanthine) is naturally found in coffee beans, cacao beans, kola nuts, guarana berries and tea leaves, of which coffee and tea are the first and second most prominent sources. The performance benefits of CF include the enhancement of mental alertness, increased levels of concentration and physical endurance, a potential reduction in fatigue and body weight and a lowering of the overall risks associated with certain metabolic syndromes [19,20]. CGA has three subclasses including5-O-caffeoylquinic acid (CQA), feruloylquinic acid (FQA) and dicaffeoylquinic acid (diCQA), of which CQA is the most common and strongest antioxidant present in coffee in the form of neochlorogenic acid (3-CQA), cryptochlorogenic acid (4-CQA) and chlorogenic acid (5-CQA) [21].

Instant coffee (regular and decaffeinated type) is a spray-dry form of coffee made from coffee extracts combined with a number of other functional ingredients (e.g., vitamins A and C, iron, inulin and oligofructose); nonetheless, the fortification of instant coffee products is necessary to improve particle size distribution, reconstitution properties, wettability and dispersibility times, as well as the overall level of satisfaction of its consumers [22]. In the manufacturing process, decaffeinated coffee is usually prepared by a treatment with waterand an organic solvent or carbon dioxide to remove intact CF from coffee beans before they are roasted and ground.The resulting coffee beveragewill then contain 1–2% of the original CF content of the regular coffee. Even after removing CF, the decaffeinated type still contains phenolic compounds (such as CGA, CA and trigonelline) and other phytochemicals. Thesecompounds and contents are similar to those found in the regular type and are known to exertcertain biological activities of interest [23–27].

Depending on the country of origin and the differing preferences of cultures and individual, coffee can be brewed by simple percolation or boiling methods, or with the use of Italian and electric coffee makers, espresso machines and French presses [5]. However, instant coffee is made by drying prepared coffee which produces a soluble powder that can be dissolved in hot water by the consumer. Different coffee preparations result in different tastes, aromas and chemical compositions [5,9,28–30]. It is likely that changes in the phenolics and CGA contents in coffee brews can affect these biological activities [30,31]. In contrast, degradation of 5-CQA that occurs during the roasting process does not affect antioxidant activity, whereas higher CF content has resulted in a greater degree of antioxidant activity indicating that the antioxidant activity may not depend only upon the CGA action [32].

Many liquid chromatographic techniques have been developed for identification of the active ingredients in coffee samples. For instance, the high-performance liquid chromatography-diode array detection (HPLC-DAD) method is often used for the simultaneous quantification of CF, trigonelline, nicotinic acid, N-methylpyridinium ion, 5-CQA, and 5-hydroxymethyl furfural. The resulting values can then be compared to the specific retention times (T_R) and concentrations of the authentic standards [33]. HPLC coupled witha mass spectrometer and a nuclear magnetic resonance spectrometer has been developed for the efficient analysis of the phenolic compounds in coffee bean extracts [34]. Recently, a fast highly-resolved sensitive ultra-high-performance liquid chromatography coupled with electrospray ionization quadrupole time-of-flight mass spectrometry (UHPLC-ESI-Q-TOF-MS), that involves a greater degree of informative structure elucidation and identification of the fragmentation patterns of the compounds, has been developed to identify polyphenols, alkaloids, diazines, and Maillard reaction products present in ground coffee samples [35]. Furthermore, a fast direct form of analysis using a real time ion source coupled with high-resolution time-of-flight mass spectrometry (DART-TOF-MS) without any prior chromatographic separationhas been developed for the quantitative analysis of CF in coffee samples [36]. In most advanced research studies, this technique can be applied for pharmaceutical, phytochemical and metabolomic analysis [37,38]. The present study aims to analyze the chemical compositions, assess the free-radical scavenging and iron-chelating activities and evaluate the toxicity of different coffee preparations.

2. Results
2.1. Information and Extraction Yield of Coffee Samples

Table 1 summarizes types and preparations of coffee samples used in this study, in which details and preparation protocols have been explained in the Materials and Methods section.

Table 1. Description of all coffee samples used in the experiments.

Coffee Samples	Source	Preparation	Condition/Instrument
Roasted Arabica coffee beans	Royal Project Foundation, Thailand	Extraction Extraction Extraction	Boiled water Automatic drip coffee machine Portable espresso coffee machine
Green Arabica coffee beans	Royal Project Foundation, Thailand	Extraction Extraction Extraction	Boiled water Automatic drip coffee machine Portable espresso coffee machine
Regular instant coffee	Tesco Supermarket, Thailand	Brewing	nd
Decaffeinated instant coffee	Tesco Supermarket, Thailand	Brewing	nd

nd = not determined.

Yields for the extracts of ground roasted coffee beans prepared from boiling and with the use of automatic coffee makers (drip and espresso methods) were found to be 17.76, 16.04 and 9.51% (w/w), respectively. The coffee extracts were further analyzed in terms of their chemical composition using the more sensitive HPLC-ESI-MS, UHPLC-ESI-Q-TOF-MS, while CA, CF and CGA concentrations were determined using HPLC/DAD. The antioxidant and iron-binding activities of the coffee products were determined and the degree of toxicity in cells and animals was evaluated.

2.2. HPLC-ESI-MS Identification of Phenolic Compounds in Coffee Bean Extracts

The phenolic constituents present in green and roasted coffee bean extracts were then analyzed using high-resolution HPLC-ESI-MS. Table 1 presents the chemical characterization of all identified phenolic compounds by peak elution order: T_R, UV absorption maxima at 270 nm from adiode array detector (DAD), exact molecular mass, molecular formula, quasimolecular ions ($[M-H]^+$, $[M-NH_4]^+$, $[M-Na]^+$ and $[M-K]^+$) with relative abundance and tentative names. As is shown in Figure 1A–F and Table 2 all analyzed roasted and green coffee extracts displayed nearly the same qualitative profiles of the bioactive substances found in the HPLC/DAD profiles at 270 nm during the course of monitoring the phenolic compounds. Their peaks were identified on the basis of UV spectra and elution/retention sequences that had been reported in previously published literature. These values were then confirmed by their mass spectrometric behavior. At least 17 phenolic compounds were detected in all roasted coffee extracts including trigonelline (peak 2), caffeic acid (peak 3), gallic acid (peak 4), epicatechin (peak 5), dicaffeoylquinic acid or dichlorogenic acid (peak 6), caffeoylquinic acid or chlorogenic acid (peak 7), caffeoyl-O-hexoside (peak 8), p-coumaroylquinic acid (peak 9), p-coumaroylglucoside (peak 10), rutin (peak 11), caffeoylquinic acid or chlorogenic acid derivative (peak 12), syringic acid (peak 13), caffeoylquinoyl-O-glucoside or chlorogenoyl-O-glucoside (peak 14), p-caffeoylquinoyldiglucoside I or chlorogenoyldiglucoside I (peak 15), p-caffeoylquinoyldiglucoside II or chlorogenoyldiglucoside II (peak 16), dicaffeoylquinoyl-O-glucoside or dichlorogenoyl-O-glucoside (peak 17) and an unidentified compound (peak 1) (Figure 1A–C and Table 1). In comparison, boiled and drip green coffee extracts contained the same 17 phenolic compounds and two additional phenolic compounds, namely dimethoxycinnamic acid (peak 10a) and caffeoylarbutin (peak 11a) (Figure 1D,E and Table 1), while espresso green coffee extract contained the same 17 phenolic compounds and three additional phenolic compounds, namely dimethoxycinnamic acid (peak 10a), caffeoylarbutin (peak 11a) and cynaroside (peak 13a) (Figure 1F and Table 1). Hence, there were no differences in the chromatographic profiles of the phenolic compounds analyzed in roasted and green coffee extracts that were prepared using the boiling, drip and espresso methods, with the exceptionof the aforementioned compounds.

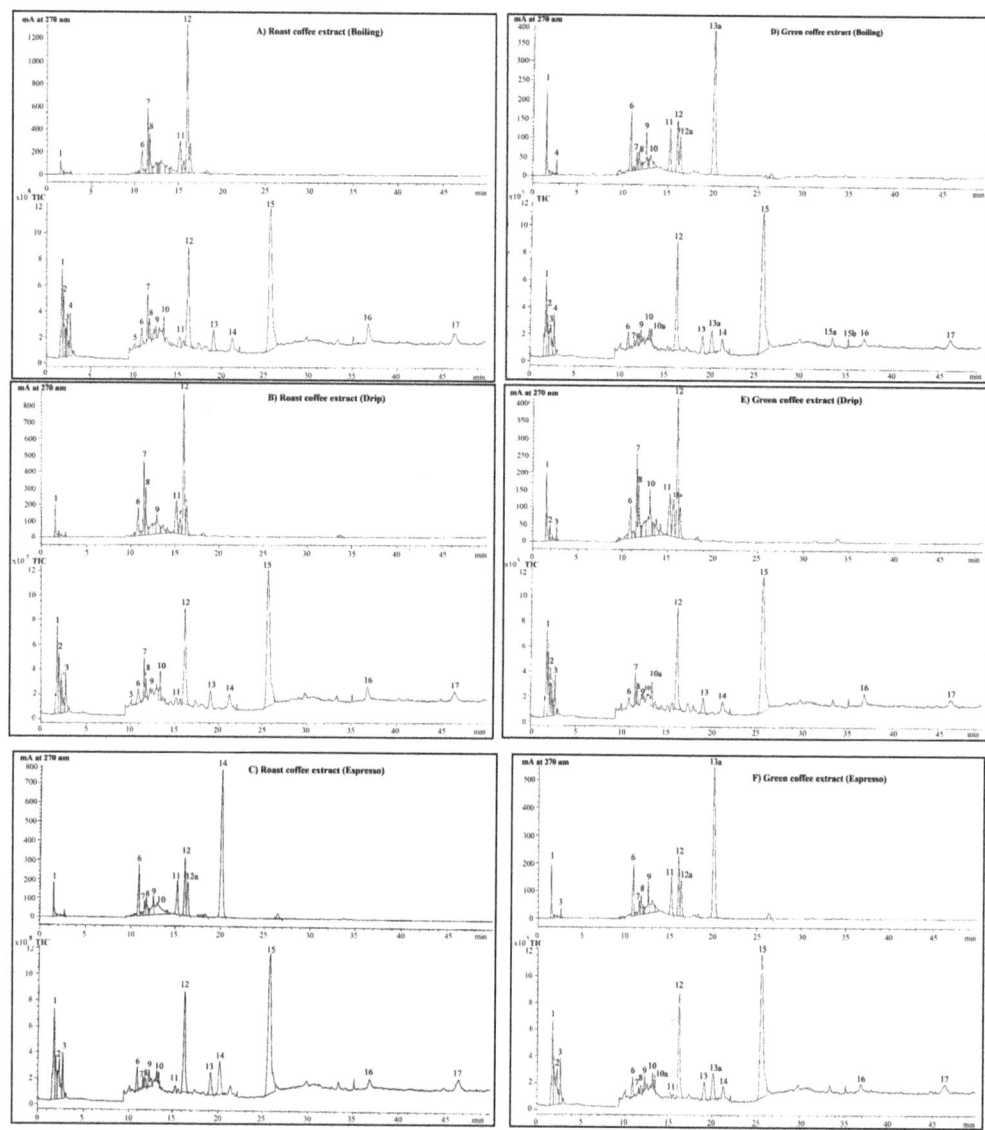

Figure 1. Chromatograms of the phenolic compounds in roasted and green coffee extracts. Roasted and green coffee were extracted with hot water using the boiling, drip and espresso methods. Phenolic compounds in the coffee extracts were identified using HPLC-ESI-MS, as has been described in Section 4.3.

Table 2. Identification of phenolic compounds in roasted and green coffee extracts. Roasted and green coffee were extracted using the boiling, drip and espresso methods. Phenolic compounds in the coffee extracts were identified using HPLC-ESI-MS, as has been described in Section 4.3.

(A) Roasted Coffee Extract (Boiling)									
Peak	OD 270 nm	TIC MS	Exact Mass	Chemical	Observed Mass (m/z)				Error	Identification
No	T_R (min)	T_R (min)	(g/mol)	Formula	$[M-H]^+$	$[M-NH_4]^+$	$[M-Na]^+$	$[M-K]^+$	(%)	
1	1.52	1.77	218.1	NA	219.1	231.1	-	262.2	0.46	Unknown
2	1.91	1.95	137.1	$C_7H_7NO_2$	138.1	154.1	-	176.1	0.73	Trigonelline
3	2.65	2.69	180.2	$C_9H_8O_4$	182.2	194.2	203.2	220.1	1.13	CA
4	ND	3.07	170.1	$C_7H_6O_5$	171.1	183.1	193.1	202.2	0.58	GA
5	10.28	10.09	290.3	$C_{15}H_{14}O_6$	291.1	303.0	314.1	325.1	0.28	Epicatechin
6	10.90	10.94	516.4	$C_{25}H_{24}O_{12}$	517.1	527.0	538.0	554.1	0.14	DiCGA
7	11.56	11.60	354.3	$C_{16}H_{18}O_9$	355.0	367.2	377.0	394.1	0.20	CGA
8	11.78	11.83	342.3	$C_{15}H_{18}O_9$	343.0	355.1	365.0	381.1	0.20	Caffeoyl-O-hexoside
9	12.49	12.54	338.3	$C_{16}H_{18}O_8$	339.1	-	361.1	377.1	0.23	p-Coumaroylquinic acid
10	13.71	13.45	327.3	$C_{15}H_{18}O_8$	328.2	341.1	351.2	-	0.27	p-Coumaroyl glycoside
11	15.22	15.25	610.5	$C_{27}H_{30}O_{16}$	611.1	-	633.0	649.0	0.10	Rutin
12	16.01	16.23	452.0	$C_{22}H_{28}O_{10}$	453.3	466.0	476.2	491.2	0.29	Unknown CGA derivative
13	ND	19.10	198.2	$C_9H_{10}O_5$	198.1	214	227.2	235.2	−0.05	Syringic acid
14	ND	21.23	516.1	$C_{22}H_{28}O_{14}$	517.0	529.0	-	547.0	0.17	CGA-O-glucoside
15	ND	25.90	678.1	$C_{28}H_{39}O_{21}$	679.4	-	-	-	0.19	CGA diglucoside I
16	ND	36.79	682.0	$C_{30}H_{34}O_{17}$	680.4	-	-	-	−0.23	CGA diglucoside II
17	ND	46.33	682.0	$C_{30}H_{34}O_{17}$	680.4	-	-	-	−0.23	DiCGA-O-glucoside
(B) Roasted Coffee Extract (Drip)										
Peak	OD 270 nm	TIC MS	Exact Mass	Chemical	Observed Mass (m/z)				Error	Identification
No	T_R (min)	T_R (min)	(g/mol)	Formula	$[M-H]^+$	$[M-NH_4]^+$	$[M-Na]^+$	$[M-K]^+$	(%)	
1	1.50	1.77	218.1	NA	219.1	231.1	-	262.2	0.46	Unknown
2	1.91	1.95	137.1	$C_7H_7NO_2$	138.1	154.1	-	176.1	0.73	Trigonelline
3	2.65	2.70	180.2	$C_9H_8O_4$	182.2	194.2	206.1	219.1	1.13	CA
6	10.90	10.93	516.4	$C_{25}H_{24}O_{12}$	517.0	527.2	538.1	555.0	0.14	DiCGA
7	11.56	11.60	354.3	$C_{16}H_{18}O_9$	355.1	367.2	377.1	394.1	0.23	CGA
8	11.78	11.83	342.3	$C_{15}H_{18}O_9$	344.1	355.1	365.0	381.1	0.53	Caffeoyl-O-hexoside
9	12.59	12.54	338.3	$C_{16}H_{18}O_8$	339.1	344.2	365.1	377.1	0.23	p-Coumaroylquinic acid
10	13.71	13.45	327.3	$C_{15}H_{18}O_8$	328.2	342.1	355.1	-	0.27	p-Coumaroyl glycoside
11	15.22	15.25	610.5	$C_{27}H_{30}O_{16}$	611.0	-	633.0	649.0	0.08	Rutin
12	16.02	16.21	452.0	$C_{22}H_{28}O_{10}$	453.3	465.0	476.1	491.2	0.29	Unknown CGA derivative
13	ND	19.12	198.2	$C_9H_{10}O_5$	198.1	214.1	227.2	235.1	−0.05	Syringic acid
14	ND	21.23	516.1	$C_{22}H_{28}O_{14}$	517.0	529.1	-	545.0	0.17	CGA-O-glucoside
15	ND	25.67	678.1	$C_{28}H_{39}O_{21}$	679.4	-	-	-	0.19	CGA diglucoside I
16	ND	36.83	682.0	$C_{30}H_{34}O_{17}$	680.4	-	-	-	−0.23	CGA diglucoside II
17	ND	46.33	682.0	$C_{30}H_{34}O_{17}$	680.4	-	-	-	−0.23	DiCGA-O-glucoside

Table 2. Cont.

(C) Roasted Coffee Extract (Espresso)

Peak No	OD 270 nm T_R (min)	TIC MS T_R (min)	Exact Mass (g/mol)	Chemical Formula	Observed Mass (m/z)				Error (%)	Identification
					$[M-H]^+$	$[M-NH_4]^+$	$[M-Na]^+$	$[M-K]^+$		
1	1.77	1.77	218.1	NA	219.1	-	-	262.1	0.46	Unknown
2	1.84	1.95	137.1	$C_7H_7NO_2$	138.1	154.1	-	176.1	0.73	Trigonelline
3	2.70	2.70	180.2	$C_9H_8O_4$	182.1	194.1	206.2	219.1	1.08	CA
6	10.91	10.95	516.4	$C_{25}H_{24}O_{12}$	517.1	-	-	-	0.14	DiCGA
7	11.58	11.62	354.3	$C_{16}H_{18}O_9$	355.0	367.2	377.0	394.1	0.20	CGA
8	11.88	11.85	354.3	$C_{16}H_{18}O_9$	355.1	367.1	377.1	393.0	0.23	Caffeoyl-O-hexoside
9	12.60	12.33	442.4	$C_{22}H_{18}O_{10}$	445.1	457.1	-	487.0	0.61	Epicatechin 3-gallate
10	13.08	13.22	313.1	$C_{15}H_{18}O_8$	314.1	327.1	337.0	347.1	0.32	p-Coumaroyl glycoside
11	15.26	15.30	610.5	$C_{27}H_{30}O_{16}$	611.0	-	633.0	649.0	0.08	Rutin
12	16.06	16.25	452.0	$C_{22}H_{28}O_{10}$	453.3	465.0	475.2	491.1	0.29	Unknown CGA derivative
13	ND	19.13	198.2	$C_9H_{10}O_5$	198.1	214.1	227.2	235.1	−0.05	Syringic acid
14	ND	21.23	516.1	$C_{22}H_{28}O_{14}$	517.0	529.1	-	545.0	0.17	CGA-O-glycoside
15	ND	25.76	678.1	$C_{28}H_{39}O_{21}$	679.4	-	-	-	0.19	CGA diglucoside I
16	ND	36.83	682.0	$C_{30}H_{34}O_{17}$	680.4	-	-	-	−0.23	CGA diglucoside II
17	ND	46.33	682.0	$C_{30}H_{34}O_{17}$	680.4	-	-	-	−0.23	DiCGA-O-glucoside

(D) Green Coffee Extract (Boiling)

Peak No	OD 270 nm T_R (min)	TIC MS T_R (min)	Exact Mass (g/mol)	Chemical Formula	Observed Mass (m/z)				Error (%)	Identification
					$[M-H]^+$	$[M-NH_4]^+$	$[M-Na]^+$	$[M-K]^+$		
1	1.50	1.77	218.1	NA	219.1	-	-	262.1	0.46	Unknown
2	1.91	1.95	137.1	$C_7H_7NO_2$	138.1	154.1	-	176.1	0.73	Trigonelline
3	2.66	2.70	180.2	$C_9H_8O_4$	182.1	194.1	206.2	219.1	1.08	CA
6	10.91	10.95	392.0	$C_{25}H_{24}O_{12}$	393.0	405.1	401.1	431.0	0.26	DiCGA
7	11.22	11.61	354.3	$C_{16}H_{18}O_9$	355.0	367.2	377.0	394.1	0.20	CGA
8	11.67	11.84	342.3	$C_{15}H_{18}O_9$	343.0	355.1	365.0	381.1	0.20	Caffeoyl-O-hexoside
9	12.09	12.32	442.4	$C_{22}H_{18}O_{10}$	445.1	457.1	-	487.0	0.61	Epicatechin 3-gallate
10	13.08	13.22	313.1	$C_{15}H_{18}O_8$	314.1	328.3	337.0	347.1	0.32	p-Coumaroyl glycoside
10a	13.46	13.47	208.2	$C_{11}H_{12}O_4$	207.1	-	-	-	−0.53	Dimethoxycinnamic acid
11	15.26	ND	610.5	$C_{27}H_{30}O_{16}$	611.1	-	633.0	649.0	0.10	Rutin
11a	15.64	ND	434.4	$C_{21}H_{22}O_{10}$	433.0	-	-	-	−0.32	Caffeoylarbutin
12	16.07	16.25	452.0	$C_{22}H_{28}O_{10}$	453.3	466.0	476.2	491.2	0.29	Unknown CGA derivative
13	ND	19.14	198.2	$C_9H_{10}O_5$	198.1	214	227.2	235.2	−0.05	Syringic acid
14	ND	21.29	516.1	$C_{22}H_{28}O_{14}$	517.0	529.1	-	545.0	0.17	CGA-O-glucoside
15	ND	25.75	678.1	$C_{28}H_{39}O_{21}$	679.4	-	-	-	0.19	CGA diglucoside I
16	ND	36.86	682.0	$C_{30}H_{34}O_{17}$	680.4	-	-	-	−0.23	CGA diglucoside II
17	ND	46.41	682.0	$C_{30}H_{34}O_{17}$	680.4	-	-	-	−0.23	DiCGA-O-glucoside

Table 2. Cont.

(E) Green Coffee Extract (Drip)

Peak No	OD 270 nm T_R (min)	TIC MS T_R (min)	Exact Mass (g/mol)	Chemical Formula	Observed Mass (m/z)				Error (%)	Identification
					$[M-H]^+$	$[M-NH_4]^+$	$[M-Na]^+$	$[M-K]^+$		
1	1.05	1.77	218.1	-	219.1	231.1	-	262.2	0.46	Unknown
2	1.91	1.95	137.1	$C_7H_7NO_2$	138.1	154.1	-	176.1	0.73	Trigonelline
3	2.65	2.70	180.2	$C_9H_8O_4$	182.2	194.2	203.2	220.1	1.13	CA
6	10.92	10.92	516.4	$C_{25}H_{24}O_{12}$	517.1	527.0	538.0	554.1	0.14	DiCGA
7	11.57	11.61	354.3	$C_{16}H_{18}O_9$	355.0	367.2	377.0	394.1	0.20	CGA
8	11.79	11.84	342.3	$C_{15}H_{18}O_9$	343.0	355.1	365.0	381.1	0.20	Caffeoyl-O-hexoside
9	ND	12.32	338.3	$C_{16}H_{18}O_8$	339.1	-	361.1	377.1	0.23	p-Coumaroylquinic acid
10	13.01	ND	313.1	$C_{15}H_{18}O_8$	314.1	328.3	337.0	347.1	0.32	p-Coumaroylglycoside
10a	13.42	13.46	208.2	$C_{11}H_{12}O_4$	207.1	-	-	-	−0.53	Dimethoxycinnamic acid
11	15.14	ND	610.5	$C_{27}H_{30}O_{16}$	611.1	-	633.0	649.0	0.10	Rutin
11a	15.63	15.67	434.4	$C_{21}H_{22}O_{10}$	433.0	-	-	-	−0.32	Caffeoylarbutin
12	16.05	16.23	452.0	$C_{22}H_{28}O_{10}$	453.3	466.0	476.2	491.2	0.29	Unknown CGA derivative
13	ND	19.11	198.2	$C_9H_{10}O_5$	198.1	214	227.2	235.2	−0.05	Syringic acid
14	ND	21.29	516.1	$C_{22}H_{28}O_{14}$	517.0	529.1	-	545.0	0.17	CGA-O-glucoside
15	ND	25.79	678.1	$C_{28}H_{39}O_{21}$	679.4	-	-	-	0.19	CGA diglucoside I
16	ND	36.64	682.0	$C_{30}H_{34}O_{17}$	680.4	-	-	-	−0.23	CGA diglucoside II
17	ND	46.40	682.0	$C_{30}H_{34}O_{17}$	680.4	-	-	-	−0.23	DiCGA-O-glucoside

(F) Green Coffee Extract (Espresso)

Peak No	OD 270 nm T_R (min)	TIC MS T_R (min)	Exact Mass (g/mol)	Chemical formula	Observed Mass (m/z)				Error (%)	Identification
					$[M-H]^+$	$[M-NH_4]^+$	$[M-Na]^+$	$[M-K]^+$		
1	1.50	1.77	218.1	NA	219.1	231.1	-	262.2	0.46	Unknown
2	ND	1.95	137.1	$C_7H_7NO_2$	138.1	154.1	-	176.1	0.73	Trigonelline
3	2.65	2.69	180.2	$C_9H_8O_4$	182.2	194.2	203.2	220.1	1.13	CA
6	10.90	10.94	516.4	$C_{25}H_{24}O_{12}$	517.1	527.0	538.0	554.1	0.14	DiCGA
7	11.56	11.61	354.3	$C_{16}H_{18}O_9$	355.0	367.2	377.0	394.1	0.20	CGA
8	11.79	11.84	342.3	$C_{15}H_{18}O_9$	343.0	355.1	365.0	381.1	0.20	Caffeoyl-O-hexoside
9	12.59	12.32	338.3	$C_{16}H_{18}O_8$	339.1	-	361.1	377.1	0.23	p-Coumaroylquinic acid
10	ND	13.21	313.1	$C_{15}H_{18}O_8$	314.1	328.3	337.0	347.1	0.32	p-Coumaroyl glycoside
10a	ND	13.46	208.2	$C_{11}H_{12}O_4$	207.1	-	-	-	−0.53	Dimethoxycinnamic acid
11	15.24	ND	610.5	$C_{27}H_{30}O_{16}$	611.1	-	633.0	649.0	0.10	Rutin
11a	ND	15.59	302.2	$C_{15}H_{10}O_7$	301.1	-	-	-	−0.38	Quercetin
12	16.04	16.23	452.0	$C_{22}H_{28}O_{10}$	453.3	466.0	476.2	491.2	0.29	Unknown CGA derivative
13	ND	19.11	198.2	$C_9H_{10}O_5$	198.1	214	227.2	235.2	−0.05	Syringic acid
13a	20.05	20.12	448.4	$C_{21}H_{20}O_{11}$	449.0	-	471.1	487.0	0.13	Cynarosideor Luteolin-7-O-glucoside
14	ND	21.26	516.1	$C_{22}H_{28}O_{14}$	517.0	529.1	-	545.0	0.17	CGA-O-glucoside
15	ND	25.67	678.1	$C_{28}H_{39}O_{21}$	679.4	-	-	-	0.19	CGA diglucoside I
16	ND	36.64	682.0	$C_{30}H_{34}O_{17}$	680.4	-	-	-	−0.23	CGA diglucoside II
17	ND	46.39	682.0	$C_{30}H_{34}O_{17}$	680.4	-	-	-	−0.23	DiCGA-O-glucoside

Abbreviations: CA = caffeic acid, CGA = chlorogenic acid, DiCGA = dichlorogenic acid, GA = gallic acid, m/z = mass to charge ratio, MS = mass spectrometry, NA = not available, ND = not detectable, TIC = total ion count, T_R = retention time.

2.3. UHPLC-ESI-Q-TOF-MS Identification of Phytochemical Compounds

According to the UHPLC-Q-TOF-MS analysis, many phytochemical compounds including 4-fluoro-L-threonine (peak 1), 3-nitroperylene (peak 2), cycloeudesmanesesquiterpenoids (peak 3), 3,4,5-tricaffeoylquinic acids (peak 4), 6-gingesulfonic acid (peak 5), phytosphingosine (peak 6), sativanine B (peak 8), sterebin E (peak 11), anofinic acid (peak 12), samandenone (alkaloids) (peak 13), sorbitan oleate (peak 14), 2-palmitoylglycerol (peak 15), citranaxanthin (peak 16), 2-stearoyl glycerol (peak 17), dodecanic acid and 12-methoxy-1-[(phosphonooxxy)methyl]1,2-ethanediyl ester (peaks 18, 19) were detected in extracts of boiled, drip and espresso roasted coffee beans and the extracts of boiled, drip and espresso green coffee beans (Figure 2 and Table 3). In addition, citronellyl butyrate, 2-[2-(4-hydroxy-3-meyhoxyphenyl)ethyl]tetrahydro-6-(4,5-dihydroxy-3methoxyphenyl)-2H-pyran-4-ylacetateand dodemorph, which correspond to peaks 7, 9 and 10, respectively, weredetected only in the boiled roasted coffee extract. However, the compounds eluted at T_R of 34.50 min (peak 20) and 36.65 min (peak 21) are unknown. The results imply that not only were two major active compounds, namely CF and CGA, found in the green and roasted Arabica coffee beans, but also that a number of minor phytochemical compounds were found in the beans. Owing to the presence of certain nutrients or the function of other important biological/pharmacological properties, these coffee extracts require further investigation.

2.4. CF, CGA and Total Phenolic Contents

HPLC-DAD profiles shown in Figure 3 have demonstrated the presence of CF and CGA, but not CA in these coffee extract samples. Stoichiometric data have revealed that the CGA contents were equal among the coffee extracts, while CF contents of the boiled and espresso coffee extracts were higher than that of the drip coffee extract. Inversely, TPC of the drip coffee extracts was higher than those of the boiled and espresso coffee extracts (Table 4). Herein, regular and decaffeinated instant coffee types were found to contain equal amounts of TPC, while their CA, CF and CGA contents were not determined in this study.

2.5. Free-Radical Scavenging Activity

In this study, the antioxidant activities of coffee samples were determined using $ABTS^{+\bullet}$ and $DPPH^{\bullet}$ methods in which Trolox was used as a standard for both experiments. Considerably, the results demonstrated that all coffee extracts and instant coffee preparations expressed free-radical scavenging properties in a concentration-dependent manner. With regard to the Trolox equivalent (TE), anti-oxidation values were assayed using the ABTS method, wherein the drip coffee extract (149.4 ± 9.2 mg TE/g) was found to be significantly more effective than the boiled coffee extract (125.8 ± 9.1 mg TE/g) and the espresso coffee extract (127.6 ± 3.0 mg TE/g) ($p < 0.05$) (Figure 4A). In this regard, the regular instant coffee (160.4 ± 4.0 mg TE/g) seemed to be as effective as the decaffeinated instant coffee (173.2 ± 8.9 mg TE/g) ($p > 0.05$).

Similarly, Trolox as well as the coffee samples were able to scavenge $DPPH^{\bullet}$ in concentration-dependent manners; however, the abilities were significantly enhanced when consecutive concentrations were increased (Figure 4B). In addition, the $DPPH^{\bullet}$ scavenging activities of the coffee samples are depicted as TE values (mean $\pm$ SD). Therefore, the espresso coffee extract (1274 ± 46 mg TE/g) and the drip coffee extract (1250 ± 38 mg TE/g) could exert stronger antioxidant activity than the boiled coffee extract (1174 ± 26 mg/g extract) ($p < 0.05$). However, regular instant coffee (2359 ± 159 mg/g) exhibited antioxidant activity equal to decaffeinated instant coffee (2358 ± 93 mg/g). The antioxidant activity of the coffee extracts and instant coffee were assayed using the $ABTS^{+\bullet}$ and $DPPH^{\bullet}$ methods and the results are summarized in Table 5.

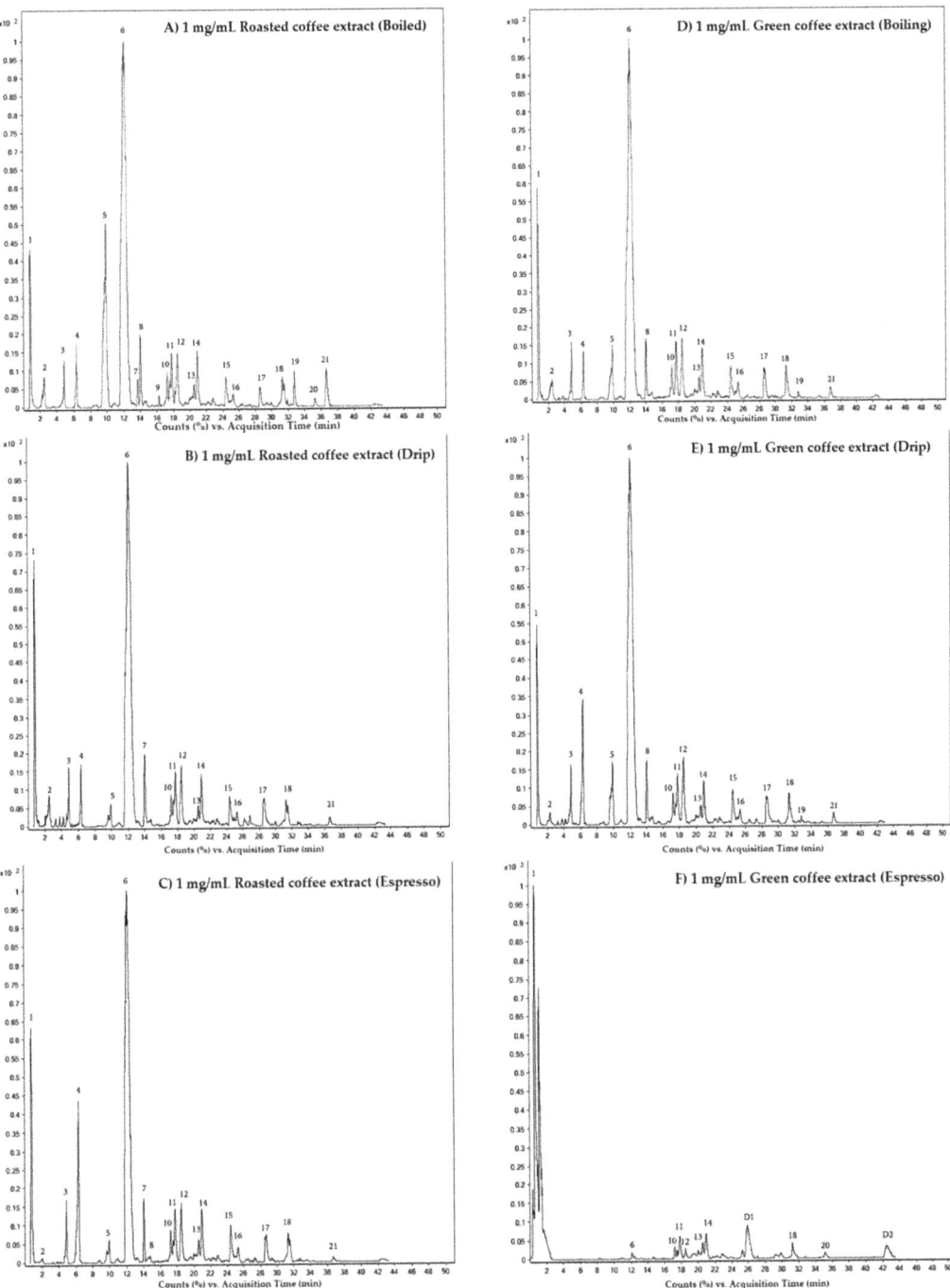

Figure 2. Chromatograms of the phytochemical compounds in roasted and green coffee extracts. Extracts of roasted coffee (**A–C**) and green coffee (**D–F**) were prepared by the boiling, drip and espresso methods, respectively. The coffee extracts were then identified using UHPLC-ESI-QTOF-MS, as has been described in Section 4.5.

Table 3. Identification of phytochemical compounds in roasted and green coffee extracts. Extracts of roasted coffee and green coffee were prepared by the boiling, drip and espresso methods (A, B, C, D, E and F, respectively). The coffee extracts were then identified using UHPLC-ESI-Q-TOF-MS, as has been described in Section 4.4.

Peak No	T_R (min)						Observed Mass $[M-H]^+$ (m/z)	Exact Mass (g/mol)	Error (%)	Chemical Formula	Possible Constitutes
	A	B	C	D	E	F					
1	0.76	0.72	0.72	0.75	0.75	0.77	137.05	137.11	0.000	$C_4H_8FNO_3$	4-Fluoro-L-threonine
2	2.49	2.48	ND	2.47	2.47	ND	297.08	297.3	0.001	$C_{20}H_{11}NO_2$	3-Nitroperylene
3	4.86	4.84	4.84	4.84	4.83	ND	490.29	490.63	0.001	$C_{28}H_{42}O_7$	Cycloeudesmane sesquiterpenoids
4	6.31	6.31	6.31	6.29	6.30	ND	678.51	678.6	0.000	$C_{34}H_{30}O_{15}$	3,4,5-Tricaffeoylquinic acids
5	9.89	9.90	9.89	9.73	9.88	ND	754.25	358.5	1.104	$C_{17}H_{26}O_6S$	6-Gingesulfonic acid
6	12.06	12.06	12.06	12.07	12.05	12.08	317.29	317.5	0.001	$C_{18}H_{39}NO_3$	Phytosphingosine
7	13.69	ND	ND	ND	ND	ND	226.19	226.35	0.001	$C_{14}H_{26}O_2$	Citronellyl butyrate
8	14.02	14.01	14.02	14.02	14.03	ND	518.29	518.6	0.001	$C_{30}H_{38}N_4O_4$	Sativanine B
9	16.23	ND	ND	ND	ND	ND	432.18	432.46	0.001	$C_{23}H_{28}O_8$	2-[2-(4-Hydroxy-3-meyhoxyphenyl)ethyl] tetrahydro-6-(4,5-dihydroxy-3methoxyphenyl)-2H-pyran-4-ylacetate
10	17.21	ND	ND	17.21	17.22	17.20	281.27	281.5	0.001	$C_{18}H_{35}NO$	Dodemorph
11	17.75	ND	ND	17.77	17.77	17.81	338.25	338.5	0.001	$C_{20}H_{34}O_4$	Sterebin E
12	18.33	18.49	18.53	18.48	18.49	18.49	204.08	204.22	0.001	$C_{12}H_{12}O_3$	Anofinic acid
13	20.52	20.55	ND	20.55	20.96	20.52	343.5	343.25	0.001	$C_{22}H_{33}NO_2$	Samandenone (alkaloids)
14	20.96	ND	20.95	20.95	20.95	20.97	428.31	428.6	0.001	$C_{24}H_{44}O_6$	Sorbitan oleate
15	24.38	24.36	24.43	24.39	24.39	ND	330.28	330.5	0.001	$C_{19}H_{38}O_4$	2-Palmitoyl glycerol
16	25.28	25.28	25.28	25.33	25.32	25.58	456.34	456.70	0.001	$C_{33}H_{44}O$	Citranaxanthin
17	28.41	28.42	28.42	28.43	28.43	28.66	358.31	358.6	0.001	$C_{21}H_{42}O_4$	2-Stearoyl glycerol
18	31.89	31.89	31.89	31.85	31.86	31.85	596.57	596.73	0.001	$C_{29}H_{57}O_{10}P$	Dodecanoic acid, 13-methoxy-1-[(phosphonooxy) methyl] 1,2-ethanediyl ester

Table 3. *Cont.*

Peak No	T$_R$ (min)						Observed Mass [M-H]$^+$ (m/z)	Exact Mass (g/mol)	Error (%)	Chemical Formula	Possible Constitutes
	A	B	C	D	E	F					
19	32.65	ND	32.52	32.76	ND	ND	596.37	596.73	0.001	$C_{29}H_{57}O_{10}P$	Dodecanoic acid, 12-methoxy-1-[(phosphono oxy)methyl]-1,2-ethanediyl ester
20	34.50	ND	ND	ND	ND	ND	585.43	ND	ND	$C_{33}H_{57}N_6OS$	Unknown
21	36.65	ND	ND	36.84	36.81	ND	641.43	ND	ND	$C_{41}H_{57}N_2O_4$	Unknown

Abbreviations: *m/z* = mass to charge ratio, ND = not detectable, TIC = total ion count, T$_R$ = retention time.

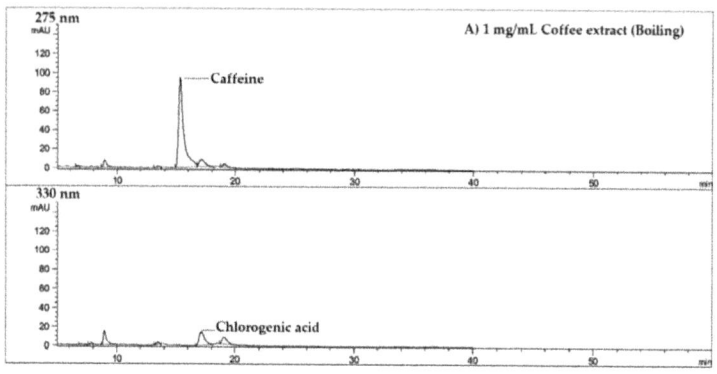
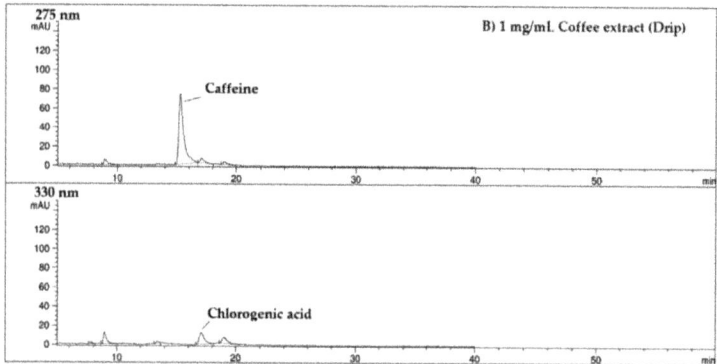
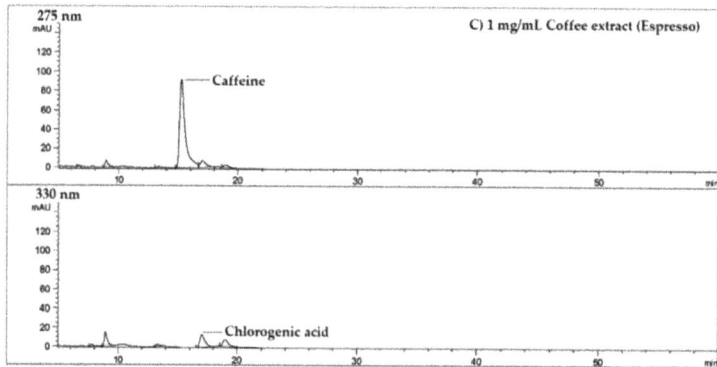

Figure 3. *Cont.*

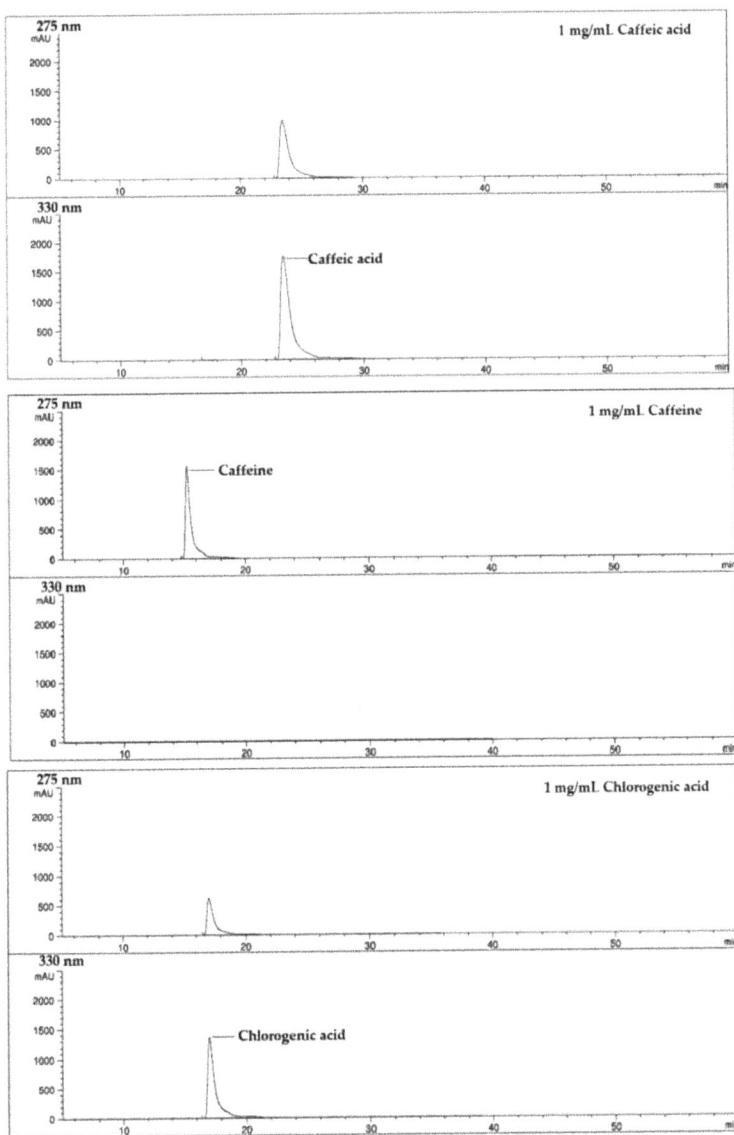

Figure 3. Chromatograms of coffee extracts, standard caffeic acid, caffeine and chlorogenic acid. Extracts of roasted coffee were prepared by the boiling, drip and espresso methods ((**A–C**), respectively). The coffee extracts and authentic standards, including caffeic acid, caffeine and chlorogenic acid, were then subjected to HPLC-DAD analysis as has been described in Section 4.5.

Table 4. Caffeine, chlorogenic acid and total phenolic contents in coffee samples. Boiled, drip and espresso coffee extracts, and the authentic standards including caffeic acid (CA), caffeine (CF) and chlorogenic acid (CGA), were subjected to HPLC-DAD analysis, as has been described in Section 4.5. Total phenolic contents in the coffee extracts and instant coffee were determined using the Folin-Ciocalteu method, as has been described in Section 4.6. Data obtained from three repetitions are expressed as mean ± standard deviation (SD) values.

Coffee Samples	CGA (mg/g)	CF (mg/g)	CA (mg/g)	TPC (mg GAE/g)
Boiled roasted coffee extract	14.47 ± 0.98	65.58 ± 9.83	ND	87.9 ± 7.9
Drip roasted coffee extract	15.67 ± 0.83	55.58 ± 10.61	ND	111.6 ± 11.4
Espresso roasted coffee extract	14.97 ± 0.89	64.25 ± 11.56	ND	97.2 ± 3.9
Regular instant coffee	nd	nd	nd	115.2 ± 22.4
Decaffeinated instant coffee	nd	nd	nd	118.1 ± 6.87

Abbreviations: CA = caffeic acid, CF = caffeine, CGA = chlorogenic acid, GAE = gallic acid equivalent, nd = not done, ND = not detectable, TPC = total phenolic content.

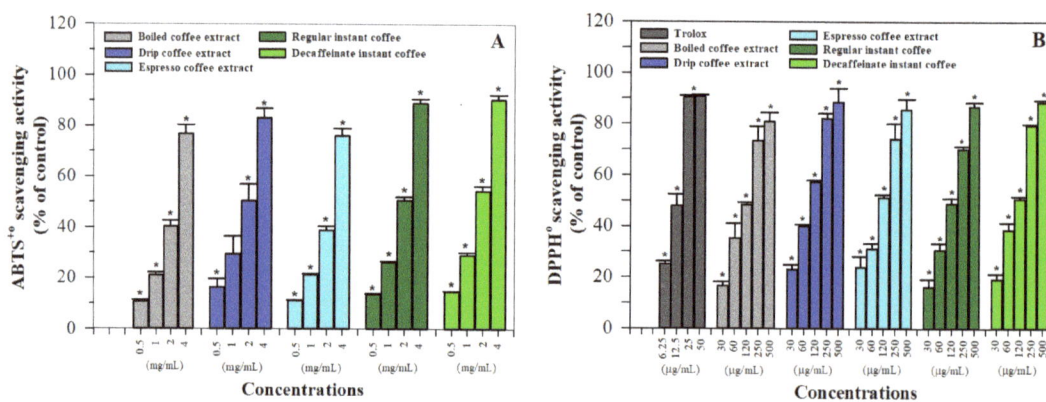

Figure 4. Free-radical scavenging activity of roasted coffee extracts and instant coffee. (A) ABTS$^{+\bullet}$ solution was incubated with deionized water (DI), standard Trolox, boiled, drip and espresso roasted coffee extracts, as well as regular and decaffeinated instant coffee products, while optical density (OD) was photometrically measured at a wavelength of 764 nm against the reagent blank. (B) DPPH$^{\bullet}$ solution was incubated at various concentrations of standard Trolox, boiled, drip and espresso coffee extracts, regular and decaffeinated instant coffee products and OD was photometrically measured at a wavelength of 517 nm against the reagent blank. Radical-scavenging activity was calculated according to the method described in Section 4.7 and is reported as percentage of ABTS$^{+\bullet}$ or DPPH$^{\bullet}$ scavenging activity. Data obtained from three repetitions are expressed as mean ± SD values.

Table 5. Antioxidant activity values of coffee samples accessed using ABTS$^{+\bullet}$ and DPPH$^{\bullet}$ methods.

Coffee Samples	Antioxidant Activity (mg TE/g)	
	ABTS$^{+\bullet}$ Method	DPPH$^{\bullet}$ Method
Roasted coffee extract (boiled)	125.8 ± 9.1	1174 ± 26
Roasted coffee extract (drip)	149.4 ± 9.2	1250 ± 38
Roasted coffee extract (espresso)	127.6 ± 3.0	1274 ± 46
Regular instant coffee	160.4 ± 4.0	2359 ± 159
Decaffeinated instant coffee	173.2 ± 9.8	2358 ± 93

ABTS$^{+\bullet}$ = 2,2′-azino-bis(3-ethylbenzothiazoline-6-sulfonic acid) diammonium salt radical, DPPH$^{\bullet}$ = 2, 2-diphenyl-1-picrylhydrazylradical, TE = Trolox equivalent.

These findings have indicated that two instant coffee products displayed significantly stronger radical-scavenging activity than all three coffee extracts in the following order:

drip coffee extract > espresso coffee extract and boiled coffee extract based on ABTS assay and drip coffee extract = espresso coffee extract > boiled coffee extract based on DPPH assay.

2.6. Iron-Binding Ability

Using scanning spectrophotometry, all the coffee samples themselves elucidated three distinct maximal absorption peaks at wavelengths of 217, 275 and 323 nm in a concentration-dependent manner (Figure 5A–C, Table 6). However, the decaffeinated instant coffee resulted in a longer absorption wavelength at 289 nm than the regular instant coffee (Figure 5D,E, Table 6), while CF itself exhibited two major absorption peaks at 217 and 275 nm and CGA itself exhibited peaks at 217 and 324 nm (Figure 5F,G, Table 6). Additionally, no peaks were exhibited by all of the samples at wavelengths of 400–600 nm.

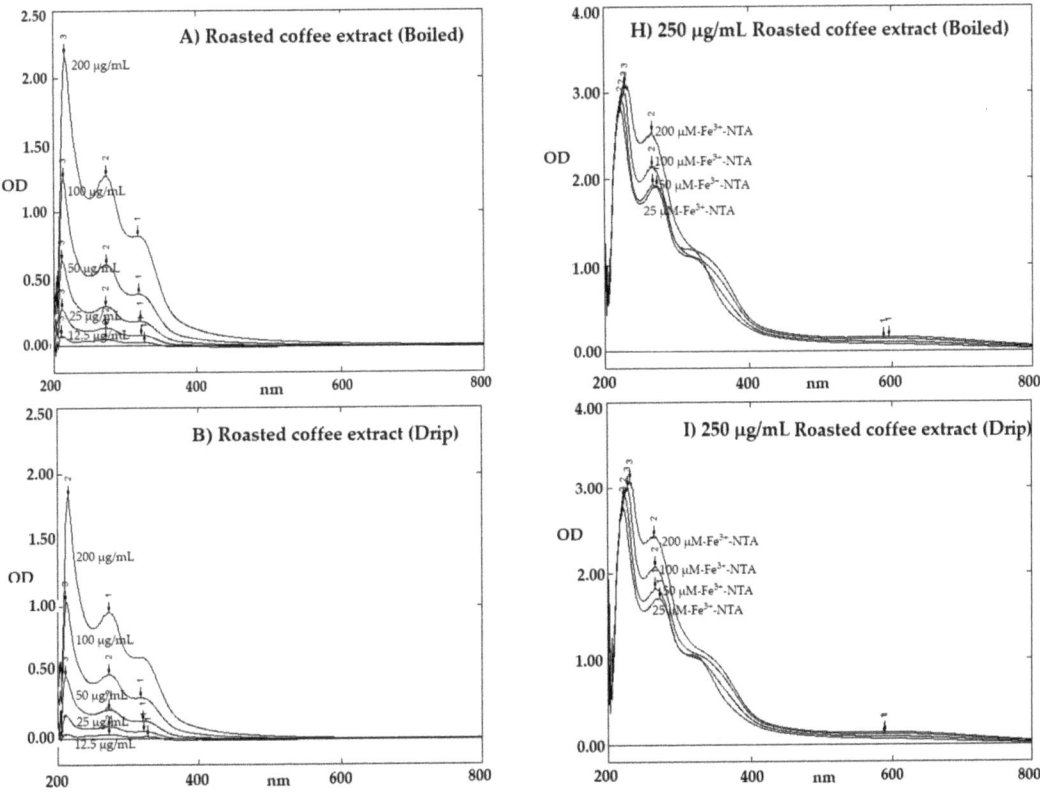

Figure 5. Cont.

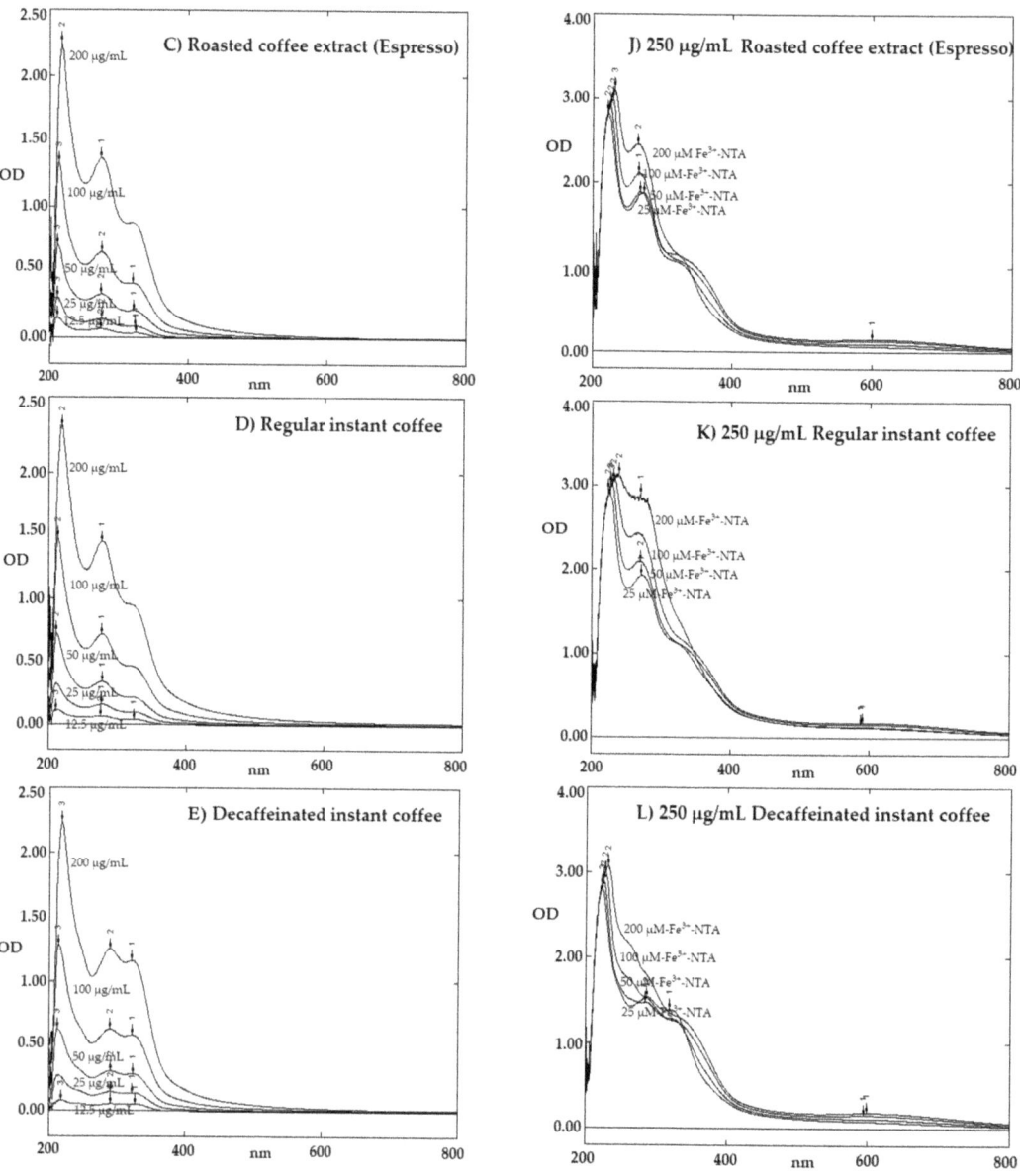

Figure 5. *Cont.*

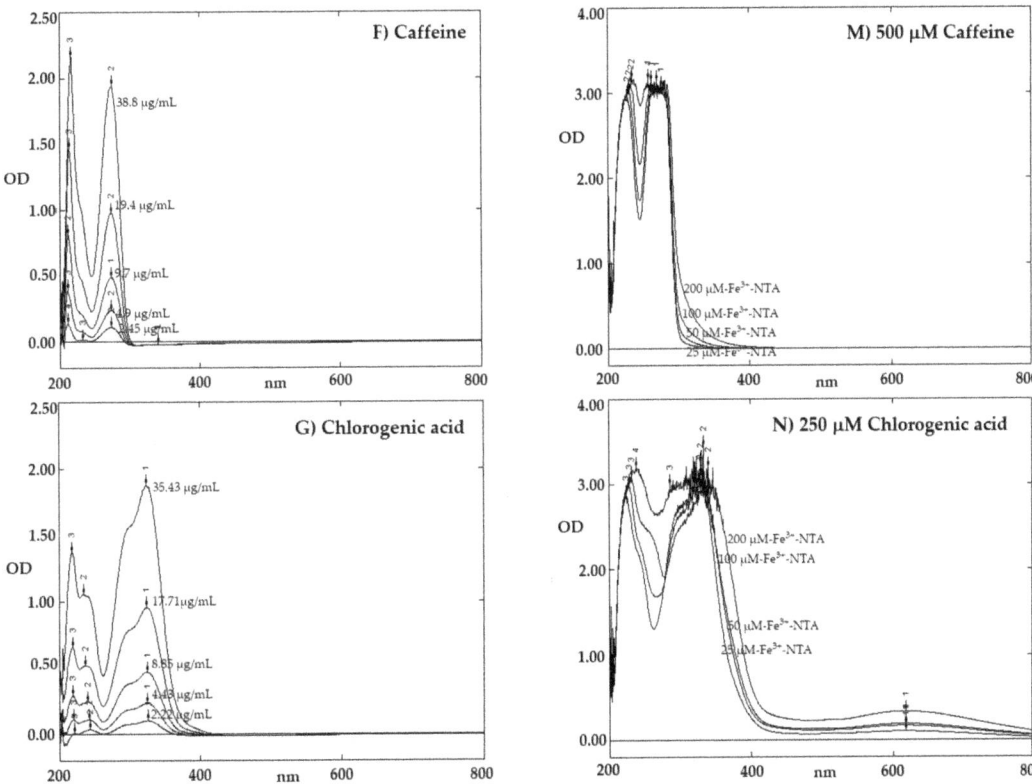

Figure 5. Spectral patterns of roasted coffee extracts, instant coffee, CF and CGA with and without the addition of iron. (**A–G**) Boiled, drip and espresso roasted coffee extracts (25–200 µg/mL), regular and decaffeinated instant coffee (12.5–200 µg/mL each), CF (2.45–38.8 µg/mL) and CGA (2.22–35.43 µg/mL) were photometrically measured at a wavelength rank of 200–800 nm against the reagent blank. (**H–N**) The coffee extracts and the instant coffees (250 µg/mL), CF (500 µM) and CGA (250 µM) were incubated with Fe-NTA (25–200 µM) for 1 h and photometrically measured at a wavelength rank of 200–800 nm against the coffee extract, instant coffee, CF and CGA solutions. Abbreviations: CF = caffeine, CGA = chlorogenic acid, Fe-NTA = ferric nitrilotriacetic acid.

Table 6. Spectral peaks from roasted coffee extracts, instant coffee, CF and CGA with and without the addition of iron.

Samples	Fe-NTA (25–200 µM)	Peak(s) (nm)
Boiled coffee extract (25–200 µg/mL)	-	217, 275, 323
Drip coffee extract (25–200 µg/mL)	-	217, 275, 323
Espresso coffee extract (25–200 µg/mL)	-	217, 275, 323
Regular instant coffee (12.5–200 µg/mL)	-	217, 275, 323
Decaffeinated instant coffee (12.5–200 µg/mL)	-	217, 289, 323
CF (2.45–38.8 µg/mL)	-	217, 275
CGA (2.22–35.43 µg/mL)	-	217, 324
Boiled coffee extract (250 µg/mL)	+	595

Table 6. *Cont.*

Samples	Fe-NTA (25–200 µM)	Peak(s) (nm)
Drip coffee extract (250 µg/mL)	+	595
Espresso coffee extract (250 µg/mL)	+	595
Regular instant coffee (250 µg/mL)	+	595
Decaffeinated instant coffee (250 µg/mL)	+	595
CF (500 µM)	+	ND
CGA (250 µM)	+	617

Abbreviations: CF = caffeine, CGA = chlorogenic acid, Fe-NTA = ferric nitrilotriacetic acid, ND = not detectable.

In comparison, CGA, but not CF, was found to bind ferric ion from Fe-NTA in a concentration-dependent manner producing a complex with a maximal absorption peak of 617 nm (Figure 5M,N, Table 6). Notably, all coffee extracts and instant coffee varieties also bound ferric ion in a concentration-dependent manner to produce a complex with a new maximal absorption peak of 595 nm (Figure 5H–L, Table 6).

2.7. Cytotoxic Effects

Herein, Arabica coffee extracts prepared by boiling, drip and espresso methods were tested in vitro for toxicity against normal human PBMC and human breast cancer (MDA-MB-231 and MCF-7) cells within 24 and 48 h of incubation and the results are shown in Figure 6. PBMC viability was found to be almost unchanged following treatment with all coffee extracts (with the exception of the 25 and 50 µg/mL espresso coffee beverages, ($p < 0.05$) for 24 h and compared to the specimens without treatment. During 48h of incubation, cell viabilities were significantly increased in a concentration-dependent manner by all the extracts, for which the drip coffee beverage was the most effective when compared to the specimens without treatment. In comparison, the viability of MDA-MB-231 cells remained unchanged following treatment with all the extracts for 24 and 48 h. When MDA-MB-231 and MCF-7 cells were treated with aggressive doses of the drip coffee extract at 31.25–1000 µg/mL, the viability of these two cells was unchanged during incubation for 24 h and tended to decrease slightly during the course of incubation up to 48 h. According to the findings, all three coffee extracts were non-toxic to normal PBMC but expressed a significantly dose-dependent increase in cell viability during 48 h of incubation based on the reducing power of the mitochondrial reductases. Likewise, no cytotoxic effect was observed in MDA-MB-231 cancer cells treated with equal concentrations of the coffee extracts for 24 and 48 h. Moreover, increasing doses of the drip coffee extract and treatment times up to 48 h were determined to be not harmful to both MDA-MB-231 and MCF-7 cells.

2.8. Acute Oral Toxicity in Rats

Acute toxicity test was conducted following OECD Guidelines for the Testing of Chemicals 425 by using initial doses of the coffee extracts at 175, 550 and 2000 mg/kg body weight (BW), respectively. However, the upper limit of the caffeine dose should not exceed 400 mg/kg BW. All clinical signs and symptoms were observed and recoded in Table 7. At the first and second dose of the drip coffee extract, neither a toxic effect nor a lethal effect was observed in the rats. One out of three rats dosed at 2000 mg/kg exhibited a sign of lethargy and drowsiness after 1 h of being given a dose of the coffee extract. After 6 h and 14 days of observation, no clinical signs of toxicity and lethality were observed. Since there were no signs of mortality and no clinical signs of toxicity at all dosing levels, the median lethal dose (LD_{50}) value of the drip coffee extract was found to be greater than 2000 mg/kg.

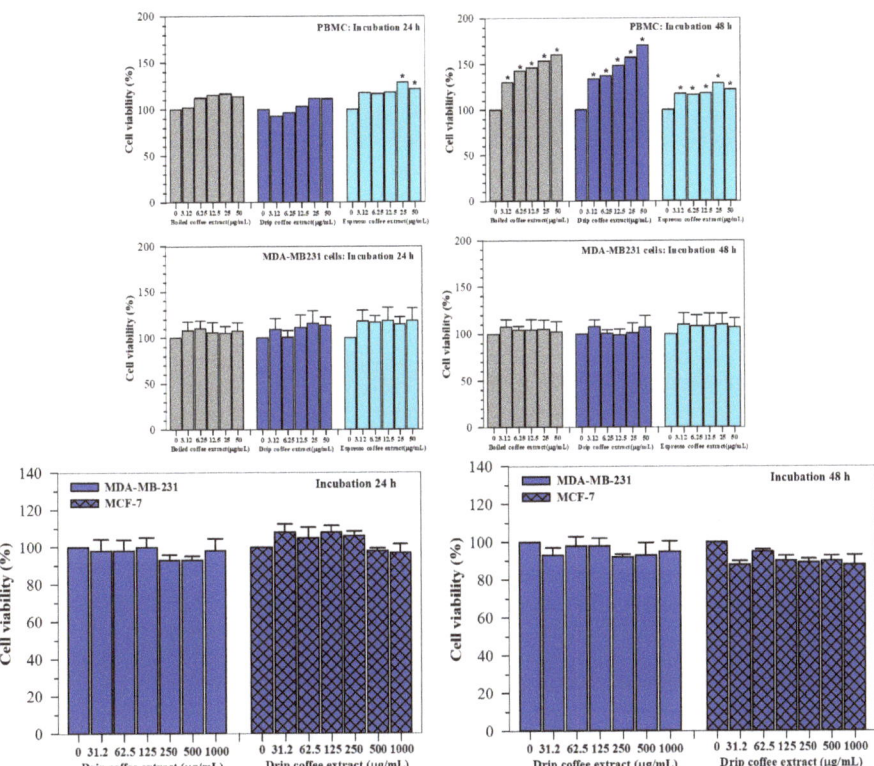

Figure 6. Viability of PBMC, MDA-MB-231 and MCF-7 cells treated with roasted coffee extracts. Normal human peripheral blood mononuclear cells (PBMC) and human breast cancer (MDA-MB-231) cell lines were treated with boiled, drip and espresso roasted coffee extracts for 24 and 48 h. The degree of viability was determined using MTT assay. Using aggressive doses, MDA-MB-231 and MCF-7 cells were treated with drip roasted coffee extract and the degree of viability was determined using MTT assay. Data are expressed as mean ± SEM values. * $p < 0.05$ when compared without treatment.

Table 7. Behavioral patterns observed in rats after single oral administration of coffee extract at doses of 175, 550 and 2000 mg/kg BW. Signs of lethargy and drowsiness were observed after 1 h of administering doses of the coffee extract.

Observations	Dose of Coffee Extract			
	175 mg/kg BW (n = 1)	550 mg/kg BW (n = 1)	2000 mg/kg BW (n = 3)	
	Days 1–14	Days 1–14	Day 1	Days 2–14
Hair falling	0/1	0/1	0/3	0/3
Ophthalmopathy	0/1	0/1	0/3	0/3
Mucous membrane	0/1	0/1	0/3	0/3
Hypersalivation	0/1	0/1	0/3	0/3
Hyperpnea	0/1	0/1	0/3	0/3
Diarrhea	0/1	0/1	0/3	0/3
Lethargy	0/1	0/1	1/3 *	0/3
Drowsiness	0/1	0/1	1/3 *	0/3
Aggressiveness	0/1	0/1	0/3	0/3
Convulsions	0/1	0/1	0/3	0/3
Tremors	0/1	0/1	0/3	0/3
Mortality	0/1	0/1	0/3	0/3

* Sigh of lethargy and drowsiness were observed after 1 h of dosing the coffee extract.

3. Discussion

Coffee is understood to be the most frequently consumed beverage in most countries worldwide. The consumption of coffee can be beneficial or may actually be detrimental to human health due to the naturally occurring active compounds that are present in coffee products. Though diterpenes (e.g., cafestol and kahweol), phenolics (e.g., CGA, CA) and heterocyclic compounds present in boiled coffee (e.g., CF and melanoidins) exhibit strong antioxidant properties, these two diterpenes have been claimed to be associated with increased serum cholesterol levels [39,40]. In the coffee trade, coffee beans are extensively used in beverage processing and are mainly comprised of a number of functional phytochemicals and nutritional carbohydrates [41]. Certain extraction factors can influence the quantity and activities of bioactive compounds that are present in coffee samples. These extraction factors include the use of solvents, the mass to volume ratio, acidity, time, temperature and pressure, as well as the application of microwaves or ultrasonic preparation methods and the specific type of coffee maker that may have been used [42–45]. Brewing is a key and final step in the production process of coffee drinks. The resulting drinks have been associated with coffee stoichiometry (known as total dissolved solids (TDS)), percent extraction and sweetness, and an inverse proportion to TDS. In addition, roasting is another key step in the preparation process of many of the most popular brewed coffee beverages. The roasting step can deliver a pleasant aroma while minimizing the bitterness of the final coffee beverage [46]. In this study, we have emphasized that all data reported were based on the use of a single-type of wet-washed Arabica coffee beans to prepare coffee beverages with and without roasting. This coffee bean was selected for the production of brewed coffee using boiled water and automatic coffee makers (drip and espresso types). Thus, we have demonstrated higher coffee extraction yields by using the boiled water and drip methods over the espresso method. A previous study reported the TDS values of the espresso and drip brewed coffees (5–10% and 1.0–1.75%, respectively), and the extraction yields of the espresso coffee (approximately 14–25%) [47]. In comparison, the extraction yields of the commercial coffee samples were found to be 81.26 ± 19.23 mg/g by methanol (100 °C), 19.12 ± 1.28 mg/g by dichloromethane (120 °C) and 1.76 ± 0.93 mg/g by n-hexane (120 °C) [48]. Consistent with the outcomes of our study, coffee drinks brewed with high temperatures and the cold dripping method exhibited the highest values in terms of TDS, extraction yields and the highest contents of CF, trigonelline, 4-CQA and 5-CQA, regardless of the roasting method that was used [49]. Moreover, Angelone and coworkers have reported on the extraction yield values assayed in espresso (13.1 ± 1.3–$22.8 \pm 1.3\%$), moka ($28.4 \pm 1.1\%$), V60 ($22.1 \pm 0.7\%$), cold brew ($23.3 \pm 0.9\%$), Aeropress ($20.4 \pm 1.2\%$) and French press ($18.7 \pm 1.1\%$) coffee beverages [50]. In roasting, galactomannan yields extracted from ground coffee with hot water were observed to increase [51]. Importantly, the amount of ochratoxin A, a naturally occurring food-borne mycotoxin produced by *Aspergillus* spp. and *Penicillium* spp. that is often found in green, roasted and brewed coffee products was reduced by the use of an espresso coffee maker (49.8%), the drip-filter method (14.5%) or the moka brewing process (32.1%) [52]. In terms of roasting coffee, CF content was not affected, CGA was degraded due to a consequence of the temperature used in brewing, CA decreased in dark roasted coffee, while melanoidins and other Maillard reaction products were developed [53]. Importantly, the roasting of green coffee beans and the brewing of ground roasted coffee are important processes that are used to make varieties of bioactive, aromatic and popular coffee drinks.

Ultraviolet (UV) detection, derivatization spectrophotometry and gas-liquid chromatography (GLC) techniques have been developed for the simple and rapid determination of CF concentrations in beverages over a long period of time [54]. At present, HPLC-ESI-MS and HPLC-ESI-MS-MS are known to be powerful techniques with high degrees of sensitivity and accuracy that can be used to determine the compound profiles of plant materials and natural products. Owing to the advantages of MS detection, coffee extracts were subjected to HPLC-ESI-MS identification by collision-induced dissociation mass spectrometer in order to identify certain phenolics, especially isomers. MS fragmentation patterns were observed after analysis by tandem MS spectra, while chromatographic retention times, relative hydrophobicity and bonding strength to quinic acid have been used to develop structure-diagnostic hierarchical keys for the identification of phenolic compounds. In this sense, the relative hydrophobicity of aglyconic phenolics can depend upon the substitution position of the phenolic ring and the number and identity of the residues. By using HPLC/ESI-MS/MS, we have identified at least 17 phenolic compounds, including trigonelline, CA, gallic acid, epicatechin, di-CGA, CGA, caffeoyl-O-hexoside, p-coumaroylquinic acid, p-coumaroylglucoside, rutin, CGA derivative, syringic acid, CGA-glucoside, CGA-diglucoside I, CGA-diglucoside II, diCGA-glucoside, and one unknown compound in the roasted coffee extracts, whereas all of the 17 compounds and an additional three compounds, namely dimethoxycinnamic acid, caffeoylarbutin and cynaroside, were detected in the green coffee extracts. Interestingly, trigonelline, which is a bitter alkaloidal ingredient that serves as an aroma generator and is responsible for certain biological activities, was detected in both green and roasted coffee extracts. Stennert and Maier demonstrated that trigonelline was degraded gradually during the roasting stage [55]. Similarly, CQAs, including 5-CQA, 3-CQA and 4-CQA; diCQAs including 3,4-diCQA, 3,5-diCQA and 4,5-diCQA; feruloylquinic acids (FQAs) including 3-FQA, 4-FQA and 5-FQA; diFQA and p-coumaroylquinic acids (p-CoQAs) including 3-p-CoQA, 4-p-CoQA and 5-p-CoQA isomers, were also found to be present inthe coffee samples [56]. Likewise, CF, CQAs, diCQAs, p-CoQAs, FQAs and caffeoylquinic acid lactone were detected in the espresso, moka, cold brewed and French Press coffee extracts [50]. This technique has been reported to be effective in identifying certain bitter compounds, such as 1,3-bis(3′,4′-dihydroxyphenyl) butane, trans-1,3-bis(3′,4′-dihydroxyphenyl)-1-butene and eight hydroxylated phenylindanes in roasted coffee at threshold concentrations of 23–178 μM [57], as well as in the detection of a carcinogenic furan in defatted, ground and constituted coffee preparations [58]. In the roasting process, free amino acids and peptides existing in green coffee beans are changed into aromatic flavors, whereas polymerization and fragmentation of proteins simultaneously generate hydrogen peroxide [59]. Unfortunately, a potential carcinogen acrylamide was also detectable in certain foods including coffee (169 ng/g) [60].

In addition, HPLC can be employed with ESI-triple quadrupole time-of-flight mass spectrometry (HPLC-ESI-QTOF-MS) in order to provide higher resolution, faster speeds and less solvent consumption, which can lead to a rapid and sensitive characterization of certain unexpected natural products. Currently, Spreng and colleagues applied this technique for the analysis of roasted coffee and reported eleven pyrazine derivatives, of which 2-(2′,3′,4′-trihydroxybutyl)-(5/6)-methyl-pyrazine and 2,(5/6)-bis(2′,3′,4′-trihydroxybutyl)-pyrazine were the most prominent compounds [61]. In the new findings, our UHPLC-ESI-QTOF-MS results have elucidated the presence of many phytochemicals, including 4-fluoro-L-threonine, 3-nitroperylene, cycloeudesmanesesquiterpenoids, 3,4,5-tri-CQAs, 6-gingesulfonic acid, phytosphingosine, sativanine B, sterebin E, anofinic acid, samandenone, sorbitan oleate, 2-palmitoylglycerol, citranaxanthin, 2-stearoyl glycerol, dodecanic acid and 12-methoxy-1-[(phosphonooxxy)methyl]1,2-ethanediyl ester, citronellyl butyrate, 2-[2-(4-hydroxy-3-meyhoxyphenyl)ethyl]tetrahydro-6-(4,5-dihydroxy-3-methoxyphenyl)-2H-pyran-4-ylacetate and dodemorph, in the extracts of the boiled, drip and espressogreen/roasted coffee beans. In addition, UHPLC-ESI-QTOF-MS analysis of the clinical specimens obtained from selected coffee consumers has indicated the appearance of CF, methylxanthines and methyluric acids in the plasma, along with eleven methylxanthine and methyluricacid metabolites, furan and methylfuran metabolites in the urine. These were mainly found in the form of sulfate, methyl derivatives, glucuronides and un-metabolized CQAs, FQAs, CoQAs, CA, ferulic acid and coumaric acids [62–64].

Through the use of HPLC-DAD analysis, we have detected the presence of CF and CGA, but not CA, in the roasted Arabica coffee extracts that were prepared by boiling and with the use of coffee makers (drip and espresso) [65]. A recent HPLC/DAD analysis of five commercial coffee samples has reported the presence of 8.35 ± 6.13 mg CF/g methanol extract [48]. Furthermore, trigonelline, CF and CGA were detected in green coffee beans (0.65 ± 0.05, 0.97 ± 0.09 and 3.13 ± 0.33 mg/g dry weight, respectively) and roasted coffee beans (0.85 ± 0.01, 1.30 ± 0.13 and 1.00 ± 0.02 mg/g dry weight, respectively), indicating a decrease in CGA and total phytochemical contents but increases in trigonelline and CF contents during the roasting process [66]. Likewise, trigonelline, CGA and CA were found in instant coffee samples [24]. Due to the thermal degradation of CGA and the relative stability of the alkaloids that occur at high temperatures during the roasting process, CF and trigonelline were the main metabolites present in roasted coffee beans [66]. During the process of roasting, CGAs and their CGA derivatives in green coffee that contribute to the acidity, astringency and bitterness of the final coffee beverages, can be isomerized and transformed to CGA lactones through a process involving water loss from quinic acid moiety and intramolecular ester bonding. This gives the coffee its flavor and determines its quality [67]. However, the amount of CGAs and their derivatives can also be used to indicate the quality of green coffee beans and to discriminate between low-quality varieties (9.1 g CGA/100 g) and commercial ones (10.4 g/100 g) [68]. Consistently, the phenolics found in our hot water/drip coffee extracts were mostly similar to those found in cold drip coffee extracts as determined by Angeloni et al. [69]. Through the use of HPLC/MS in our study, the phenolic and alkaloid profiles identified for the espresso coffee extracts were not totally the same as those assayed by Aves and colleagues [70]. However, the analyzed compounds were mostly comprised of trigonelline, CQAs, FQAs, coumaroylquinic acids, diCGAs, diFQA and caffeoylferuloylquinic acid. In the present study, when using UHPLC-ESI-QTOF-MS, 19 of 21 phytochemical compounds were detected consistently in all of the green coffee and roasted coffee extracts. Consistently, Angelino and colleagues used the UHPLC-MS-MS technique and have revealed the presence of trigonelline, CF, CGA, FQA, coumaroylquinic acids, hydroxycinnamate dimers, caffeoylshikimic acids and caffeoylquinic lactones in regular espresso coffee varieties [71].

With regard to antioxidant activity, food processing generally affects the content and activities of persisting phytochemicals and the corresponding antioxidant capacities in functional foods. Coffee beans are the main sources of anti-oxidative compounds and require roasting and drying before utilization. Consequently, the two processes may give rise to significant alterations in the antioxidant compositions and properties of theresultant coffee products, while the Strecker and Maillard reactions may increasefree-radical scavenging capacities [41]. Functional phenolic compounds contain hydroxyl components on the aglyconic phenolic ring and glyconic ring that scavenge ROS. Nonetheless, the antioxidant capacity that is related to human health benefits is dependent upon the bioavailability of the phytochemicals after consumption, which is subsequently dependent upon the soluble parts of the coffee known as the extraction yields [17]. In accordance with this finding, all roast coffee extracts and instant coffee doses dependently scavenged $ABTS^{\bullet+}$ and $DPPH^{\bullet}$ at different potencies depending on the coffee type and the preparation method used, in which the drip coffee extract exhibited the most efficient activity. With regard to the relevant technical factors, decaffeinated espresso coffee exerted slightly higher $DPPH^{\bullet}$ scavenging activity than regular espresso coffee (32% and 38%, respectively), which was directly related to the phenolic contents [70]. Iwai and colleagues previously elucidated that the potency order of superoxide anion radical scavenging activity was diCGA > CA, CGA > FQA, for which the activities of the diCGA were twice as effective as those of CGA and four times as effective as those of 5-FQA [68]. Importantly, total phenolic content, $ABTS^{\bullet+}$- and $DPPH^{\bullet}$-scavenging capacities of coffee samples were increased during the roasting process [66]. More importantly, the order of antioxidant capacity of all coffee brews was espresso > mocha > plunger > drip-filtered; herein, CGAs scavenged Fremy's salt (potassium nitrosodisulfonate) radicals, while melanoidins scavenged 2,2,6,6-tetramethyl-1-piperidin-1-oxyl radicals [72].

Notably, the consumption of a cup of drip coffee or instant coffee resulted in approximately a 39% decrease in dietary non-heme iron. Additionally, when coffee is consumed along with a meal, ferric-ethylenediamine tetraacetic acid or ferric chloride decreases the degree of dietary iron absorption from 5.88% to 1.64% depending upon the coffee dose [6]. For the preparation of coffee extracts, water was found to be the optimum solvent in terms of producing higher yields, better protective effects against lipid peroxidation, and effective ROS-scavenging and iron-chelating properties when compared to methanol, ethanol and n-hexane [73]. It was noted that certain phenolic compounds, such as CGAs, hydrolysable multi-galloyl tannin and galloylcatechin, did not condense the tannins or catechins present in the coffee utilized diol groups onto the phenolic ring that would bind iron at different affinity levels and interfere with dietary iron absorption [74,75]. In addition, the CGAs abundant in roasted coffee beans utilized the hydroxyl groups to bind thiol-molecules [76]. Moreover, coffee grounds were found to adsorb divalent metal ions in the following order: Cu < Pb < Zn < Cd, whereas tea leaves exhibited a similar outcome in the opposite order [77].

It has been reported that coffee constituents (e.g., kahweol, cafestol and CGAs) induced antioxidant-response element gene expression and activated the production of anti-oxidative enzymes in peripheral blood lymphocytes [78]. Herein, our roasted coffee extracts significantly increased the MTT-assayed viability of normal PBMC at 48 h, but did not influence the viability of breast cancer MDA-MB-231 and MCF-7 cells at any time period. Consistently, the consumption of coffee containing cafestol and kahweol induced antioxidant enzymes by an increase of 38% of superoxide dismutase activity [79]. Taken together, the treatment of PBMC with kahweol, cafestol and CGAs-rich roasted coffee can increase and empower mitochondrial reducing enzyme systems and subsequently intensify blue formazan products, which can result in higher percentages of cell viability. Ambiguously, a recent placebo-controlled intervention trial in healthy subjects has demonstrated that the consumption of coffee at up to 5 cups per day had no detectable beneficial or harmful effects on human health [80]. In contrast, the biocompatible copper sulfate-oxidized nanoparticles of Arabica coffee bean extracts showed anti-proliferative activity against MCF-7 breast cancer cells [81]. In some studies, CGAs and the natural extracts of green and roasted coffee beans could be employed as chemoprotective dietary supplements against the proliferation of Ras-dependent breast cancer MDA-MB-231 cell lines [82]. On the other hand, trigonelline, a niacin-related natural constituent of coffee (1%), was found to stimulate MCF-7 cell growth by acting as a phytoestrogen through the mediation of the estrogen receptor [83]. However, there is no evidence to support a relationship of either caffeinated or decaffeinated coffee intake with breast cancer risk [84]. Despite the fact that they are the two major active flavor ingredients in coffee, CA (6 and 10 mg/plate) and CGA (19 and 28 mg/plate), have been reported for their mutagenic properties in the L5178Y mouse lymphoma TK$^{+/-}$ assay. This was possibly due to the oxidative degradation and transformation of the two compounds to hydrogen peroxide at a neutral pH in the presence of a metal ion^{2+} (such as Mn^{2+}) [85]. Additionally, green and light roasted coffee extracts were found to promote higher inhibitory effects on the viability of PC-3 and DU-145 metastatic prostate cancer cell lines that had been assayed by the MTT test [53]. Moreover, green coffee beans containing CGAs and their derivatives showed anti-proliferative effects against U937, KB, MCF7 and WI38-VA cancer cell lines [68]. In controversy, diterpenes (e.g., cafestol and kahweol) are claimed to promote an increase in plasma cholesterol concentrations as a risk factor of cardiovascular diseases and type 2 diabetic mellitus [86].

4. Materials and Methods

4.1. Chemicals and Reagents

In terms of materials, 2,2′-azino-bis (3-ethylbenzothiazoline-6-sulfonic acid)diammonium salt (ABTS), chlorogenic acid (CGA), caffeic acid (CA), caffeine (CF), gallic acid (GA), 2,2-diphenyl-1-picrylhydrazyl (DPPH), formic acid, 3-(N-morpholino) propanesulfonic acid (MOPS), 3-[4,5-dimethylthiazol-2-yl]-2,5-diphenyl tetrazolium bromide (MTT), nitrilotriacetate disodium salt, nitrilotriacetate trisodium salt, phosphate-buffered saline (PBS) pH 7.0, phosphoric acid, 6-hydroxy-2,5,7,8-tetramethylchroman-2-carboxylic acid

(Trolox) and potassium thiosulfate were purchased from Sigma-Aldrich, Chemical Company (St. Louis, MO, USA). Ferric nitrate nonahydrate, sodium carbonate and Folin-Ciocalteu's phenol reagent were obtained from the Merck-Millipore Company (Merck KGaA, Damstadt, Germany). Organic solvents, such as acetonitrile, were all of HPLC grade and all chemicals were highly pure and of analytical grade. Commercial instant coffee (regular and decaffeinated types) were purchased from a Tesco Supermarket in Chiang Mai Province, Thailand.

4.2. Preparation of Coffee Extracts

Green coffee beans (*Coffea arabica*) were harvested from coffee fields belonging to the Royal Project Foundation at Mae Hair Village, Mae Wang District, Chiang Mai Province, Thailand. Whole beans were roasted and ground (known as milling) with the use of a coffee-grinding machine in a Coffee Factory of the Royal Project Foundation, Chiang Mai, Thailand. Extracts of ground green and roasted coffee were prepared using the boiled water, drip and espresso coffee maker methods, as has been previously described by Hutachok and colleagues [65]. Boiled coffee extract was manually prepared by placing ground roasted coffee (10 g) into hot water (90 °C) (100 mL), allowing the coffee to dissolve for 10 min and then allowing it to cool down. In pressurized percolation (espresso coffee extract), hot water (100 mL) was forced through fine coffee (10 g) under pressure using a manual portable Espresso maker (Minipresso GR, Wacaco Company, Pagewood NSW, Australia) with a specification of 8g ground coffee capacity, 120-mL water capacity and 8-bar pump pressure according to the manufacturer's instructions. During the brewing of drip coffee, hot water (100 mL) was forced through fine coffee (10 g) and filtered on high-quality paper under pressure using an automated drip coffee maker (Delonghi, Thiptanaporn, Company Limited, Bangna, Bangkok, Thailand) with a specification of 1.4-L tank capacity and 15-bar pump pressure according to the manufacturer's instructions. All the coffee brewed was filtered through a clean filter cloth and centrifuged at 6000 rpm for 20 min, while the supernatant was lyophilized to dryness. Coffee extracts were then stored in separate brown bottles at −20 °C until further analysis.

4.3. HPLC/ESI-MS Identification of Phenolic Compounds

All of the coffee extracts were analyzed in terms of their polar phenolic compounds at the Central Laboratory (North Branch), Department of Land Development, Ministry of Agriculture and Cooperation, Chiang Mai using the HPLC/MS method established by Cuyckens and colleagues with slight modifications [65,87]. The HPLC system (Agilent Technologies 1100 Series, Deutschland GmbH, Waldbronn, Germany) consisted of a quaternary pump (G1311A), an online vacuum degasser (G1322A), an autosampler (G1313A), a thermostated column compartment (G1316A) and a photodiode array (PDA) detector (G1315A). The outlet of the PDA was coupled directly to the atmospheric pressure ESI interface of the mass spectrometer (MS) detector (Agilent Technologies 1100 LC/MSD SL, Palo Alto, CA, USA) through a flow splitter (1:1). In terms of the analysis, coffee extracts were constituted in 1.0 mL of a mixture comprised of solvent A (acetonitrile) and solvent B (10 mM formate buffer pH 4.0) (1:1, *v/v*) and filtered through a syringe filter (polytetrafluoroethylene membrane, 25 mm diameter, 0.45-µm pore size, Corning®) before being used. The mixture was then injected (20 µL) into the system. Chromatographic separation was carried out on a column (LiChroCART RP-18e, 150 mm × 4.6 mm, 5 µm particle size; Purospher STAR, Merck, Darmstadt, Germany) operated at 40 °C. The mobile phases A and B were run at a flow rate of 1.0 mL/min under the gradient program of 100% B (0% A) for an initial period of 5 min, 0–20% A from 5 to 10 min, 20% A from 10 to 20 min, 20–40% A from 20 to 60 min, 40% A for 3 min, and followed by an initial step of 100% B for 5 min. PDA detection was set at 270 nm. MS analysis was done in positive ESI mode and spectra were acquired within the mass to a charge ratio (*m/z*) ranging from 100 to 700. For the single quadrupole MS system, the ESI energy was set at 70 eV, while the temperatures of the ion source and the interface were set at 150 °C and 230 °C, respectively. Nitrogen was used as the nebulizing, drying and collision gas. The capillary temperature was set to 320 °C, the nebulizer pressure was set to 60 psi and the drying gas flow rate was set to 13 L/min. Capillary voltages were set to 3500 V (positive) and 150 V (negative). The oven temperature was programmed as follows: 80 °C (held for 3 min), ramped to 110 °C at 10 °C/min (held for 5 min), increased to 190 °C (held for 3 min), ramped to 220 °C at 10 °C/min (held for 4 min) and increased to 280 °C at 15 °C/min (held for 13 min). Accurate mass measurements were performed by employing the auto mass calibration method using an external mass calibration solution (ESI-L Low Concentration Tuning Mix; Agilent calibration solution B). Herein, the limit of detection (LOD), limit of quantitation (LOQ) and recovery value were found to be 0.5 mg/kg, 1.20 mg/kg and 70–110%, respectively. The chromatographic and mass spectrometric analysis and prediction of the chemical formula, including the exact mass calculation, were performed by Mass Hunter software version B.04.00 built to 4.0.479.0 (Agilent Technology). Available authentic phenolics (1 mg/mL each), such as gallic acid, catechin, tannic acid, rutin, isoquercetin, hydroquinine, eriodictyol and quercetin, were also analyzed and used to generate a database. In addition, MS data were searched for in published literature repositories.

4.4. UHPLC-ESI-QTOF-MS Analysis of Phytochemical Compounds

Phytochemical compounds of the coffee extracts were analyzed at the Central Laboratory, Faculty of Agriculture, Chiang Mai University, Chiang Mai, Thailand using the HPLC-ESI-QTOF-MS method [88]. The HPLC instrument was equipped with an ESI-QTOF-MS machine (Agilent Acquisition SW version 6200 series TOF/6545 series LC/Q-TOF) and QTOF Firmware Version 25.698 (Agilent Technologies, Santa Clara, CA, USA). Mobile phase A (acetonitrile) and mobile phase B (0.1% formic acid) were degassed at 25 °C for 15 min. The GSO (20 mg) was constituted in 1.0 mL of the A:B mixture (1:1, v/v). It was then filtered using a syringe filer (polyvinylidene fluoride type, 0.45 μm pore size, Millipore, MA, USA) and put into HPLC vials. The flow rate was set to 0.35 mL/min, the injection volume was measured at 10 μL for each sample and the running time was 60 min. Chromatographic separation was carried out on a column (InfinityLabPoroshell 120 EC-C18 type, 2.1 mm × 100 mm, 2.7 μm, Agilent Technologies, Santa Clara, CA, USA) that was regulated thermally at 40 °C. The ESI-MS-MS spectra were recorded using an Agilent Q-TOF mass spectrometer. In the MS system, nitrogen gas nebulization was set at 45 pounds per inch^2 with a flow rate of 5 L/min at 300 °C, while the sheath gas was set at 11 L/min at 250 °C. In addition, the capillary and nozzle voltage values were set at 3.5 kV and 500 V, respectively. A complete mass scan was conducted as a mass-to-charge ratio (m/z) ranging from 200 to 3200. All the operations, acquisition and analysis of the data were monitored using Agilent LC-Q-TOF-MS MassHunter Acquisition Software Version B.04.00 (Agilent Technologies) and operated under MassHunter Qualitative Analysis Software Version B.04.00 (Agilent Technologies). Peak identification was performed in positive modes using the library database, while the identification scores were further selected for characterization and m/z verification.

4.5. HPLC-DAD Analysis of Caffeic Acid, Caffeine and Chlorogenic acid Contents

Owing to their popularity and the general preferences of consumers, we focused our investigation on the biological properties of roasted coffee extracts in comparison with instant coffee. Thus, CA, CGA and CF contents were analyzed in boiled, drip and espresso roasted coffee extracts using the HPLC-DAD method [65]. The conditions included a column (C18-type, 4.6 × 250 mm, 5-μm particle size, Agilent Technologies, Santa Clara, California, United States), isocratic mobile-phase solvent containing 0.2% phosphoric acid and acetonitrile (90:10, v/v), a flow rate of 1.0 mL/min and a wavelength detection of 275 nm for CF and 330 nm for CA and CGA. Data were recorded and integrated using Millenium 32 HPLC Software. CGA and CF were identified by comparison with the specific T_R of the authentic standards. Determined concentrations were then established from the standard curves constructed from different concentrations.

4.6. Determination of Total Phenolic Content

Total phenolic compounds (TPC) of the roasted coffee extracts and instant coffee were determined using the Folin-Ciocalteu method [89]. Briefly, coffee samples (100 µL) were incubated with 10% Folin-Ciocalteu reagent (200 µL) at room temperature for 4 min and then incubated with 700 mM Na_2CO_3 (800 µL) for 30 min. The optical density (OD) was then measured at 765 nm against the reagent blank. GA (6.25–200 µg/mL) was used to generate a calibration curve that was then used to calculate TPC and identify the gallic acid equivalent (GAE).

4.7. Determination of Free-Radical Scavenging Activity

Antioxidant activity was determined by using the ABTS radical cation ($ABTS^{\bullet+}$) decolorization method [89] and the DPPH radical ($DPPH^{\bullet}$) scavenging method as was previously established by Hatano and colleagues with slight modifications [89]. Ten microliters of DI (control), coffee extracts (50–4000 µg/mL), instant coffee (50–4000µg/mL) or Trolox (6.25–800 µg/mL) was incubated with the freshly prepared solution consisting of 3.5 mM $ABTS^{\bullet+}$ and 1.22 mM $K_2S_8O_8$ in the dark at room temperature for exactly 6 min. The substance was then photometrically measured at a wavelength of 764 nm. Results are expressed as percentage of inhibition of $ABTS^{\bullet+}$ production and reported in mg Trolox equivalent antioxidant capacity as (TEAC)/g dry weight of the extracts. With the use of the $DPPH^{\bullet}$ scavenging method, DI (control), coffee extracts (30–500 µg/mL), instant coffee (30–500µg/mL) or Trolox (6.25–50 µg/mL) were incubated with of 0.4 mM $DPPH^{\bullet}$ solution in the dark at room temperature for 30 min and photometrically measured at a wavelength of 517 nm. Antioxidant activity was determined using the following equation:

$$\text{Radical scavenging activity (\%)} = [(OD_{DI} - OD_{\text{coffee or Trolox}})/OD_{DIt}] \times 100 \quad (1)$$

Results were reported as percentage of $ABTS^{\bullet}$ or $DPPH^{\bullet}$ scavenging activity (y-axis) and compared to DI plotted against coffee extract, instant coffee or Trolox concentration (x-axis), while the concentration of the substance that provided 50% reduction of $DPPH^{\bullet}$ concentration (IC_{50}) value was calculated from the graph.

4.8. Determination of Iron-Binding Ability

Iron-binding activity of the coffee extracts and the instant coffee was determined with the method established by Srichairatanakool and colleagues [90]. Nitrilotriacetate (NTA) pH 7.0 solution (80 mM) was prepared by mixing 80 mM nitrilotriacetate disodium salt solution with 80 mM nitrilotriacetate trisodium salt solution until the pH reached 7.4. Stock ferric nitrilotriacetate (Fe-NTA) solution (1 mM) was freshly prepared by dissolving ferric nitrate nonahydrate in 80 mM NTA pH 7.0 solution and the mixture was incubated at room temperature for 1 h. Coffee extracts (250 µg/mL) were diluted in 10 mM MOPS buffer pH 7.0 to reach designated concentrations. The coffee extracts, CGA and CF, were incubated with 1 mM Fe-NTA solution at room temperature for 1 h and were photometrically measured within a wavelength range of 200–800 nm using a scanning double-beam UV-VIS spectrophotometer against the reagent blank. In addition, the coffee extracts (250 µg/mL) were incubated with different concentrations of Fe-NTA solution at room temperature for 30 min. Finally, the produced coffee-iron complex was photometrically measured within a wavelength range of 200–800 nm using a scanning double-beam UV-VIS spectrophotometer against the reagent blank.

4.9. Cell Toxicity Study Using MTT Assay

Colorimetric MTT assay was used for the determination of cell viability based on mitochondrial reductases presented in live cells, in which yellow MTT dye was reduced to purple-colored formazan product in a direct proportion to the number of viable cells [91]. Normal human peripheral blood mononuclear cells (PBMC) and two human breast cancer cell lines were used in the toxicity study. The blood collection protocol was approved of by the director of Maharaj Nakorn Chiang Mai, Faculty of Medicine, ChiangMai University, Chiang Mai, Thailand and the Ethical Committee for Human Study (Reference Number 8393 (8).9/436). Informed consent was provided by healthy blood volunteers. Blood samples (10 mL) obtained from healthy donors were collected into tubes (BD Vacutainer®, USA) containing 158 USP units of sodium heparin and diluted with sterile PBS, pH 7.4(1:1, v/v). Ficoll-Paque (Histopaque®-1077) solution was then overlaid with heparinized blood (at a ratio of 1:2, v/v) in 50-mL sterile conical tubes and centrifuged at 1500 rpm and at a temperature of 25 °C for 30 min. After centrifugation, the upper layer was aspirated and mononuclear cells at the interphase were collected into 15-mL sterile centrifuge tubes. The suspension was washed twice with 10 mL of sterile PBS, then centrifuged at 6000 rpm and at a temperature of 4 °C for 5 min, while the supernatant was aspirated and the cell pellets were resuspended in complete culture medium. The suspension was then incubated overnight in a humidified environment containing 5%CO_2 at 37 °C. In the assay, PBMC (7×10^4 cells/well) were incubated with boiled, drip and espresso coffee extracts (3.125–50 µg/mL) or PBS for 24 and 48 h in a humidified 5%CO_2, 37 °C incubator. After the treatment, MTT reagent (20 µL) was added into each well and the cells were incubated in the dark at 37 °C for 4 h. Finally, the formazan crystal product was dissolved with DMSO (200 µL) and the OD was measured at wavelengths of 570 and 630 nm. The percentage of cell viability was calculated using the following formula: $OD_{sample}/OD_{control} \times 100$.

The human breast cancer cell lines, MCF-7 (ATCC HTB-22) and MDA-MB-231 (ATCC HTB-26), were kindly provided by Professor Dr. Masami Suganuma, PhD. at the Graduate School of Science and Engineering, Saitama University, Saitama, Japan and Professor Dr. Ratana Banjerdpongchai, MD., PhD. at the Department of Biochemistry, Faculty of Medicine, Chiang Mai University, Thailand. MDA-MB-231 cells were cultured incomplete RPMI-1640 medium supplemented with 10% (v/v) heat-inactivated fetal bovine serum (FBS) and penicillin-streptomycin solution (10,000 U/mL). They were then treated with boiled, drip and espresso coffee extracts (3.12–50 µg/mL) for 24 and 48 h. Viability was then determined using the MTT assay. In another study, MDA-MB-231 and MCF-7 cells were treated with drip coffee extracts (31.25–1000µg/mL) in an incubator (5% CO_2, 37 °C) for 24 and 48 h. Viability was then determined using the MTT assay.

4.10. Acute Toxicity Study

Acute toxicity of the drip coffee extract was investigated in female rats for 14 days using the Standard Operating Procedure (SOP) according to OECD Guidelines 2008 established by van den Heuvel and colleagues [92]. Animals were purchased from the National Laboratory Animal Center, Mahidol University, Salaya Campus, Thailand. The study protocol was approved of by the Animal Ethics Committee, Faculty of Medicine, Chiang Mai University, Thailand (Reference Number 25/2561). Male Sprague–Dawley rats at approximately 8–10 weeks of age and weighing around 180–200 g were used for the acute toxicity study. Rats were orally fed with a single dose of drip coffee extract using an oral dosing needle. Animals were observed after being dosed for 30 min, 1, 4 and 6 h and daily for 14 d. In addition, rat body weights were recorded before treatment, then once a week for the next 2 weeks or at the time of death. The initial dose was estimated from the median lethal dose (LD_{50}) of caffeine content, while the starting dose was 175 mg/kg BW. The dose for the next animal was then adjusted up or down (1.3 of dose progression factor) depending upon the outcome for the first animal. If an animal survived, the dose for the next animal was increased; if the animal died, the dose for the next animal was decreased. The upper and lower limits of the dose should not exceed 2000 mg/kg BW or be lower than 5 mg/kg BW, respectively. Remark: The caffeine content in the upper limit dose should not exceed 200 mg/kg BW. Visual observations including behavioral patterns (trembling, diarrhea, breathing, impairment in food intake, water consumption, postural abnormalities, hair loss, sleep, lethargy and restlessness) and physical appearance (eye color, mucous membrane, salivation, skin/fur effects, body weight and injury) were recorded daily for 14 d.

4.11. Statistical Analysis

The experiments were performed in at least triplicate. Data were analyzed using the SPSS Statistics Program (IBM SPSS® Software version 22, IBM Corporation, Armonk, NY, USA, with a shared license assigned to Chiang Mai University, Thailand) and are expressed as mean±standard deviation (SD) or mean±standard error of the mean (SEM) values. Statistical significance was analyzed using one-way analysis of variance (ANOVA) followed by Tukey HSD's post-test. A p-value < 0.05 was considered significant.

5. Conclusions

We have conducted a comparative analysis of the phytochemical compounds, radical scavenging activities and also the iron binding capacity of roasted and green Arabica coffee extracts and instant coffee. The profiles of the phenolics and phytochemicals were practically the same in all green and roasted coffee bean extracts prepared using the boiling, drip-filtered and espresso methods. An exception of this would be the existence of additional dimethoxycinnamic acid, caffeoylarbutin and cynaroside that were present in green coffee. Similar to instant coffee, besides an abundance of caffeine, all the roasted coffee extracts were determined to be rich phenolic compounds; particularly, chlorogenic acid, which is known to exert dose dependent antioxidant and iron-binding properties. Among them, the drip coffee extract contained higher total phenolic content which would reflect stronger free-radical scavenging activity. The coffee extracts were not harmful to breast cancer cells and normal white blood cells, while they did support viability of the white blood cells at longer incubation times. Taken together, an automatic drip-filter coffee maker would be the most efficient way of brewing roasted coffee to yield the highest free-radical capacity, as well as the greatest amounts of phenolic compounds.

Author Contributions: N.H. designed and conducted the experiments, analyzed the data and shared in the discussion of the data; P.K. analyzed the caffeine and phenolics using an HPLC-DAD and HPLC-MS machine, and shared in the discussion; T.P. analyzed the phytochemical compounds using an HPLC-QTOF-MS machine; P.A. verified all coffee information and shared in the discussion; C.C. supplied roasted coffee beans and blended the roasted coffee; S.F. offered advice on the experiments and shared in the discussion; S.S. conceived of the study and experiments, contributed to the discussion, wrote and revised the manuscript. All authors have read and agreed to the published version of the manuscript.

Funding: This study was financially supported by the Research and Researchers for Industries (RRI) Ph.D. Scholarship, Thailand Science Research and Innovation (PHD60I0020), the Royal Project Foundation, Chiang Mai and the Medical Faculty Endowment Fund, Faculty of Medicine, Chiang Mai University, Thailand (Code 113/2563).

Institutional Review Board Statement: Not applicable.

Informed Consent Statement: Not applicable.

Data Availability Statement: Data available in a publicly accessible repository.

Acknowledgments: The authors wish to thank the Royal Project Foundation, Chiang Mai, Thailand for kindly supplying fresh and roasted coffee beans, the Ethics Committee for Animal Study at the Faculty of Medicine, Chiang Mai University for consideration and approval of the study protocol and the Animal Unit of the Faculty of Medicine, Chiang Mai University for granting access to the facilities used for animal housing. We express our appreciation to John B. Porter, at the UCL Biomedical Research Centre, Cardiometabolic Programme, University College London Medical School, London, United Kingdom for his helpful comments and suggestions.

Conflicts of Interest: The authors declare that they hold no conflict of interest.

Sample Availability: Samples of the green and roasted coffee beans, the boiled, drip and espresso coffee extracts are available from the authors.

Abbreviations

3-CQA = neochlorogenic acid; 4-CQA = cryptochlorogenic acid; 5-CQA = 5-O-caffeoylquinic acid; 5-FQA = 5-feruloylquinic acid; ABTS = 2,2′-Azino-bis (3-ethylbenzothiazoline-6-sulfonic acid) diammonium salt; ABTS$^{+\bullet}$ = 2,2′-azino-bis(3-ethylbenzothiazoline-6-sulfonic acid) diammonium salt radical cation; ANOVA = analysis of variance; BW = body weight; CA = caffeic acid; Cd = cadmium; CF = caffeine; CGA = chlorogenic acid; CQA = caffeoylquinic acid; Cu = copper; DAD = diode-array detector; DI = deionized water; diCGA = dichlorogenic acid; diCQA = dicaffeoylquinic acid; diFQA = diferuloylquinic acid; DMSO = dimethyl sulfoxide; DPPH = 2, 2-diphenyl-1-picrylhydrazyl; DPPH$^{\bullet}$ = 2, 2-diphenyl-1-picrylhydrazyl radical cation; ESI = electrospray ionization; Fe^{3+}-NTA = ferric-nitrilotriacetic acid; Fe-NTA = ferric nitrilotriacetic acid; FQA = feruloylquinic acid; GA = gallic acid; GAE = gallic acid equivalent; GLC = gas-liquid chromatography; HPLC-DAD = high-performance liquid chromatography-diode array detection; HPLC-ESI-MS = high-performance liquid chromatography-electrospray ionization-mass spectrometry; HPLC-MS = high-performance liquid chromatography-mass spectrometry; HPLC-ESI-MS-MS = high-performance liquid chromatography-electrospray ionization-tandem mass spectrometry; IC$_{50}$ = half-maximal inhibitory concentration; K$_2$S$_8$O$_8$ = potassium thiosulfate; LD$_{50}$ = median lethal dose; LOD = limit of detection; LOQ = limit of quantitation; m/z = mass to charge ratio; Mn^{2+} = Manganese ion; MOPS = 3-(N-morpholino) propanesulfonic acid; MS = mass spectrometry; MTT = 3-[4,5-dimethylthiazol-2-yl]2,5-diphenyl tetrazolium bromide; NA = not available; ND = not detectable; NTA = nitrilotriacetate; OD = optical density; OECD = organization for economic co-operation and development; Pb = lead; PBMC = peripheral blood mononuclear cells; PBS = phosphate-buffered saline; PDA = photodiode array; p-CoQAs = para-coumaroylquinic acids; ROS = reactive oxygen species; rpm = revolutions per minute; SD = standard deviation; SEM = standard error of the means; SOP = standard operating procedure; TDS = total dissolved solid; TE = Trolox equivalent; TEAC = Trolox equivalent antioxidant capacity; TIC = total ion count; TPC = total phenolic content; T$_R$ = retention time; UHPLC-ESI-QTOF-MS = ultra-high performance liquid chromatography-electrospray ionization-quadrupole time-of-flight-mass spectrometry; UHPLC-QTOF-MS = ultra-high performance liquid chromatography-quadrupole time-of-flight-mass spectrometry; UV = ultraviolet; v/v = volume by volume; w/w = weight by weight; Zn = zinc.

References

1. Jeszka-Skowron, M.; Zgoła-Grześkowiak, A.; Grześkowiak, T. Analytical methods applied for the characterization and the determination of bioactive compounds in coffee. *Eur. Food Res. Technol.* **2015**, *240*, 19–31. [CrossRef]
2. Clarke, R.J. Coffee: Green coffee/roast and ground. In *Encyclopedia of Food Science and Nutrition*; Caballero, B., Ed.; Elsevier Science Ltd.: West Sussex, UK, 2003; Volume 3, pp. 187–191.
3. Park, J.B. NMR Confirmation and HPLC quantification of javamide-I and javamide-II in green coffee extract products available in the market. *Int. J. Anal. Chem.* **2017**, *2017*, 1927983. [CrossRef] [PubMed]
4. Ramirez-Coronel, M.A.; Marnet, N.; Kolli, V.S.; Roussos, S.; Guyot, S.; Augur, C. Characterization and estimation of proanthocyanidins and other phenolics in coffee pulp (*Coffea arabica*) by thiolysis-high-performance liquid chromatography. *J. Agric. Food Chem.* **2004**, *52*, 1344–1349. [CrossRef] [PubMed]
5. Farah, A. Coffee constituents. In *Coffee*; Chu, Y.F., Ed.; John Wiley & Sons: Hoboken, NJ, USA, 2012.
6. Morck, T.A.; Lynch, S.R.; Cook, J.D. Inhibition of food iron absorption by coffee. *Am. J. Clin. Nutr.* **1983**, *37*, 416–420. [CrossRef]
7. Cavalcante, J.W.; Santos, P.R., Jr.; Menezes, M.G.; Marques, H.O.; Cavalcante, L.P.; Pacheco, W.S. Influence of caffeine on blood pressure and platelet aggregation. *Arq. Bras. Cardiol.* **2000**, *75*, 97–105. [CrossRef]
8. Franks, A.M.; Schmidt, J.M.; McCain, K.R.; Fraer, M. Comparison of the effects of energy drink versus caffeine supplementation on indices of 24-hour ambulatory blood pressure. *Ann. Pharmacother.* **2012**, *46*, 192–199. [CrossRef]
9. Gonzalez de Mejia, E.; Ramirez-Mares, M.V. Impact of caffeine and coffee on our health. *Trends Endocrinol. Metab.* **2014**, *25*, 489–492. [CrossRef]
10. Von Ruesten, A.; Feller, S.; Bergmann, M.M.; Boeing, H. Diet and risk of chronic diseases: Results from the first 8 years of follow-up in the EPIC-Potsdam study. *Eur. J. Clin. Nutr.* **2013**, *67*, 412–419. [CrossRef]
11. Hang, D.; Zeleznik, O.A.; He, X.; Guasch-Ferre, M.; Jiang, X.; Li, J.; Liang, L.; Eliassen, A.H.; Clish, C.B.; Chan, A.T.; et al. Metabolomic signatures of long-term coffee consumption and risk of type 2 diabetes in women. *Diabetes Care* **2020**, *43*, 2588–2596. [CrossRef]
12. Hong, C.T.; Chan, L.; Bai, C.H. The effect of caffeine on the risk and progression of Parkinson's disease: A meta-analysis. *Nutrients* **2020**, *12*, 1860. [CrossRef]

13. Hu, G.; Bidel, S.; Jousilahti, P.; Antikainen, R.; Tuomilehto, J. Coffee and tea consumption and the risk of Parkinson's disease. *Mov. Disord.* **2007**, *22*, 2242–2248. [CrossRef]
14. Nkondjock, A. Coffee consumption and the risk of cancer: An overview. *Cancer Lett.* **2009**, *277*, 121–125. [CrossRef]
15. Hashibe, M.; Galeone, C.; Buys, S.S.; Gren, L.; Boffetta, P.; Zhang, Z.F.; La Vecchia, C. Coffee, tea, caffeine intake, and the risk of cancer in the PLCO cohort. *Br. J. Cancer* **2015**, *113*, 809–816. [CrossRef]
16. Monteiro, M.; Farah, A.; Perrone, D.; Trugo, L.C.; Donangelo, C. Chlorogenic acid compounds from coffee are differentially absorbed and metabolized in humans. *J. Nutr.* **2007**, *137*, 2196–2201. [CrossRef]
17. Farah, A.; Monteiro, M.; Donangelo, C.M.; Lafay, S. Chlorogenic acids from green coffee extract are highly bioavailable in humans. *J. Nutr.* **2008**, *138*, 2309–2315. [CrossRef]
18. Higdon, J.V.; Frei, B. Coffee and health: A review of recent human research. *Crit. Rev. Food Sci. Nutr.* **2006**, *46*, 101–123. [CrossRef]
19. Heckman, M.A.; Weil, J.; Gonzalez de Mejia, E. Caffeine (1, 3, 7-trimethylxanthine) in foods: A comprehensive review on consumption, functionality, safety, and regulatory matters. *J. Food Sci.* **2010**, *75*, R77–R87. [CrossRef]
20. Baspinar, B.; Eskici, G.; Ozcelik, A.O. How coffee affects metabolic syndrome and its components. *Food Funct.* **2017**, *8*, 2089–2101. [CrossRef]
21. Stalmach, A.; Mullen, W.; Nagai, C.; Crozier, A. On-line HPLC analysis of the antioxidant activity of phenolic compounds in brewed, paper-filtered coffee. *Braz. J. Plant Physiol.* **2006**, *18*, 253–262. [CrossRef]
22. Benkovic, M.; Srecec, S.; Spoljaric, I.; Mrsic, G.; Bauman, I. Fortification of instant coffee beverages—influence of functional ingredients, packaging material and storage time on physical properties of newly formulated, enriched instant coffee powders. *J. Sci. Food Agric.* **2014**, *95*, 2607–2618. [CrossRef]
23. Olechno, E.; Puscion-Jakubik, A.; Markiewicz-Zukowska, R.; Socha, K. Impact of brewing methods on total phenolic content (TPC) in various types of coffee. *Molecules* **2020**, *25*, 5274. [CrossRef]
24. Arai, K.; Terashima, H.; Aizawa, S.; Taga, A.; Yamamoto, A.; Tsutsumiuchi, K.; Kodama, S. Simultaneous determination of trigonelline, caffeine, chlorogenic acid and their related compounds in instant coffee samples by HPLC using an acidic mobile phase containing octanesulfonate. *Anal. Sci.* **2015**, *31*, 831–835. [CrossRef]
25. Buijs, M.M.; Kobaek-Larsen, M.; Kaalby, L.; Baatrup, G. Can coffee or chewing gum decrease transit times in Colon capsule endoscopy? A randomized controlled trial. *BMC. Gastroenterol.* **2018**, *18*, 95. [CrossRef]
26. Cho, E.S.; Jang, Y.J.; Hwang, M.K.; Kang, N.J.; Lee, K.W.; Lee, H.J. Attenuation of oxidative neuronal cell death by coffee phenolic phytochemicals. *Mutat. Res.* **2009**, *661*, 18–24. [CrossRef]
27. Hoelzl, C.; Knasmuller, S.; Wagner, K.H.; Elbling, L.; Huber, W.; Kager, N.; Ferk, F.; Ehrlich, V.; Nersesyan, A.; Neubauer, O.; et al. Instant coffee with high chlorogenic acid levels protects humans against oxidative damage of macromolecules. *Mol. Nutr. Food Res.* **2010**, *54*, 1722–1733. [CrossRef]
28. Gloess, A.N.; Schönbächler, B.; Klopprogge, B.; D'Ambrosio, L.; Chatelain, K.; Bongartz, A.; Strittmatter, A.; Rast, M.; Yeretzian, C. Comparison of nine common coffee extraction methods: Instrumental and sensory analysis. *Eur. Food Res. Technol.* **2013**, *236*, 607–627. [CrossRef]
29. Zlotek, U.; Karas, M.; Gawlik-Dziki, U.; Szymanowska, U.; Baraniak, B.; Jakubczyk, A. Antioxidant activity of the aqueous and methanolic extracts of coffee beans (*Coffea arabica* L.). *Acta Sci. Pol. Technol. Aliment.* **2016**, *15*, 281–288. [CrossRef]
30. Niseteo, T.; Komes, D.; Belščak-Cvitanović, A.; Horžić, D.; Budeč, M. Bioactive composition and antioxidant potential of different commonly consumed coffee brews affected by their preparation technique and milk addition. *Food Chem.* **2012**, *134*, 1870–1877. [CrossRef]
31. Budryn, G.; Nebesny, E.; Podsędek, A.; Żyżelewicz, D.; Materska, M.; Jankowski, S.; Janda, B. Effect of different extraction methods on the recovery of chlorogenic acids, caffeine and Maillard reaction products in coffee beans. *Eur. Food Res. Technol.* **2009**, *228*, 913–922. [CrossRef]
32. Vignoli, J.A.; Bassoli, D.G.; Benassi, M.T. Antioxidant activity, polyphenols, caffeine and melanoidins in soluble coffee: The influence of processing conditions and raw material. *Food Chem.* **2011**, *124*, 863–868. [CrossRef]
33. Gant, A.; Leyva, V.E.; Gonzalez, A.E.; Maruenda, H. Validated HPLC-diode array detector method for simultaneous evaluation of six quality markers in coffee. *J. AOAC Int.* **2015**, *98*, 98–102. [CrossRef] [PubMed]
34. Morishita, H.; Iwahashi, H.; Osaka, N.; Kido, R. Chromatographic separation and identification of naturally occurring chlorogenic acids by 1H nuclear magnetic resonance spectroscopy and mass spectrometry. *J. Chromatogr.* **1984**, *315*, 253–260. [CrossRef]
35. Rocchetti, G.; Braceschi, G.P.; Odello, L.; Bertuzzi, T.; Trevisan, M.; Lucini, L. Identification of markers of sensory quality in ground coffee: An untargeted metabolomics approach. *Metabolomics* **2020**, *16*, 127. [CrossRef] [PubMed]
36. Danhelova, H.; Hradecky, J.; Prinosilova, S.; Cajka, T.; Riddellova, K.; Vaclavik, L.; Hajslova, J. Rapid analysis of caffeine in various coffee samples employing direct analysis in real-time ionization-high-resolution mass spectrometry. *Anal. Bioanal. Chem.* **2012**, *403*, 2883–2889. [CrossRef] [PubMed]
37. Farooq, M.U.; Mumtaz, M.W.; Mukhtar, H.; Rashid, U.; Akhtar, M.T.; Raza, S.A.; Nadeem, M. UHPLC-QTOF-MS/MS based phytochemical characterization and anti-hyperglycemic prospective of hydro-ethanolic leaf extract of *Butea monosperma*. *Sci. Rep.* **2020**, *10*, 3530. [CrossRef]
38. Qiao, X.; Qu, C.; Luo, Q.; Wang, Y.; Yang, J.; Yang, H.; Wen, X. UHPLC-qMS spectrum-effect relationships for *Rhizomaparidis* extracts. *J. Pharm. Biomed. Anal.* **2020**, *194*, 113770.

39. Borrelli, R.C.; Fogliano, V. Bread crust melanoidins as potential prebiotic ingredients. *Mol. Nutr. Food Res.* **2005**, *49*, 673–678. [CrossRef]
40. Gross, G.; Jaccaud, E.; Huggett, A.C. Analysis of the content of the diterpenes cafestol and kahweol in coffee brews. *Food Chem. Toxicol.* **1997**, *35*, 547–554. [CrossRef]
41. De Melo Pereira, G.V.; de Carvalho Neto, D.P.; Magalhaes Junior, A.I.; Vasquez, Z.S.; Medeiros, A.B.P.; Vandenberghe, L.P.S.; Soccol, C.R. Exploring the impacts of postharvest processing on the aroma formation of coffee beans—A review. *Food Chem.* **2018**, *272*, 441–452. [CrossRef]
42. Nzekoue, F.K.; Angeloni, S.; Navarini, L.; Angeloni, C.; Freschi, M.; Hrelia, S.; Vitali, L.A.; Sagratini, G.; Vittori, S.; Caprioli, G. Coffee silver skin extracts: Quantification of 30 bioactive compounds by a new HPLC-MS/MS method and evaluation of their antioxidant and antibacterial activities. *Food Res. Int.* **2020**, *133*, 109128. [CrossRef]
43. Lopes, G.R.; Passos, C.P.; Rodrigues, C.; Teixeira, J.A.; Coimbra, M.A. Impact of microwave-assisted extraction on roasted coffee carbohydrates, caffeine, chlorogenic acids and coloured compounds. *Food Res. Int.* **2020**, *129*, 108864. [CrossRef]
44. Kim, J.H.; Ahn, D.U.; Eun, J.B.; Moon, S.H. Antioxidant effect of extracts from the coffee residue in raw and cooked meat. *Antioxidants* **2016**, *5*, 21. [CrossRef]
45. Rebollo-Hernanz, M.; Canas, S.; Taladrid, D.; Benitez, V.; Bartolome, B.; Aguilera, Y.; Martin-Cabrejas, M.A. Revalorization of coffee husk: Modeling and optimizing the green sustainable extraction of phenolic compounds. *Foods* **2021**, *10*, 653. [CrossRef]
46. Cotter, A.R.; Batali, M.E.; Ristenpart, W.D.; Guinard, J.X. Consumer preferences for black coffee are spread over a wide range of brew strengths and extraction yields. *J. Food Sci.* **2020**, *86*, 194–205. [CrossRef]
47. Frost, S.C.; Ristenpart, W.D.; Guinard, J.X. Effects of brew strength, brew yield, and roast on the sensory quality of drip brewed coffee. *J. Food Sci.* **2020**, *85*, 2530–2543. [CrossRef]
48. Ahmad, R.; Ahmad, N.; Al-Anaki, W.S.; Ismail, F.A.; Al-Jishi, F. Solvent and temperature effect of accelerated solvent extraction (ASE) coupled with ultra-high-pressure liquid chromatography (UHPLC-PDA) for the determination of methyl xanthines in commercial tea and coffee. *Food Chem.* **2019**, *311*, 126021. [CrossRef]
49. Cordoba, N.; Moreno, F.L.; Osorio, C.; Velasquez, S.; Ruiz, Y. Chemical and sensory evaluation of cold brew coffees using different roasting profiles and brewing methods. *Food Res. Int.* **2021**, *141*, 110141. [CrossRef]
50. Angeloni, G.; Guerrini, L.; Masella, P.; Bellumori, M.; Daluiso, S.; Parenti, A.; Innocenti, M. What kind of coffee do you drink? An investigation on effects of eight different extraction methods. *Food Res. Int.* **2019**, *116*, 1327–1335. [CrossRef]
51. Simoes, J.; Nunes, F.M.; Domingues, M.R.; Coimbra, M.A. Extractability and structure of spent coffee ground polysaccharides by roasting pre-treatments. *Carbohydr. Polym.* **2013**, *97*, 81–89. [CrossRef]
52. Perez De Obanos, A.; Gonzalez-Penas, E.; Lopez De Cerain, A. Influence of roasting and brew preparation on the ochratoxin A content in coffee infusion. *Food Addit. Contam.* **2005**, *22*, 463–471. [CrossRef]
53. Montenegro, J.; Dos Santos, L.S.; de Souza, R.G.G.; Lima, L.G.B.; Mattos, D.S.; Viana, B.; da Fonseca Bastos, A.C.S.; Muzzi, L.; Conte-Junior, C.A.; Gimba, E.R.P.; et al. Bioactive compounds, antioxidant activity and antiproliferative effects in prostate cancer cells of green and roasted coffee extracts obtained by microwave-assisted extraction (MAE). *Food Res. Int.* **2021**, *140*, 110014. [CrossRef]
54. Abdel-Moety, E.M. First-derivative spectrophotometric and gas-liquid chromatographic determination of caffeine in foods and pharmaceuticals. I. Rapid determination of caffeine in coffee, tea and soft drinks. *Z. Lebensm. Unters. Forsch.* **1988**, *186*, 412–416. [CrossRef]
55. Stennert, A.; Maier, H.G. Trigonelline in coffee. II. Content of green, roasted and instant coffee. *Z. Lebensm. Unters. Forsch.* **1994**, *199*, 198–200. [CrossRef]
56. Fujioka, K.; Shibamoto, T. Quantitation of volatiles and nonvolatile acids in an extract from coffee beverages: Correlation with antioxidant activity. *J. Agric. Food Chem.* **2006**, *54*, 6054–6058. [CrossRef]
57. Frank, O.; Blumberg, S.; Kunert, C.; Zehentbauer, G.; Hofmann, T. Structure determination and sensory analysis of bitter-tasting 4-vinylcatechol oligomers and their identification in roasted coffee by means of LC-MS/MS. *J. Agric. Food Chem.* **2007**, *55*, 1945–1954. [CrossRef]
58. Van Lancker, F.; Adams, A.; Owczarek, A.; De Meulenaer, B.; De Kimpe, N. Impact of various food ingredients on the retention of furan in foods. *Mol. Nutr. Food Res.* **2009**, *53*, 1505–1511. [CrossRef]
59. Montavon, P.; Mauron, A.F.; Duruz, E. Changes in green coffee protein profiles during roasting. *J. Agric. Food Chem.* **2003**, *51*, 2335–2343. [CrossRef]
60. Murkovic, M. Acrylamide in Austrian foods. *J. Biochem. Biophys. Methods* **2004**, *61*, 161–167. [CrossRef]
61. Spreng, S.; Schaerer, A.; Poisson, L.; Chaumonteuil, M.; Mestdagh, F.; Davidek, T. Discovery of polyhydroxyalkyl pyrazine generation upon coffee roasting by in-bean labeling experiments. *J. Agric. Food Chem.* **2021**, *69*, 6636–6649. [CrossRef]
62. Martinez-Lopez, S.; Sarria, B.; Baeza, G.; Mateos, R.; Bravo-Clemente, L. Pharmacokinetics of caffeine and its metabolites in plasma and urine after consuming a soluble green/roasted coffee blend by healthy subjects. *Food Res. Int.* **2014**, *64*, 125–133. [CrossRef]
63. Gomez-Juaristi, M.; Martinez-Lopez, S.; Sarria, B.; Bravo, L.; Mateos, R. Bioavailability of hydroxycinnamates in an instant green/roasted coffee blend in humans. Identification of novel colonic metabolites. *Food Funct.* **2017**, *9*, 331–343. [CrossRef] [PubMed]
64. Stegmuller, S.; Beissmann, N.; Kremer, J.I.; Mehl, D.; Baumann, C.; Richling, E. A New UPLC-qTOF approach for elucidating furan and 2-methylfuran metabolites in human urine samples after coffee consumption. *Molecules* **2020**, *25*, 5104. [CrossRef] [PubMed]

65. Hutachok, N.; Angkasith, P.; Chumpun, C.; Fucharoen, S.; Mackie, I.J.; Porter, J.B.; Srichairatanakool, S. Anti-platelet aggregation and anti-cyclooxygenase activities for a range of coffee extracts (*Coffea arabica*). *Molecules* **2020**, *25*, 2474.
66. Acidri, R.; Sawai, Y.; Sugimoto, Y.; Handa, T.; Sasagawa, D.; Masunaga, T.; Yamamoto, S.; Nishihara, E. Phytochemical profile and antioxidant capacity of coffee plant organs compared to green and roasted coffee beans. *Antioxidants* **2020**, *9*, 93. [CrossRef]
67. Farah, A.; de Paulis, T.; Trugo, L.C.; Martin, P.R. Effect of roasting on the formation of chlorogenic acid lactones in coffee. *J. Agric. Food Chem.* **2005**, *53*, 1505–1513. [CrossRef]
68. Iwai, K.; Kishimoto, N.; Kakino, Y.; Mochida, K.; Fujita, T. In vitro antioxidative effects and tyrosinase inhibitory activities of seven hydroxycinnamoyl derivatives in green coffee beans. *J. Agric. Food Chem.* **2004**, *52*, 4893–4898. [CrossRef]
69. Angeloni, G.; Guerrini, L.; Masella, P.; Innocenti, M.; Bellumori, M.; Parenti, A. Characterization and comparison of cold brew and cold drip coffee extraction methods. *J. Sci. Food Agric.* **2018**, *99*, 391–399. [CrossRef]
70. Alves, R.C.; Costa, A.S.; Jerez, M.; Casal, S.; Sineiro, J.; Nunez, M.J.; Oliveira, B. Antiradical activity, phenolics profile, and hydroxymethylfurfural in espresso coffee: Influence of technological factors. *J. Agric. Food Chem.* **2010**, *58*, 12221–12229. [CrossRef]
71. Angelino, D.; Tassotti, M.; Brighenti, F.; Del Rio, D.; Mena, P. Niacin, alkaloids and (poly)phenolic compounds in the most widespread Italian capsule-brewed coffees. *Sci. Rep.* **2018**, *8*, 17874. [CrossRef]
72. Perez-Martinez, M.; Caemmerer, B.; De Pena, M.P.; Cid, C.; Kroh, L.W. Influence of brewing method and acidity regulators on the antioxidant capacity of coffee brews. *J. Agric. Food Chem.* **2010**, *58*, 2958–2965. [CrossRef]
73. Yen, W.J.; Wang, B.S.; Chang, L.W.; Duh, P.D. Antioxidant properties of roasted coffee residues. *J. Agric. Food Chem.* **2005**, *53*, 2658–2663. [CrossRef]
74. Brune, M.; Rossander, L.; Hallberg, L. Iron absorption and phenolic compounds: Importance of different phenolic structures. *Eur. J. Clin. Nutr.* **1989**, *43*, 547–557.
75. Tamilmani, P.; Pandey, M.C. Iron binding efficiency of polyphenols: Comparison of effect of ascorbic acid and ethylenediaminetetraacetic acid on catechol and galloyl groups. *Food Chem.* **2015**, *197*, 1275–1279. [CrossRef]
76. Muller, C.; Hofmann, T. Screening of raw coffee for thiol binding site precursors using "in bean" model roasting experiments. *J. Agric. Food Chem.* **2005**, *53*, 2623–2629. [CrossRef]
77. DjatiUtomo, H.; Hunter, K.A. Adsorption of divalent copper, zinc, cadmium and lead ions from aqueous solution by waste tea and coffee adsorbents. *Environ. Technol.* **2006**, *27*, 25–32.
78. Volz, N.; Boettler, U.; Winkler, S.; Teller, N.; Schwarz, C.; Bakuradze, T.; Eisenbrand, G.; Haupt, L.; Griffiths, L.R.; Stiebitz, H.; et al. Effect of coffee combining green coffee bean constituents with typical roasting products on the Nrf2/ARE pathway in vitro and in vivo. *J. Agric. Food Chem.* **2012**, *60*, 9631–9641. [CrossRef]
79. Bichler, J.; Cavin, C.; Simic, T.; Chakraborty, A.; Ferk, F.; Hoelzl, C.; Schulte-Hermann, R.; Kundi, M.; Haidinger, G.; Angelis, K.; et al. Coffee consumption protects human lymphocytes against oxidative and 3-amino-1-methyl-5H-pyrido[4,3-b]indole acetate (Trp-P-2) induced DNA-damage: Results of an experimental study with human volunteers. *Food Chem. Toxicol.* **2007**, *45*, 1428–1436. [CrossRef]
80. Shaposhnikov, S.; Hatzold, T.; Yamani, N.E.; Stavro, P.M.; Lorenzo, Y.; Dusinska, M.; Reus, A.; Pasman, W.; Collins, A. Coffee and oxidative stress: A human intervention study. *Eur. J. Nutr.* **2018**, *57*, 533–544. [CrossRef]
81. Sunoqrot, S.; Al-Shalabi, E.; Al-Bakri, A.G.; Zalloum, H.; Abu-Irmaileh, B.; Ibrahim, L.H.; Zeno, H. Coffee bean polyphenols can form biocompatible template-free antioxidant nanoparticles with various sizes and distinct colors. *ACS Omega* **2021**, *6*, 2767–2776. [CrossRef]
82. Palmioli, A.; Ciaramelli, C.; Tisi, R.; Spinelli, M.; De Sanctis, G.; Sacco, E.; Airoldi, C. Natural compounds in cancer prevention: Effects of coffee extracts and their main polyphenolic component, 5-O-caffeoylquinic acid, on oncogenic Ras proteins. *Chem. Asian J.* **2017**, *12*, 2457–2466. [CrossRef]
83. Allred, K.F.; Yackley, K.M.; Vanamala, J.; Allred, C.D. Trigonelline is a novel phytoestrogen in coffee beans. *J. Nutr.* **2009**, *139*, 1833–1838. [CrossRef]
84. Gierach, G.L.; Freedman, N.D.; Andaya, A.; Hollenbeck, A.R.; Park, Y.; Schatzkin, A.; Brinton, L.A. Coffee intake and breast cancer risk in the NIH-AARP diet and health study cohort. *Int. J. Cancer* **2011**, *131*, 452–460. [CrossRef]
85. Fung, V.A.; Cameron, T.P.; Hughes, T.J.; Kirby, P.E.; Dunkel, V.C. Mutagenic activity of some coffee flavor ingredients. *Mutat. Res.* **1988**, *204*, 219–228. [CrossRef]
86. Ranheim, T.; Halvorsen, B. Coffee consumption and human health—beneficial or detrimental?—Mechanisms for effects of coffee consumption on different risk factors for cardiovascular disease and type 2 diabetes mellitus. *Mol. Nutr. Food Res.* **2005**, *49*, 274–284. [CrossRef]
87. Cuyckens, F.; Claeys, M. Optimization of a liquid chromatography method based on simultaneous electrospray ionization mass spectrometric and ultraviolet photodiode array detection for analysis of flavonoid glycosides. *Rapid Commun. Mass Spectrom.* **2002**, *16*, 2341–2348. [CrossRef]
88. Prommaban, A.; Utama-Ang, N.; Chaikitwattana, A.; Uthaipibull, C.; Porter, J.B.; Srichairatanakool, S. Phytosterol, lipid and phenolic composition, and biological activities of guava seed oil. *Molecules* **2020**, *25*, 2474. [CrossRef]
89. Hatano, T.; Kagawa, H.; Yasuhara, T.; Okuda, T. Two new flavonoids and other constituents in licorice root: Their relative astringency and radical scavenging effects. *Chem. Pharm. Bull.* **1988**, *36*, 2090–2097. [CrossRef]
90. Srichairatanakool, S.; Ounjaijean, S.; Thephinlap, C.; Khansuwan, U.; Phisalpong, C.; Fucharoen, S. Iron-chelating and free-radical scavenging activities of microwave-processed green tea in iron overload. *Hemoglobin* **2006**, *30*, 311–327. [CrossRef]

91. Mosmann, T. Rapid colorimetric assay for cellular growth and survival: Application to proliferation and cytotoxicity assays. *J. Immunol. Methods* **1983**, *65*, 55–63. [CrossRef]
92. Van den Heuvel, M.J.; Clark, D.G.; Fielder, R.J.; Koundakjian, P.P.; Oliver, G.J.; Pelling, D.; Tomlinson, N.J.; Walker, A.P. The international validation of a fixed-dose procedure as an alternative to the classical LD50 test. *Food Chem. Toxicol.* **1990**, *28*, 469–482. [CrossRef]

Article

Development and Characterization of Functional Starch-Based Films Incorporating Free or Microencapsulated Spent Black Tea Extract

Surakshi Wimangika Rajapaksha [1] and Naoto Shimizu [2,*]

[1] Laboratory of Agricultural Bio-System Engineering, Graduate School of Agriculture, Hokkaido University, Hokkaido 060-8589, Japan; surakshi.wima@gmail.com

[2] Research Faculty of Agriculture/Field Science Center for Northern Biosphere, Hokkaido University, Hokkaido 060-8589, Japan

* Correspondence: shimizu@bpe.agr.hokudai.ac.jp; Tel.: +81-11706-3848; Fax: +81-11706-3848

Abstract: Antioxidant polyphenols in black tea residue are an underused source of bioactive compounds. Microencapsulation can turn them into a valuable functional ingredient for different food applications. This study investigated the potential of using spent black tea extract (SBT) as an active ingredient in food packaging. Free or microencapsulated forms of SBT, using a pectin–sodium caseinate mixture as a wall material, were incorporated in a cassava starch matrix and films developed by casting. The effect of incorporating SBT at different polyphenol contents (0.17% and 0.34%) on the structural, physical, and antioxidant properties of the films, the migration of active compounds into different food simulants and their performance at preventing lipid oxidation were evaluated. The results showed that adding free SBT modified the film structure by forming hydrogen bonds with starch, creating a less elastic film with antioxidant activity (173 and 587 µg(GAE)/g film). Incorporating microencapsulated SBT improved the mechanical properties of active films and preserved their antioxidant activity (276 and 627 µg(GAE)/g film). Encapsulates significantly enhanced the release of antioxidant polyphenols into both aqueous and fatty food simulants. Both types of active film exhibited better barrier properties against UV light and water vapour than the control starch film and delayed lipid oxidation up to 35 d. This study revealed that starch film incorporating microencapsulated SBT can be used as a functional food packaging to protect fatty foods from oxidation.

Keywords: biodegradable; edible film; release; polyphenols; antioxidant activity; pectin; sodium caseinate; oil oxidation; waste biomass

Citation: Rajapaksha, S.W.; Shimizu, N. Development and Characterization of Functional Starch-Based Films Incorporating Free or Microencapsulated Spent Black Tea Extract. *Molecules* **2021**, *26*, 3898. https://doi.org/10.3390/molecules26133898

Academic Editors: Severina Pacifico and Simona Piccolella

Received: 4 June 2021
Accepted: 23 June 2021
Published: 25 June 2021

Publisher's Note: MDPI stays neutral with regard to jurisdictional claims in published maps and institutional affiliations.

Copyright: © 2021 by the authors. Licensee MDPI, Basel, Switzerland. This article is an open access article distributed under the terms and conditions of the Creative Commons Attribution (CC BY) license (https://creativecommons.org/licenses/by/4.0/).

1. Introduction

Lipid oxidation or oxidative rancidity, one of the main causes of food deterioration alongside microbial spoilage, generates harmful compounds in foods as well as degrading their colours and nutrients [1]. Synthetic antioxidants such as butylated hydroxyanisole (BHA), butylated hydroxy toluene (BHT), and propyl gallate have often been used to overcome rancidity. However, undesirable side-effects associated with these synthetic antioxidants have directed scientific studies towards investigating natural antioxidants [2,3]. Therefore, phenolic compounds originating from plants, the main category of natural antioxidants, have become of growing scientific interest.

The large amount of waste arising from food manufacturing is a potential source of underused polyphenols which possess antioxidant power [4,5]. In this sense, using this waste is a sustainable and economically attractive way of obtaining natural antioxidants.

Tea has been a popular beverage since ancient times. One type, black tea, is consumed throughout the world and contains many types of polyphenols such as theaflavin, thearubigin, and catechin which are responsible for its antioxidant activity [6,7]. Tea is prepared by infusing tea leaves in boiling water for a short time. After black tea is brewed during the industrial beverage manufacturing process, the remaining residue, spent black tea (SBT), is usually discarded as waste. SBT still contains a significantly high and recoverable quantity of antioxidant phenolic compounds and our previous study has shown that subcritical water extraction with ethanol as a co-solvent is an efficient process for extracting phenolic compounds from SBT [8]. This method, subcritical solvent extraction (SSE), uses a pressurized mixture of water and ethanol in the liquid state above its boiling point to enhance the solubility of the compounds and the mass transfer rate during the extraction process, thereby improving the extract yield while decreasing the time and solvent consumption during the process.

The phenolic extract from SBT can be incorporated directly into foods or in food packaging to control lipid oxidation. However, adding phenolic extracts directly to food may neutralize their antioxidant activity after they react with the food and reduce its quality. As the lipid oxidation process is induced from the surface of the foodstuff, designing a functional food film incorporating phenolic extracts has recently received more research attention [9,10]. Antioxidants incorporated into a film can, not only prevent oxidative damage in fatty foods, but also act as a functional additive by migrating from the packaging into the food product. However, the stability of polyphenols can deteriorate during food processing and storage because of their sensitivity to oxygen, light, and heat. Encapsulating phenolic extracts in various types of wall material can efficiently reduce this deterioration during the preparation of the films. Encapsulation can also help to regulate the kinetics of the active compounds as they are released into food products while maintaining the physical properties of the film.

Spray drying is the most widely used technique for encapsulating active and heat-labile compounds because of its short thermal contact time and suitability for industrial application. The type of wall material is also crucial when producing microencapsulates by spray drying [11]. Of the various types of wall material, conjugated mixtures of protein and polysaccharides, such as sodium caseinate and pectin, are of great interest because of their complementary effects on the stability of the core material [12,13]. Sodium caseinate is derived from casein, the principal protein in the milk of bovine and other ruminant animals. Its emulsifying properties combined with its heat stability promote its suitability as a wall material for encapsulation [14]. Pectin is a negatively charged plant polysaccharide which has a wide range of food applications including as a protective carrier because of its gel-forming and stabilization properties. In our recent study, polyphenols obtained from SBT were successfully encapsulated in a mixture of pectin and sodium caseinate to produce a functional food ingredient [8]. These encapsulated polyphenols can be used as stable active compounds in different food applications. However, few studies have reported on the application of these prepared functional microencapsulates to food products and active food packaging incorporating SBT extract microencapsulates has not yet been developed to the best of our knowledge. For producing food packaging, the use of biobased polymers rather than synthetic material is gaining popularity because they are biodegradable, cheap, abundant, and edible. Starch from a variety of plant sources is considered as one of the most promising biopolymers. In particular, cassava starch has been reported to be an excellent raw material for food packaging because of its characteristics of being odourless, tasteless, colourless, non-toxic, and with a high amylopectin content and high viscosity [15–18]. It has also been used as a carrier for antioxidants and antimicrobials in active food packaging and improved the physical properties of the film [19–21].

Hence, the present study aimed to develop a functional film incorporating SBT extract and to investigate the effect of incorporating free and encapsulated forms of the extract on the physico-chemical properties and migration behaviour of active compounds into food simulants. The performance of antioxidants in preventing lipid oxidation of soybean oil will also be studied.

2. Results and Discussion

2.1. Fourier Transform Infrared (FTIR) Analysis of Films and SBT Extract Powders

FTIR spectroscopy is used for analysing the functional bonds and intermolecular interactions between compounds by identifying their molecular vibrations. The infrared spectra of free and encapsulated spent black tea extract powders (SBT, SBT$_{en}$ respectively) are shown in Figure 1a. The FTIR spectrum of SBT extract reflects the main functional groups in polyphenols, amino acids, and alkaloids. The broad band at 3318 cm^{-1} is related to the O–H and N–H stretching modes in tea extract [22,23]. The peaks observed at 2922, 2858, and 1034 cm^{-1} have been attributed to C–H stretching, O–H stretching in alkanes and carboxylic acid, and C=C bond stretch in aromatic rings, respectively [24]. The 1234 cm^{-1} band probably arises from the C–O group in polyols such as hydroxyflavones and catechins [25]. The absorption bands at 1630 and 1530 cm^{-1} have been attributed to amide I (C=O stretching) and amide II (N–H bending) in the amino acids present in black tea [26].

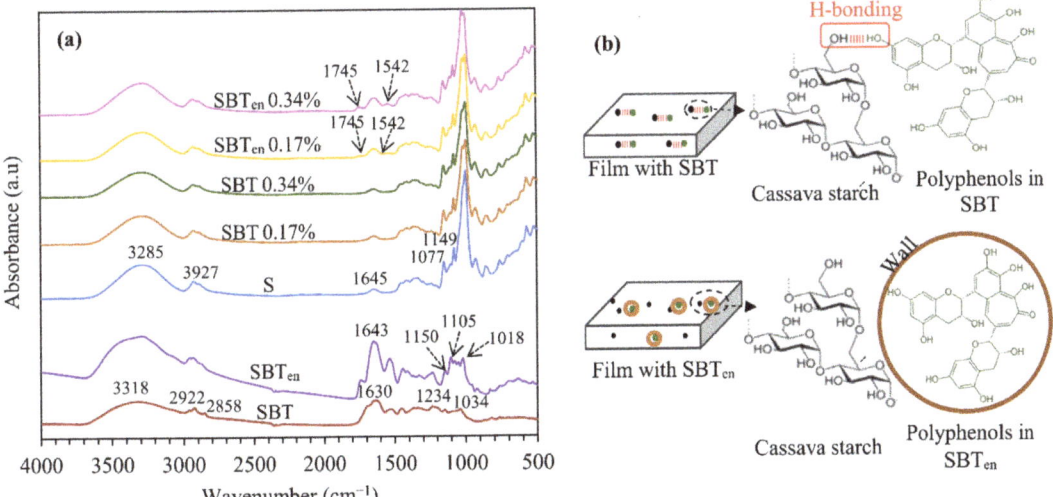

Figure 1. (a) FTIR spectra of spray-dried SBT and SBT$_{en}$, starch-based films without (S), and with SBT or SBT$_{en}$; (b) Schematic illustration of interaction between starch and SBT polyphenols in active films with SBT or SBT$_{en}$.

For SBT$_{en}$ powder, a strong peak observed at 1643 cm^{-1} could be caused by the migration of the –COO stretching vibration at 1630 cm^{-1} from the spectrum of pectin and the C=O band at 1648 cm^{-1} from the spectrum of casein (Figure S1). These spectral changes have been documented as characteristics of the pectin–casein bonding which may be created due to the Maillard conjugation [27–29]. The results from X-ray diffraction and thermal stability in our previous study have also confirmed the formation of a pectin–caseinate complex [8]. Moreover, emerged new peaks at 1105 and 1018 cm^{-1} in SBT$_{en}$ presented a band at 1150 cm^{-1} (C–O stretching) in both SBT and SBT$_{en}$ and a mild shift in the peak of O–H stretching at 3290 cm^{-1}, and indicated that conjugation had occurred between the polyphenols and sodium caseinate–pectin during the encapsulation process (Figure S1). This result was consistent with Jin et al.'s study [30], which revealed FTIR spectra of conjugation between tea polyphenols, pectin, and soy protein.

In the IR spectra of all films (Figure 1a), the wide band between 3000 and 3600 cm^{-1} associated with O–H stretching and the peak at 2927 cm^{-1} could be attributed to the C–H from alkyl groups [31]. The peaks at 1149 and 1077 cm^{-1} were attributed to C–O bond stretching of the C–O–H group. It has been reported that the vibrational band at around 1645 cm^{-1} is related to the O–H bending of the adsorbed water in the amorphous regions of cassava starch [32]. In films incorporating SBT$_{en}$, this peak overlapped with C=O stretching (amide I; at 1635 cm^{-1}) in the sodium caseinate wall materials. The films with SBT$_{en}$ exhibited bands at 1542 cm^{-1} (amide II) and 1745 cm^{-1}, corresponding to typical peaks for SBT$_{en}$ powder, thus confirming the successful incorporation of the encapsulated powder.

After adding intact SBT extract to the film, a flattening and a slight red shift of the O–H stretching band from 3285 cm^{-1} to 3291 cm^{-1} can be observed, indicating that there was no chemical interaction, but hydrogen bonds between active groups of the starch matrix and the phenolic hydroxy groups in the SBT. In contrast, the related peak in films with SBT$_{en}$ showed less flattening than with SBT and shifted to the lower wavenumber. This behaviour indicated the higher availability of O–H in SBT$_{en}$ films than in the other formulations owing to having unbound surface O–H groups in SBT$_{en}$ film [33]. The ratio of the intensities of the peaks at 3300 and 1149 cm^{-1} (I_{3300}/I_{1149}), associates to the stretching vibration of 'C–O' in the 'C–O–H' group, was also calculated to compare the number of hydroxyl groups available in the different formulations. The films containing SBT$_{en}$ exhibited higher ratios indicating a higher number of available hydroxyl groups and the film containing SBT showed a lower number than the other formulations [19]. It can thus be presumed that phenolic hydroxyl groups can interact with the cassava starch matrix by creating hydrogen bonds, but encapsulating SBT in wall materials can interrupt these bonds, as illustrated in Figure 1b.

2.2. Morphology of Control and Active Films

The effect of incorporating free and biopolymer-encapsulated SBT extract into starch films on their superficial morphology was analysed using scanning electron microscopy (SEM) images (Figure 2).

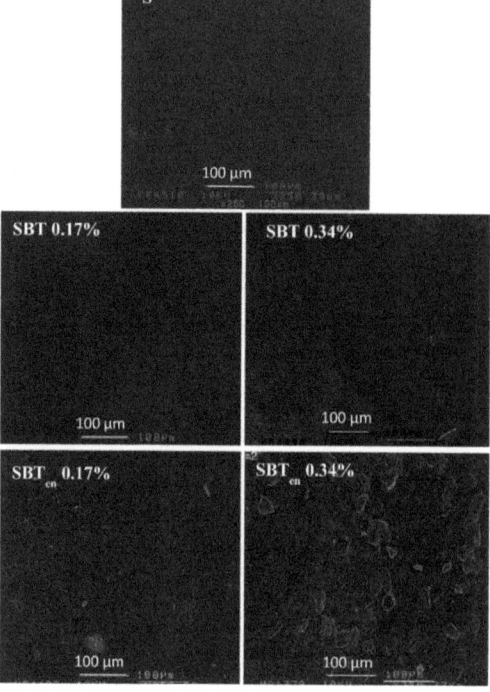

Figure 2. SEM micrographs of the surface of starch films incorporating free or encapsulated SBT extracts: control film (S) and active films incorporating SBT or SBT$_{en}$ at 0.17% or 0.34%.

The cassava starch films exhibited a smooth, homogenous, and flat surface, but adding SBT generated the rough surface on films, which interfered with the regular starch matrix. This damaging of matrix could be ground for the poor mechanical properties of the active films. Films incorporating the free SBT extract showed no significant differences in their surfaces regardless of the polyphenol concentration, suggesting that the SBT extract had spread evenly throughout the starch matrix and interacted via hydrogen bonding during the processing of the film. Similar results have been observed during the development of active starch films incorporating tea polyphenols [34]. After adding SBT_{en}, white particles were observed on the surface of the films, possibly related to the partial dissolving of SBT_{en} resulting in the entrapment of SBT extract within the core of the microencapsulates. The SBT_{en} 0.34% films exhibited a very rough surface with a less homogenous surface, suggesting that the polymer chains had been interrupted by the microcapsules. This phenomenon has also been reported where undissolved particles were observed on the surface of cassava starch films incorporating anthocyanin microencapsulates [35].

2.3. Tensile Properties, Thickness, Young's Modulus, and Viscosity of Film-Forming Solutions

The mechanical behaviour of films is related to their internal structure and can be described in terms of tensile strength (TS), elongation at break (EAB), Young's modulus (YM), and film thickness. For the different film formulations, Table 1 shows the mean values of their mechanical parameters and Figure 3, their stress–strain curves.

Table 1. Viscosity of FFD, thickness, mechanical properties, and water vapor transmission rates (WVTR) of starch films incorporating free or encapsulated SBT extracts.

Film	Viscosity of FFD (mPa.s)	Thickness (mm)	Tensile Strength (MPa)	Young's Modulus (MPa)	WVTR (g mm²/m² 24 h)
S	195.5 ± 2.5 [a]	0.1133 ± 0.0036 [a]	13.43 ± 0.141 [b]	224.45 ± 86.3 [c]	0.61 ± 0.19 [a]
SBT 0.17%	119.0 ± 5.0 [b]	0.1002 ± 0.0120 [ab]	17.54 ± 4.535 [ab]	231.94 ± 63.7 [c]	0.29 ± 0.04 [b]
SBT 0.34%	104.3 ± 5.1 [bc]	0.0957 ± 0.0028 [ab]	25.33 ± 3.706 [a]	1282.39 ± 84.4 [a]	0.54 ± 0.12 [a]
SBT_{en} 0.17%	90.5 ± 1.5 [bc]	0.0906 ± 0.0043 [b]	25.17 ± 4.578 [a]	955.05 ± 74.0 [b]	0.38 ± 0.05 [ab]
SBT_{en} 0.34%	82.5 ± 3.5 [c]	0.0862 ± 0.0032 [b]	23.47 ± 5.301 [ab]	1088.19 ± 72.34 [ab]	0.52 ± 0.13 [a]

Starch films incorporating free or encapsulated SBT extracts: control film (S) and active films incorporating SBT or SBT_{en} at 0.17% or 0.34%. Different letters in the same column (a–c) indicate a statistically significant difference ($p < 0.05$) between mean values. Values represent the mean ± standard deviation of three individual runs.

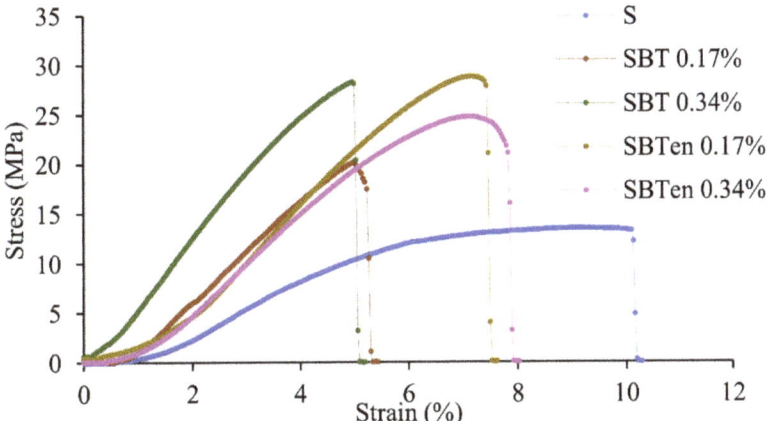

Figure 3. Stress–strain curves of starch films incorporating free or encapsulated SBT extracts: control film (S) and active films incorporating SBT or SBT_{en} at 0.17% or 0.34%.

Incorporating free and encapsulated SBT decreased the thickness of the active films associated with the decreasing starch mass. The control film with the highest mass of cassava starch exhibited the highest thickness influenced by the viscosity of the film-forming dispersion (FFD) (Table 1), an important parameter to be evaluated for the film casting process. This can affect the spread ability, thickness, uniformity of the casting layer, and the mechanical properties of the biopolymer film [36]. Table 1 shows that the tensile strength (TS) of the films incorporating SBT was significantly higher than that of the control film ($p < 0.05$) with the film incorporating 0.34% SBT exhibiting the highest value of 25.33 MPa. Simultaneously, the strain at break of the SBT films decreased significantly. This behaviour can be attributed to the generation of hydrogen bond interactions between the black tea extract and the cassava starch molecules. The spent black tea extract comprises polyphenols with hydroxyl groups, particularly thearubigins, theaflavin, and catechin, which influence their linking with the functional groups in cassava starch. Similar results have been found in starch films incorporating tea polyphenols and rosemary extracts [19,34]. The incorporation of encapsulated SBT into film also significantly increased both the value of TS and Young's modulus (YM) ($p < 0.05$). This may have been caused by the microencapsulates acting as a reinforcement filler in the starch matrix, thus developing a less continuous microstructure. However, the values of TS and YM in films with SBT_{en} were lower than those with free SBT at 0.34%, possibly influenced by the sodium caseinate and pectin used as wall materials in the microencapsulates. High levels of rigidity and brittleness are seriously disadvantageous limitations in active biobased packaging films and thus affect their use in the food industry. Consequently, incorporating encapsulated SBT into the starch matrix can overcome these disadvantages by increasing the elasticity and positively affecting the mechanical behaviour of the film.

2.4. Water Vapour Transmission Rate

The water vapor transmission rate (WVTR) is a critical property of packaging films which affects the moisture level of the food products by preventing the loss or gaining moisture. The WVTR of the different film formulations with free and encapsulated SBT was assessed (Table 1). The results indicated that the significant effect on WVTR of adding SBT at 0.17% could be attributed to the formation of hydrogen bonds between the starch and polyphenols, as discussed previously. This can reduce the quantity of hydrophilic functional groups, thus lowering the WVTR value. However, this value increased again with the addition of SBT at a concentration of 0.34% owing to the hygroscopic nature of phenolic extracts. The WVTR of a film can be influenced by several factors such as the hydrophobic/hydrophilic nature of the materials used for the preparation of film, the presence of cracks or voids, and steric hindrance and tortuosity in the structure [37]. The WVTR values of active films with encapsulated SBT increased, but not above those of the control and the SBT 0.34% films, possibly because of the water-soluble nature of pectin in an aqueous environment [38] and the hydrophilic groups of sodium caseinate in the wall of the microencapsulates [39]. However, the decrease in the WVTR indicated the adequate cohesion of the polymeric chain formed, thus illustrating that strong intermolecular interactions had created barriers to the diffusion of water vapor through the matrix, whereas the presence of microcapsules as fillers can also create a tortuous path for water molecules to pass through.

2.5. Light Transmission and Transparency of Films

The light transparency of packaging materials is an important parameter, because it affects the quality of the protection to foods while also influencing its appearance and attractiveness. Figure 4a shows that the control film based only on cassava starch was clear and transparent, but adding SBT or SBT_{en} made the films slightly brown in colour. The control film (S) exhibited the highest transparency value (-1.783), but after adding the SBT and SBT_{en} powder, the transparency of the resulting active films (average: -2.98 and -3.15, respectively) decreased significantly ($p < 0.05$).

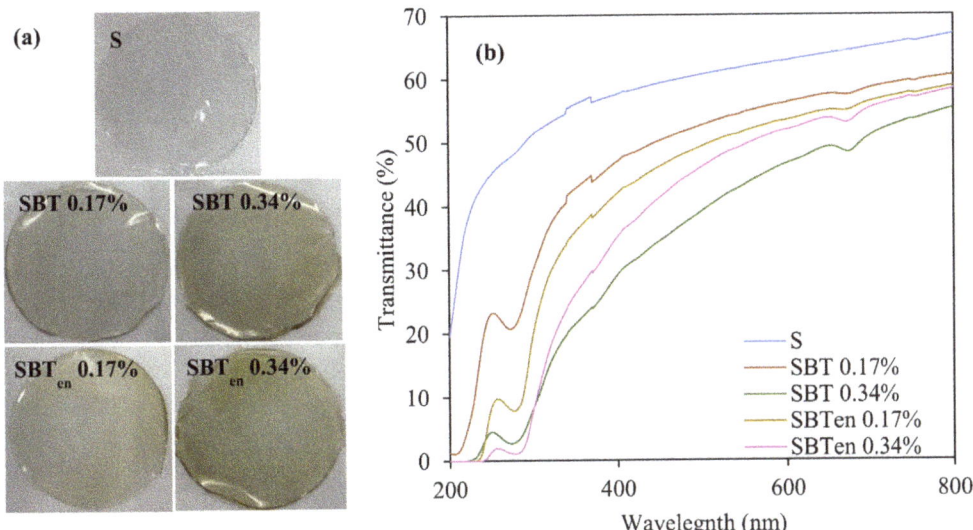

Figure 4. The visual appearance (a) and light transmittance (b) of the control film (S) and starch films incorporating free or encapsulated SBT extracts at 0.17% or 0.34%.

Light can catalyse many reactions causing food to deteriorate, influencing lipid oxidation, off-flavour development, the generation of undesired colours, and the degradation of nutritional compounds. UV light, with a higher energy than visible light, has a great potential for breaking chemical bonds. Therefore, blocking out UV light is a desirable property for a packaging film, particularly when used on foods with a high lipid content which are prone to photosensitized oxidation.

The wavelength of ultraviolet (UV) light ranges from 10 to 380 nm and visible light from 400 to 780 nm [40]; thus, the transmission of UV and visible light was determined at a wavelength between and 200 to 800 nm (Figure 4b). All the active films provided a lower transmission of UV light than the control film (S), possibly due to the fact that UV light was absorbed after incorporating the phenolic extracts into the starch matrix. It was shown to increase the obstruction to UV light with increasing SBT or SBT_{en} content. This behaviour agreed with previous studies on active cassava starch film incorporating Chinese berry anthocyanins [21], and cassava starch/chitosan active film with Pitanga leaf extract [41]. In the case of the SBT_{en} 0.34% film, the light transmission in the range of 400–800 nm was 35–58%, whereas for the SBT 0.34% film, it was 29–55%, possibly because the polymeric wall materials in SBT_{en} had covered the active compounds.

2.6. Antioxidant Content of Films and Their Migration into Food Simulants

Biopolymer films incorporating antioxidant compounds are manufactured for active food packaging applications. The incorporated antioxidant compounds can help reduce oxidative reactions in foods and thus significantly increase their shelf life. The DPPH radical scavenging activity assay is commonly used to evaluate the activity of antioxidant compounds in food by quantifying their ability to quench the DPPH radical. The dark purple colour of DPPH disappears when it is reduced to its non-radical form. The degree of fading of the free radical scavenger can be quantified by measuring the absorbance at 517 nm. The antiradical effects of all active film formulations measured by DPPH are shown in Table 2. The results showed that the concentration of polyphenols and encapsulation significantly affected the antiradical effect of the films ($p < 0.05$). The total antioxidant activity of the SBT_{en} 0.34% film was the highest at 630 µg GAE/g film, whereas that of the SBT 0.17% film was significantly lower than that of the other formulations at 173 µg GAE/g film. This can be attributed to the loss of antioxidant polyphenols during the preparation and drying of films incorporating free SBT. However, the microencapsulation of polyphenols could preserve the antioxidant activity of the films, which agreed with a report on the use of microencapsulation for preserving green tea polyphenols during different food applications [42]. Incorporating a higher level of polyphenols (0.34%), whether encapsulated or free, also significantly increased the antioxidant activity of the films.

Table 2. Antioxidant content and their migration into food simulants (water and 95% ethanol), expressed in terms of µg (gallic acid equivalent; GAE)/g film.

Film	Total Antioxidant µg (GAE)/g Film	Migration (Water) µg (GAE)/g Film	Migration (95% Ethanol) Mg (GAE)/gFilm
SBT 0.17%	173.14 ± 6.88 [d]	9.84 ± 3.00 [c]	10.85 ± 7.22 [c]
SBT 0.34%	587.06 ± 6.98 [b]	30.03 ± 4.00 [c]	53.31 ± 5.11 [b]
SBT$_{en}$ 0.17%	276.13 ± 6.88 [c]	105.63 ± 10.11 [b]	35.08 ± 4.44 [bc]
SBT$_{en}$ 0.34%	629.70 ± 20.80 [a]	391.22 ± 24.40 [a]	118.53 ± 5.12 [a]

Starch films incorporating free or encapsulated SBT extracts: control film (S) and active films incorporating SBT or SBT$_{en}$ at 0.17% or 0.34%. Different letters in the same column (a–d) indicate a statistically significant difference ($p < 0.05$) between mean values. Values represent the mean ± standard deviation of three individual runs.

The migration or release test is important for providing information on the affinity of food products for active materials, thus making it feasible to select the most suitable active material for each type of food. The release completely depends on the compatibility of the antioxidant compound with the food product or food simulant [43]. Various studies have quantified the compounds which are released into recommended food simulants such as water (representing aqueous foods) and 95% ethanol (representing fatty foods) [44,45]. The results of the present study showed a significantly higher release of antioxidant compounds from films incorporating SBT microcapsules (SBT$_{en}$) into water than into 95% ethanol (Table 2, Figure 5a).

The three main factors affecting the migration of active compounds into food simulants are the liquid diffusion into the film network, the solubility of the film in the simulant, and the diffusion of the film into the simulant [44]. The liquid diffusion into the film matrix also depends on the polarity of the simulant, as indicated by the swelling degree of the film. Figure 5b,c shows that the swelling degree of all the active films in water was higher than those in ethanol and that the films incorporating encapsulated SBT exhibited a higher swelling degree in water than those incorporating free SBT. This could be the effect of the hydrophilic nature of the films, thus making them swell more in water than in ethanol. In particular, films incorporated the encapsulated SBT were more hydrophilic than the film incorporated free SBT, and thus led to the higher migration of antioxidant compounds into water. In contrast, films with free SBT released more active compounds into 95% ethanol than into water, possibly because of the greater affinity and swelling of SBT films in 95% ethanol (Figure 5c). This could also be attributed to the hydrophobic nature of polyphenols in their free form, but when covered with a polymer wall material, the resulting microcapsules exhibited a more hydrophilic nature. Even though the release from SBT films into 95% ethanol was higher than into water, it was always lower than that from the SBT$_{en}$ films. This behaviour could also be attributed to the presence of hydrogen bonds between the polyphenols and starch molecules in the SBT film, but their absence in the SBT$_{en}$ films promotes the release of microcapsules. Thus, active films incorporating encapsulated SBT released more into both types of food simulant, but after release, the hydrophilic nature of the microcapsules seemed to limit the complete release of active compounds more into ethanol than into water. A similar observation has been reported for the release of microencapsulated eugenol from thermoprocessed starch films [46].

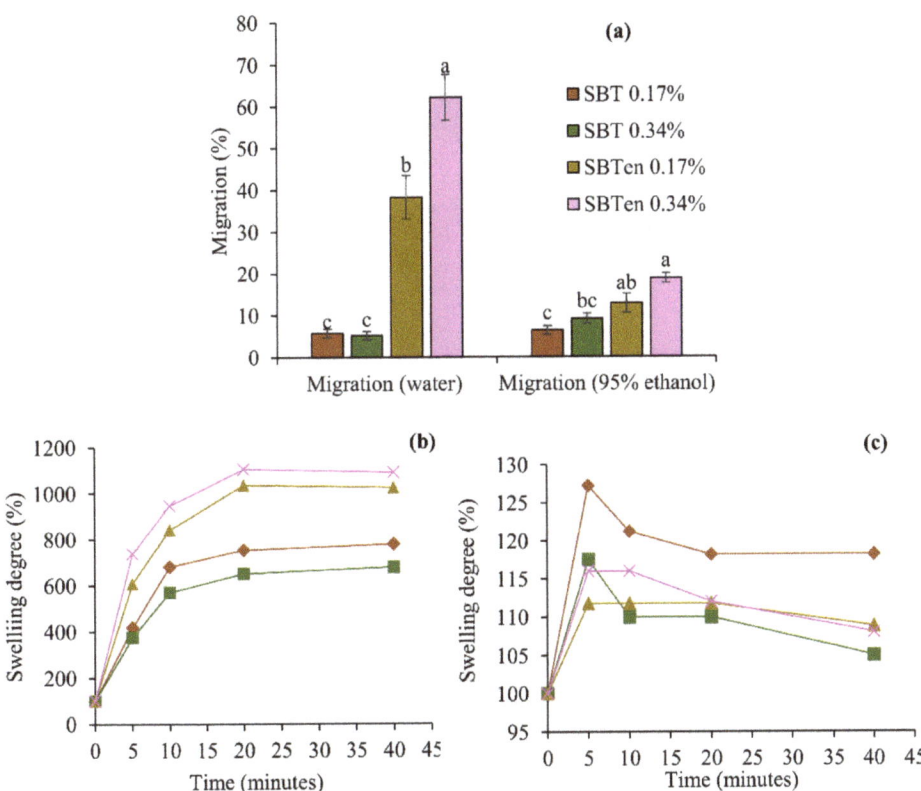

Figure 5. Migration percentage of antioxidant compounds into water and 95% ethanol, (**a**) Swelling degree in water (**b**) and Swelling degree in 95% ethanol (**c**) of starch films incorporating free or encapsulated SBT extracts at 0.17% or 0.34%. Different letters in each food simulant (a–c) indicated a statistically significant difference ($p < 0.05$).

2.7. Effect of Antioxidant Activity on Preventing Lipid Oxidation

Lipid oxidation is a detrimental process in food systems, because it can reduce the nutritional value and sensory quality of foods and produces toxic compounds hazardous to human health. The degree of lipid oxidation is commonly measured by evaluating the peroxide value (PV) of foods. This is also an important test for measuring the effectiveness of active packaging films placed in direct contact with foods. The PV determines the concentration of hydroperoxide, the primary oxidation products of foods, with a high peroxide value indicating a higher level of lipid oxidation or food spoilage [47]. In the present study, the PV of soybean oil samples in contact with the different formulations of films was measured during storage for 35 d. The soybean oil in the open vial reached the highest PV of 68.4 ± 3.04 (meq O2/Kg) after 35 d of storage (Figure 6) because of its exposure to light and oxygen.

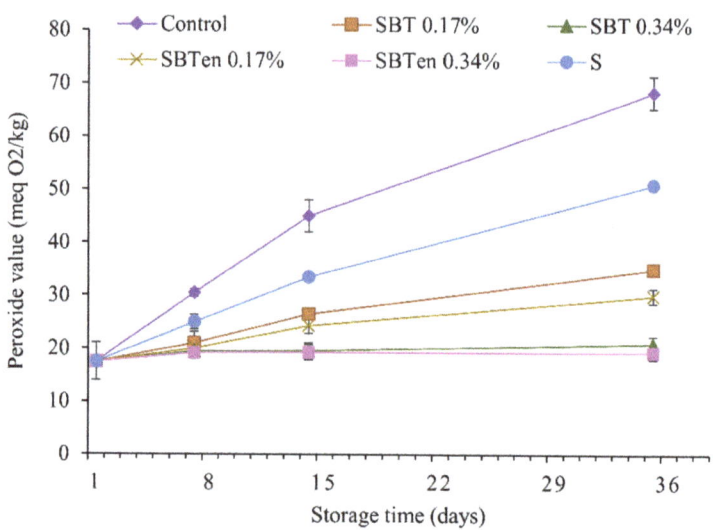

Figure 6. The changes of peroxide value of soybean oil during 35 days of storage at 27 °C, open glass vial (control) and starch films without and with incorporating free or encapsulated SBT extracts at 0.17% or 0.34%.

The all-active films exhibited a lower PV than the open control sample and the sample sealed with the starch film. As the concentration of polyphenols increased, the PV of the oil decreased. This suggested that adding SBT extract to the film in either the free or encapsulated form could retard the formation of peroxides due to lipid oxidation. Theoretically, lipid oxidation can be induced by oxygen in the presence of initiators such as heat, free radicals, light, and metal ions [48]. Thus, the polyphenols in the SBT extract could scavenge the free radicals in oil by donating hydrogen from the phenolic hydroxyl group, which helped to stop the chain of radical propagation.

The PV of oil sealed with SBT_{en} films was lower than that of oil sealed with SBT films, possibly because of its higher release of antioxidant compounds. However, the PV of oil samples sealed with either SBT or SBT_{en} films at 0.34% were similar, possibly because the lower transparency of the SBT films at 0.34% would remove the initiators needed for oil oxidation.

3. Materials and Methods

3.1. Materials

The cassava starch was obtained from a local supermarket in Sapporo (Hokkaido, Japan). Unblended black tea was supplied by New Vithanakande Tea Factory (Pvt) Ltd., (Ratnapura, Sri Lanka). Glycerol, gallic acid, 2,2-diphenyl-1-picrylhydrazyl (DPPH), Folin and Ciocalteu phenol reagent, and casein sodium salt from bovine milk were provided by Sigma-Aldrich (St. Louis, MO, USA), and pectin from citrus, sodium carbonate, magnesium nitrate, ethanol, and methanol were provided by Fujifilm Wako Pure Chemical Corp. (Osaka, Japan). Distilled water was used in all experiments. All other chemicals and solvents used were of analytical grade.

3.2. Preparation of SBT Powders

3.2.1. Preparation of SBT

The SBT was produced as described previously [8]. Briefly, black tea leaves were brewed in boiling water for 6 min (2 g/100 mL), then filtered to obtain the residue. The SBT was prepared after air-drying the filtered residue in an oven at 45 °C.

3.2.2. Extraction and Encapsulation of SBT Polyphenols

The antioxidant phenolic compounds from SBT were extracted by subcritical solvent extraction (SSE) using the conditions optimized in a previous study [8]. Briefly, a mixture of pure water and ethanol (71% concentration) and SBT (20 mL/g) were mixed in an 11-mL reactor with an agitator (Chemi-station PPV 3000, Tokyo Rikakikai Co. Ltd., Tokyo, Japan). The extraction reactor was purged with nitrogen gas and 2.0 MPa of initial pressure was applied. The sample in the reactor was then heated to 180 °C and maintained there for 10 min. During extraction, the agitation speed was kept at 1000 rpm to prevent any local overheating and to increase the mass transfer. After the extraction, the reactor was cooled in a cold-water bath; then, the sample was filtered through filter paper (6 μm) to recover the SBT extract. The recovered SBT extract was concentrated using a rotary evaporator (N-1210 and SB-1300 water bath, EYELA Tokyo Rikakikai Co., Ltd., Tokyo, Japan), then stored at 4 °C until encapsulation.

The SBT extract was microencapsulated as described previously [8]. A 50%:50% combination of pectin and sodium caseinate was used as the wall or coating material. First, the coating material was dissolved in distilled water at 90 °C (3 g/100 mL), then stirred until a clear dispersion was obtained. The polymer solution was then kept in a refrigerator overnight to allow complete hydration. On the next day, the concentrated SBT extract was added dropwise to the prepared biopolymer solution (1 g SBT extract: 20 g solution) after it had been heated to 40 °C; then, the mixed solution was stirred for 20 min. The prepared feed solutions were sonicated for 20 min, then homogenized for 30 min (HERACLES-16g, Koike Precision Instruments, Tokyo, Japan). The samples with a 3% solids concentration were then spray-dried with a laboratory-scale spray dryer OSK 55MO102 (Osaka Seimitsu Kikai Co. Ltd., Osaka, Japan) using a spray nozzle (diameter 0.5 mm), an inlet air temperature of 140 °C, an outlet air temperature of 85 °C, an atomization pressure of 0.4 MPa, and a feed flow rate of 5 mL/min. The non-encapsulated SBT extract was spray dried under similar conditions to the other samples with each experiment performed in duplicate. The resulting SBT extract powders with or without encapsulation were packed in zip-lock bags, covered with aluminium foil, then stored in a refrigerator until further use.

3.3. Preparation of the Films

The films were prepared by the solvent casting process, as described in previous studies [19,35,46]. Cassava starch control films were produced by blending 4.0 g of starch, 1.2 g of glycerol, and 84.0 g of distilled water. Active starch films were prepared by replacing the starch component in the formulations with the same amount of free or encapsulated SBT extract powder with predetermined phenolic content by Folin–Ciocalteu method [8] (total phenolics content of powders: 359.90 and 17.14 μg of gallic acid equivalent/mg, respectively), to obtain films with two different concentrations of total polyphenols (Table 3). The total solids concentration of all the control and active films were set at 4% (m/m). First, an aqueous starch dispersion was heated at 95 °C for 30 min under constant stirring until the starch was gelatinized. Glycerol was then added to the solution at a total solid: glycerol ratio of 10:3 (m/m) and agitated for 30 min. After cooling the solutions to 40 °C in a cold-water bath, the free and microencapsulates of SBT were added to the starch solution, then homogenized for 15 min. The film-forming dispersions (FFD) obtained were degassed by sonication then poured into polypropylene Petri dishes while maintaining the amount of FFD in the dish constant at 0.28 g/cm². After drying at 35 °C for 24 h, the films were conditioned at 25 °C and 53% relative humidity over saturated $Mg(NO_3)_2$ before characterization.

Table 3. Formulations of starch films incorporating free or encapsulated SBT extracts.

	Types	Polyphenols %	Starch (g)	SBT/SBT$_{en}$ (g)	Glycerol (g)	Water (g)
Control films	S		4.000	-	1.2	84.0
Films with SBT extract	SBT 0.17%	0.17%	3.981	0.0190	1.2	84.0
	SBT 0.34%	0.34%	3.962	0.0380	1.2	84.0
Films with encapsulated SBT extract	SBT$_{en}$ 0.17%	0.17%	3.600	0.4000	1.2	84.0
	SBT$_{en}$ 0.34%	0.34%	3.200	0.8000	1.2	84.0

3.4. Measurement of Viscosity of FFD

The viscosity of all the FFD was measured using a Sinewave Vibro Viscometer SV-10 (A&D Co. Ltd., Tokyo, Japan). All measurements were carried out at room temperature in triplicate, and the average value was taken as the final value.

3.5. Characterization of the Films

3.5.1. Fourier Transform Infrared (FT-IR) Spectroscopy of Powders and Films

The preliminary structures of the SBT and SBT$_{en}$ powders were characterized using an FTIR spectrophotometer (JASCO FTIR660plus, JASCO Corp., Tokyo, Japan). The samples were prepared by pressing a mixture of the powder sample and KBr into pellets. The functional groups of the films were identified by the FTIR spectrophotometer equipped with an attenuated reflection accessory, ATR (Spectrum 100, PerkinElmer Co., Ltd., Shelton, CT, USA). A spectral resolution of 4 cm^{-1} was used and 32 scans were acquired for each spectrum in the range of 650–4000 cm^{-1}. This experiment was performed in a room at 23 °C and 50% relative humidity.

3.5.2. Scanning Electron Microscopy (SEM)

The surface morphology of the films was observed using a field emission scanning electron microscope (SEM, JSM-6301F, JEOL Ltd., Tokyo, Japan). The samples were dried at 35 °C for 12 h, mounted on aluminium stubs using double-sided carbon tape, then sputter-coated with gold. The micrographs were captured at an accelerating voltage of 10 kV.

3.5.3. Film Thickness and Tensile Properties

The thickness of the films was measured by a digital micrometer (ABSOLUTE Digimatic Micrometer, Mitutoyo Corp., Kawasaki, Japan) to the nearest 0.001 mm. The average thickness was based on measurements at three random positions on the film.

The tensile properties, tensile strength (TS), elongation at break (%EAB), and Young's modulus (YM) of the films were analysed using a universal testing machine system (AG-100kNXplus, Shimadzu Corp., Kyoto, Japan) according to the ISO 527-3 (2018) test method [49]. All the films were initially cut into a dumbbell shape using a cutter (JIS K7113 No.1$\frac{1}{2}$, Shimadzu Corp.) and preconditioned for 5 d at 23 °C and 50% relative humidity before the mechanical testing. The initial grip separation and the testing speed were set at 50 mm and 50 mm/min, respectively. The results are based on three measurements for each formulation.

3.5.4. Water Vapour Transmission Rate (WVTR)

The water vapour transmission rate of the samples was analysed as described previously [50], with minor modifications. First, the film samples were equilibrated in a desiccator containing a saturated solution of NaCl (75% relative humidity). The films were then fixed using elastic bands over the top of a weighing bottle containing 5.0 g of anhydrous CaCl$_2$, then weighed immediately (W_1). All the bottles were then placed in a chamber for 24 h (75% relative humidity at 25 °C) then weighed again (W_2). The WVTR was calculated as follows:

$$\text{WVTR} \left(\text{g mm/m}^2\ 24\ \text{h}\right) = \frac{(W2 - W1) \times \text{Film thickness}}{\text{Area} \times \text{Holding time}} \quad (1)$$

3.5.5. Light Transmission

The light transmission of the films was measured using a UV-Vis spectrophotometer (JASCO V-560, JASCO Corp., Tokyo, Japan) in the wavelength range of 200–800 nm. First, the films were cut into rectangular strips (3.5 cm × 0.5 cm), then placed directly into a cuvette to allow the light beam to pass through the films. The analyses were performed in triplicate and an empty cell was used as the reference. The film transparency for visible light was determined at 600 nm [51]. The percentage transparency was calculated as follows:

$$\text{Transparency value} = (\log T_{600})/x \quad (2)$$

where T is the fractional transmission at 600 nm and x is the thickness of the film sample (mm).

3.5.6. DPPH Radical Scavenging Assay

The antiradical effect of the films was determined by the 2,2-diphenyl-1-picrylhydrazyl (DPPH) scavenging activity method [52] with slight modifications. First, a piece of film (30 mg) was completely dissolved in distilled water, then 3 mL of ethanol was added. The resulting solution was centrifuged at 6000× g for 10 min. Two millilitres of the supernatant and 5 mL of 0.1 mM DPPH solution were mixed, kept in the dark for 30 min, then the absorbance was measured at 517 nm using a UV–vis spectrophotometer (JASCO V-560, JASCO Corp., Tokyo, Japan). Gallic acid was used as the standard for preparation of the standard curve (0.5–6.0 μg/mL, R^2 = 0.99). The DPPH scavenging capacity was expressed as micrograms of gallic acid equivalent/g (dry weight) film (μg GAE)/g film).

3.5.7. Migration Test

Pieces of film sample (2 cm × 2 cm) were placed in 8 mL of water (representing an aqueous food) and 95% ethanol (representing a fatty food) in separate glass vials. The vials were then purged with nitrogen gas before closing and kept at 25 °C for 7 d. The migration of the antioxidant compounds into each food simulant was then tested using the DPPH method, as described earlier (Section 3.5.6). The results were expressed as micrograms of gallic acid equivalent/g (dry weight) film (μg GAE)/g film).

3.5.8. Peroxide Value (PV) of Soybean Oil

Soybean oil (5 mL) was poured into dark glass vials, then sealed with films of each formulation (S, SBT 0.17%, SBT 0.34%, SBT_{en} 0.17% and SBT_{en} 0.34%). The sealed vials containing the oil were then placed upside-down to ensure the oil was in contact with the films. An open dark glass vial containing 5 mL of soybean oil was used as the control. All vials were stored at 27 ± 2 °C and 50 ± 5% relative humidity under fluorescent light, then samples were taken after 0, 7, 14, and 35 d of storage to determine the PV of the oil. The peroxide value of the soybean oil was determined by iodometric titration. The oil sample (1 g) was mixed with 10 mL of glacial acetic acid and chloroform (3:2, v/v) in an Erlenmeyer flask. The mixture was shaken vigorously to dissolve the oil completely, then mixed with 0.5 mL of saturated potassium iodide solution. The mixture was then kept in the dark for 1 min before dilution with 30 mL of distilled water and titrated against 0.01 M sodium thiosulphate in the presence of a starch indicator (5 mL). The volume of sodium thiosulphate consumed was recorded and the PV expressed as millimolar equivalents of free iodine per kilogram of oil (meq of oxygen/kg). All the analyses were performed in triplicate.

3.6. Statistical Analysis

Statistical analyses were carried out using Minitab 19.1.1. (Minitab Inc., State College, PA, USA) to determine the significance of differences of factors and levels. One-way analysis of variance (ANOVA) and Tukey's multiple comparison test were used to compare and identify significant differences ($p < 0.05$) between group means. The results were reported as the mean value of three repeated experimental data.

4. Conclusions

In the present study, a functional film has been developed using free and microencapsulated SBT extract as the active ingredient. The results showed that adding SBT either in the free or encapsulated form significantly affected the mechanical, barrier (water and light), and antioxidant properties of the films. Non-encapsulated SBT extract could modify the starch matrix by forming hydrogen bonds with cassava starch, thus inducing a reduction in the mechanical properties of the film and in antioxidant migration from the film. The film incorporating microencapsulated SBT extract exhibited a heterogenous film surface and enhanced physical properties compared with the SBT film. The microencapsulation of SBT protected the antioxidant compounds during the film processing and caused a significant increase in the migration of the active compounds into both the aqueous and fatty food simulants. The SBT_{en} 0.34% film exhibited the highest free radical scavenging activity and prevented lipid oxidation of soybean oil samples for more than 35 d.

Supplementary Materials: The following are available online, Figure S1: FTIR spectra of spray dried SBT, SBT_{en}, sodium caseinate, and pectin.

Author Contributions: Conceived and performed experiments, analyzed data, prepared the manuscript draft, writing, S.W.R.; review, editing, supervision, N.S. Both authors have read and agreed to the published version of the manuscript.

Funding: This work was partially supported by Moonshot Agriculture, Forestry and Fisheries Research and Development Program (MS 508), Government of Japan.

Institutional Review Board Statement: Not applicable.

Informed Consent Statement: Not applicable.

Data Availability Statement: Data of the compounds are available from the authors.

Acknowledgments: The FTIR analysis of powders was performed at the Global Facility Centre of Hokkaido University, and the scanning microscopy was provided by the Electron Microscope Laboratory in the Research Faculty of Agriculture, Hokkaido University. We thank Philip Creed, from Edanz Group (https://en-author-services.edanz.com/) (accessed on 2 April 2020) for editing a draft of this manuscript.

Conflicts of Interest: The authors declare no conflict of interest.

Sample Availability: Samples of the compounds are not available from the authors.

References

1. Vieira, S.A.; Zhang, G.; Decker, E.A. Biological Implications of Lipid Oxidation Products. *J. Am. Oil Chem. Soc.* **2017**, *94*, 339–351. [CrossRef]
2. Caleja, C.; Barros, L.; Antonio, A.L.; Oliveira, M.B.P.P.; Ferreira, I.C.F.R. A Comparative Study between Natural and Synthetic Antioxidants: Evaluation of Their Performance after Incorporation into Biscuits. *Food Chem.* **2017**, *216*, 342–346. [CrossRef] [PubMed]
3. Pires, M.A.; Munekata, P.E.S.; Villanueva, N.D.M.; Tonin, F.G.; Baldin, J.C.; Rocha, Y.J.P.; Carvalho, L.T.; Rodrigues, I.; Trindade, M.A. The Antioxidant Capacity of Rosemary and Green Tea Extracts to Replace the Carcinogenic Antioxidant (BHA) in Chicken Burgers. *J. Food Qual.* **2017**, 2409527. [CrossRef]
4. Kumar, K.; Srivastav, S.; Sharanagat, V.S. Ultrasound Assisted Extraction (UAE) of Bioactive Compounds from Fruit and Vegetable Processing By-Products: A Review. *Ultrason. Sonochem.* **2021**, *70*, 105325. [CrossRef] [PubMed]
5. Dueñas, M.; García-Estévez, I. Agricultural and Food Waste: Analysis, Characterization and Extraction of Bioactive Compounds and Their Possible Utilization. *Foods* **2020**, *9*, 817. [CrossRef]

6. Butt, M.S.; Imran, A.; Sharif, M.K.; Ahmad, R.S.; Xiao, H.; Imran, M.; Rsool, H.A. Black Tea Polyphenols: A Mechanistic Treatise. *Crit. Rev. Food Sci. Nutr.* **2014**, *54*, 1002–1011. [CrossRef]
7. Łuczaj, W.; Skrzydlewska, E. Antioxidative Properties of Black Tea. *Prev. Med.* **2005**, *40*, 910–918. [CrossRef] [PubMed]
8. Rajapaksha, D.S.W.; Shimizu, N. Valorization of FFD by Recovery of Antioxidant Polyphenolic Compounds: Subcritical Solvent Extraction and Microencapsulation. *Food Sci. Nutr.* **2020**, *8*, 4297–4307. [CrossRef]
9. Ceballos, R.L.; Ochoa-Yepes, O.; Goyanes, S.; Bernal, C.; Famá, L. Effect of Yerba Mate Extract on the Performance of Starch Films Obtained by Extrusion and Compression Molding as Active and Smart Packaging. *Carbohydr. Polym.* **2020**, *244*, 116495. [CrossRef] [PubMed]
10. Valdés García, A.; Juárez Serrano, N.; Beltrán Sanahuja, A.; Garrigós, M.C. Novel Antioxidant Packaging Films Based on Poly(ε-Caprolactone) and Almond Skin Extract: Development and Effect on the Oxidative Stability of Fried Almonds. *Antioxidants* **2020**, *9*, 629. [CrossRef] [PubMed]
11. Ray, S.; Raychaudhuri, U.; Chakraborty, R. An Overview of Encapsulation of Active Compounds Used in Food Products by Drying Technology. *Food Biosci.* **2016**, *13*, 76–83. [CrossRef]
12. Baracat, M.M.; Nakagawa, A.M.; Casagrande, R.; Georgetti, S.R.; Verri, W.A., Jr.; de Freitas, O. Preparation and Characterization of Microcapsules Based on Biodegradable Polymers: Pectin/Casein Complex for Controlled Drug Release Systems. *AAPS PharmSciTech* **2012**, *13*, 364–372. [CrossRef]
13. Nooshkam, M.; Varidi, M. Maillard Conjugate-Based Delivery Systems for the Encapsulation, Protection, and Controlled Release of Nutraceuticals and Food Bioactive Ingredients: A Review. *Food Hydrocoll.* **2020**, *100*, 105389. [CrossRef]
14. Augustin, M.A.; Oliver, C.M.; Hemar, Y. Casein, Caseinates, and Milk Protein Concentrates. In *Dairy Ingredients for Food Processing: Chandan/Dairy Ingredients for Food Processing*, 1st ed.; Chandan, R.C., Kilara, A., Eds.; Blackwell Publishing Ltd.: Ames, IA, USA, 2011; pp. 161–178.
15. Luchese, C.L.; Garrido, T.; Spada, J.C.; Tessaro, I.C.; de la Caba, K. Development and Characterization of Cassava Starch Films Incorporated with Blueberry Pomace. *Int. J. Biol. Macromol.* **2018**, *106*, 834–839. [CrossRef]
16. Lim, W.S.; Ock, S.Y.; Park, G.D.; Lee, I.W.; Lee, M.H.; Park, H.J. Heat-Sealing Property of Cassava Starch Film Plasticized with Glycerol and Sorbitol. *Food Packag. Shelf Life* **2020**, *26*, 100556. [CrossRef]
17. dos Santos Caetano, K.; Almeida Lopes, N.; Haas Costa, T.M.; Brandelli, A.; Rodrigues, E.; Hickmann Flôres, S.; Cladera-Olivera, F. Characterization of Active Biodegradable Films Based on Cassava Starch and Natural Compounds. *Food Packag. Shelf Life* **2018**, *16*, 138–147. [CrossRef]
18. Qin, Y.; Liu, Y.; Yong, H.; Liu, J.; Zhang, X.; Liu, J. Preparation and Characterization of Active and Intelligent Packaging Films Based on Cassava Starch and Anthocyanins from Lycium Ruthenicum Murr. *Int. J. Biol. Macromol.* **2019**, *134*, 80–90. [CrossRef]
19. Piñeros-Hernandez, D.; Medina-Jaramillo, C.; López-Córdoba, A.; Goyanes, S. Edible Cassava Starch Films Carrying Rosemary Antioxidant Extracts for Potential Use as Active Food Packaging. *Food Hydrocoll.* **2017**, *63*, 488–495. [CrossRef]
20. Luchese, C.L.; Spada, J.C.; Tessaro, I.C. Starch Content Affects Physicochemical Properties of Corn and Cassava Starch-Based Films. *Ind. Crops Prod.* **2017**, *109*, 619–626. [CrossRef]
21. Yun, D.; Cai, H.; Liu, Y.; Xiao, L.; Song, J.; Liu, J. Development of Active and Intelligent Films Based on Cassava Starch and Chinese Bayberry (Myrica Rubra Sieb. et Zucc.) Anthocyanins. *RSC Adv.* **2019**, *9*, 30905–30916. [CrossRef]
22. Brza, M.A.; Aziz, S.B.; Anuar, H.; Ali, F.; Dannoun, E.M.A.; Mohammed, S.J.; Abdulwahid, R.T.; Al-Zangana, S. Tea from the Drinking to the Synthesis of Metal Complexes and Fabrication of PVA Based Polymer Composites with Controlled Optical Band Gap. *Sci. Rep.* **2020**, *10*, 18108. [CrossRef] [PubMed]
23. Moosa, A.; Ridha, A.M.; Allawi, M.H. Green Synthesis of Silver Nanoparticles Using Spent Tea Leaves Extract with Atomic Force Microscopy. *Int. J. Curr. Eng. Sci. Res.* **2015**, *5*, 3233–3241. Available online: http://inpressco.com/category/ijcet (accessed on 19 May 2020).
24. Ali, A.; Bilal, M.; Khan, R.; Farooq, R.; Siddique, M. Ultrasound-Assisted Adsorption of Phenol from Aqueous Solution by Using Spent Black Tea Leaves. *Environ. Sci. Pollut. Res. Int.* **2018**, *25*, 22920–22930. [CrossRef] [PubMed]
25. Rengga, W.D.P.; Yufitasari, A.; Adi, W. Synthesis of Silver Nanoparticles from Silver Nitrate Solution Using Green Tea Extract (Camelia Sinensis) as Bioreductor. *J. Bahan Alam Terbarukan* **2017**, *6*, 32–38. [CrossRef]
26. Thummajitsakul, S.; Samaikam, S.; Tacha, S.; Silprasit, K. Study on FTIR Spectroscopy, Total Phenolic Content, Antioxidant Activity and Anti-Amylase Activity of Extracts and Different Tea Forms of Garcinia Schomburgkiana Leaves. *Lebenson. Wiss. Technol.* **2020**, *134*, 110005. [CrossRef]
27. Ghazi, A. Extraction of Beta-Carotene from Orange Peels. *Nahrung* **1999**, *43*, 274–277. [CrossRef]
28. Ren, J.-N.; Hou, Y.-Y.; Fan, G.; Zhang, L.-L.; Li, X.; Yin, K.; Pan, S.-Y. Extraction of Orange Pectin Based on the Interaction between Sodium Caseinate and Pectin. *Food Chem.* **2019**, *283*, 265–274. [CrossRef] [PubMed]
29. Abd El-Salam, M.H.; El-Shibiny, S. Preparation and Potential Applications of Casein-Polysaccharide Conjugates: A Review: Casein-Polysaccharide Conjugates. *J. Sci. Food Agric.* **2020**, *100*, 1852–1859. [CrossRef] [PubMed]
30. Jin, B.; Zhou, X.; Liu, Y.; Li, X.; Mai, Y.; Liao, Y.; Liao, J. Physicochemical Stability and Antioxidant Activity of Soy Protein/Pectin/Tea Polyphenol Ternary Nanoparticles Obtained by Photocatalysis. *Int. J. Biol. Macromol.* **2018**, *116*, 1–7. [CrossRef]
31. Lei, Y.; Wu, H.; Jiao, C.; Jiang, Y.; Liu, R.; Xiao, D.; Lu, J.; Zhang, Z.; Shen, G.; Li, S. Investigation of the Structural and Physical Properties, Antioxidant and Antimicrobial Activity of Pectin-Konjac Glucomannan Composite Edible Films Incorporated with Tea Polyphenol. *Food Hydrocoll.* **2019**, *94*, 128–135. [CrossRef]

32. Edhirej, A.; Sapuan, S.M.; Jawaid, M.; Zahari, N.I. Effect of Various Plasticizers and Concentration on the Physical, Thermal, Mechanical, and Structural Properties of Cassava-Starch-Based Films: Effect of Various Plasticizers. *Starke* **2017**, *69*, 1500366. [CrossRef]
33. Nebahani, L.; Jaisingh, A. Polymer Science and Innovative Applications. In *Chemical Analysis of Polymers: Materials, Techniques, and Future Developments*, 1st ed.; Almaadeed, M.A.A., Ponnamma, D., Carignano, M.A., Eds.; Elsevier: Cambridge, MA, USA, 2020; pp. 69–117.
34. Feng, M.; Yu, L.; Zhu, P.; Zhou, X.; Liu, H.; Yang, Y.; Zhou, J.; Gao, C.; Bao, X.; Chen, P. Development and Preparation of Active Starch Films Carrying Tea Polyphenol. *Carbohydr. Polym.* **2018**, *196*, 162–167. [CrossRef]
35. Stoll, L.; Costa, T.M.H.; Jablonski, A.; Flôres, S.H.; de Oliveira Rios, A. Microencapsulation of Anthocyanins with Different Wall Materials and Its Application in Active Biodegradable Films. *Food Bioproc. Tech.* **2016**, *9*, 172–181. [CrossRef]
36. Manshor, N.M.; Jai, J.; Hamzah, F.; Somwangthanaroj, A.; Ongdeesoontorn, W.T. Rheological Properties of Film Solution from Cassava Starch and Kaffir Lime Oil. *J. Phys. Conf. Ser.* **2019**, *1349*, 012045. [CrossRef]
37. Moghadam, M.; Salami, M.; Mohammadian, M.; Khodadadi, M.; Emam-Djomeh, Z. Development of Antioxidant Edible Films Based on Mung Bean Protein Enriched with Pomegranate Peel. *Food Hydrocoll.* **2020**, *104*, 105735. [CrossRef]
38. Chen, P.-H.; Kuo, T.-Y.; Kuo, J.-Y.; Tseng, Y.-P.; Wang, D.-M.; Lai, J.-Y.; Hsieh, H.-J. Novel Chitosan–Pectin Composite Membranes with Enhanced Strength, Hydrophilicity and Controllable Disintegration. *Carbohydr. Polym.* **2010**, *82*, 1236–1242. [CrossRef]
39. Chew, S.C.; Tan, C.P.; Nyam, K.L. Microencapsulation of Refined Kenaf (*Hibiscus Cannabinus* L.) Seed Oil by Spray Drying Using β-Cyclodextrin/Gum Arabic/Sodium Caseinate. *J. Food Eng.* **2018**, *237*, 78–85. [CrossRef]
40. Plakett, D.; Siro, I. Developments in Packaging Materials. In *Emerging food Packaging Technologies: Principles and Practice*, 1st Ed.; Yam, K.L., Lee, D.S., Eds.; Woodhead Publishing Limited: Philadelphia, PA, USA, 2012; pp. 237–359.
41. Chakravartula, S.S.N.; Lourenço, R.V.; Balestra, F.; Bittante, A.M.Q.B.; Sobral, P.J.A.; Rosa, M.D. Influence of Pitanga (*Eugenia uniflora* L.) Leaf Extract and/or Natamycin on Properties of Cassava Starch/Chitosan Active Films. *Food Packag. Shelf Life* **2020**, *24*, 100498. [CrossRef]
42. Massounga Bora, A.F.; Ma, S.; Li, X.; Liu, L. Application of Microencapsulation for the Safe Delivery of Green Tea Polyphenols in Food Systems: Review and Recent Advances. *Food Res. Int.* **2018**, *105*, 241–249. [CrossRef]
43. Gómez-Estaca, J.; López-de-Dicastillo, C.; Hernández-Muñoz, P.; Catalá, R.; Gavara, R. Advances in Antioxidant Active Food Packaging. *Trends Food Sci. Technol.* **2014**, *35*, 42–51. [CrossRef]
44. Adilah, Z.A.M.; Jamilah, B.; Hanani, Z.A.N. Functional and Antioxidant Properties of Protein-Based Films Incorporated with Mango Kernel Extract for Active Packaging. *Food Hydrocoll.* **2018**, *74*, 207–218. [CrossRef]
45. Fasihnia, S.H.; Peighambardoust, S.H.; Peighambardoust, S.J.; Oromiehie, A.; Soltanzadeh, M.; Peressini, D. Migration Analysis, Antioxidant, and Mechanical Characterization of Polypropylene-Based Active Food Packaging Films Loaded with BHA, BHT, and TBHQ. *J. Food Sci.* **2020**, *85*, 2317–2328. [CrossRef]
46. Talón, E.; Vargas, M.; Chiralt, A.; González-Martínez, C. Eugenol Incorporation into Thermoprocessed Starch Films Using Different Encapsulating Materials. *Food Packag. Shelf Life* **2019**, *21*, 100326. [CrossRef]
47. Wang, S.; Xia, P.; Wang, S.; Liang, J.; Sun, Y.; Yue, P.; Gao, X. Packaging Films Formulated with Gelatin and Anthocyanins Nanocomplexes: Physical Properties, Antioxidant Activity and Its Application for Olive Oil Protection. *Food Hydrocoll.* **2019**, *96*, 617–624. [CrossRef]
48. Laguerre, M.; Lecomte, J.; Villeneuve, P. Evaluation of the Ability of Antioxidants to Counteract Lipid Oxidation: Existing Methods, New Trends and Challenges. *Prog. Lipid Res.* **2007**, *46*, 244–282. [CrossRef]
49. Babaghayou, M.I.; Mourad, A.-H.I.; Lorenzo, V.; Chabira, S.F.; Sebaa, M. Anisotropy Evolution of Low Density Polyethylene Greenhouse Covering Films during Their Service Life. *Polym. Test.* **2018**, *66*, 146–154. [CrossRef]
50. Akhter, R.; Masoodi, F.A.; Wani, T.A.; Rather, S.A. Functional Characterization of Biopolymer Based Composite Film: Incorporation of Natural Essential Oils and Antimicrobial Agents. *Int. J. Biol. Macromol.* **2019**, *137*, 1245–1255. [CrossRef] [PubMed]
51. Loo, C.P.Y.; Sarbon, N.M. Chicken Skin Gelatin Films with Tapioca Starch. *Food Biosci.* **2020**, *35*, 100589. [CrossRef]
52. Dou, L.; Li, B.; Zhang, K.; Chu, X.; Hou, H. Physical Properties and Antioxidant Activity of Gelatin-Sodium Alginate Edible Films with Tea Polyphenols. *Int. J. Biol. Macromol.* **2018**, *118*, 1377–1383. [CrossRef]

Development and Characterization of Novel Biopolymer Derived from *Abelmoschus esculentus* L. Extract and Its Antidiabetic Potential

Abd Elmoneim O. Elkhalifa [1], Eyad Al-Shammari [1], Mohd Adnan [2], Jerold C. Alcantara [3], Khalid Mehmood [4], Nagat Elzein Eltoum [1], Amir Mahgoub Awadelkareem [1], Mushtaq Ahmad Khan [5] and Syed Amir Ashraf [1,*]

[1] Department of Clinical Nutrition, College of Applied Medical Sciences, University of Hail, Hail P.O. Box 2440, Saudi Arabia; ao.abdalla@uoh.edu.sa (A.E.O.E.); eyadhealth@hotmail.com (E.A.-S.); nagacademic0509@gmail.com (N.E.E.); mahgoubamir22@gmail.com (A.M.A.)
[2] Department of Biology, College of Science, University of Hail, Hail P.O. Box 2440, Saudi Arabia; drmohdadnan@gmail.com
[3] Department of Clinical Laboratory Sciences, College of Applied Medical Sciences, University of Hail, Hail P.O. Box 2240, Saudi Arabia; jerold.alcantara@yahoo.com
[4] Department of Pharmaceutics, College of Pharmacy, University of Hail, Hail P.O. Box 81481, Saudi Arabia; adckhalid@gmail.com
[5] Department of Microbiology and Immunology, College of Medicine and Health Sciences, UAE University, Al Ain 15551, United Arab Emirates; mushtaq.khan@uaeu.ac.ae
* Correspondence: s.amir@uoh.edu.sa or amirashrafy2007@gmail.com; Tel.: +966-591491521 or +966-165358298

Abstract: *Abelmoschus esculentus* (Okra) is an important vegetable crop, widely cultivated around the world due to its high nutritional significance along with several health benefits. Different parts of okra including its mucilage have been currently studied for its role in various therapeutic applications. Therefore, we aimed to develop and characterize the okra mucilage biopolymer (OMB) for its physicochemical properties as well as to evaluate its in vitro antidiabetic activity. The characterization of OMB using Fourier-transform infrared spectroscopy (FT-IR) revealed that okra mucilage containing polysaccharides lies in the bandwidth of 3279 and 1030 cm^{-1}, which constitutes the fingerprint region of the spectrum. In addition, physicochemical parameters such as percentage yield, percentage solubility, and swelling index were found to be 2.66%, 96.9%, and 5, respectively. A mineral analysis of newly developed biopolymers showed a substantial amount of calcium (412 mg/100 g), potassium (418 mg/100 g), phosphorus (60 mg/100 g), iron (47 mg/100 g), zinc (16 mg/100 g), and sodium (9 mg/100 g). The significant antidiabetic potential of OMB was demonstrated using α-amylase and α-glucosidase enzyme inhibitory assay. Further investigations are required to explore the newly developed biopolymer for its toxicity, efficacy, and its possible utilization in food, nutraceutical, as well as pharmaceutical industries.

Keywords: okra mucilage; okra polysaccharides; biopolymer; α-amylase activity; α-glucosidase activity; nutraceuticals; antidiabetic activity

1. Introduction

Abelmoschus esculentus (L.) Moench is a popular vegetable crop cultivated throughout the world mostly in tropical and subtropical regions. The cultivation of okra vegetable is globally known for its palatability [1]. Okra plant and its derived products have been studied for various therapeutic purposes, such as antidiabetic, antioxidant, anticancer, immunomodulatory potentials, as well as its ability to ease constipation [1]. Currently, the mucilage or latex present in okra has drawn attention of the scientific community for its application as an intervention for new therapeutic purposes, as the infusion of okra mucilage has been earlier used in traditional Indian ethnomedicine for treating dysentery, diarrhea, and many more [2]. Previous studies indicated that okra mucilage could have a potential role in the management of diabetes [3]. Mucilaginous substances present in the pod walls of okra containing significant amount of protein, carbohydrate, neutral sugars, minerals, and other complex polysaccharides [4,5]. Polysaccharides are a very important class of biopolymers and represent a structurally diverse class of macromolecules. Furthermore, natural polymers derived from plant or animal sources have high molecular weight along with increased polarity, as these polymers are made up of monosaccharide units and joined by glycosides linkage [6]. Moreover, polysaccharides are highly diverse in structure and biological functions including serving as structural components of cell walls, cell recognition, cell proliferation, energy storage, cell differentiation, regulation of signaling, and immune responses [7]. This enormous potential variability in polysaccharide structure gives the necessary flexibility for the precise regulatory mechanisms of various cell–cell interactions in higher organisms. Recent research is focused on polysaccharides isolated from natural sources, because of its low side effect or with minimal toxicity. Meanwhile, several naturally occurring polysaccharides such as cellulose, starch, pectin, acacia gum, gum arabic, arabinogalactan, xylan, beta-glucan, and karaya gum has been reported [8,9]. Furthermore, the main bioactive component of okra mucilage is okra polysaccharide, which is reported to be comprising of pectic polysaccharides [10]. Meanwhile, the compositions of water-extractable polysaccharides were reported to be galacturonic acid, rhamnose, arabinose, xylose, mannose, galactose, glucose, xylan, starch, and uronic acid [11,12]. In addition, okra polysaccharides have been reported for their antioxidant activity, immunomodulatory activity, ability to improve metabolic disorders and intestinal function, hypoglycemic activity, and antifatigue activities. Rhamnogalacturonan, a polysaccharide extracted from okra, has been reported to have an antidiabetic effect [11].

Okra polysaccharides have been seen as a promising bioactive component considering its future prospective in food and pharmaceutical purposes for the development of novel polymer [10,11,13,14]. Additionally, okra polysaccharides could also become a source for the development of antidiabetic biopolymer, since it is considered to be very economical, non-toxic, and biodegradable [6]. Diabetes is currently one of the most prevalent epidemics worldwide; it represents an increase in socioeconomic burden, affecting about 382 million people globally, and each year, around 1.3 million people die from diabetes. By 2045, an estimate of 629 million people will be diabetic worldwide, as reported by the International Diabetes Federation in 2017 [15]. Moreover, the etiology of different types of diabetes varies, but complications related to high blood glucose are common in both types of diabetes. Meanwhile, drug or diet ability to delay the production or absorption of glucose by inhibiting carbohydrate-hydrolyzing enzymes such as α-amylase and α-glucosidase is one of the most common therapeutic approaches used for the treatment of hyperglycemia [16]. Additionally, α-glucosidase inhibitors are considered to be more effective categories of antidiabetic agents used in hyperglycemia, especially in case of postprandial hyperglycemia over α-amylase inhibitors. The membrane bound α-glucosidase enzymes speed up the digestion of oligosaccharides and disaccharides into simple monosaccharides, after which they get absorbed and enter into the bloodstream. The inhibition of α-glucosidase as well as α-amylases enzyme can help in delaying the digestion of carbohydrates, thereby reducing the levels of glucose in blood [17]. At present, the use of carbohydrate digesting enzyme inhibitors plays a vital role in controlling hyperglycemia by reducing the intestinal absorption of glucose [16].

However, several reports suggest that pharmacological agents are usually associated with some side effects, adverse effects, and even sometimes, their efficacies are controversial. Hence, attention has been shifted toward traditional and alternative medicines or food-derived products rich in antidiabetic phytoconstituents. The bioactive components present in plants and plants derived products such as alkaloids, flavonoids, glycosides, gum, carbohydrates, triterpenes, and different types of peptides are usually responsible for their therapeutic importance [18–21]. Recently, okra has been recognized for its potential therapeutic purposes due to the presence of various important phytochemicals including polysaccharides [1,22]. Therefore, despite having various therapeutic applications of okra fruits, seeds, pods, its mucilage has not been much explored toward its promising potential. Hence, a novel biopolymer derived from okra mucilage has been developed along with its physicochemical characterization, and its antidiabetic properties are studied.

2. Results and Discussion

2.1. Characterization of Okra Mucilage Biopolymer by FT-IR

Okra mucilage biopolymer (OMB) characterization using FT-IR analysis identifies several functional groups representing characteristic bands of polysaccharides lying between 3279 and 1030 cm^{-1}, which constitute the fingerprint region of the spectrum, as presented in Figure 1. The broad band at 3279 cm^{-1} is mainly due to the presence of hydrogen-bonded hydroxyl groups, which give rise to the complex vibrational bands associated with free intermolecular and intramolecular bound hydroxyl groups, which leads to the gross structure of carbohydrates [23,24]. Our FT-IR data also revealed the characteristic of polysaccharide consisting of galactose, rhamnose, and galacturonic acid represented by the broad-spectrum peak at 3279 cm^{-1}, suggesting the presence of aromatic sugar with O-H as the principle functional groups. Meanwhile, the presence of an O-H functional group in the broad peak characterizes hydrophilic nature of the polysaccharides. The hydroxyl groups in carbohydrate have intermolecular and intramolecular hydrogen bonding that give broad band at 3279 cm^{-1} [24]. In addition, the band present at 2938 cm^{-1} is also a characteristic of methyl C–H bonding associated with benzene rings. In cellulose and hemicellulose components, the characteristic C–H stretching corresponds to the band at 2942 cm^{-1} [25]. In complex polysaccharides spectra, 1245 cm^{-1} was assigned for the C–O stretching band, whereas 1030 cm^{-1} was assigned for the C–O–C group, indicating the presence of aromatic bonds present in galactose, galacturonic acid, as well as in rhamnose.

The amide I band corresponds to the band at 1625 cm^{-1}, which comes in the most sensitive spectral region, depicting the secondary organizational units of proteins [26]. This band is purely due to peptide linkages and CO stretch vibrations, indicating the presence of protein. As the biopolymer was not subject to deproteinization, the protein bands were observed [27]. The absorption at 1732 cm^{-1} was observed because of ester carbonyl, which has also been reported in a previous study on okra mucilage [28]. Moreover, the presence of carbonyl, methyl, as well as hydroxyl functional groups in okra are representative of polysaccharides molecules, which is considered to be the backbone of the developed polymer. Our results confirmed that OMB is composed of polysaccharides. However, the polysaccharides were not in the form of cellulose or starch, but few functional bands indicated the presence of peptide cross-link along with some amino sugars, and our results are in line with earlier studies [24,29].

2.2. Percentage Yield, Solubility, and Swelling Index of Okra Mucilage Biopolymer

Okra mucilage extraction and its percentage yield were found to be 2.66% w/w using water as the extraction liquid, and according to previous studies, the percentage yield of okra mucilage was reported to range from 0.5% to 11% [28,30,31]. These variations in the percentage yield of okra mucilage could be due to several factors, such as the physical state of pods (dried or hydrated), cultivation region of pods, breed of okra, parts of okra (crown or pulp), and maturation state of okra pods. In addition, the percentage yield could also be affected by the extraction method used [25]. It has been reported that acetone is more effective when compared to methanol as an extraction solvent; the yield rises by 31 times [28]. Percentage yield was recorded on a dry weight basis, and the moisture content of extracted mucilage was found to be 9.6%. Currently, scientific communities are working more on these factors to determine the optimal conditions of mucilage extraction, such as extraction time, temperature, extraction cycle number, and raw material-to-solvent ratio. Okra mucilage extraction by the water extraction method gave the yield of 2.66%, and the conditions were temperature of 70 °C for 2 h and agitating at 200 rpm [32]. The solution obtained was caramel-colored, viscous, and slippery. We found that the solubility of extracted mucilage polymer at neutral pH was 96.9%. Earlier studies have also indicated that okra mucilage is partially soluble in cold water and soluble in warm water. Our results are also consistent with the previous reported studies. All the characteristics obtained were similar to those reported previously [31,33].

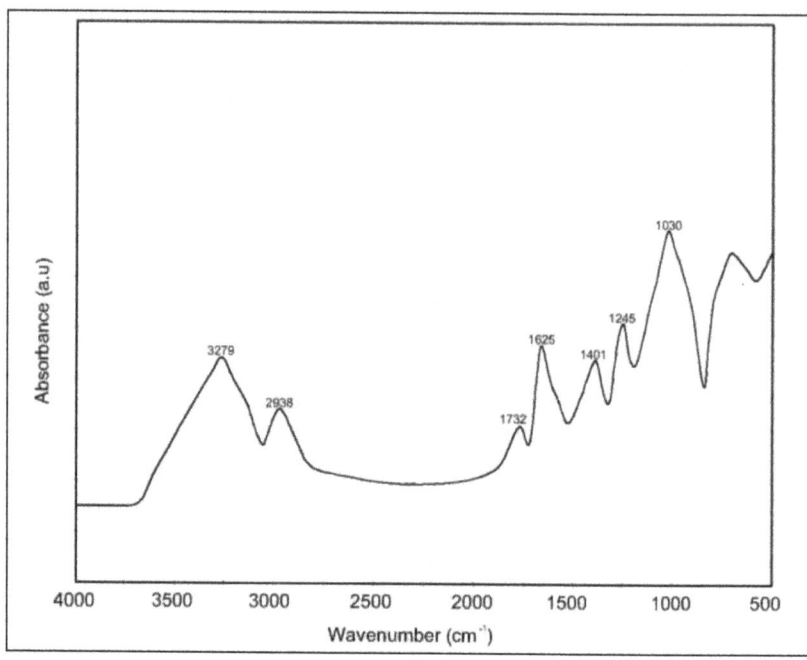

Figure 1. FT-IR spectrum of okra mucilage.

Additionally, we found that the swelling index of OMB at pH 7.5 in deionized water was 5.0. Moreover, the swelling index has been found to be increasing with a gradual increase of pH, as presented in Figure 2. This increase in the swelling index of okra mucilage could be due to its increase in intra-ionic repulson when the ionization of carboxyl groups (-COOH) is high. In addition, we also found that at pH 5.5, the swelling index was the lowest; this could be due to the low ionization of the major functional groups present in mucilage biopolymer. The pH-responsive behavior shown by OMB could present a favorable condition for the controlled release of a bioactive active component in pharmaceutical as well as nutraceutical industries [24]. In addition, the controlled release of active agents at different pH indicates a unique polymer behavior, which is indispensable in the packaging of food. Earlier studies have indicated that the polymer prepared for drug development shows reduced swelling onwards of pH > 7.4, which could be due to carboxyl groups ionization, and polymer dissolution occurs [34]. However, further studies are required for a better analysis of this reported behavior.

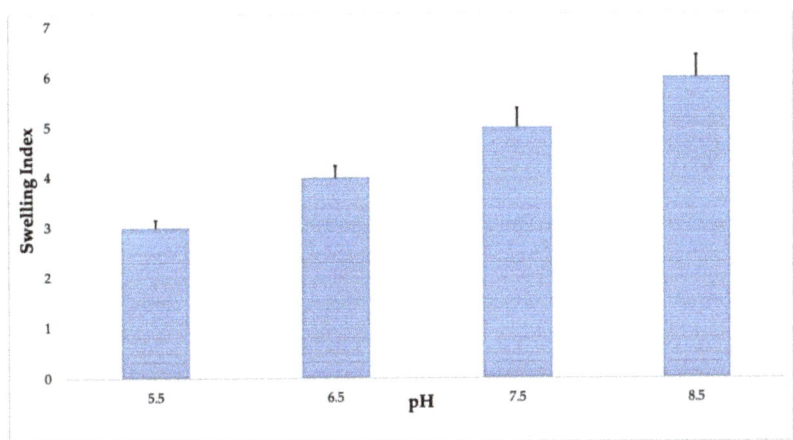

Figure 2. Swelling index of okra mucilage biopolymer at different pH. All experiments were performed in triplicate, and data represent mean ±standard deviation.

2.3. Mineral Composition

Minerals are considered as one of the important parts of human nutrition, which help to promote a healthy physical and mental state. The minerals are the principal elements of the bones, teeth, tissues, muscles, blood, and nerve cells [35]. They play an important role in acid–base balance maintenance, nerve response to physiological stimulation, and in the clotting of blood [36]. The major component of bone is calcium, which is essential in teeth development, blood coagulation, and intracellular cement substance integrity [37]. The mineral constitution of OMB is summarized in Figure 3. The calcium concentration of OMB was found to be 412 mg/100 g. One of the key trace elements required is iron, which is helpful in the synthesis of hemoglobin, central nervous system functioning, and oxidation of carbohydrates, fats, and proteins [38,39], which is further important in preventing diabetes. The low iron content in the body leads to infection in the gastrointestinal pathway, epistaxis (nose bleeding), and myocardial infection [40]. Figure 3 presents the iron content of OMB, which was found to be 47 mg/100 g. Zinc is another required trace element having important roles in many cellular processes such as normal body growth, development of the brain, behavioral responses, formation of bone, and smooth wound healing [41]. Protein and carbohydrate metabolism requires zinc. The hepatic stellate cells (HSCs) of the liver are the storage sites of vitamin A, and zinc is involved in its mobility to other body parts from the liver. Zinc metalloenzymes are involved in DNA and RNA biosynthesis [42]. The deficiency of zinc is commonly seen in people suffering from Crohn's disease, hypothyroidism, and gum disease (Periodontitis). People with zinc inadequacy are vulnerable to viral infections and diabetes mellitus. Zinc plays a favorable role in treating viral infections such as AIDS, prostate gland enlargement, rheumatoid arthritis, laceration, acne, eczema, and stress [43]. Figure 3 presents the zinc content of OMB, which was found to be 16.31 mg/100 g. Other minerals detected in okra mucilage in significant amount were phosphorus (60 mg/100 g), potassium (418 mg/100 g), and sodium (9 mg/100 g). A high potassium level in the body causes an increased utilization of iron in the body. Potassium is also beneficial to patients administered with diuretics for hypertension control and those having excessive potassium excretion via body fluid [44].

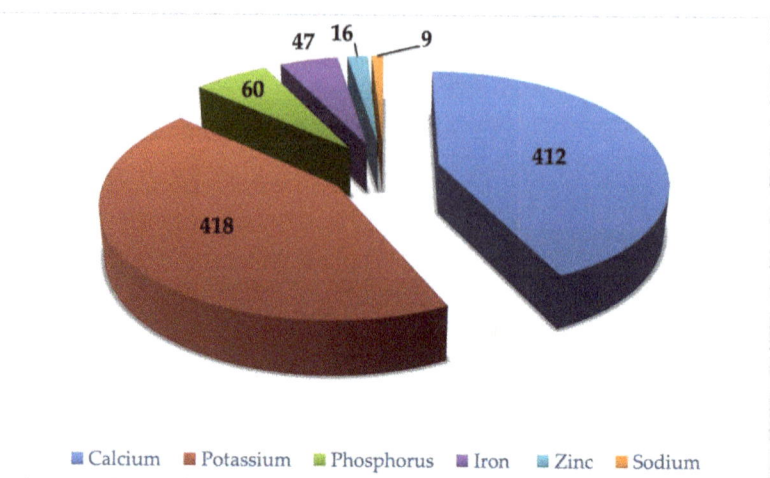

Figure 3. Mineral contents (mg/100 g) of okra mucilage biopolymer.

2.4. Antidiabetic Activity

The use of plant-based approaches in the existing modern medications system for the treatment of chronic disease such as diabetes is gaining recognition. Majorly, two of the α-amylase and α-glucosidase enzymes are considered to be responsible for diabetic conditions: α-amylase begins the process of carbohydrate digestion by hydrolysis of 1, 4-glycosidic linkages of polysaccharides (starch, glycogen) to disaccharides, and α-glucosidase catalyzes the disaccharides to monosaccharides, which leads to postprandial hyperglycemia. Hence, inhibitors of α-amylase and α-glucosidase are useful for the control of high glucose level, as they delay carbohydrate digestion, which consequently reduces the postprandial plasma glucose level [16].

To study the antidiabetic activity of OMB, α-amylase enzyme assay (Figure 4a) and α-glucosidase inhibitory assay (Figure 4b) were performed. The α-amylase inhibition enzyme assay of okra mucilage revealed that the new developed okra biopolymer had significant anti-diabetic property with the increase in the concentration of biopolymer, and its inhibitory activity was highest. On the other hand, α-glucosidase inhibitory assay showed that OMB had concentration-dependent inhibitory effects. The mucilage biopolymer at different concentration of 1, 2, 3, 4, and 5 mg/mL showed 9.9, 16.5, 24.5, 28.2, and 49.8 α-amylase inhibitory activity as well as 30, 41.5, 50.5, 62.2, and 69.7 α-glucosidase inhibitory activity, respectively. Figure 4a,b showed that the higher the concentration, the higher the inhibition of α-amylase and α-glucosidase inhibition. This could be due to high concentrations representing more solutes in the form of secondary metabolites from okra polymer, which had the ability to inhibit the action of both antidiabetic enzymes.

Earlier, the antidiabetic activity of okra has been reported by Ahmad et al. (2016), and they found that aqueous extract of okra at different increasing concentrations (50, 100, 150, 200, and 250 μg/mL) has reportedly increased the inhibition for both percentage of α-amylase and α-glucosidase enzyme. This reported result was consistent with our results [45,46]. In both the assays, the mucilage polymer showed significant results, indicating that OMB has potential, and it needs to be further explored for toxicity studies and clinical trials, and by virtue of that, it could become an important anti-diabetic agent.

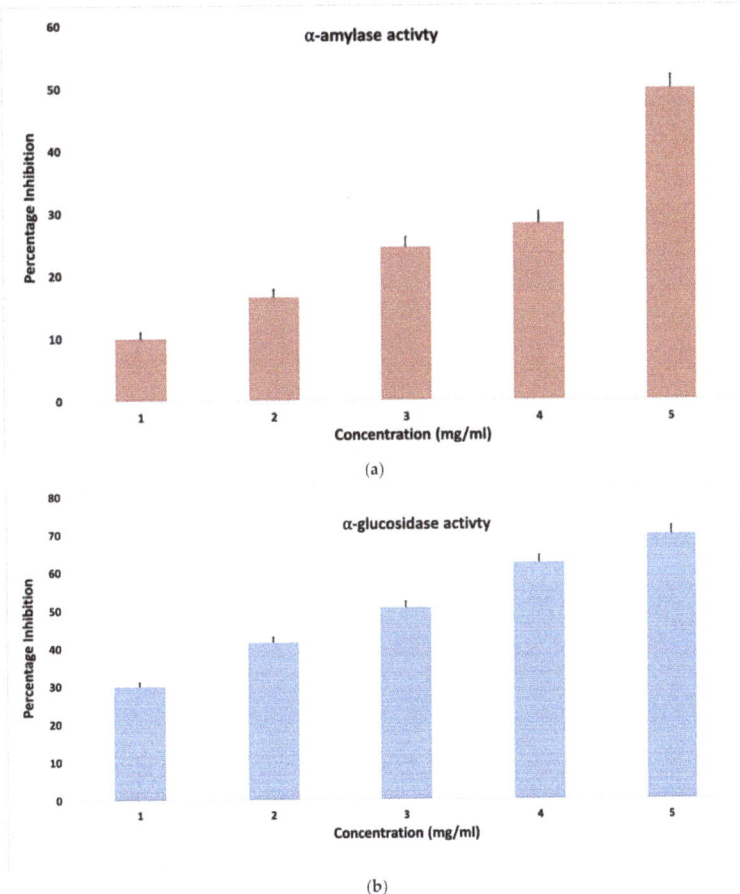

Figure 4. Screening of (**a**) α-amylase inhibitory assay, (**b**) α-glucosidase inhibitory assay of okra mucilage biopolymer at various concentrations. All experiments were performed in triplicate, and data are expressed as mean ±standard deviation.

3. Materials and Methods

3.1. Materials

Mature okra pods were collected during the month of September 2020 from the local markets of Hail city, Saudi Arabia. Samples collected from the local market were carefully assorted, cleaned, washed from the earthy or waste material, and further dried under the shade for 1 day. The kits used for α-amylase enzyme, α-glucosidase inhibitory assays, and ethanol was bought from Sigma-Aldrich® (St. Louis, MO, USA). No additional purification was done in these kits before using them for the experiments. All other reagents used to carry out the experiments in this study were of analytical grade.

3.2. Extraction of Okra Mucilage

The okra pods were sliced, deseeded, and then water-soaked at ambient temperature. Liquid fraction or filtrate was separated from the solid content after 12 h using a muslin cloth. To the filtrate, triple the volume of ethanol was added. The liquid was stirred slowly by hand until the mucilage was fully precipitated. The mucilage was left to dry in an oven for about 12 h at 30 °C. Later, the dried mucilage was ground into powdered form evenly using a grinder and then passed through sieves. The finely pulverized polymer was kept in polyethylene pouch bags in the dark until further analyses [47].

3.3. Characterization of Okra Mucilage Biopolymer

The FT-IR experiments were carried out using the Thermo Scientific® Nicolet TM 6700 FT-IR spectrophotometer. The attenuated total reflection was performed to obtain the spectra. In this technique, Zn-Se crystals were used to press the samples, and 4000–650 cm^{-1} was the range of collection. An average of 16 scans having a resolution of 4 cm^{-1} was considered.

3.4. Quantitative Yield Determination

The percentage yield was calculated by the previously reported method of Jouki et al. (2014) [48,49]. The weight of okra pods without seeds in grams (m_s) was used to calculate the yield. After the extraction, the weight of mucilage polymer was converted into milligrams (m). The percentage yield was calculated using Equation (1) as mentioned below.

$$\text{Yield} = \frac{m}{m_s} \tag{1}$$

3.5. Swelling Index

The swelling index (SI) was determined according to the method reported by Cotrim et al. (2016) with certain modifications. In a graduated cylinder of 10 mL, 0.1 g of okra mucilage biopolymer was transferred. After measuring the initial dried sample volume, ionized water was added, and a final volume of 10 mL was made. The mucilage polymer swelled up to form a viscous gel. After 24 h, the volume of swelled up mucilage was measured. The swelling index was calculated using Equation (2) mentioned below [49].

$$SI = \frac{V_f}{V_b} \tag{2}$$

where V_f = final volume 24 h later, and V_b = dry mucilage initial volume.

3.6. Solubility of Okra Mucilage Biopolymer

Dry okra mucilage biopolymer (0.1 g) was taken in a graduated cylinder of 10 mL to study its solubility in water. Deionized water was added to make the volume up to 10 mL, and the sample solutions were left for 6 h. Furthermore, the dispersed samples were subjected to magnetic stirring at 60 °C for 1 h. Later on, the sample solution was centrifuged at 4000 rpm for 30 min. After the centrifugation process, the insoluble matrix was separated and dried at 105 °C until it achieved constant weight. The solubility percentage of the sample was obtained as per Equation (3), as mentioned below [31]

$$S = \frac{w_1 - w_2}{w_1} \times 100 \tag{3}$$

where S = percentage solubility, w_1 = initial mass of dry mucilage, and w_2 = mass of insoluble dry matter obtained after centrifugation.

3.7. Determination of Mineral Content

The mineral analysis of OMB was carried out according to the procedure mentioned in the AOAC (2016). Additionally, sample preparation was made by taking 1 g of the pre-dried sample in crucible and burned over a hot plate followed by ashing inside the muffle furnace at 550 °C. Once the sample cools down, it was mixed 20% supra pure nitric acid (50 mL) in an Erlenmeyer flask and heated at 70–85 °C for 6 h. Meanwhile, during the digestion of the sample, supra pure nitric acid was added intermittently to maintain the volume. Furthermore, the sample was filtered cool and the volume was made up to 100 mL in a volumetric flask using deionized water. Afterwards, the prepared samples were placed for analysis of selected elements [50]. The standard flame emission photometer (PerkinElmer, Waltham, MA, USA) was used to determine the concentrations of sodium (Na) and potassium (K) present in the okra mucilage. Phosphorus (P) content was quantified by the vanadomolybdate method AOAC (2016). Atomic absorption spectrophotometer (PerkinElmer, Waltham, MA, USA) was used to measure calcium (Ca), iron (Fe), and zinc (Zn) concentrations [51].

3.8. Antidiabetic Activity

To investigate the in vitro antidiabetic activity, α-amylase inhibitory assay and α-glucosidase inhibitory assay were performed.

3.8.1. α-Amylase Inhibitory Assay

α-Amylase inhibitory activity of OMB was carried out as per the method presented by Oliviya et al. (2018) with slight modification. Non-identical aliquots of mucilage biopolymer (1, 2, 3, 4, and 5 mg/mL), were taken separately. Subsequently, 250 µL of 0.02 N sodium phosphate buffer containing α-amylase solution (0.5 mg/mL) was added to it. Starch solution (1%, 250 µL) prepared in 0.02 M sodium phosphate was added to the solution. The incubation of 10 min at 25 °C was provided to this mixture. With the addition of 500 µL of dinitrosalicylic acid, the reaction was terminated. Then, it was kept for 5 min in a boiling water bath. Then, the solution was cooled down, and absorbance (Abs) was recorded at 540 nm [52]. Negative control was prepared using distilled water, and the percentage inhibition was calculated by using following formula [53].

$$\% \text{ Inhibition} = \frac{\text{Abs.}(540)_{control} - \text{Abs.}(540)_{sample}}{\text{Abs.}(540)_{control}} \times 100$$

3.8.2. α-Glucosidase Inhibitory Assay

α-Glucosidase inhibitory activity of okra mucilage biopolymer was carried out as per the method presented by Oliviya et al. (2018) with slight modification. Pre-incubation of 100 µL of glucosidase with 50 µL at different concentration (1, 2, 3, 4, 5 mg/mL) of mucilage biopolymer extract were taken separately. To start the reaction, 50 µL of 3.0 mM P-nitrophenyl-α-D-glucopyranoside was added in 20 mM phosphate buffer. The solution was incubated for 20 min at a temperature of 37 °C. At this stage, the reaction was terminated by adding 0.1 M of Na_2CO_3. The absorbance was measured at 405 nm [52]. Negative control was prepared using distilled water, and the percentage inhibition was calculated by using the following formula [53].

$$\% \text{ Inhibition} = \frac{\text{Abs.}(405)_{control} - \text{Abs.}(405)_{sample}}{\text{Abs.}(405)_{control}} \times 100$$

3.9. Statistical Analysis

The data of all the experiments were expressed as mean ±standard deviation of triplicate measurements. All the analysis were carried out using IBM SPSS software version 23.0 (IBM Corp. Armonk, NY, USA).

4. Conclusions

Okra, an easily available nutritive vegetable crop, has been found very interesting in terms of mineral content. Mucilage present in okra has been found to be a rich source of polysaccharides. Our study showed that the newly developed biopolymer had key physicochemical characteristics. Key parameters such as swelling index, solubility, as well as mineral analysis indicated that the newly developed biopolymer could become a potential source for food and pharmaceutical industries as packaging materials (edible polymer), emulsion stabilizers, binding agent, water retention agent, thickeners, etc. The high swelling index of biopolymer at different pH could open the door of possibilities for the development of edible polymer as well as for the development of nutraceuticals by incorporating bioactive components. In addition, FT-IR analysis suggested that it contains polysaccharides majorly composed of galactose, rhamnose, and galacturonic acid. Furthermore, mucilage biopolymer was evaluated for in vitro antidiabetic activity, and both α-amylase and α-glucosidase inhibitory assay showed positive response against the okra mucilage biopolymer, indicating its possible application for the advancement of anti-diabetic agent. Therefore, based upon our results, we can conclude that the newly developed okra mucilage biopolymer is enriched with nutritional content as well as it could become an important natural source for various food, nutraceutical as well as pharmaceutical applications.

Author Contributions: Conceptualization, A.E.O.E., E.A.-S. and S.A.A.; methodology, M.A., J.C.A., K.M. and N.E.E.; validation, M.A., M.A.K. and S.A.A.; formal analysis, E.A.-S., M.A., J.C.A. and A.M.A.; investigation, S.A.A., M.A., M.A.K. and K.M.; data curation, K.M., A.E.O.E., S.A.A., N.E.E. and M.A.; writing—original draft preparation, S.A.A. and M.A.; writing—review and editing, A.E.O.E., M.A. and M.A.K.; visualization, K.M., A.M.A. and S.A.A.; supervision, M.A., A.E.O.E. and M.A.K.; project administration, S.A.A., A.E.O.E. and M.A. All authors have read and agreed to the published version of the manuscript.

Funding: This research has been funded by Scientific Research Deanship at University of Ha'il Saudi Arabia through project number RG-191333.

Institutional Review Board Statement: Not applicable.

Informed Consent Statement: Not applicable.

Data Availability Statement: All data generated or analyzed during this study are included in this article.

Conflicts of Interest: The authors declare no conflict of interest. The funders had no role in the design of the study; in the collection, analyses, or interpretation of data; in the writing of the manuscript, or in the decision to publish the results.

Sample Availability: Samples of the compounds are not available from the authors.

References

1. Elkhalifa, A.E.O.; Alshammari, E.; Adnan, M.; Alcantara, J.C.C.; Awadelkareem, A.M.; Eltoum, N.E.; Mehmood, K.; Panda, B.P.; Ashraf, S.A. Okra (*Abelmoschus Esculentus*) as a Potential Dietary Medicine with Nutraceutical Importance for Sustainable Health Applications. *Molecules* **2021**, *26*, 696. [CrossRef] [PubMed]
2. Roy, A.; Shrivastava, S.L.; Mandal, S.M. Functional properties of Okra *Abelmoschus esculentus* L. (Moench): Traditional claims and scientific evidences. *Plant Sci. Today* **2014**, *1*, 121–130. [CrossRef]
3. Khosrozadeh, M.; Heydari, N.; Abootalebi, M. The Effect of Abelmoschus Esculentus on Blood Levels of Glucose in Diabetes Mellitus. *Iran. J. Med. Sci.* **2016**, *41*, S63. [PubMed]
4. Ahiakpa, K. Mucilage Content of 21 Accessions of Okra (*Abelmoschus* spp L.). *Scientia Agriulturae* **2014**, *2*, 96–101. [CrossRef]
5. Gemede, H.F.; Ratta, N.; Haki, G.D.; Woldegiorgis, A.Z. Nutritional Quality and Health Benefits of Okra (*Abelmoschus Esculentus*): A Review. *Glob. J. Med Res.* **2014**, *14*. [CrossRef]
6. Pala, K. Synthesis, characterization of Okra mucilage as a potential new age therapeutic intervention. In *MOL2NET, International Conference Series on Multidisciplinary Sciences*; USINEWS-04: US-IN-EU Worldwide Science Workshop Series; University of Minnesota Duluth: Duluth, MN, USA, 2020. [CrossRef]
7. Wang, H.J.; Liu, Y.M.; Qi, Z.M.; Wang, S.Y.; Liu, S.X.; Li, X.; Xia, X.C. An Overview on Natural Polysaccharides with Antioxidant Properties. *Curr. Med. Chem.* **2013**, *20*, 2899–2913. [CrossRef]
8. Mohammed, A.S.A.; Naveed, M.; Jost, N. Polysaccharides; Classification, Chemical Properties, and Future Perspective Applications in Fields of Pharmacology and Biological Medicine (A Review of Current Applications and Upcoming Potentialities). *J. Polym. Environ.* **2021**, 1–13. [CrossRef]
9. Li, Q.; Niu, Y.; Xing, P.; Wang, C. Bioactive polysaccharides from natural resources including Chinese medicinal herbs on tissue repair. *Chin. Med.* **2018**, *13*, 1–11. [CrossRef]
10. Liao, Z.; Zhang, J.; Liu, B.; Yan, T.; Xu, F.; Xiao, F.; Wu, B.; Bi, K.; Jia, Y. Polysaccharide from Okra (*Abelmoschus esculentus* (L.) Moench) Improves Antioxidant Capacity via PI3K/AKT Pathways and Nrf2 Translocation in a Type 2 Diabetes Model. *Molecules* **2019**, *24*, 1906. [CrossRef]
11. Zhu, X.-M.; Xu, R.; Wang, H.; Chen, J.-Y.; Tu, Z.-C. Structural Properties, Bioactivities, and Applications of Polysaccharides from Okra [*Abelmoschus esculentus* (L.) Moench]: A Review. *J. Agric. Food Chem.* **2020**, *68*, 14091–14103. [CrossRef]

12. Xu, K.; Guo, M.; Du, J. Molecular characteristics and rheological properties of water-extractable polysaccharides derived from okra (*Abelmoschus esculentus* L.). *Int. J. Food Prop.* **2017**, *20*, S899–S909. [CrossRef]
13. Gao, H.; Zhang, W.; Wang, B.; Hui, A.; Du, B.; Wang, T.; Meng, L.; Bian, H.; Wu, Z. Purification, characterization and anti-fatigue activity of polysaccharide fractions from okra (*Abelmoschus esculentus* (L.) Moench. *Food Funct.* **2018**, *9*, 1088–1101. [CrossRef] [PubMed]
14. Nie, X.-R.; Fu, Y.; Wu, D.-T.; Huang, T.-T.; Jiang, Q.; Zhao, L.; Zhang, Q.; Lin, D.-R.; Chen, H.; Qin, W. Ultrasonic-Assisted Extraction, Structural Characterization, Chain Conformation, and Biological Activities of a Pectic-Polysaccharide from Okra (Abelmoschus esculentus). *Molecules* **2020**, *25*, 1155. [CrossRef] [PubMed]
15. International Diabetes Federation (IDF). *International Diabetes Atlas*, 8th ed.; International Diabetes Federation: Brussels, Belgium, 2017; ISBN 78-2-930229-87-4. Available online: www.diabetesatlas.org (accessed on 30 March 2021).
16. Hullatti, K.; Telagari, M. In-vitro α-amylase and α-glucosidase inhibitory activity of Adiantum caudatum Linn. and Celosia argentea Linn. extracts and fractions. *Indian J. Pharmacol.* **2015**, *47*, 425–429. [CrossRef] [PubMed]
17. Bhatia, A.; Singh, B.; Arora, R.; Arora, S. In vitro evaluation of the α-glucosidase inhibitory potential of methanolic extracts of traditionally used antidiabetic plants. *BMC Complementary Altern. Med.* **2019**, *19*, 1–9. [CrossRef]
18. Ashraf, S.A.; ElKhalifa, A.E.O.; Siddiqui, A.J.; Patel, M.; Awadelkareem, A.M.; Snoussi, M.; Ashraf, M.S.; Adnan, M.; Hadi, S. Cordycepin for Health and Wellbeing: A Potent Bioactive Metabolite of an Entomopathogenic Cordyceps Medicinal Fungus and Its Nutraceutical and Therapeutic Potential. *Molecules* **2020**, *25*, 2735. [CrossRef]
19. Siddiqui, A.J.; Danciu, C.; Ashraf, S.A.; Moin, A.; Singh, R.; Alreshidi, M.; Patel, M.; Jahan, S.; Kumar, S.; Alkhinjar, M.I.M.; et al. Plants-Derived Biomolecules as Potent Antiviral Phytomedicines: New Insights on Ethnobotanical Evidences against Coronaviruses. *Plants* **2020**, *9*, 1244. [CrossRef]
20. Noumi, E.; Snoussi, M.; Anouar, E.; Alreshidi, M.; Veettil, V.; Elkahoui, S.; Adnan, M.; Patel, M.; Kadri, A.; Aouadi, K.; et al. HR-LCMS-Based Metabolite Profiling, Antioxidant, and Anticancer Properties of *Teucrium polium* L. Methanolic Extract: Computational and In Vitro Study. *Antioxidants* **2020**, *9*, 1089. [CrossRef]
21. Reddy, M.N.; Adnan, M.; Alreshidi, M.M.; Saeed, M.; Patel, M. Evaluation of Anticancer, Antibacterial and Antioxidant Properties of a Medicinally Treasured Fern Tectaria coadunata with its Phytoconstituents Analysis by HR-LCMS. *Anti Cancer Agents Med. Chem.* **2020**, *20*, 1845–1856. [CrossRef]
22. Durazzo, A.; Lucarini, M.; Novellino, E.; Souto, E.B.; Daliu, P.; Santini, A. *Abelmoschus esculentus* (L.): Bioactive Components' Beneficial Properties—Focused on Antidiabetic Role—For Sustainable Health Applications. *Molecules* **2018**, *24*, 38. [CrossRef]
23. Ahmed, M.A.; Holz, M.; Woche, S.K.; Bachmann, J.; Carminati, A. Effect of soil drying on mucilage exudation and its water repellency: A new method to collect mucilage. *J. Plant Nutr. Soil Sci.* **2015**, *178*, 821–824. [CrossRef]
24. Emeje, M.; Isimi, C.; Byrn, S.; Fortunak, J.; Kunle, O.; Ofoefule, S. Extraction and Physicochemical Characterization of a New Polysaccharide Obtained from the Fresh Fruits of Abelmoschus Esculentus. *Iran. J. Pharm. Res. IJPR* **2011**, *10*, 237–246.
25. Lengsfeld, C.; Titgemeyer, F.; Faller, G.; Hensel, A. Glycosylated Compounds from Okra Inhibit Adhesion of Helicobacter pylori to Human Gastric Mucosa. *J. Agric. Food Chem.* **2004**, *52*, 1495–1503. [CrossRef]
26. Kong, J.; Yu, S. Fourier Transform Infrared Spectroscopic Analysis of Protein Secondary Structures. *Acta Biochim. Biophys. Sin.* **2007**, *39*, 549–559. [CrossRef]
27. Haq, M.A.; Alam Jafri, F.; Hasnain, A. Effects of plasticizers on sorption and optical properties of gum cordia based edible film. *J. Food Sci. Technol.* **2016**, *53*, 2606–2613. [CrossRef]
28. Archana, G.; Sabina, K.; Babuskin, S.; Radhakrishnan, K.; Fayidh, M.A.; Babu, P.A.S.; Sivarajan, M.; Sukumar, M. Preparation and characterization of mucilage polysaccharide for biomedical applications. *Carbohydr. Polym.* **2013**, *98*, 89–94. [CrossRef]
29. Zaharuddin, N.D.; Noordin, M.I.; Kadivar, A. The Use of *Hibiscus esculentus* (Okra) Gum in Sustaining the Release of Propranolol Hydrochloride in a Solid Oral Dosage Form. *BioMed Res. Int.* **2014**, *2014*, 1–8. [CrossRef]
30. Antoniou, J.; Liu, F.; Majeed, H.; Qazi, H.J.; Zhong, F. Physicochemical and thermomechanical characterization of tara gum edible films: Effect of polyols as plasticizers. *Carbohydr. Polym.* **2014**, *111*, 359–365. [CrossRef]
31. Kaur, G.; Singh, D.; Brar, V. Bioadhesive okra polymer based buccal patches as platform for controlled drug delivery. *Int. J. Biol. Macromol.* **2014**, *70*, 408–419. [CrossRef]
32. Lee, C.S.; Chong, M.F.; Robinson, J.; Binner, E. Optimisation of extraction and sludge dewatering efficiencies of bio-flocculants extracted from *Abelmoschus esculentus* (okra). *J. Environ. Manag.* **2015**, *157*, 320–325. [CrossRef]
33. Kale, P. Extraction and Characterization of Okra Mucilage as a Pharmaceutical Aid. *Int. J. Sci. Dev. Res.* **2020**, *5*, 189–193.
34. Rizwan, M.; Yahya, R.; Hassan, A.; Yar, M.; Azzahari, A.D.; Selvanathan, V.; Sonsudin, F.; Abouloula, C.N. pH Sensitive Hydrogels in Drug Delivery: Brief History, Properties, Swelling, and Release Mechanism, Material Selection and Applications. *Polymers* **2017**, *9*, 137. [CrossRef]
35. Soetan, K.O.C.; Oyewole, O. The importance of mineral elements for humans, domestic animals and plants: A review. *Afr. J. Food Sci.* **2010**, *4*, 200–222.
36. Hanif, R.I.Z.; Iqbal, M.; Hanif, S.; Rasheed, M. Use of vegetables as nutritional food: Role in human health. *J. Agric. Biol. Sci.* **2006**, *1*, 18–22.
37. Oka, Y.; Udaka, K.; Tsuboi, A.; Elisseeva, O.A.; Ogawa, H.; Aozasa, K.; Kishimoto, T.; Sugiyama, H. Cancer Immunotherapy Targeting Wilms' Tumor Gene WT1 Product. *J. Immunol.* **2000**, *164*, 1873–1880. [CrossRef]

38. Chitturi, R.; Baddam, V.R.; Prasad, L.; Prashanth, L.; Kattapagari, K. A review on role of essential trace elements in health and disease. *J. Dr. NTR Univ. Health Sci.* **2015**, *4*, 75. [CrossRef]
39. National Research Council (US) Committee on Diet and Health. *Diet and Health: Implications for Reducing Chronic Disease Risk*; National Academies Press: Washington, DC, USA, 1989; ISBN 978-0-309-07474-2.
40. Gasche, C.; Lomer, M.; Cavill, I.; Weiss, G. Iron, anaemia, and inflammatory bowel diseases. *Gut* **2004**, *53*, 1190–1197. [CrossRef]
41. Chasapis, C.T.; Ntoupa, P.-S.A.; Spiliopoulou, C.A.; Stefanidou, M.E. Recent aspects of the effects of zinc on human health. *Arch. Toxicol.* **2020**, *94*, 1443–1460. [CrossRef]
42. Jabeen, S.; Shah, M.; Khan, S.; Hayat, M.Q. Determination of major and trace elements in ten important folk therapeutic plants of Haripur basin, Pakistan. *J. Med. Plants Res.* **2010**, *4*, 559–566. [CrossRef]
43. Read, S.A.; Obeid, S.; Ahlenstiel, C.; Ahlenstiel, G. The Role of Zinc in Antiviral Immunity. *Adv. Nutr.* **2019**, *10*, 696–710. [CrossRef]
44. Kardalas, E.; Paschou, S.A.; Anagnostis, P.; Muscogiuri, G.; Siasos, G.; Vryonidou, A. Hypokalemia: A clinical update. *Endocr. Connect.* **2018**, *7*, R135–R146. [CrossRef] [PubMed]
45. Ahmed, B.T.; Kumar, S.A. Antioxidant and Antidiabetic properties of Abelmoschus esculentus extract—An in vitro assay. *Res. J. Biotechnol.* **2016**, *11*, 34–41.
46. Aligita, W.; Muhsinin, S.; Susilawati, E.; Dahlia; Pratiwi, D.S.; Aprilliani, D.; Artarini, A.; Adnyana, I.K. Antidiabetic Activity of Okra (Abelmoschus esculentus L.) Fruit Extract. *Rasayan J. Chem.* **2019**, *12*, 157–167. [CrossRef]
47. Ameena, K.; Dilip, C.; Saraswathi, R.; Krishnan, P.; Sankar, C.; Simi, S. Isolation of the mucilages from Hibiscus rosasinensis linn. and Okra (Abelmoschus esculentus linn.) and studies of the binding effects of the mucilages. *Asian Pac. J. Trop. Med.* **2010**, *3*, 539–543. [CrossRef]
48. Jouki, M.; Yazdi, F.T.; Mortazavi, S.A.; Koocheki, A. Quince seed mucilage films incorporated with oregano essential oil: Physical, thermal, barrier, antioxidant and antibacterial properties. *Food Hydrocoll.* **2014**, *36*, 9–19. [CrossRef]
49. Cotrim, M.D.A.P.; Mottin, A.C.; Ayres, E. Preparation and Characterization of Okra Mucilage (*Abelmoschus esculentus*) Edible Films. *Macromol. Symp.* **2016**, *367*, 90–100. [CrossRef]
50. Alghamdi, A.A.; Awadelkarem, A.M.; Hossain, A.B.M.S.; Ibrahim, N.A.; Fawzi, M.; Ashraf, S.A. Nutritional Assessment of Different Date Fruits (*Phoenix Dactylifera* L.) Varieties Cultivated in Hail Province, Saudi Arabia. *Biosci. Biotechnol. Res. Commun.* **2018**, *11*, 263–269. [CrossRef]
51. Latimer, G.W.; AOAC International. *Official Methods of Analysis*; AOAC International: Rockville, MD, USA, 2016; Volume 20, ISBN 0935584870.
52. Christynal Oliviya, R.; Krishnakumari, S. Anti-diabetic activity of Abelmoschus esculentus and its nano particle—A Comparative Study. *Int. J. Res. Anal. Rev.* **2018**, *5*, 415–421.
53. Mechchate, H.; Es-Safi, I.; Louba, A.; Alqahtani, A.S.; Nasr, F.A.; Noman, O.M.; Farooq, M.; Alharbi, M.S.; Alqahtani, A.; Bari, A.; et al. In Vitro Alpha-Amylase and Alpha-Glucosidase Inhibitory Activity and In Vivo Antidiabetic Activity of *Withania frutescens* L. Foliar Extract. *Molecules* **2021**, *26*, 293. [CrossRef]

Article

Fermentation of Jamaican Cherries Juice Using *Lactobacillus plantarum* Elevates Antioxidant Potential and Inhibitory Activity against Type II Diabetes-Related Enzymes

Andri Frediansyah [1,*], Fitrio Romadhoni [2], Suryani [3], Rifa Nurhayati [1] and Anjar Tri Wibowo [4,5,*]

1. Research Division for Natural Products Technology (BPTBA), Indonesian Institute of Sciences (LIPI), Yogyakarta 55861, Indonesia; rifa004@gmail.com
2. Department of Chemistry, Faculty of Mathematic and Natural Science, Islamic University of Indonesia (UII), Yogyakarta 55584, Indonesia; kimiajogja2@gmail.com
3. Department of Biology, Faculty of Science and Technology, Sunan Kalijaga Islamic State University (UIN Sunan Kalijaga), Yogyakarta 55281, Indonesia; lactobacilluspentosus@gmail.com
4. Departement of Biology, Faculty of Science and Technology, Airlangga University, Kampus C, Mulyorejo, Surabaya 60115, Indonesia
5. Biotechnology of Tropical Medicinal Plants Research Group, Airlangga University, Kampus C, Mulyorejo, Surabaya 60115, Indonesia
* Correspondence: andri.frediansyah@lipi.go.id (A.F.); anjar.tri@fst.unair.ac.id (A.T.W.); Tel.: +62-274-392-570 (A.F.); +62-315-936-501 (A.T.W.)

Citation: Frediansyah, A.; Romadhoni, F.; Suryani; Nurhayati, R.; Wibowo, A.T. Fermentation of Jamaican Cherries Juice Using *Lactobacillus plantarum* Elevates Antioxidant Potential and Inhibitory Activity against Type II Diabetes-Related Enzymes. *Molecules* **2021**, *26*, 2868. https://doi.org/10.3390/molecules26102868

Academic Editors: Severina Pacifico and Simona Piccolella

Received: 22 April 2021
Accepted: 10 May 2021
Published: 12 May 2021

Publisher's Note: MDPI stays neutral with regard to jurisdictional claims in published maps and institutional affiliations.

Copyright: © 2021 by the authors. Licensee MDPI, Basel, Switzerland. This article is an open access article distributed under the terms and conditions of the Creative Commons Attribution (CC BY) license (https://creativecommons.org/licenses/by/4.0/).

Abstract: Jamaican cherry (*Muntinga calabura* Linn.) is tropical tree that is known to produce edible fruit with high nutritional and antioxidant properties. However, its use as functional food is still limited. Previous studies suggest that fermentation with probiotic bacteria could enhance the functional properties of non-dairy products, such as juices. In this study, we analyze the metabolite composition and activity of Jamaican cherry juice following fermentation with *Lactobacillus plantarum* FNCC 0027 in various substrate compositions. The metabolite profile after fermentation was analyzed using UPLC-HRMS-MS and several bioactive compounds were detected in the substrate following fermentation, including gallic acid, dihydrokaempferol, and 5,7-dihydroxyflavone. We also found that total phenolic content, antioxidant activities, and inhibition of diabetic-related enzymes were enhanced after fermentation using *L. plantarum*. The significance of its elevation depends on the substrate composition. Overall, our findings suggest that fermentation with *L. plantarum* FNCC 0027 can improve the functional activities of Jamaican cherry juice.

Keywords: Jamaican cherry; fermentation; *Lactobacillus plantarum*; antioxidant; antidiabetic; food nutrition improvement

1. Introduction

Jamaican cherry (*Muntingia calabura* Linn.) is a plant species belonging to Elaeocarpaceae. This plant is indigenous to Southern Mexico, Northern South America, Central America, Trinidad, St. Vincent, and the Greater Antilles [1]. It is also spread and commonly found in other tropical countries including Indonesia, Malaysia, the Philippines, and India [2]. The flower, roots, and barks are used as folk remedies for various medical conditions including fever, liver disease, incipient cold, and antiseptic agents in Southeast Asia [1]. The leaves are known to contain phenolic compounds that exhibits various biological activities such as antimicrobial [3], antioxidant [4], anti-cancer [5], hepatoprotective [6], and hypotensive activities [7]. In addition, Jamaican cherry fruit is edible; they can be eaten raw or processed as juice. Morphologically, the fruit is relatively small, weighted around 1.50 g, round smooth-shape, and the color turns from green to red when ripened. It has a sweet taste due to low titratable acidity and high content of soluble solids. The fruit is also known to have high nutritional value [8] and contain high concentration of flavonoid and phenolic compounds [9]. Due to its taste, nutritional value, and high content of antioxidant compounds, Jamaican cherry fruit has a great potential to be used as a functional food in the food industry.

In previous work, we reported the isolation of lactic acid bacteria from Jamaican cherry fruit [10]. Despite their abundance and diversity, little is known about lactic acid bacteria natural physiological function. Nevertheless, these bacteria are widely known in the food industry due to their probiotic properties [11]. Fermentation using lactic acid bacteria was reported to enhance the flavor [12], shelf life [13], bioavailability [14], and functional quality of the fermented substrates [15]. Our previous study showed that fermentation of black grape juice using *Lactobacilli* could increase juices' functional properties in vitro [16]. The use of *L. pentosus* and *Leuconostoc* in the fermentation of carrot juice also shown to improve the bioavailability of iron in the juice [17]. Another study showed that lactic bacteria and other associated microorganisms in kefir could elevate the α-glucosidase inhibitory effect of *Aronia melanocarpa* fruit juice [18]. The market interest and demands for functional food are continuously increasing [19], aforementioned studies suggest that probiotics bacteria could be used to enhance the functional properties of non-dairy products such as juices. In this study, we use *L. plantarum* strain FNCC 0027 for the fermentation of Jamaican cherry juice. Six different substrate compositions were tested and functional properties of the fermented juices in each of the substrate compositions were reported, including phenolic content, antioxidant capacity, and the inhibition of several type II diabetes-related enzymes. Further, we also provided the metabolite profile of the fermented juice. Here we showed that Jamaican cherry juice fermentation using *L. plantarum* FNCC 0027 can effectively improve functional properties of the juice.

2. Results and Discussion

2.1. Fermentation of Jamaican Juice by L. plantarum

Fermentation using various substrate compositions containing different ratios of Jamaican cherry juice and MRS broth (M1 to M6) was monitored for 48 h. As shown in Figure 1A. the initial viable count of *L. plantarum* was 7.02 log CFU/mL and the initial pH was 6.51. The rapid growth of *L. plantarum* was observed after 4 h of fermentation while the pH of the substrate continuously declined in an opposite trend. The viable count of *L. plantarum* reached a maximum at 24 h and was maintained until 48 h. The highest viable count observed was 9.3 logs CFU/mL at 24 h in M6, followed by M5, M4, M2, M1, and M3 with 8.9, 8.6, 8.4, 8.5, and 8.5 log CFU/mL, respectively (Figure 1A). Our data showed that the growth rate of *L. plantarum* was positively correlated with the amount of MRSB in the substrate. Propagation of viable cells and growth pattern are crucial for the production of desired metabolite in the fermented substrate. Østlie, et al. [20] reported that the production of organic acid, carbon dioxide, and volatile compounds was influenced by the total cell count of probiotic *L. acidophilus*, *L. rhamnosus*, *L. reuteri*, and *Bifidobacterium animalis* in the fermented substrate. On the other hand, *L. plantarum* viable count was negatively correlated with substrate pH, while the viable count is increasing the pH is decreasing over the fermentation period ($p < 0.05$) (Figure 1B). The pH value of the substrate is an essential parameter to assess the progress and endpoint of the fermentation; it was reported to influence the flavor of the final product [21]. After 48 h of fermentation, the pH of the fermented juices has decreased from 6.5 to 3.3 (M1), 3.4 (M2), 4 (M3), 3.7 (M4), 3.4 (M5), and 3.9 (M6). This pattern is quite similar to the 48-h-fermentation of Indonesian black grape juice using *L. plantarum* [16].

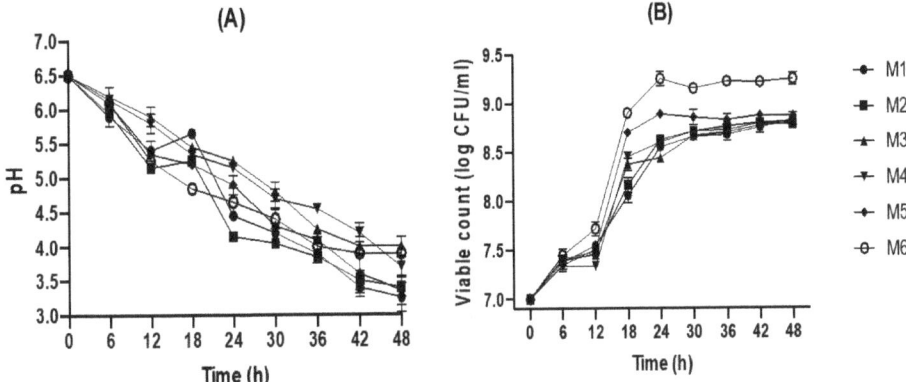

Figure 1. Effect of fermentation on pH and viable count (**A**) the change of pH (**B**) viable count.

2.2. Total Phenolic Content after Fermentation

The fruit of Jamaican cherry are reported to be rich in phenol, flavonoid, and other bioactive substances [8]. Phenols are essential antioxidants in fruit and vegetables, therefore, in this study we evaluated total phenolic content in different substrates formulation (M1 to M6) before and after fermentation. Our data revealed that fermentation using *L. plantarum* can significantly increase total phenolic compounds in M1 and M2 formulation to about 1.3- and 1.4-folds ($p < 0.05$), respectively (Figure 2). This result is in consonance with published studies about fruit fermentation using *L. plantarum*, including in the fermentation of blueberry [22], mulberry [23], and kiwi juice [24]. Changes in total phenolic content in the substrate might occur through multiple factors. For example, changes in substrate pH during fermentation period could affect the structure of the phenolic compound [22]. Alternatively, a macromolecular form of phenol can be disintegrated into small phenols due to specific metabolic activity of the lactic acid bacteria, such as de-glycosylation [23]. *Lactobacillus* fermentation does not always have a positive effect on phenol concentration. Liao, et al. [25] reported that fermentation using *L. brevis* MPL39 can significantly reduce phenolic content in mango juice. Therefore, it should be considered that different lactic acid bacteria strains and substrate matrices could have different effects on the phenolic content.

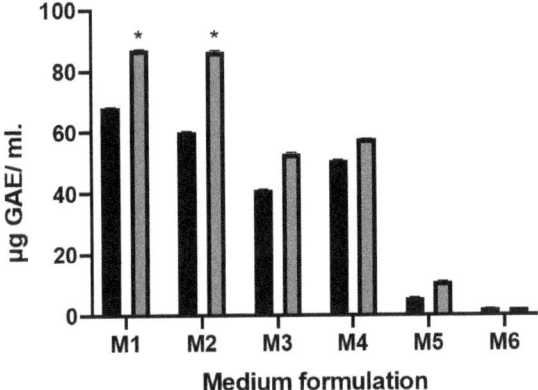

Figure 2. Effect fermentation of *L. plantarum* on the total phenol content. Black bar represent phenol content before fermentation and gray bar represent phenol content after fermentation. Each value represents the mean ± SE (n = 3). The star (*) means significantly different at $p < 0.05$.

2.3. Antioxidative Activity of the Fermented Juices

Antioxidants that present in fermented substrates, such as polyphenol and flavonoid, can eliminate free radicals through electron donation or hydrogen atom transfer. Various types of antioxidative compounds may act against oxidizing agents through diverse mechanisms [26,27]. The antioxidant activity of Jamaican cherry fruit has been reported in previous study [8], which attributed to the presence of polyphenol, flavonoid, and vitamin C in the fruit [28]. Fermentation with lactic acid bacteria could improve the bioavailability of fruits' natural constituents [14]. Here we performed four assays for assessing the antioxidant capacity of fermented Jamaican cherry juice (M1–M6), including DPPH radical, ABTS cation radical, FRAP, and ORAC assay (Figure 3).

The highest DPPH scavenging activity was observed in fermented M1 that contains 100% of Jamaican cherry juice, showing 77.81% DPPH activity, comparable to gallic acid (76.77%) and ascorbic acid (85.57%) that were used as positive control in the experiment. The fermented M1 to M3 substrates exhibited enhanced DPPH scavenging activity compared to their respective non-fermented controls; it was 1.7, 1.8, and 2.2 times higher in fermented M1 to M3, respectively (Figure 3A). Overall, fermented M1 to M3 showed more than 50% of radical scavenging activity after 48 h of fermentation using *L. plantarum*. Similar enhancement in DPPH radical scavenging following fermentation has been reported in previous studies [16,23]. The ABTS radical scavenging activity is also elevated after 48 h fermentation. Similar to DPPH activity the highest ABTS scavenging activity was observed in fermented M1 that contains 100% Jamaican cherry juice, showing 69.25% ABTS activity. Gallic acid and ascorbic acid exhibited 75.43% and 73.45% ABTS activity, respectively. Comparing fermented and non-fermented group, significant increase in ABTS activity was observed in the fermented M1 to M4 ($p \leq 0.05$), where 1.70; 1.46; 1.47; and 1.76-fold increase in ABTS activity was recorded, respectively (Figure 3B). FRAP (Figure 3C) and ORAC (Figure 3D) values also showed a similar trend. M1 and M2 antioxidative activity significantly elevated ($p \leq 0.05$) after fermentation using *L. plantarum*. The FRAP and ORAC values for both M1 and M2 are 1.3 times higher following 48 h fermentation

Generally, we showed that *L. plantarum* could elevate antioxidant activity of Jamaican cherry juice, additionally we observed that substrate containing 100% (M1) and 80% (M2) Jamaican cherry juice could significantly increase ($p \leq 0.05$) antioxidant activity in all tested assays. Suggesting that higher juice content in the substrate correlates with higher antioxidative activity. This results was consistent with our previous study about black grape juices fermentation with *L. plantarum* [16]. Several recent studies also showed the enhancement of antioxidant capacity in fruit juices fermented using lactic acid bacteria, such as in kiwifruit [24], dragon fruit [29], apple [30], and bergamot [31]. Fermentation using lactic acid bacteria could increase the concentration of functional components like phenolic and flavonoids compounds through hydrolysis mechanisms [32]. In addition, the pH change during fermentation could also influence the phytochemical structure of antioxidant compounds [33]. Further, changes in pH could alter the activity of various metabolic enzymes and increase the bioavailability of various compounds [16].

2.4. Enzyme Inhibition Ability

Controlling postprandial hyperglycemia is an essential part of diabetes treatment. This could be achieved by reducing or inhibiting the absorption of glucose during the digestion step. There are numerous pathways to prevent the absorption of glucose, one strategy is by inhibiting α-glucosidase, α-amylase, and amyloglucosidase activity. These three enzymes are essential hydrolase that could metabolize carbohydrates and regulate the blood sugar level in the human body. Various extracts from bacteria, fungi, and plants, including fruit, leaves, roots, and roots were reported to be able to inhibit diabetes-related enzymes activity [16,34]. Recent study showed that fermentation using probiotic bacteria, including lactic acid bacteria, could increase the antidiabetic potential of blueberry juice [22]. Such study is essential for the development of fermentation based functional beverages. Here we reported α-glucosidase, α-amylase, and amyloglucosidase inhibition by different formulations (M1–M6) of Jamaican cherry juice fermentation using *L. plantarum*. Overall, significant increase in the inhibition of α-glucosidase, α-amylase, and amyloglucosidase activity ($p < 0.05$) was observed in M1 to M4 following fermentation for 48 h (Figure 4). The highest increase in inhibition against α-glucosidase (1.9 times) and amyloglucosidase (1.6 times) was recorded in fermented M2, while the highest increase in inhibition against α-amylase (1.9 times) was observed in fermented M1. Acarbose, a well-known antidiabetic drug, a positive control in this study, showed higher but comparable inhibitory properties against those three enzymes (Figure 4). These results indicate that fermentation of Jamaican cherry fruit juice using *L. plantarum* can improve the inhibitory activities of the juice and their antidiabetic potential. The increase in inhibitory activities of plant extract against diabetes-related enzymes following *L. plantarum* fermentation is in agreement with previous studies [16,22,35]. Further, the cell-free filtrate of *L. plantarum* X1 was also reported to be able to inhibit α-glucosidase activity [36].

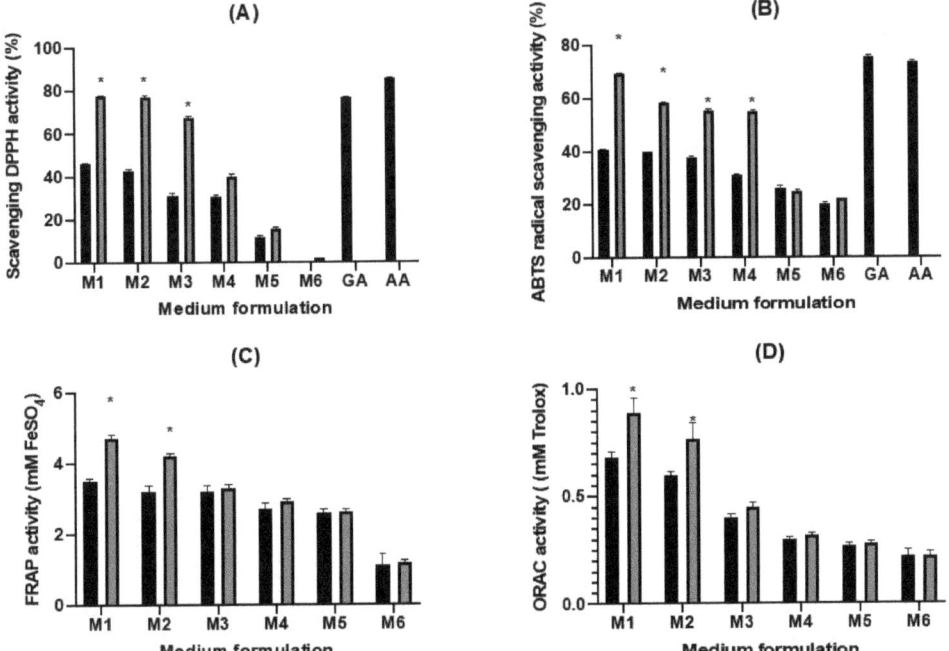

Figure 3. The effect of fermentation by *L. plantarum* on the antioxidant activities of Jamaican cheery juices. (**A**) Scavenging DPPH activity (**B**) ABTS Scavenging radical activity (**C**) FRAP activity (**D**) ORAC activity. Black bar represent antioxidant activities before fermentation and gray bar represent antioxidant activities after fermentation. Each value represents the mean ± SE (n = 3). The star (*) means significantly different at $p < 0.05$.

Structural changes of indigenous compounds following fermentation might contribute to the inhibition activity. It is reported that fermentation of *Momordica charantia* (bitter melon) juice using *L. plantarum* BET003 results in the transformation of aglycone compounds and improved antidiabetic potential of the fermented juice [37]. In addition, fermentation might enhance the activity various proteolytic enzymes that can inhibit amylase and amyloglucosidase activity, such as protamex, alcalase, neutrase, and flavourzyme [16].

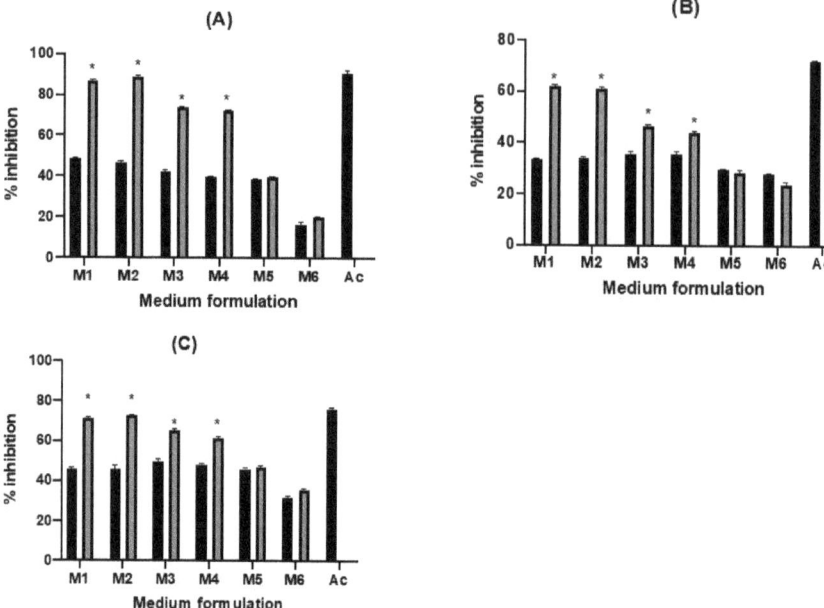

Figure 4. Effect fermentation of L. plantarum to the inhibition of (**A**) α-glucosidase, (**B**) α-amylase, and (**C**) amyloglucosidase. Black bar represent inhibitory activities before fermentation and gray bar represent inhibitory activities after fermentation M1 to M6 means medium formulation and Ac means Acarbose 5 mg/mL. Each value represents the mean ± SE (n = 3). The star (*) means significantly different at $p < 0.05$.

2.5. Mass Spectrometry Analysis of M1 Substrate

Our results showed that total phenolic content, antioxidant activities, and inhibitory activity against diabetes-related enzymes was highest in substrate with 100% Jamaican cherry juice. To evaluate whether those activities are correlated with changes in metabolite composition of the fermented juice, we analyzed the metabolite profile of the fermented and unfermented M1 using UHPLC-QTOF-HRMS/MS. The electrospray ionization mode was set into a negative mode, three more peaks were detected in the fermented M1 namely A, B, and C (Figure 5). We then annotate the putative metabolites A, B, and C using molecular formula and the fragmentation pattern against database, aided with spectral library search together with the suggested fragmentation trees using SIRIUS (https://bio.informatik.uni-jena.de/software/sirius/ [accessed 20 June 2020]). We identified the three putative metabolites of fermented Jamaican fruit juice, including gallic acid (A), dihydrokaempferol (B), and 5,7-dihydroxyflavone (C) (Table 1 and Figures S1–S3).

A previous study reported that L. plantarum strain CIR1 could produce gallic acid during fermentation. This bacteria could produce tannase, an enzyme that could facilitate bioconversion of tannic acid to gallic acid [38]. L. plantarum strain FNCC 0027 used in this study might also produce tannase that can hydrolyze tannic acid presence in Jamaican fruits [8,39]. Recent study reported that fermentation of Solanum retroflexum leaf using L. plantarum strain 75 could induce the production of various bioactive compounds, such as gallic acid, vanillic, coumaric, ellagic acid, quercetin, and catechin [40]. Fermentation with L. plantarum FNCC 0027 also induces the production of bioactive compounds, including gallic acid, dihydrokaempferol, and 5,7-dihydroxyflavone. The production of dihydrokaempferol possibly due to the L. plantarum shikimic pathway that can convert flavonoid naringenin presents in Jamaican fruit into dihydrokaempferol [8]. Hydroxylation of naringenin by hydroxylase enzyme known to be produced by L. plantarum could produce intermediate compounds from dihydroflavonols class, such as dihydrokaempferol and dihydroxyflavone [41–43]. Gallic acid and flavonoids such as dihydrokaempferol and 5,7-dihydroxyflavone is known to have antidiabetic and antioxidant properties [44–46]. Thus, their presence in the fermented substrate can influence the functional properties of the Jamaican cherry juice.

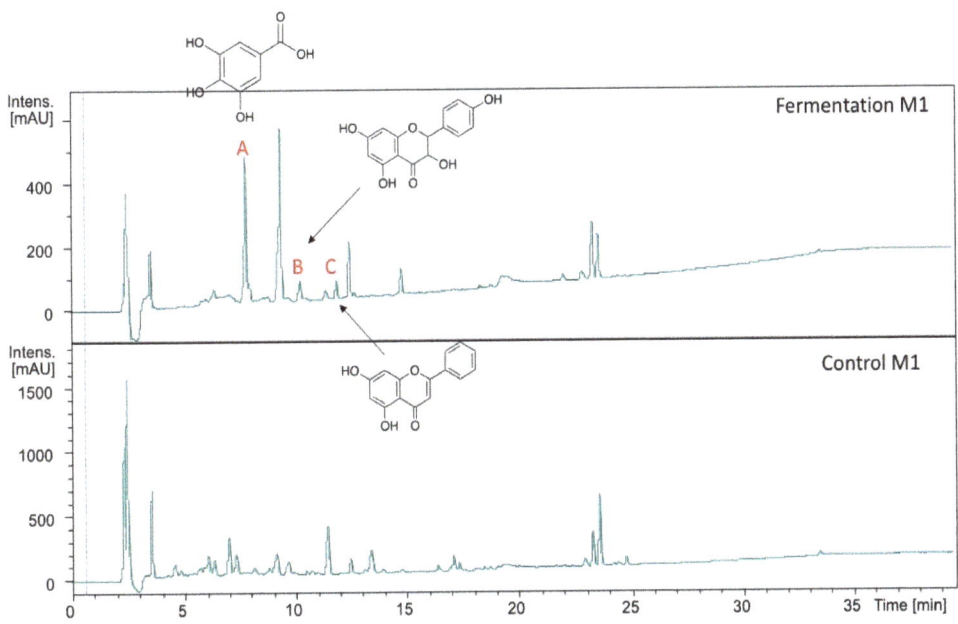

Figure 5. The UV chromatogram of M1 containing *L. plantarum* (**above**) and M1 without *L. plantarum* (**below**).

Table 1. Chemical constituent identified after fermentation of M1 using *L. plantarum*.

Peak	Putative Compound	Experiment (m/z)	Theoretical (m/z)	Adduct	Error (ppm)	rdb
A	gallic acid	169.0141	169.0142	[M-H]⁻	0.7	5
B	dihydrokaempferol	287.0561	287.0561	[M-H]⁻	−0.1	10
C	5,7-dihydroxyflavone	253.0505	23.0506	[M-H]⁻	0.5	11

rdb means ring double bond.

3. Materials and Methods

3.1. Chemical and Reagents

The chemicals and reagents used are p-nitrophenyl-α-D-glucopyranoside (Calbiochem®, San Diego, MA, USA), starch (Merck, Darmstadt, Germany), acarbose (Fluka Analytical, Sigma–Aldrich, Laramie, WY, USA), Man Rogosa Sharpe/MRS broth (Sigma–Aldrich, St. Louis, MO, USA), α-glucosidase from *Saccharomyces cerevesiae* (Sigma–Aldrich, St. Louis, MO, USA), α-amylase from *Aspergillus oryzae* (Sigma–Aldrich, St. Louis, MO, USA), amyloglucosidase from *Aspergillus niger* (Sigma–Aldrich, St. Louis, MO, USA), 2,2-Diphenyl-1-picrylhydrazyl (Sigma–Aldrich, St. Louis, MO, USA), 2,4,6-Tris(2-pyridyl)-s-triazine (Fluka Analytical, Sigma–Aldrich, St. Louis, MO, USA), 3,5-dinitrosalicylic acid (Sigma–Aldrich, St. Louis, MO, USA), 2,2′-Azino-bis(3-ethylbenzothiazoline-6-sulfonic acid) (Sigma–Aldrich, St. Louis, MO, USA), Folin–Ciocalteu's phenol reagent (Merck, Darmstadt, Germany), gallic acid, peroxidase. Other reagents were analytical grade or better.

3.2. Bacterial Strain and Fruit Material

Lactobacillus plantarum FNCC 0027 was obtained from the culture collection of Biotechnology Laboratory, Graduate School of Biotechnology, Universitas Gadjah Mada, Yogyakarta, Indonesia. Jamaican cherry fruits were collected from the research field at Research Division for Natural Product Technology (BPTBA), the Indonesian Institute of Sciences (LIPI), Yogyakarta, Indonesia in March 2016. The fruits (250 g) were washed and kept at 4 °C before further use. Jamaican cherry juice was produced from the fruit using sterilized-commercial food juicer at room temperature.

3.3. Preparation of Inoculant

L. plantarum FNCC 0027 were propagated in MRS broth at 37 °C for 48 h under anaerobic conditions. The cells were harvested, pelletized, and re-suspended in sterilized phosphate buffer saline at 7.02 log CFU/mL for inoculation.

3.4. Fermentation Procedure

First, Jamaican cherry juice was produced by homogenizing Jamaican cherry fruits with sterile water in a commercial blender at the concentration of 1 g/mL. The homogenized juices were filtered and pasteurized for 3 min at 85 °C [20]. Fermentation was performed in 250 mL sterilized Erlenmeyer. Erlenmeyer was filled with pasteurized Jamaican cherry juice (J) and sterilized MRS broth (M) with a total reaction volume of 100 mL. Six different J to M ratios were used in this study (in a total of 100 mL), including M1 (1:0), M2 (4:1), M3 (3:1), M4 (1:1), M5 (1:4), and M6 (0:1). Each formulation was subjected to two treatments: control non-fermented group and group fermented using L. plantarum (with an initial starter of 7.02 log CFU/mL). All samples were then incubated aerobically at 37 °C, 100 rpm, for 48 h before harvested for subsequent analysis.

3.5. Bacterial Cell Separation

To obtain cell-free supernatant, the fermented juices were centrifuged for 15 min at 10,000× g [47]. The resulting supernatant was then filtered through a 0.22 µm membrane filter (Millipore, Burlington, MA, USA) and kept at −20 °C.

3.6. Bacterial Viable Count

Viable count was performed at different intervals over the course of 48 h fermentation (0, 6, 12, 18, 24, 30, 36, 42, and 48 h). To perform viable count, 10 sterile 15mL test tubes were filled with 9 mL of sterilized phosphate buffer saline (PBS). Serial 10-fold dilutions in PBS were then prepared in the test tubes, using 1 mL of fermented juice as the starter. Viable count was next performed using pour plate method in duplicate, 1 mL of solution from each dilution series was mixed with 25 mL tempered (47 °C) Plate Count Agar (OXOID®, Basingstoke, England). The plate was incubated for 72 h at 30 °C ± 2 °C, and colony numbers was calculated using a colony counter. Plates with 15 to 300 colonies were considered for colony forming unit calculation. The viable colonies were converted into weighted mean colony forming units per milliliter (CFU/mL) using the following Equation (1):

$$N = \sum C/[(n_1 + 0.1 n_2)d] \tag{1}$$

N is the number of colonies in the plate; $\sum C$ is the sum of plates containing 15 to 300 colonies; n_1 is the number of plates retained in the first dilution; n_2 is the number of plates retained in the second dilution; and d is the first dilution factor. The viable colonies were then converted into log CFU/mL.

3.7. pH Measurement

The pH of the samples was measured using Eutech PC 700 pH meter (Thermo Scientific, Waltham, MA, USA). The pH measurement was performed at different intervals over the course of 48 h fermentation (0, 6, 12, 18, 24, 30, 36, 42, and 48 h).

3.8. Total Phenolic Content Measurement

Total phenolic content was determined using methods developed by Zhou, Wang, Zhang, Yang, Sun, Zhang, and Yang [24], with minor modifications. Briefly, 25 μL of 0.2 N Folin–Ciocalteu phenol reagents was added into 96-well plates containing 5 μL of each supernatant and 195 μL of distilled water. The mixture was then incubated for 6 min at room temperature in darkness, followed by addition of 75 μL of 7% sodium carbonate. The mixture was then incubated further for 2 h at room temperature in darkness. Blank solution was prepared by the same steps as described above except that supernatant was substituted with 5 μL of water. For all samples, the absorbance at 765 nm was recorded using Multiskan® Go microplate spectrophotometer (Thermo Scientific, Vantaa, Finland). Standard curve was produced using Gallic acid, the equation for the standard curve is $y = 0.0104x - 0.0159$ ($R^2 = 0.9988$) where y is the absorbance at 735 nm and x is the concentration of Gallic acid in μg/mL. Total phenol was expressed as μg of Gallic acid equivalent (GAE) per mL of supernatant (μg GAE/mL).

3.9. 2,2-Diphenyl-1-picrylhydrazyl (DPPH) Radical Scavenging Activity

The DPPH scavenging assay was performed using 96-well plates following the method by Kosem, et al. [48], with modifications. Briefly, 50 μL of filtered supernatant was mixed with 70 μL of methanol and the absorbance of the pre-plate reading was recorded at 517 nm. About 80 μL of 0.5 mM DPPH solution in methanol was then added to the well. The degree of purple (from DPPH) decolorization to yellow represents the scavenging efficiency of the supernatants. After an incubation period of 30 min at room temperature (25 ± 2 °C) in the darkness, the decrease in the absorbance was recorded at 517 nm using Multiskan® Go microplate spectrophotometer (Thermo Scientific, Vantaa, Finland). Lower absorbance represents higher free radical-scavenging activity. The scavenging activity against DPPH was calculated using following Equation (2):

$$\text{DPPH scavenging rate (\%)} = [1 - (Abs_1 - Abs_0)] \times 100\% \qquad (2)$$

Abs_0 was absorbance of control and Abs_1 was the absorbance in the presence of supernatant.

3.10. 2,2′-Azino-bis(3-ethylbenzothiazoline-6-sulfonic acid) (ABTS) Assay

The ABTS assay was performed according to Re et al. [49]. In brief, radical cation was generated by reacting 5 mL of 7 mM ABTS with 5 mL of 2.45 mM of potassium persulfate. The reaction was performed by incubating the mixture for 16 h in the dark. Working solution for radical cation was obtained by diluting the reacted solution at the OD 0.7 ± 0.02 at 734 nm. Subsequently, 300 μL of the diluted radical cation solution was mixed with 3 μL of supernatant. The ODs was then recorded at 734 nm after 10 min incubation at room temperature using Multiskan® Go microplate spectrophotometer (Thermo Scientific, Vantaa, Finland). The ABTS radical scavenging activity was calculated using Equation (2).

3.11. Ferric Reducing-Antioxidant Power (FRAP) Assay

The FRAP assay was conducted according to Cecchini and Fazio [50], which was originally described by Benzie and Strain [51], with minor modifications. In brief, FRAP reagent was prepared by mixing 300 mM sodium acetate buffer (pH 3.6), 10 mM of 2,4,6-Tris(2-pyridyl)-s-triazine (TPTZ) in 40 mM HCl, and 20 mM $FeCl_3 \cdot 6H_2O$ at 10:1:1 (*v/v/v*) ratio. Subsequently, 10 μL of supernatant was added to the 96-well plate containing 300 μL FRAP solution as an oxidizing reagent. After 5 min incubation in dark at 37 °C, the absorbance was measured at 593 nm using Tecan Infinite® 200 Pro microplate reader. The experiment was calibrated with $FeSO_4 \cdot 7H_2O$ and the results were expressed in terms of $FeSO_4 \cdot 7H_2O$ equivalents (μM).

3.12. Oxygen Radical Absorbance Capacity (ORAC) Assay

ORAC were measured according to Zulueta, et al. [52]. The 2,2′-Azobis(2-amidinopropane) dihydrochloride (AAPH) radical stock solution was prepared freshly by adding 434 mg of AAPH to 10 mL of phosphate buffer (75 mM) to obtain final concentration of 161 mmol/L. Fluorescein stock solution (1.03 mmol/L) was prepared using phosphate buffer solution. Subsequently, the supernatant (25 µL) was added to the 96-well plate and mixed with 150 µL of fluorescein solution (40 nm/L). Mixture was then incubated for 5 min at 37 °C. Subsequently, 25 µL AAPH solutions was added and the fluorescein was recorded immediately at an excitation wavelength of 485 nm and emission wavelength of 535 nm. The fluorescein was monitored using Tecan Infinite® 200 Pro microplate reader every minute for 30 min, ORAC values were expressed in term of Trolox equivalents (µM).

3.13. In Vitro Inhibiting Activity of α-Glucosidase

The α-glucosidase inhibitory activity of the fermented supernatant was determined according to the chromogenic method, in a 96-well plate, according to our previous published method [16]. First, 20 µL of supernatant was mixed with 10 µL of a 1.0 U/mL α-glucosidase and 50 µL of 0.1 M sodium phosphate buffer (pH 6.9). The mixed solution was incubated at 37 °C for 15 min. After pre-incubation, the enzymatic reaction was initiated by adding 20 µL of 5 mM p-nitrophenyl-α-D-glucopyranoside solution in 0.1 M sodium phosphate buffer (pH 6.9). The mixture was incubated for 20 min at 37 °C. The absorbance was subsequently measured at 405 nm using Multiskan® Go microplate spectrophotometer (Thermo Scientific, Vantaa, Finland). Percent inhibition was calculated relative to the diabetes drug acarbose as the reference. Reaction system without supernatant was used as negative control while reaction system without α-glucosidase was used as a blank for correcting the background absorbance. The percentage of enzymatic inhibition activity was calculated using following Equation (3):

$$\% \text{ inhibition activity} = [(\text{Abs}_A - \text{Abs}_B)/\text{Abs}_A] \times 100\% \quad (3)$$

Abs_A is the absorbance of the control and Abs_B is the absorbance of the tested supernatant.

3.14. In Vitro Inhibiting Activity of α-Amylase

The α-amylase inhibitory activities of the fermented supernatant were carried out according to procedure reported by Telagari and Hullatti [53]. The assay system was carried out in a 96-well plate. First, a reaction mixture containing 10 µL of a 2.0 unit/mL α–amylase, 50 µL sodium phosphate buffer (0.1 M, pH = 6.9), and 20 µL of supernatant was prepared. The mixed solution was incubated at 37 °C for 20 min. After pre-incubation, 50 µL of 1% soluble starch (Merck, Darmstadt, Germany) in 0.1 M sodium phosphate buffer pH 6.9 was added as a substrate and the mixture was incubated further for 30 min at 37 °C. Next, 100 µL of the 3.5-dinitrosalicylic acid solution was added and heated at 100 °C in a water bath for 10 min. The absorbance was subsequently measured at 540 nm using Multiskan® Go microplate spectrophotometer (Thermo Scientific, Vantaa, Finland). Acarbose was used as a positive reference standard. The percentage of α-amylase inhibition was calculated using Equation (2).

3.15. In Vitro Inhibiting Activity of Amyloglucosidase

The amyloglucosidase (exo-1,4-α-glucosidase) inhibitory activities of the fermented supernatant was carried out in 96-well plate according to Saul et al. [54] and Warren et al. [55], with modifications. First, a mixture containing 10 µL of a 1.0 U/mL α-amyloglucosidase, 10 µL of 0.1 M sodium acetate (pH 5.0), and 25 µL of supernatant was prepared. The mixed solution was incubated at 37 °C for 20 min. After pre-incubation, the enzymatic reaction was initiated by adding 5 µL of 5 mM p-nitrophenyl-α-D-glucopyranoside solution in 0.1 M sodium phosphate buffer (pH 6.9). At the end of the incubation, 200 µL of 0.4 mM glycine buffer (pH 10.4) was added to each well to stop reaction. The p-nitrophenil released was then measured at 410 nm using Tecan Infinite® 200 Pro microplate reader.

3.16. UPLC-HRMS-MS Analysis

High resolution MS was carried out using Bruker maXis 4G ESI time of flight mass-spectrometer (Bruker Daltonics, Bremen, Germany) attached to an Ultimate 3000 HPLC (Thermo Fisher Scientific, Bremen, Germany). The UHPLC-method was performed using Reprosil 3 μm C18 100 Å, 10 × 3.3 mm (flow rate of 0.3 mL/min, monitored at 210 and 240 nm) with linear gradient starts from 90% to 0% of A (A: ddH$_2$O, B: Acetonitrile, both solvents containing 0.01% formic acid) for 30 min and held constant of 100% B for 10 min. The parameter was set in a capillary voltage of 4500 V, nebulizer nitrogen pressure of 2 bars, the dry gas flow of 9 L/min source temperature, ion source temperature 200 °C, and spectral rate of 3 Hz. The MS data subsequently analyzed using Bruker Compass Data Analysis 4.4 SR1(x64).

3.17. Statistical Analysis

All assays were conducted in triplicate. The mean and standard error (mean ± SE) was determined of each data. The analysis of variance (ANOVA) and student t-test was employed using SPSS 16 to determine the level of statistical differences between control and Jamaican juice formulation fermented with *L. plantarum*. Differences at $p < 0.05$ were considered statistically significant.

4. Conclusions

Our results demonstrated that fermentation of Jamaican cherry juice using *L. plantarum* FNCC 0027 could significantly improve the functional activities of the substrates, including higher polyphenol content, antioxidant capacity (DPPH, ABTS, FRAP, and ORAC), and inhibitory activity against diabetic-related enzymes (α-glucosidase, α-amylase, and amyloglucosidase). Various fermentation formulations with different ratios of Jamaican cherry juice and MRSB volume were studied in this study and we found that M1 formulation with 100% Jamaican cherry juice showed the highest functional activities compared to the other composition. This result indicates that the addition of MRSB is not required to facilitate effective fermentation and improvement of Jamaican cherry juice functional properties. In summary, the functional properties and beneficial activity of Jamaican cherry juice can be enhanced by fermentation with *L. plantarum*, one of "generally regarded as safe" (GRAS) strain. This work can be used as the basis for the development of new functional beverages using fermented Jamaican cherry juice as the main component.

Supplementary Materials: The following are available online, Figure S1: Peak A found in fermented Jamaican cherry juice with a negative HR-ESI-TOF-MS m/z 235.0506 [M-H]$^-$; Figure S2: Peak B found in fermented Jamaican cherry juice with a negative HR-ESI-TOF-MS m/z 169.0140 [M-H]$-$; Figure S3: Peak C found in fermented Jamaican cherry juice with a negative HR-ESI-TOF-MS m/z 287.0561 [M-H]$^-$.

Author Contributions: Conceptualization, formal analysis, writing—original draft, investigation, validation A.F.; methodology, data curation, data analysis F.R.; methodology, visualization, software S.; project administration, resources, R.N.; project administration, funding acquisition, supervision, writing—review and editing, A.T.W. All authors have read and agreed to the published version of the manuscript.

Funding: This research was funded by the DIPA -Indonesian Institute of Science (LIPI) and Research Group Grant from Universitas Airlangga (Grant numbers 458 343/UN3.14/PT/2020).

Institutional Review Board Statement: Not applicable.

Informed Consent Statement: Not applicable.

Data Availability Statement: Publicly available datasets were analyzed in this study. This data can be found at: https://www.kaggle.com/microbiologii/dataset-plantarum (accessed on 20 June 2020).

Acknowledgments: The authors thank Gross for provided us equipment and facilities in the Pharmaceutical Institute, Eberhard Karls Universität Tübingen, Germany, to conduct some assays and mass spectrometry.

Conflicts of Interest: We declare that we have no known competing financial interests or personal relationships that could have appeared to influence the work reported in this paper. We also declare that there is no conflict of interest in the submission of the manuscript and the manuscript is approved by all authors for publication.

Sample Availability: Samples of fermented extracts are not available from the authors.

References

1. Mahmood, N.; Nasir, N.; Rofiee, M.; Tohid, S.M.; Ching, S.; Teh, L.; Salleh, M.; Zakaria, Z. *Muntingia calabura*: A review of its traditional uses, chemical properties, and pharmacological observations. *Pharm. Biol.* **2014**, *52*, 1598–1623. [CrossRef]
2. Harshini, V.; Gayathri, H.S.; Padmaja, A. Development of *Muntingia calabura* Fruit Based Squash. *Asian J. Dairy Food Res.* **2020**, *9*, 256–260.
3. Buhian, W.P.C.; Rubio, R.O.; Valle, D.L., Jr.; Martin-Puzon, J.J. Bioactive metabolite profiles and antimicrobial activity of ethanolic extracts from *Muntingia calabura* L. leaves and stems. *Asian Pac. J. Trop. Biomed.* **2016**, *6*, 682–685. [CrossRef]
4. Zolkeflee, N.K.Z.; Isamail, N.A.; Maulidiani, M.; Abdul Hamid, N.A.; Ramli, N.S.; Azlan, A.; Abas, F. Metabolite variations and antioxidant activity of *Muntingia calabura* leaves in response to different drying methods and ethanol ratios elucidated by NMR-based metabolomics. *Phytochem. Anal.* **2021**, *32*, 69–83. [CrossRef] [PubMed]
5. Sufian, A.S.; Ramasamy, K.; Ahmat, N.; Zakaria, Z.A.; Yusof, M.I.M. Isolation and identification of antibacterial and cytotoxic compounds from the leaves of *Muntingia calabura* L. *J. Ethnopharmacol.* **2013**, *146*, 198–204. [CrossRef]
6. Zakaria, Z.A.; Mahmood, N.D.; Mamat, S.S.; Nasir, N.; Omar, M.H. Endogenous Antioxidant and LOX-Mediated Systems Contribute to the Hepatoprotective Activity of Aqueous Partition of Methanol Extract of *Muntingia calabura* L. Leaves against Paracetamol Intoxication. *Front. Pharmacol.* **2018**, *8*, 982. [CrossRef]
7. Shih, C.-D.; Chen, J.-J.; Lee, H.-H. Activation of nitric oxide signaling pathway mediates hypotensive effect of *Muntingia calabura* L. (Tiliaceah) leaf extract. *Am. J. Chin. Med.* **2006**, *34*, 857–872. [CrossRef]
8. Pereira, G.A.; Arruda, H.S.; de Morais, D.R.; Eberlin, M.N.; Pastore, G.M. Carbohydrates, volatile and phenolic compounds composition, and antioxidant activity of calabura (*Muntingia calabura* L.) fruit. *Food Res. Int.* **2018**, *108*, 264–273. [CrossRef]
9. Recuenco, M.C.; Lacsamana, M.S.; Hurtada, W.A.; Sabularse, V.C. Total phenolic and total flavonoid contents of selected fruits in the Philippines. *Philipp. J. Sci.* **2016**, *145*, 275–281.
10. Frediansyah, A.; Nurhayati, R.; Sholihah, J. Lactobacillus pentosus isolated from *Muntingia calabura* shows inhibition activity toward alpha-glucosidase and alpha-amylase in intra and extracellular level. In *IOP Conference Series: Earth and Environmental Science*; IOP Publishing: Bristol, UK, 2019; p. 012045.
11. Zielińska, D.; Kołożyn-Krajewska, D. Food-origin lactic acid bacteria may exhibit probiotic properties. *Biomed. Res. Int.* **2018**, *2018*. [CrossRef]
12. Chen, C.; Lu, Y.; Yu, H.; Chen, Z.; Tian, H. Influence of 4 lactic acid bacteria on the flavor profile of fermented apple juice. *Food Biosci.* **2019**, *27*, 30–36. [CrossRef]
13. Parada, J.L.; Caron, C.R.; Medeiros, A.B.P.; Soccol, C.R. Bacteriocins from lactic acid bacteria: Purification, properties and use as biopreservatives. *Braz. Arch. Biol. Technol.* **2007**, *50*, 512–542. [CrossRef]
14. Sharma, N.; Angural, S.; Rana, M.; Puri, N.; Kondepudi, K.K.; Gupta, N. Phytase producing lactic acid bacteria: Cell factories for enhancing micronutrient bioavailability of phytate rich foods. *Trends Food Sci. Technol.* **2020**, *96*, 1–12. [CrossRef]
15. Szutowska, J. Functional properties of lactic acid bacteria in fermented fruit and vegetable juices: A systematic literature review. *Eur. Food Res. Technol.* **2020**, *246*, 357–372. [CrossRef]
16. Frediansyah, A.; Nurhayati, R.; Romadhoni, F. Enhancement of antioxidant activity, α-glucosidase and α-amylase inhibitory activities by spontaneous and bacterial monoculture fermentation of Indonesian black grape juices. In *AIP Conference Proceedings*; AIP Publishing LLC: Melville, NY, USA, 2017; p. 020022.
17. Bergqvist, S.W.; Andlid, T.; Sandberg, A.-S. Lactic acid fermentation stimulated iron absorption by Caco-2 cells is associated with increased soluble iron content in carrot juice. *Br. J. Nutr.* **2006**, *96*, 705–711.
18. Du, X.; Myracle, A.D. Fermentation alters the bioaccessible phenolic compounds and increases the alpha-glucosidase inhibitory effects of aronia juice in a dairy matrix following in vitro digestion. *Food Funct.* **2018**, *9*, 2998–3007. [CrossRef] [PubMed]
19. Siro, I.; Kápolna, E.; Kápolna, B.; Lugasi, A. Functional food. Product development, marketing and consumer acceptance—A review. *Appetite* **2008**, *51*, 456–467. [CrossRef]
20. Østlie, H.M.; Helland, M.H.; Narvhus, J.A. Growth and metabolism of selected strains of probiotic bacteria in milk. *Int. J. Food Microbiol.* **2003**, *87*, 17–27. [CrossRef]
21. McFeeters, R. Fermentation microorganisms and flavor changes in fermented foods. *J. Food Sci.* **2004**, *69*, FMS35–FMS37. [CrossRef]
22. Zhang, Y.; Liu, W.; Wei, Z.; Yin, B.; Man, C.; Jiang, Y. Enhancement of functional characteristics of blueberry juice fermented by Lactobacillus plantarum. *LWT* **2021**, *139*, 110590. [CrossRef]

23. Kwaw, E.; Ma, Y.; Tchabo, W.; Apaliya, M.T.; Wu, M.; Sackey, A.S.; Xiao, L.; Tahir, H.E. Effect of lactobacillus strains on phenolic profile, color attributes and antioxidant activities of lactic-acid-fermented mulberry juice. *Food Chem.* **2018**, *250*, 148–154. [CrossRef] [PubMed]
24. Zhou, Y.; Wang, R.; Zhang, Y.; Yang, Y.; Sun, X.; Zhang, Q.; Yang, N. Biotransformation of phenolics and metabolites and the change in antioxidant activity in kiwifruit induced by Lactobacillus plantarum fermentation. *J. Sci. Food Agric.* **2020**, *100*, 3283–3290. [CrossRef]
25. Liao, X.-Y.; Guo, L.-Q.; Ye, Z.-W.; Qiu, L.-Y.; Gu, F.-W.; Lin, J.-F. Use of autochthonous lactic acid bacteria starters to ferment mango juice for promoting its probiotic roles. *Prep. Biochem. Biotechnol.* **2016**, *46*, 399–405. [CrossRef] [PubMed]
26. Pisoschi, A.M.; Pop, A.; Cimpeanu, C.; Predoi, G. Antioxidant capacity determination in plants and plant-derived products: A review. *Oxidative Med. Cell. Longev.* **2016**, *2016*. [CrossRef]
27. Santos-Sánchez, N.F.; Salas-Coronado, R.; Villanueva-Cañongo, C.; Hernández-Carlos, B. *Antioxidant Compounds and Their Antioxidant Mechanism*; IntechOpen: London, UK, 2019.
28. Preethi, K.; Vijayalakshmi, N.; Shamna, R.; Sasikumar, J. In vitro antioxidant activity of extracts from fruits of *Muntingia calabura* Linn from India. *Pharmacogn. J.* **2010**, *2*, 11–18. [CrossRef]
29. Muhialdin, B.J.; Kadum, H.; Zarei, M.; Hussin, A.S.M. Effects of metabolite changes during lacto-fermentation on the biological activity and consumer acceptability for dragon fruit juice. *LWT* **2020**, *121*, 108992. [CrossRef]
30. Wu, C.; Li, T.; Qi, J.; Jiang, T.; Xu, H.; Lei, H. Effects of lactic acid fermentation-based biotransformation on phenolic profiles, antioxidant capacity and flavor volatiles of apple juice. *LWT* **2020**, *122*, 109064. [CrossRef]
31. Hashemi, S.M.B.; Jafarpour, D. Fermentation of bergamot juice with Lactobacillus plantarum strains in pure and mixed fermentations: Chemical composition, antioxidant activity and sensorial properties. *LWT* **2020**, *131*, 109803. [CrossRef]
32. Hur, S.J.; Lee, S.Y.; Kim, Y.-C.; Choi, I.; Kim, G.-B. Effect of fermentation on the antioxidant activity in plant-based foods. *Food Chem.* **2014**, *160*, 346–356. [CrossRef]
33. Friedman, M.; Jürgens, H.S. Effect of pH on the stability of plant phenolic compounds. *J. Agric. Food Chem.* **2000**, *48*, 2101–2110. [CrossRef]
34. Nurhayati, R.; Frediansyah, A.; Rachmah, D.L. Lactic Acid Bacteria Producing Inhibitor of Alpha Glucosidase Isolated from Ganyong (Canna Edulis) and Kimpul (Xanthosoma sagittifolium). In *IOP Conference Series: Earth and Environmental Science*; IOP Publishing: Bristol, UK, 2017; p. 012009.
35. Wang, Z.; Hwang, S.H.; Lee, S.Y.; Lim, S.S. Fermentation of purple Jerusalem artichoke extract to improve the α-glucosidase inhibitory effect in vitro and ameliorate blood glucose in db/db mice. *Nutr. Res. Pract.* **2016**, *10*, 282. [CrossRef] [PubMed]
36. Li, X.; Wang, N.; Yin, B.; Fang, D.; Zhao, J.; Zhang, H.; Wang, G.; Chen, W. Lactobacillus plantarum X1 with α-glucosidase inhibitory activity ameliorates type 2 diabetes in mice. *RSC Adv.* **2016**, *6*, 63536–63547. [CrossRef]
37. Mazlan, F.A.; Annuar, M.S.M.; Sharifuddin, Y. Biotransformation of Momordica charantia fresh juice by Lactobacillus plantarum BET003 and its putative antidiabetic potential. *PeerJ* **2015**, *3*, e1376. [CrossRef] [PubMed]
38. Aguilar-Zarate, P.; Cruz, M.A.; Montañez, J.; Rodriguez-Herrera, R.; Wong-Paz, J.E.; Belmares, R.E.; Aguilar, C.N. Gallic acid production under anaerobic submerged fermentation by two bacilli strains. *Microb. Cell Factories* **2015**, *14*, 1–7. [CrossRef] [PubMed]
39. Krishnaveni, M.; Dhanalakshmi, R. Qualitative and quantitative study of phytochemicals in *Muntingia calabura* L. leaf and fruit. *World J. Pharm. Res.* **2014**, *3*, 1687–1696.
40. Degrain, A.; Manhivi, V.; Remize, F.; Garcia, C.; Sivakumar, D. Effect of lactic acid fermentation on color, phenolic compounds and antioxidant activity in African nightshade. *Microorganisms* **2020**, *8*, 1324. [CrossRef] [PubMed]
41. Zha, J.; Koffas, M.A. Production of anthocyanins in metabolically engineered microorganisms: Current status and perspectives. *Synth. Syst. Biotechnol.* **2017**, *24*, 259–266. [CrossRef]
42. Rodríguez, H.; Landete, J.M.; de las Rivas, B.; Muñoz, R.J.F.C. Metabolism of food phenolic acids by Lactobacillus plantarum CECT 748T. *Food Chem.* **2008**, *107*, 1393–1398. [CrossRef]
43. Zha, J.; Wu, X.; Gong, G.; Koffas, M.A. Pathway enzyme engineering for flavonoid production in recombinant microbes. *Metab. Eng. Commun.* **2019**, *9*, e00104. [CrossRef]
44. Adefegha, S.A.; Oboh, G.; Ejakpovi, I.I.; Oyeleye, S.I. Antioxidant and antidiabetic effects of gallic and protocatechuic acids: A structure–function perspective. *Comp. Clin. Pathol.* **2015**, *24*, 1579–1585. [CrossRef]
45. Al-Salih, R.M. Clinical experimental evidence: Synergistic effect of Gallic acid and tannic acid as Antidiabetic and antioxidant agents. *Thi-Qar Med. J.* **2010**, *4*, 109–119.
46. Punithavathi, V.R.; Prince, P.S.M.; Kumar, R.; Selvakumari, J. Antihyperglycaemic, antilipid peroxidative and antioxidant effects of gallic acid on streptozotocin induced diabetic Wistar rats. *Eur. J. Pharmacol.* **2011**, *650*, 465–471. [CrossRef] [PubMed]
47. Motevaseli, E.; Shirzad, M.; Raoofian, R.; Hasheminasab, S.-M.; Hatami, M.; Dianatpour, M.; Modarressi, M.-H. Differences in vaginal lactobacilli composition of Iranian healthy and bacterial vaginosis infected women: A comparative analysis of their cytotoxic effects with commercial vaginal probiotics. *Iran. Red Crescent Med. J.* **2013**, *15*, 199. [CrossRef] [PubMed]
48. Kosem, N.; Han, Y.-H.; Moongkarndi, P. Antioxidant and cytoprotective activities of methanolic extract from Garcinia mangostana hulls. *Sci. Asia* **2007**, *33*, 283–292. [CrossRef]
49. Re, R.; Pellegrini, N.; Proteggente, A.; Pannala, A.; Yang, M.; Rice-Evans, C. Antioxidant activity applying an improved ABTS radical cation decolorization assay. *Free Radic. Biol. Med.* **1999**, *26*, 1231–1237. [CrossRef]

50. Cecchini, S.; Fazio, F. Assessment of Total Antioxidant Capacity in Serum of Heathy and Stressed Hens. *Animals* **2020**, *10*, 2019. [CrossRef]
51. Benzie, I.F.; Strain, J.J. The ferric reducing ability of plasma (FRAP) as a measure of "antioxidant power": The FRAP assay. *Anal. Biochem.* **1996**, *239*, 70–76. [CrossRef]
52. Zulueta, A.; Esteve, M.J.; Frígola, A. ORAC and TEAC assays comparison to measure the antioxidant capacity of food products. *Food Chem.* **2009**, *114*, 310–316. [CrossRef]
53. Telagari, M.; Hullatti, K. In-Vitro α-amylase and α-glucosidase inhibitory activity of Adiantum caudatum Linn and Celosia argentea Linn extracts and fractions. *Indian J. Pharmacol.* **2015**, *47*, 425.
54. Saul, R.; Molyneux, R.; Elbein, A. Studies on the mechanism of castanospermine inhibition of α-and β-glucosidases. *Arch. Biochem. Biophys.* **1984**, *230*, 668–675. [CrossRef]
55. Warren, F.J.; Zhang, B.; Waltzer, G.; Gidley, M.J.; Dhital, S. The interplay of α-amylase and amyloglucosidase activities on the digestion of starch in in vitro enzymic systems. *Carbohydr. Polym.* **2015**, *117*, 192–200. [CrossRef] [PubMed]

Review

Okra (*Abelmoschus Esculentus*) as a Potential Dietary Medicine with Nutraceutical Importance for Sustainable Health Applications

Abd Elmoneim O. Elkhalifa [1], Eyad Alshammari [1], Mohd Adnan [2], Jerold C. Alcantara [3], Amir Mahgoub Awadelkareem [1], Nagat Elzein Eltoum [1], Khalid Mehmood [4], Bibhu Prasad Panda [5] and Syed Amir Ashraf [1,*]

[1] Department of Clinical Nutrition, College of Applied Medical Sciences, University of Hail, Hail 2440, Saudi Arabia; ao.abdalla@uoh.edu.sa (A.E.O.E.); eyadhealth@hotmail.com (E.A.); mahgoubamir22@gmail.com (A.M.A.); nagacademic0509@gmail.com (N.E.E.)

[2] Department of Biology, College of Science, University of Hail, Hail 2440, Saudi Arabia; drmohdadnan@gmail.com

[3] Department of Clinical Laboratory Sciences, College of Applied Medical Sciences, University of Hail, Hail 2440, Saudi Arabia; jerold.alcantara@yahoo.com

[4] Department of Pharmaceutics, College of Pharmacy, University of Hail, Hail 2440, Saudi Arabia; adckhalid@gmail.com

[5] Microbial and Pharmaceutical Biotechnology Laboratory, Centre for Advanced Research and Pharmaceutical Sciences, School of Pharmaceutical Education and Research, Jamia Hamdard, New Delhi 110062, India; bppanda@jamiahamdard.ac.in

* Correspondence: amirashrafy2007@gmail.com; Tel.: +966-591-491-521 or +966-165-358-298

Citation: Elkhalifa, A.E.O.; Alshammari, E.; Adnan, M.; Alcantara, J.C.; Awadelkareem, A.M.; Eltoum, N.E.; Mehmood, K.; Panda, B.P.; Ashraf, S.A. Okra (*Abelmoschus Esculentus*) as a Potential Dietary Medicine with Nutraceutical Importance for Sustainable Health Applications. *Molecules* **2021**, *26*, 696. https://doi.org/10.3390/molecules26030696

Academic Editors: Severina Pacifico and Simona Piccolella

Received: 7 January 2021
Accepted: 22 January 2021
Published: 28 January 2021

Publisher's Note: MDPI stays neutral with regard to jurisdictional claims in published maps and institutional affiliations.

Copyright: © 2021 by the authors. Licensee MDPI, Basel, Switzerland. This article is an open access article distributed under the terms and conditions of the Creative Commons Attribution (CC BY) license (https://creativecommons.org/licenses/by/4.0/).

Abstract: Recently, there has been a paradigm shift from conventional therapies to relatively safer phytotherapies. This divergence is crucial for the management of various chronic diseases. Okra (*Abelmoschus esculentus* L.) is a popular vegetable crop with good nutritional significance, along with certain therapeutic values, which makes it a potential candidate in the use of a variety of nutraceuticals. Different parts of the okra fruit (mucilage, seed, and pods) contain certain important bioactive components, which confer its medicinal properties. The phytochemicals of okra have been studied for their potential therapeutic activities on various chronic diseases, such as type-2 diabetes, cardiovascular, and digestive diseases, as well as the antifatigue effect, liver detoxification, antibacterial, and chemo-preventive activities. Moreover, okra mucilage has been widely used in medicinal applications such as a plasma replacement or blood volume expanders. Overall, okra is considered to be an easily available, low-cost vegetable crop with various nutritional values and potential health benefits. Despite several reports about its therapeutic benefits and potential nutraceutical significance, there is a dearth of research on the pharmacokinetics and bioavailability of okra, which has hampered its widespread use in the nutraceutical industry. This review summarizes the available literature on the bioactive composition of okra and its potential nutraceutical significance. It will also provide a platform for further research on the pharmacokinetics and bioavailability of okra for its possible commercial production as a therapeutic agent against various chronic diseases.

Keywords: antidiabetic; cardioprotective; functional foods; nutraceuticals; okra; phytotherapy

1. Introduction

Okra (*Abelmoschus esculentus* L.), belonging to the family *Malvaceae*, is commonly known as Lady's finger, as well as by several vernacular names, including okra, bhindi, okura, quimgombo, bamia, gombo, and lai long ma, in the different geographical regions of its cultivation [1]. Okra is believed to have originated near Ethiopia, where it was frequently cultivated by the Egyptians during the 12th century, and thereafter spread throughout the Middle East and North Africa [2,3]. Okra is an annual shrub that is cultivated mostly within tropical and subtropical regions across the globe and represents a popular garden crop, as well as a farm crop. It is a widely cultivated food crop and is globally known for its palatability. The immature green pods of okra are usually consumed as vegetables, while the extract of the pods also serves as a thickening agent in numerous recipes for soups, as well as sauces, to augment their viscosity [4,5]. Another noteworthy application of okra fruit is their wide use in the pickle industry. The polysaccharides present in okra are used in sweetened frozen foods such as icecreams, as well as bakery products, due to their health benefits and longer shelf-lives [6–8]. Anatomically, the fruits, stem, and leaves of okra are covered with minute soft, hairy structures. Although the flowering of the okra plant is perennial, it is highly dependent on various biotic and abiotic factors. The leaves of okra are polymorphous, characterized by hairy upper and lower surfaces, whereas the petioles are around 15 cmlong. The flowers of okra can be easily recognized due to their slight yellowish color with a crimson center. The edible part of okra or its capsule (pod) measures approximately 15–20 cm in length and has a pyramidal-oblong, pentagonal, hispid appearance. Historically, okra pods were utilized for various purposes, such as in food, appetite boosters, astringents, and as an aphrodisiac. Furthermore, okra pods have also been recommended to cure dysentery, gonorrhea, and urinary complications [9]. Extracts of young okra pods have also been reported to display moisturizing and diuretic properties, whereas the seeds of this plant have been reported to possess anticancer and fungicidal properties [10].

Recently, okra has been used not only for its nutritional values but, also, for its nutraceutical and therapeutic properties, owing to the presence of various important bioactive compounds and their associated bioactivities. This review presents a summary of the nutritional significance of okra, as well as the possible pharmacological applications of okra bioactive components, and to explore the possible characteristics forthe development and formulation of nutraceuticals andfunctional food. In addition, this review also focuses on the nutraceutical potential of *Abelmoschus esculentus* for various therapeutic purposes, as well as to demonstrate the benefit of okra-based nutraceuticals and their consumption.

2. Nutritional and Bioactive Constituents in Okra

Okra is probably a proficient dietary constituent rather than a staple food crop. Small industries (Surajbala Exports Private Limited, New Delhi, India and Hunan QiyiXinye culture media, Hunan, China) have utilized okra seeds for oil extraction [11,12]. Additionally, the lipid content of any food item is considered as one of the important aspects of itsnutritional value, andseveral food types have different levels oflipid contents, including triacylglycerols, polar lipids, free fatty acids, or diacylglycerols. Among these constituents, fatty acids are largely responsible for determining the stability and nutritional value of food types. Triacylglycerols are biomolecules that are composed of unsaturated and saturated fatty acids with subtle differences in the number of associated acyl group repeats, along with the repetitions and positions of double bonds. Importantly, these lipids are naturally destined to be energy reservoirs [13].

Earlier, Savello et al. reported that the seeds of the okra plant represent a rich source of oil, constituting 20 to 40% of the total composition, which varies with the extraction procedure [14]. Linoleic acid, a well-known representative of polyunsaturated fatty acids (PUFA), is the dominant constituent of the oil content (47.4%) of okra seeds [14]. Other important dietary constituents essential for human growth are the amino acids and their polymers, viz., proteins [15,16]. Okra seeds have been reported to have different protein compositions from cereals and pulses, as their protein ingredients are modified to bear a balance of characteristic amino acids, namely lysine and tryptophan. Thus, owing to their rich content of essential amino acids, okra seeds represent an important constituent of the human diet [17]. Okra also serves as a potentially rich source of vitamins and carbohydrates, which are also regarded as a vital nutritional components of the diet [18]. Okra pods are also reported for their rich nutritional compositions and are frequently consumed either after boiling, frying, or cooking [19]. The nutritional values of different edible parts of okra (per a 100 g serving) are mentioned in Table 1.

Table 1. Nutritional composition of raw okra per 100 g of serving. Data reported from USDA (United States Department of Agriculture) SR-21.

S. No.	Dietary Constituents	Amount Per Serving	%DV *
1.	Total calories	130 kJ	2
2.	Total carbohydrates	7 g	2
3.	Total protein	2.0 g	4
4.	Dietary fiber	3.2 g	13
5.	Starch	0.3 g	-
6.	Sugar	1.2 g	-
7.	Total fat	0.1 g	-
8.	Trans-fat	-	-
9.	Saturated fat	0.0 g	0
10.	Cholesterol	0.0 mg	0
11.	Total omega-3 fatty acids	0.001 g	-
12.	Total omega-6 fatty acids	0.026 g	-
13.	Phytosterols	0.024 g	-

* Indicates the daily limit percentage for adults and children aged ≤ four years [20].

Okra also provides a rich supply of minerals required for maintaining normal homeostasis at the cellular level. The edible plant parts have been documented to possess calcium (Ca), phosphorus (P), and iron (Fe) at different amounts of 84, 90, and 1.20 mg, respectively. It also possesses a β-carotene, riboflavin, and vitamin B complex at the approximate concentrations of 185 µg, 0.08 mg, and 0.04 mg, respectively [21]. The other compositional vitamins of okra plants are mentioned in Table 2.

Table 2. Mineral intake per serving of 100 g of okra. Data reported from USDA SR-21.

S. No.	Minerals	Amount Per Serving	%DV *
1.	Potassium	303 mg	9
2.	Calcium	81.0 mg	8
3.	Phosphorus	63.0 mg	6
4.	Magnesium	57.0 mg	14
5.	Copper	0.1 mg	5
6.	Selenium	0.7 µg	1
7.	Manganese	1.0 mg	50
8	Zinc	0.6 mg	4
9.	Sodium	8.0 mg	0
10.	Iron	0.8 mg	4

* Indicates the daily limit percentage for adults and children aged ≤ 4 years [20].

The mucilage from the okra plant is primarily constituted by carbohydrates [22]. Furthermore, the young pods of okra plants are composed of polysaccharides (Mw~170,000), along with 11% amino acids. Moreover, okra pods are primarily constituted by equivalent amounts of galactose and galacturonic acid (25 and 27%, respectively), along with 22% rhamnose. In some parts of the world, especially West Africa, okra pods are also consumed in a dried form after the addition of other ingredients. However, one major nutritional drawback of such consumption is the absence of β-carotene or retinol (vitamin A) in a dried form [23]. Moreover, fresh pods of okra also provide viscous dietary fiber, which has been reported to minimize cholesterol levels. In previous studies, the maximum concentrations of nutrients were reported from okra pods that were aged only up to seven days [24,25].

3. The Pharmacological and Potential Applications of Okra-Derived Biomolecules

3.1. Antidiabetic Efficacy

The incidences of metabolic disorders, particularly diabetes, have tremendously increased in populations throughout the world. Although diabetes is caused by several factors, oxidative imbalance and inflammatory responses have been identified as their most common repercussions [26,27]. Recently, type 2 diabetes patients have also been reported to be associated with obesity and high lipid profiles [28]. Okra plant parts have been widely reported to reduce hyperglycemic levels. Okra mucilage, along with ethanolic and aqueous extracts of the pods, have been reported to lower the glucose levels in the blood of alloxan-induced diabetes models [29,30]. Diabetes nephropathy (DN) is a common complication of diabetes that has become a serious threat to human health and is expected to become the common cause of end-stage renal disease and cardiovascular events [31–33].

In clinical practice, a Chinese single plant-based drug extracted from the dry flowers of *Abelmoschus manihot*, named the Huangkui capsule (HKC), is used for the treatment of CKD (chronic kidney disease), DN, chronic glomerulonephritis, membranous nephropathy, and other inflammatory diseases. It is a patented drug approved by the State Food and Drug Administration of China (Z19990040) in 1999 for diabetes-related complications [34–36].

In a rat model of unilateral nephrectomy and doxorubicin-induced nephropathy, a HKC dose of 0.5 and 2 g/kg is administered via an intragastric (IG) manner for 28 days. It was found that the general status of the rat improved, as alleviated renal histological changes, proteinuria, albuminuria, glomerulosclerosis, and a decreased infiltration of ED1+ and ED3+ macrophages into the glomeruli were noticed. An inhibition of the protein expression of tumor necrosis factor (TNF)-α in the kidney was found. Further studying of the mechanism shows that, in a rat model of doxorubicin-induced nephropathy, HKC can downregulate the protein expression of transforming growth factor (TGF)-β1 and the p38mitogen-activated protein kinase (MAPK) by suppressing the p38/ MAPK signaling pathway [37,38].

Additionally, within a rat model of DN induced by unilateral nephrectomy and streptozotocin (STZ) injections rather than lipoic acid, it was noted that, when HKC was administered (IG) at doses of 0.75 and 2.0 g/kg for 56 days, the urinary albumin levels were reduced. It also improves renal function, as it decreases the blood urea nitrogen (BUN) and serum uric acid levels. The number of cells and the extracellular matrix of glomeruli is reduced to alleviate kidney fibrosis by HKC and reverse the increase in the markers of oxidative stress, such as malondialdehyde (MDA), 8-hydroxy-2′-deoxyguanosine, total superoxide dismutase (SOD), and nicotinamide adenine dinucleotide phosphate oxidase-4 [39]. Further studies of the mechanism proved that HKC simultaneously decreased the protein expression of pp38MAPK, p-Akt, TGF-β1, and TNF-α by inhibiting the p38MAPK and Akt signaling pathways in the kidney in a rat model of DN.

Later, in vitro and in vivo studies indicated that an increase in the mRNA expression of peroxisome proliferator-activated receptor (PPAR)-α and PPARG in the livers and kidneys of rats with DN was observed when HKC was administered (IG) at doses of 75, 135, and 300 mg/kg for 84 days. It was also found that HKC administration also increased the serum albumin levels, while the serum triglycerides, cholesterol, and total fats levels were lowered in a dose-dependent manner. This result was seen in the livers of rats with DN as compared to irbesartan [33]. Moreover, HKC decreased the expression of interleukin (IL)-1, IL-2, IL-6, and TNF-α by suppressing the inflammatory reaction in the kidneys of rats with DN. Strikingly, HKC alleviated endoplasmic reticulum stress and decreased the activation of the c-Jun NH2-terminal kinase in the livers and kidneys of rats with DN and, subsequently, reduced renal injury [33].

The results obtained from the above studies demonstrated that HKC can be a potential agent for DN treatment in humans. It was found that, when HKC was administered via IG at a dose of 0.75 g/Kg for 28 days, a significant decrease in the levels of BUN, serum creatinine, and urine protein in the plasma was found. The molecular mechanisms demonstrated that HKC notably downregulated the protein expression of NADPH oxidase

(NOX)-1, NOX-2, NOX-4, α-smooth muscle actin (αSMA), and the p-extracellular signal-regulated kinase (ERK)1/2 by inhibiting the NADPH oxidase/reactive oxygen species (ROS)/ERK signaling pathways in renal tissue in rats with chronic renal failure induced by adenine in vivo [40]. Subsequent phytochemical investigations have shownthat the main bioactive components of HKC are quercetin, quercetin-3'-O-glucoside, isoquercitrin, and hyperoside. A 100-μM concentration of gossypetin-8-O-β-D-glucuronide can inhibit the protein expression of smooth muscle actin, p-ERK1/2, NOX-1, NOX-2, and NOX-4 in HK-2 cells, which are induced by high glucose levels in the same way as the NOX inhibitor diphenyleneiodonium [40].

Furthermore, aqueous extracts from the pods of okra plants co-administered with metformin also resulted in hypoglycemia in Long Evans rats [41]. The enzyme α-amylase, which acts by breaking polysaccharides, resulting in the availability of glucose, has been considered as a vital enzyme for fulfilling the energy requirements of the human body. Water-soluble seed and peel extracts from okra plants have also been previously reported to inhibit the activities of both α-glucosidase and α-amylase [42]. Moreover, Lu and coworkers reported that the α-glucosidase and α-amylase inhibitory activities of premature seeds of the okra plant were due to oligomeric proanthocyanidins [43]. Further research on the antidiabetic efficacy of the okra plant suggested that rhamnogalacturonan also mediated the antidiabetic activity [44].

3.2. Antioxidant Efficacy

Okra pods are immature fruits of okra that are consumed in nearly all parts of the world as a vegetable. Previous studies have reported that immature okra pods have antioxidant potential [45,46]. In elaboration of this, a recent study elucidated that the antioxidant efficacy of okra pods may be due to the large amounts of polyphenols (29.5%) present within the seeds of immature pods. After the ingestion of food or a beverage, flavonoids in the ingested matrix must pass from the gut lumen into the circulatory system in order to be absorbed. Since, in plants, almost all flavonoids are in the form of glycosides, the attached sugar must be removed following consumption before absorption can take place [47]. These polyphenols from immature okra pods carry out the antioxidant activity by lowering the MDA level and increasing the SOD and glutathione peroxidase (GSH-Px) levels [48,49]. Flavonoids are a large group of secondary plant metabolites and occur as either aglycones or conjugates with glycosides and acyl groups, wherein around 8000 different types have been identified so far [50]. The major phenolic compounds found in okra fruits are quercetin-3-O-gentiobioside, quercetin-3-O-glucoside (isoquercitrin), rutin, a quercetin derivative, protocatechuic acid, and a catechin derivative, of which quercetin-3-O-gentiobioside was the most abundant phenolic compound [51,52]. The major contributor of the antioxidant capacity is quercetin-3-O-gentiobioside. It also exhibits inhibitory effects on digestive enzymes like lipase, α-glucosidase, and α-amylase [48,51].

The free radicalscavenging and ferric-reducing capabilities of okra pods have been documented in previous studies. In one study, okra extract obtained by cold extraction and boiling the fruit in water showed notable antioxidant activity [53]. Seeds from the okra plant are also a rich source of phenols, namely procyanidin B1 and B2, both of which are involved in DPPH (1,1-diphenyl-2-picrylhydrazyl) and ABTS (2,2'-azino-bis(3-ethylbenzothiazoline-6-sulfonic acid) free radicalscavenging activities [54]. Liao et al. also reported the antioxidant and ferric-reducing activities of okra pods, although they identified two specific glucopyranoside compounds, 5,7,3',4'-tetrahydroxy-4"-O-methylflavonol-3-O-β-D-glucopyranoside and 5,7,3',4'-tetrahydroxy flavonol–3-O-[β-D-glucopyranosyl-(1→6)]-β-D-glucopyranoside, to be the responsible bioactive compounds [55]. Subsequently, it was also reported that alternative parts of okra—namely, the flowers, leaf, seed, and pods—also had substantial antioxidant efficacy [55]. In a desirability study, a powder of the seed and peel of okra augmented the levels of hepatic, renal, and pancreatic SOD and glutathione peroxidase in streptozotocin models of diabetes. A similar treatment regime also resulted in reduced levels of glutathione and thiobarbituric acid [41]. Similar observations were also recorded by Doreddula et al., in which extracts from okra seeds at concentrations in the range of 100–250 µg/mL showed substantial antioxidant effects through ferric reduction, β-carotene-linoleic assay, and DPPH. Moreover, different fractions of okra plants have also been reported to reduce malondialdehyde and elevate the levels of glutathione peroxidase and superoxide dismutase [51].

3.3. Anticancer Effect

Cancer is the second-leading cause of death globally, and despite the advances in drug development, it is still necessary to develop new plant-derived medicines. There is an urgent need for new anticancer drugs, because cancerous cells are developing resistances against the currently available drugs, like vinca alkaloids and taxanes [56,57].

The term cancer represents a broad group of malignancies showing the key characteristic of uncontrolled proliferation, aided by various regulatory and functional changes, which, in turn, ensure a systemic spread throughout the body [58]. Although scientific cancer research communities have made remarkable progress in understanding the mechanisms responsible for such a debilitating disease, an effective treatment strategy without accompanying toxic effects has yet to be made available. The therapeutic management of such life-threatening modalities should focus on the increased exploration of a chemopreventive approach through plant-based natural agents from different sources. To this end, plant products have attracted several researchers across the world because of their selective toxicity against malignant cells [59–61]. It is also reported that the flowers of okra contain substantial amounts of flavonoids and phenols, as compared to the pods, peel, leaves, and seeds [62,63]. A recent report elucidated that purified fractions of flavonoids from the flowers of okra plants had a significant antitumor effect on colorectal malignancy both in-vitro, as well as in vivo, exerting a strong antioxidant potency concomitantly with substantial antiproliferative effects on tumor growth. The antiproliferative effect of flavonoids within okra flowers induced the activation of p53, culminating into the ceasing of mitochondrial functions within colorectal tumor cells, ultimately resulting in apoptosis and restraining the autophagy [64].

Subsequently, the anticancerous effect of okra seed extracts have also been documented in vitro through several other cell lines. The flavonoid constituents of seed extracts showed enhanced cytotoxic effects on human-derived breast cancer cells (MCF-7) in comparison with hepatoma cells of human origin (HepG2) and human cervical cancer (HeLa) cells in a dose-dependent manner. These observations affirmed that the flavonoid isoquercitrin, in a synergistic association with other flavonoids, inhibited vascular endothelial growth factor (VEGF), resulting in the apoptosis of cancerous cells [65]. Additionally, Hyperin—also known as quercetin-3-O-β-D-galactoside—is an important flavonoid constituent of okra. Hyperin has also been reported for its anticancerous potential in gastric cancer cells (CHI) by establishing an antiproliferative, antimigratory, and anti-invasive environment resulting in apoptosis by blocking the Wnt/β-catenin signal pathway [66].

Carbohydrate-binding proteins such as lectins have been widely investigated for their anticancer effects [67,68]. Lectins from okra have been documented to activate caspase-mediated downstream signaling in MCF7 cells and normal fibroblasts (CCD-1059 SK) [69]. Furthermore, the extract of okra pulp has also been attached to gold nanoparticles and observed to exhibit a significant induction of oxidative stress, followed by the depolarization of mitochondrial membrane potential and apoptosis in Jurkat cells, thereby indicating its anticancer potential [70]. However, studies focusing on the capability of okra and its parts to reduce cancer progression and its associated effects are rare.

3.4. Immunomodulatory Potential

The immune system acts as a well-defined protective shield against noxious external or internal intervening agents. The immune system plays a critical role in protecting the human body from infectious diseases. Its two main contributors include innate and acquired immunity responses. The most important feature of innate immunity is its lack of specific recognition. This type of immune system responds to all pathogens, regardless of their nature. Innate immunity is composed of immune and nonimmune components, whilst acquired immunity has only immune elements. Phytochemicals are the naturally occurring secondary metabolites present in abundance in fruits and vegetables. They do not have any nutritional importance, but they are essential for the growth and maintenance of plants, and with evolution, humans have learnt the ways to harvest and manipulate these phytochemicals for their own benefits [71,72].

The biologically potent constituents of *A. esculentus* have also been reported to modulate the complex immune system. The administration of lectin from okra in mice at low concentrations (0.01, 0.1, and 1 mg/kg) has demonstrated significant inflammatory effects [45]. Recently, the administration of the ethanolic extract of okra lowered inflammation in Wistar rats subjected to acute gastric mucosal injury [73]. The polysaccharide constituents of an aqueous okra extract were demonstrated to augment the hemoglobin content and expression of major histocompatibility complex (MHC) II and CD80/89 within the bone marrow hematopoietic cells derived from rats, as well as reduce endocytosis. Furthermore, the aqueous extracts also upregulate the expression of interleukin-12 and interferon-γ, along with the simultaneous reduction of the anti-inflammatory cytokine interleukin-10 [74]. Macrophages are key components of innate immunity and act as a prerequisite for the effective functioning of the innate immune systems. The immunomodulatory effects of polysaccharides from okra have also been evaluated on macrophage cell lines by Chen et al. (2016), who reported an increase in nitric oxide (NO) synthesis, inducible NO synthase (iNOS) expression, and the levels of tumor necrosis factor-α and cytokines in RAW264.7 cells following treatment [75].

3.5. Microbicidal Action

The lipid content—namely, palmitic and stearic acids of lyophilized extracts from okra pods—along with its aqueous counterpart, have earlier been reported to inhibit *Rhodococcus opacus*, *Rhodococcus erythropolis*, *Mycobacterium aurum*, *Escherichia coli*, *Staphylococcus aureus*, *Pseudomonas aeruginosa*, and *Xanthobacter* Py2 [76]. Moreover, the gold nanoparticles synthesized from the okra extract (pulp) also displayed substantial microbicidal efficacy against *Bacillus cereus*, *Bacillus subtilis*, *E. coli*, *P. aeruginosa*, and *M. luteus* [77]. Furthermore, fractionsof the pods from okra that are rich in carbohydrates were also documented for their activity against *Helicobacter pylori* [78]. Apart from the above-mentioned therapeutic effects of okra fruits, some therapeutic effects and their mechanisms of action are presented below in Table 3, alongwith the presentation of various bioactive components present in okra and their chemical structures (Figure 1).

Table 3. Different bioactive components derived from okra showing their therapeutic benefits on human health, along with their mechanisms of action.

Bioactive Components	Therapeutic Benefits	Mechanisms of Action	Reference
Polysaccharide	Antidiabetic	It helps to lower body weight and glucose levels, improve glucose tolerance, and decrease the total serum cholesterol levels in high-fat diet-fed C57BL/6 mice.	[79]
Rhamnogalacturonan	Antidiabetic	Hypoglycemic effect,	[80]
Lectins	Anticancer	Arrest the cell cycle and activate the caspase cascades.	[81]
		Inhibit cellular proliferation in human breast cancer in vitro.	[70]

Table 3. Cont.

Bioactive Components	Therapeutic Benefits	Mechanisms of Action	Reference
Pectin	Anticancer	Involved in cell adhesion, growth, and survival, as well as tumor development and cancer prevention therapy.	[82–84]
	Antiproliferative and proapoptotic	Induce apoptosis and inhibit cellular proliferation.	[85]
Pectin	Lower bad cholesterol	Okra promotes cholesterol degradation and inhibits the production of fat in the body. It lowers bad cholesterol by altering the bile production in the intestines. This helps in eliminating the clots and deposited cholesterol.	[86]
Polyphenolic compounds	Antioxidant	Extract exhibits a strong DPPH radical scavenging activity and reducing power.	[87]
Quercetin 3-O-glucosyl (1→6) glucoside (QDG) and quercetin 3-O-glucoside (QG)	Antioxidant	Excellent reducing power and free radical scavenging capabilities, including DPPH, superoxide anions, and hydroxyl radicals.	[88]
Vitamin C, calcium, iron, manganese, and magnesium	Antioxidant	Eliminating free radicals.	[86]
Quercetin derivatives and epigallocatechin	Antioxidant	Inhibitory effects on the generation of reactive oxygen species (ROS).	[54]
Polysaccharide	Metabolic disorders	Inhibition of LXR and PPAR signaling.	[79]
Polyphenolic compounds,	Antioxidant	Perform the function of capturing free radicals and stopping the chain reactions.	[89]
Vitamin A; B vitamins (B1, B2, B6); and vitamin C and traces of zinc, calcium, folic acid, and fiber	Pregnancy benefits	Folates prevent miscarriages. They are also beneficial in the formation of the neural tube of the fetus and protect these tubes, preventing defects. This helps prevent birth defects like spina bifida and can even stop constipation during pregnancy.	[86]
Polyphenols like catechin and flavonoids like quercetin possess	Antifatigue effects	Decreased the levels of blood lactic acid (BLA) and BUN in the blood; MDA in the liver; and increased the levels of HG, SOD, and GSH in the liver during fatigue recovery, which proved that OSD could alleviate physical fatigue and promote recovery.	[51]
Probiotics	Gut bacteria-friendly	Biosynthesis of the vitamin B complex.	[86]
Glutathione	Detoxify liver, antioxidant	The slimy substance in okra contains substances that bind bile acid and cholesterol to detoxify the liver.	[86]
Mucilaginous	Ulcer treatment	The slimy stuff in okra is alkaline. This helps in neutralizing the acid. Additionally, it provides a protective coating within the digestive tract, which speeds up the healing process of peptic ulcers.	[86]
Mucilaginous with fiber	Relieves and prevents constipation	Bind toxins and lubricates the large intestines. This ensures effortless and normal bowel movement due to its natural laxative property.	[86]

Table 3. Cont.

Bioactive Components	Therapeutic Benefits	Mechanisms of Action	Reference
Vitamin K and C	Bone health and essential for the blood-clotting process. It also helps restore bone density and prevent osteoporosis.	Several mechanisms are suggested by which vitamin K can modulate bone metabolism. Besides the gamma-carboxylation of osteocalcin, a protein believed to be involved in bone mineralization, there is increasing evidence that vitamin K also positively affects the calcium balance, a key mineral in bone metabolism.	[86,90]
Vitamin A, along with antioxidant contents such as lutein, xanthein, and carotenes	Improves vision	Okra contains beta-carotenes (precursor of vitamin A), xanthin, and lutein, all with antioxidant properties preventing eye problems like cataract and glaucoma.	[86]
Glycosylated compounds	Antibacterial activity	Inhibit the adhesion of *Helicobacter pylori* to the human gastric mucosa.	[78]
Rhamnogalacturonan Polysaccharides	Antiadhesive properties	Interrupt the adhesion of *H. pylori* to human stomach tissues via interfering with the outer membrane proteins.	[78,91–93]
Polyphenols and flavonoids (okra seeds)	Antifatigue	Reduce the levels of BLA and BUN, enhancing hepatic glycogen storage and the promoting antioxidant ability by lowering the MDA level and increasing the SOD and GSH-PX levels.	[51]

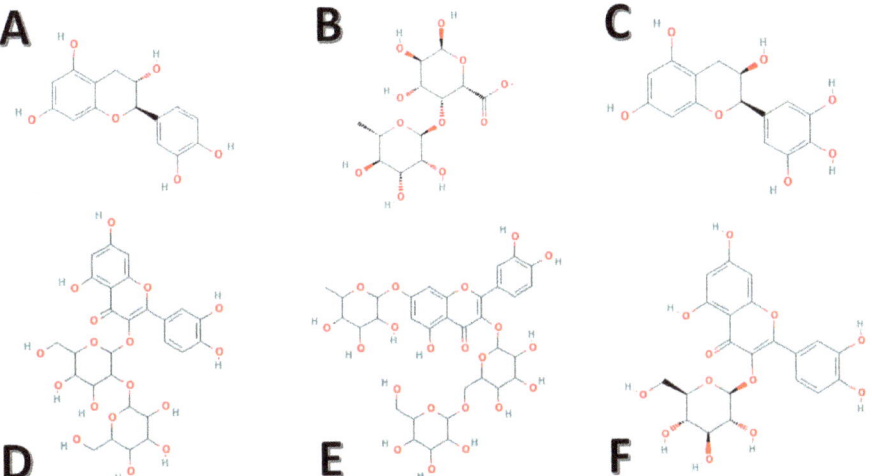

Figure 1. Chemical structures of the identified potent bioactive components derived from okra. (**A**) Catechin, (**B**) rhamnogalacturonan, (**C**) epigallocatechin, (**D**) quercetin-3-O-sophoroside, (**E**) Quercetin-3-O-[glucosyl(1->6) glucoside]-7-O-rhamnoside, and (**F**) isoquercitrin.

4. In Vivo Studies on the Health Benefits of Okra and Its Components

As stated earlier in this review, okra and its bioactive components are reported to have potential beneficial health effects. A significant reduction in the blood glucose level, along with an increase in body weight, was reported by Dubey and Mishra (2017) and Sabitha et al. (2011) in streptozotocin-induced diabetic rats when fed the peel and seed powder of okra.Sabitha et al. also reported a significant increased level of hemoglobin and the total protein level and a decrease in HbA1c. They also reported that okra peel and seed powder at a dose of 200 mg/kg showed a significant reduction inblood glucose compared to a 100 mg/kg dose. Additionally, the treatment of both the doses of okra seed powders significantly produced a greater blood glucose reduction when compared to okra peel powderat a 100 or 200-mg/kg dose. Meanwhile, there several studies suggested multiple mechanisms of antidiabetic plants to exert their blood glucose-lowering effects, such as the inhibition of carbohydrate metabolizing enzymes, enhancement of insulin sensitivity, regeneration of damaged pancreatic islet β-cells, and enhancement of insulin secretion and release [41,94]. Okra polysaccharides were also reported to lower the body weight and glucose levels, improve the glucose tolerance, and decrease the total serum cholesterol levels in mice fed with a high-fat diet. Different studies reported the effects of different parts of okra on alloxan-induced diabetic rats and showed that there was a significant reduction in the blood glucose level and glycated hemoglobin and an improvement in the lipid profile compared with the diabetic nontreated control rats and comparable with the metformin-positive control group [95,96]. Different parts of okra fruits can stimulate glycogen synthesis in the liver and delay the intestinal absorption of glucose in alloxan-induced diabetic rats [97]. In the same study, the histopathological examination of the pancreatic tissue after the administration of okra fruits revealed the evidence of pancreatic islet cell regeneration [97]. Okra supplementation statistically reduced the highlevels of fasting blood sugar, total cholesterol, and triglycerides and decreased the homeostasis model assessment of the basal insulin resistance index in diabetic rats, as reported by Majd et al. (2018) [98].

Nguekouo et al. (2018), in an in vivo study, showed that boiling and roasting do not change the antidiabetic potential of okra fruits and seeds [99]. Ortaç et al. (2018) reported in an in vivo study that okra has a gastroprotective effect against ethanol and could decrease a gastric ulcer, as indicated by the biochemical and histopathological data. They concluded that okra could be a possible therapeutic antiulcer agent [74]. Hossen et al. (2013) showed that the methanol extract of okra had a good central nervous system depressant activity, along with a high painkiller activity, on Swiss albino mice [100]. Wang et al. (2104) demonstrated that mice fed an okra diet execrated more cholesterol in their stools and had lower total blood cholesterol levels compared to the control mice group [101]. A four-year study conducted on 1100 people showed that people who ate a diet rich in polyphenols had lower inflammatory markers associated with heart disease, and as okra is one of the polyphenol-rich diets, okra may therefore help to protect from cardiovascular diseases [102]. Monte et al. (2014) reported that the lectin available in okra can stop cancer cell growth by up to 63% when they did a testtube study on breast cancer cells. Additionally, Vayssade et al. (2010), in a testtube study on metastatic mouse melanoma cells, showed that okra extract leads to cancer cell death [85]. Doreddula et al. (2014) revealed that the seed extracts of okra have an antioxidant, antistress effect in the bloodstream of mice [103].

5. Therapeutic Prospects of Okra as Dietary Medicine/Nutraceuticals

Around two-thirds of the world population (7.8 billion) is dependent on plant-based materials for their medicinal and healing properties, mainly because of their easy availability, accessibility, affordability, and safety, as well as the traditional beliefs of the consumers [104]. A very old quote by Hippocrates stated, "Let food be thy medicine and medicine be thy food", which described the significance of food and its nutritional, as well as therapeutic, values for the prevention, treatment, and management of diseases [105]. Thereafter, DeFelice coined the term nutraceutical by merging "nutrition" and "pharmaceutical" and defined it as food or part of a food that not only imparts health benefits but, also, contributes to the prevention or treatment of various diseases [106]. Importantly, nutraceuticals have been formulated in such a way that they could benefit or facilitate the management of human health without instigating any harm due to their natural occurrence. Nutraceuticals derived from plants, animals, or live microorganisms possess great potential for use by scientific communities, food researchers, and food-processing industries to produce unique foods or food components for the forthcoming needs of human beings to stay healthy without any side effects. Currently, the rapid rise in demand for nutraceutical products has been largely observed because of their therapeutic value in various diseases, such as diabetes, hypertension, arthritis, inflammatory bowel disease, the common cold, dyslipidemia, heart disease, and cancer. Nutraceutical products may also increase the lifespan by delaying aging, promoting the integrity of the body, and sustaining smooth normal functioning [107]. Moreover, based upon various pharmacological potentials of okra-derived molecules, okra has been seen as one of the potential sources of nutraceuticals.

It was observed that there was an increase in the number of studies investigating the therapeutic value of okra (Figure 2). The phytochemicals present in okra have been suggested to have potential applications for the treatment and management for various diseases (Tables 3 and 4) [9]. However, the potential of this extraordinary, cost-effective, cheap vegetable crop is still not fully used for its therapeutic or nutraceutical potential effects [9]. Therefore, there is an urgent need for such an easily available cheap vegetable crop like okra to be used in a nutraceutical formulation. Hence, okra-basednutraceuticals could play an essential role in the prevention and management of health, along with health improvements. Therefore, *A. esculentus*, an edible vegetable, could be an ideal source of nutraceuticals, since it contains both nutritionally active chemicals, as well as a source for various physiological advantages, as presented in Table 4. Few countries have traditionally used okra in folk medicine for various therapeutic purposes, such as gastroprotective, antiulcerogenic, and as a diuretic. The recent urbanization and changes in lifestyles, eating habits, and other factors have exposed the global population to a variety of chronic diseases. Since okra is an easily available and low-cost vegetable, it could potentially become an important nutraceutical product for populations in countries, irrespective of their stage of development. Particularly, okra-based nutraceuticals could become an ideal source of nutrition for people suffering from malnutrition in lesser-developed countries. However, the available studies in the literature have not specifically analyzed the therapeutic potential of okra as a nutraceutical.

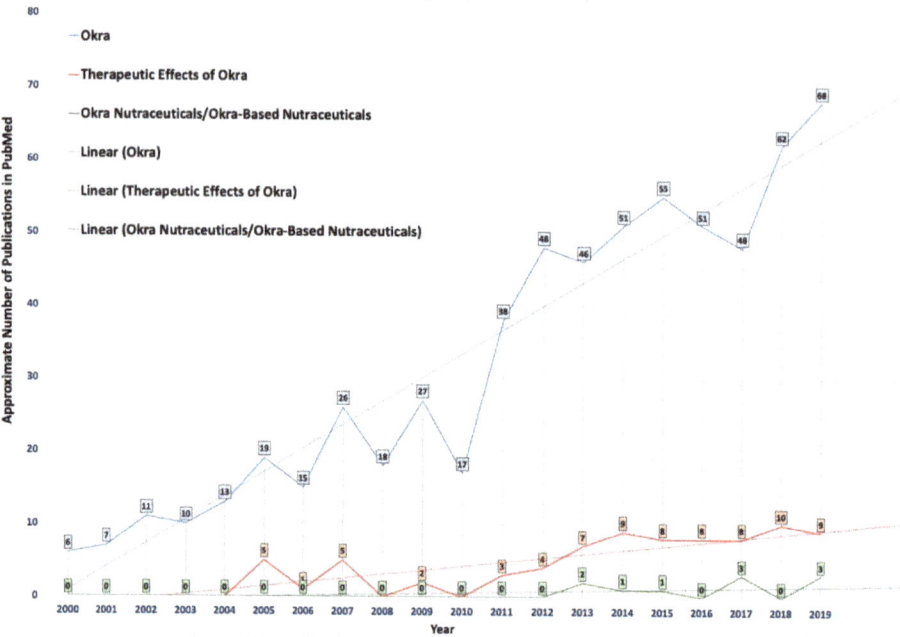

Figure 2. Statistical data showing the number of publications on the use of okra in pharmaceuticals indexed in PubMed from 2000 to 2019. The colors indicate publications available in PubMed after using the following keywords/phrases: (1) okra (blue), (2) therapeutic effect of okra (orange), and (3) okra nutraceuticals/okra-based nutraceuticals (green). The average trendlines show the importance and urgent need for research concerning the development of easily available okra (or its by-products)-based nutraceuticals and functional foods.

Table 4. Health benefits of various phytochemicals present in okra and its various parts.

Bioactive Components	Part	Health Benefits	Reference
Polyphenols Carotene		Important for eyesight, along with healthy skin.	[78,108]
Folic acid		Beneficial for fetus development.	[1,109]
Thiamine	Pod	Improves the nervous system, brain, heart, stomach, muscles, and intestine functions.	[8,110]
Riboflavin		Needed for growth and overall good health.	[111,112]
Niacin		Keeps our nervous system, digestive system, and skin healthy.	[108,113]
Vitamin C		Helps in the overall growth of the body and tissue repair.	[1,114]
Oligomeric catechin		Prevents and is used for treating chronic ailments, e.g., cardiovascular diseases and cancer.	[10,115]
Flavonol derivatives		Improves vascular health, leading to a reduced risk of diseases.	[116,117]
Lysine	Seed	Improves calcium absorption and retention.	[118,119]
Palmitic acid		An important constituent of the cell membrane, with a critical role in protein palmitoylation and palmitoylated signal molecules.	[120,121]
Oleic acid		Decreases the cholesterol levels and prevents heart diseases.	[122,123]
Linoleic acid		Improves cardiovascular health.	[88,124]
Carbohydrate	Roots	Prime energy source and fuel for the brain, kidney, heart, and muscles.	[125,126]
Flavonoids		Exhibits substantial anticancer, antioxidant, anti-inflammatory, and hepatoprotective activities.	[10,127]
Minerals	Leaves	Helps in overall growth and body development.	[128,129]
Tannins		Accelerate blood clotting, reduces the serum lipid and blood pressure, and modulates the immune responses.	[130–132]

6. Formulation and Development of Okra-Based Nutraceuticals

Nutraceuticals are broadly described as food or parts of food that provide incremental health benefits. Okra-based nutraceuticals represent popular health foods, owing to its intrinsic nutritional and other bioactive components, which show health-associated beneficial properties (Figure 3) [133]. Several efforts are being made to improve the well-known hypoglycemic outcomes of okra fruit by formulating different proportions of seeds and peels of Ex-maradi Okra fruit in the ratio of (10:90, 20:80, 30:70, 40:60, 50:50%, and so on), which is subsequently followed by investigating the antidiabetic and antioxidant efficacy of these formulations in vitro. Recent findings have led to the conclusion that seeds and peels at the ratio of 10:90% are the most efficient in exhibiting substantial in vitro antidiabetic and antioxidant efficacy [134,135]. Subsequently, it was recommended that the nutraceutical formulation of peel and Ex-maradi okra seeds in the ratio (10:90) exhibits substantial hypoglycemic and hypolipidemic activity in alloxan models of diabetes (rodents) and was thus appropriate for further improvements for the formulation of okra-based nutraceutical interventions in diabetes mellitus [135]. On the other hand, okra polysaccharides have also been reported to inhibit human cancer cell proliferation [136]. This possibly indicates their potential usage as anticancer nutraceutical formulations. However, an individual's susceptibility to any disease also largely depends upon genetic predisposition and lifestyle habits, such as smoking and high alcohol consumption. Therefore, the efficacy of nutraceuticals can vary from person to person. Nutraceuticals have proven health benefits, and their consumption (within their acceptable recommended dietary intakes) may help in the prevention of disease and allow humans to maintain overall good health. Therefore, since various parts of okra (fruit, seed, pulp, and mucilage) carry several therapeutic purposes, it can be considered to be an important vegetable crop for nutraceutical purposes.

Figure 3. Illustrative representation of the okra-mediated beneficial effects that have been scientifically established to date.

7. Global Okra Production and Possible Nutraceutical Market

Okra, being an inexpensive popular vegetable crop, is consumed by several populations globally and is a local staple food in low-income countries. Nowadays, due to its nutritional and health benefits, there is a growing demand for okra, and different okra products are available for purchase on online marketplaces. Recently, the agency Market Research Future estimated that the global okra seed market could earn a revenue of USD 352.7 million and register a 9.8% compound annual growth rate during the period 2018–2023 [137]. Globally, the market of okra seeds is geographically largely divided into Europe, Asia-Pacific, and North America, followed by the remaining countries. In 2018, the largest accreditation for the contribution of the okra market share (63.77%) was recorded by the Asia-Pacific region. It is estimated that the okra-based nutraceutical market will reach a worth of 222.9 million USD by the end of the year 2023. Small-scale manufacturers are a major cause for the disintegration of the okra market in the Asia-Pacific region. Pakistan, Malaysia, India, and the Philippines are regarded as the dominant producers of okra seeds [19,137–139]. In recent times, India has been the prominent producer of okra globally, followed by the remaining countries mentioned above. Since 2018, these remaining countries have held a 33.0% share of the global okra market. This enhanced expansion within the local market is attributed to increased cultivation, as well as the development of genetically modified (GM) seeds. Furthermore, the acceptance of hybrid and disorder-resistant seeds within the region has also facilitated the noticeable expansion of the okra market. Africa is now predicted to globally dominate the market for the consumption of okra seeds. It represents approximately 69% of the territorial market share due to increased accessibility to more arable croplands within the country. On the other hand, during 2017, North America accounted for only 2.2% of the okra market share, whereas Europe accounted for only 1.0%. At the same time, Mexico is known to be a dominant producer of okra in North America because of the high cultivation of okra within the country [109]. The global okra seed-mediated market (OSM) is divided categorically and regionally. Based on the category, the OSM is further divided into conventional and organic seeds of okra plants. The conventional OSM (cOSM) is more prominent, with a market share of 90.5% since 2018. The market dominance of cOSM could be attributed to the exploitation of different varieties, namely open-pollinated and traditional. In contrast, organic OSM is estimated to show a high growth rate of 10.7%, which could be attributed to a shift in consumer awareness resulting in an increased preference for organic plant produce [17]. Thus, the high production of okra the world over should be utilized to some extent in the large-scale production of okra-based nutraceuticals, which could also be used to alleviate the problem of malnutrition in underdeveloped countries.

8. Safety and Efficacy

Okra is a food crop, and its elongated, edible pods are mostly harvested during the immature stages and eaten primarily as a vegetable dish. Other parts of okra—namely, the flowers and buds—are also palatable [112]. The premature pods are commonly ingested in vegetable dishes and are also consumed dried, marinated in salads, fried, raw, or boiled, along with various other ingredients [10]. An average fresh okra pod is estimated to contain approximately 740 IU of vitamin A. Okra seeds also provide a rich source of edible oil, constituting up to 22% of the biomass [10]. Toxicological reports have suggested that the fruit and seeds of okra are nontoxic at normal levels of consumption. Few reports have substantiated the safety and efficacy of okra extracts in controlled human trials. *Abelmoschus manihot*, a single medicament of traditional Chinese medicine widely used to treat kidney disease, was evaluated in the form of a huangkui capsule, 2.5 g, three times per day, losartan potassium, 50mg/d, or a combined treatment—a huangkui capsule at 2.5 g, three times per day, was combined with losartan potassium, 50mg/d. The duration of the intervention was 24 weeks. The results were evaluated in the form of changes within the mean baseline urine protein excretion and changes in the estimated glomerular filtration rate (eGFR) from the baseline after treatment. The study proved the efficacy of

Abelmoschus manihot as a promising therapeutic for patients with primary kidney disease (chronic kidney disease stages 1 and 2) with moderate proteinuria [34].

Similarly, IQP-AE-103 is a combination of dehydrated powder of okra (*A. esculentus* (L.) Moench) pods and inulin; a heterogeneous mixture of fructose polymers extracted from chicory roots was used to evaluate the efficacy and tolerability/safety of IQP-AE-103 on body weight reduction in overweight-to-moderately obese adults. A beneficial effect of IQP-AE-103 on the lipid metabolism was also demonstrated in the subgroup of subjects with baseline total cholesterol levels above 6.2 mmol/L. The study indicated that IQP-AE-103 could be an effective and safe weight loss intervention; the trial was registered with NCT03058367 [140]. However, currently, a trial to evaluate the glycemic evaluation of okra seed noodles is currently active, although not recruiting; see clinicaltrials.gov (ClinicalTrials.gov Identifier: NCT03990844). Data on the safety and toxicity of okra fruits are limited. Therefore, further clinical studies and research is required on this edible medicinal plant in the context of nutraceutical and functional food development, food excipients, drug discovery, and development [9,117].

9. Future Perspectives and Conclusions

The attributes of the different edible okra parts discussed in this review highlight the nutritional relevance of okra and its nutraceutical potential for providing health benefits to humans. Okra is a cost-effective and economically affordable natural source with ample reservoirs of carbohydrates, proteins, fatty acids, vitamins, fiber, and minerals, with various other bioactive phytochemicals that are important for human well-being. Although the potential benefits of okra on different chronic disorders have been scientifically explored to some extent, several aspects, such as the pharmacokinetics and bioavailability of okra, as well as its specific mechanisms of action on different diseases, need further investigation. Such a knowledge gap may have arisen due to the complex etiology of diseases, along with different related factors aiding the diseases. Owing to the nutritional composition of okra, it can be potentially used to compensate for the problem of malnutrition in underdeveloped countries across the globe. Moreover, the formulation of okra-based nutraceuticals could prove to be beneficial, since it is easily available and inexpensive. Therefore, further studies should focus on the development of functional foods, nutraceuticals, or drugs from okra components.

Author Contributions: Conceptualization, A.E.O.E. and S.A.A.; methodology, E.A., J.C.A., K.M., B.P.P., and A.M.A.; investigation, N.E.E., E.A., and A.M.A.; data curation, B.P.P., J.C.A., and K.M.; writing—original draft preparation, A.E.O.E. and S.A.A.; writing—review and editing, M.A.; visualization, M.A., and S.A.A.; supervision, M.A. and S.A.A.; and project administration, A.E.O.E. and S.A.A. All authors have read and agreed to the published version of the manuscript.

Funding: This research was funded by the Scientific Research Deanship at the University of Ha'il—Saudi Arabia through project number RG-191333.

Institutional Review Board Statement: Not applicable.

Informed Consent Statement: Not applicable.

Data Availability Statement: All data generated or analyzed during this study are included in this article.

Conflicts of Interest: The authors declare no conflict of interest. The funders had no role in the design of the study; in the collection, analyses, or interpretation of data; in the writing of the manuscript, or in the decision to publish the results.

References

1. Jain, N.; Jain, R.; Jain, V.; Jain, S. A review on: *Abelmoschus esculentus*. *Pharmacia* **2012**, *1*, 84–89.
2. Lamont, W.J. Okra—A versatile vegetable crop. *Hort Technol.* **1999**, *9*, 179–184. [CrossRef]
3. Tindall, H.D. *Vegetables in the Tropics*; Macmillan Publishers Limited: London, UK, 1983. [CrossRef]
4. Dhaliwal, M.S. *Okra (Abelmoschus esculentus) L. (Moench)*; Kalyani Publishers: New Delhi, India, 2010.

5. Kumar, A.; Kumar, P.; Nadendla, R. A review on: *Abelmoschus esculentus* (Okra). *Int. Res. J. Pharm. Appl. Sci.* **2013**, *3*, 129–132.
6. Archana, G.; Babu, P.A.S.; Sudharsan, K.; Sabina, K.; Raja, R.P.; Sivarajan, M.; Sukumar, M. Evaluation of Fat Uptake of Polysaccharide Coatings on Deep-Fat Fried Potato Chips by Confocal Laser Scanning Microscopy. *Int. J. Food Prop.* **2015**, *19*, 1583–1592. [CrossRef]
7. Costantino, A.; Romanchik-Cerpovicz, J. Physical and sensory measures indicate moderate fat replacement in frozen dairy dessert is feasible using okra gum as a milk-fat ingredient substitute. *J. Am. Diet. Assoc.* **2004**, *104*, 44. [CrossRef]
8. Yuennan, P.; Sajjaanantakul, T.; Kung, B. Effect of Okra Cell Wall and Polysaccharide on Physical Properties and Stability of Ice Cream. *J. Food Sci.* **2014**, *79*, E1522–E1527. [CrossRef]
9. Islam, M.T. Phytochemical information and pharmacological activities of Okra (*Abelmoschus esculentus*): A literature-based review. *Phytother. Res.* **2019**, *33*, 72–80. [CrossRef]
10. Durazzo, A.; Lucarini, M.; Novellino, E.; Souto, E.B.; Daliu, P.; Santini, A. *Abelmoschus esculentus* (L.): Bioactive Components' Beneficial Properties-Focused on Antidiabetic Role-For Sustainable Health Applications. *Molecules* **2018**, *24*, 38. [CrossRef]
11. S.E.P. Limited. 2021. Available online: https://www.sbepl.com/ (accessed on 18 January 2021).
12. Hunan, H. QiyiXinyeCulture Media, China. 2021. Available online: https://www.globalsources.com/si/AS/Hunan-Qiyi/6008850060925/Showroom/3000000149681/ALL.htm (accessed on 19 January 2021).
13. Vermerris, W.; Nicholson, R. Families of Phenolic Compounds and Means of Classification. *Phenolic Compd. Biochem.* **2006**. [CrossRef]
14. Savello, P.A.; Martin, F.W.; Hill, J.M. Nutritional composition of okra seed meal. *J. Agric. Food Chem.* **1980**, *28*, 1163–1166. [CrossRef]
15. Cieślik, E.; Gębusia, A.; Filipiak-Florkiewicz, A.; Mickowska, B. The content of protein and of amino acids in Jerusalem artichoke tubers (*Helianthus tuberosus* L.) of red variety Rote Zonenkugel. *Acta Sci. Pol. Technol. Aliment.* **2012**, *10*, 433–441.
16. Alghamdi, A.A. Nutritional Assessment of Different Date Fruits (*Phoenix Dactylifera* L.) Varieties Cultivated in Hail Province, Saudi Arabia. *Biosci. Biotechnol. Res. Commun.* **2018**, *11*, 263–269. [CrossRef]
17. Gemede, H.F.; Haki, G.D.; Beyene, F.; Woldegiorgis, A.Z.; Rakshit, S.K. Proximate, mineral, and antinutrient compositions of indigenous Okra (*Abelmoschus esculentus*) pod accessions: Implications for mineral bioavailability. *Food Sci. Nutr.* **2015**, *4*, 223–233. [CrossRef] [PubMed]
18. Aykroyd, W.R.; Gopalan, C.; Balasubramanian, S.C. The Nutritive Value of Indian Foods and the Planning of Satisfactory Diets. *Spec. Rep. Ser. Indian Counc. Med. Res.* **1963**, *42*, 1–255. [PubMed]
19. Akintoye, H.; Adebayo, A.; Aina, O. Growth and Yield Response of Okra Intercropped with Live Mulches. *Asian J. Agric. Res.* **2011**, *5*, 146–153. [CrossRef]
20. Available online: https://nutritiondata.self.com/facts/vegetables-and-vegetable-products/2497/2 (accessed on 10 November 2020).
21. Sharma, K.; Gupta, A.; Kumar, M.; Singh, M.K.; Malik, S.; Singh, B.; Chaudhary, V. Effect of Integrated Nutrient Management and Foliar Spray of Bioregulators on Growth and Yield of Okra. *Int. J. Curr. Microbiol. Appl. Sci.* **2020**, *9*, 344–354.
22. Kumar, R.; Patil, M.; Patil, S.; Paschapur, M. Evaluation of Abelmoschus esculentus mucilage as suspending agent in paracetamol suspension. *Int. J. Pharm. Tech. Res.* **2009**, *1*, 658–665.
23. Avallone, S.; Tiemtore, T.-W.E.; Mouquet-Rivier, C.; Trèche, S. Nutritional value of six multi-ingredient sauces from Burkina Faso. *J. Food Compos. Anal.* **2008**, *21*, 553–558. [CrossRef]
24. Kendall, C.; Jenkins, D.J.A. A Dietary portfolio: Maximal reduction of low-density lipoprotein cholesterol with diet. *Curr. Atheroscler. Rep.* **2004**, *6*, 492–498. [CrossRef]
25. Agbo, A.E.; Gnakri, D.; Beugrz, G.M.; Fondio, L.; Kouamz, C.M. Maturity degree of four Okra fruits varieties and their nutrients composition. *Electoron. J. Food Plants Chem.* **2008**, *5*, 1–4.
26. Langenberg, C.; Lotta, L.A. Genomic insights into the causes of type 2 diabetes. *Lancet* **2018**, *391*, 2463–2474. [CrossRef]
27. Ripon, S.S.; Mahmood, A.; Chowdhury, M.M.; Islam, M.T. A Possible Anti-Atherothrombosis Activity via Cytoprotective Trait of the Clerodenrumviscosum Leaf Methanol Extract. *Insights Biomed.* **2016**, *1*, 1–6. [CrossRef]
28. Tomoda, M.; Shimizu, N.; Gonda, R.; Kanari, M.; Yamada, H.; Hikino, H. Anticomplementary and hypoglycemic activity of Okra and Hibiscus mucilages. *Carbohydr. Res.* **1989**, *190*, 323–328. [CrossRef]
29. Saha, D.; Jain, B.; Jain, V.K. Phytochemical evaluation and characterization of hypogylcemic activity of various extracts of Abelmoschus esculentus Linn. Fruit. *Int. J. Pharm. Pharm. Sci.* **2011**, *3*, 183–185.
30. Khatun, H.; Rahman, A.; Biswas, M.; Islam, A.U. Water-soluble Fraction of *Abelmoschus esculentus* L. Interacts with Glucose and Metformin Hydrochloride and Alters Their Absorption Kinetics after Coadministration in Rats. *ISRN Pharm.* **2011**, *2011*, 260537. [CrossRef]
31. Balakumar, P.; Kadian, S.; Mahadevan, N. Are PPAR alpha agonists a rational therapeutic strategy for preventing abnormalities of the diabetic kidney? *Pharmacol. Res.* **2012**, *65*, 430–436. [CrossRef]
32. Lv, M.; Chen, Z.; Hu, G.-Y.; Li, Q. Therapeutic strategies of diabetic nephropathy: Recent progress and future perspectives. *Drug Discov. Today* **2015**, *20*, 332–346. [CrossRef]
33. Ge, J.; Miao, J.-J.; Sun, X.-Y.; Yu, J.-Y. Huangkui capsule, an extract from *Abelmoschus manihot* (L.) medic, improves diabetic nephropathy via activating peroxisome proliferator–activated receptor (PPAR)-α/γ and attenuating endoplasmic reticulum stress in rats. *J. Ethnopharmacol.* **2016**, *189*, 238–249. [CrossRef]

34. Zhang, L.; Li, P.; Xing, C.-Y.; Zhao, J.-Y.; He, Y.; Wang, J.-Q.; Wu, X.-F.; Liu, Z.; Zhang, A.; Lin, H.; et al. Efficacy and Safety of *Abelmoschus manihot* for Primary Glomerular Disease: A Prospective, Multicenter Randomized Controlled Clinical Trial. *Am. J. Kidney Dis.* **2014**, *64*, 57–65. [CrossRef]
35. Du, L.; Tao, J.; Jiang, S.; Qian, D.; Guo, J.; Duan, J. Metabolic profiles of the Flos *Abelmoschus manihot* extract by intestinal bacteria from the normal and CKD model rats based on UPLC-Q-TOF/MS. *Biomed. Chromatogr.* **2016**, *31*. [CrossRef]
36. Li, P.; Chen, Y.; Lin, H.; Ni, Z.; Zhan, Y.; Wang, R.; Yang, H.; Fang, J.-A.; Wang, N.; Li, W.-G.; et al. *Abelmoschus manihot*—A traditional Chinese medicine versus losartan potassium for treating IgA nephropathy: Study protocol for a randomized controlled trial. *Trials* **2017**, *18*, 170. [CrossRef] [PubMed]
37. Tu, Y.; Sun, W.; Wan, Y.-G.; Che, X.-Y.; Pu, H.-P.; Yin, X.-J.; Chen, H.-L.; Meng, X.-J.; Huang, Y.-R.; Shi, X.-M. Huangkui capsule, an extract from *Abelmoschus manihot* (L.) medic, ameliorates adriamycin-induced renal inflammation and glomerular injury via inhibiting p38MAPK signaling pathway activity in rats. *J. Ethnopharmacol.* **2013**, *147*, 311–320. [CrossRef]
38. Zhao, Q.; Wan, Y.G.; Sun, W.; Wang, C.J.; Wei, Q.X.; Chen, H.L.; Meng, X.J. Effects of Huangkui Capsule on renal inflammatory injury by intervening p38MAPK signaling pathway in rats with adriamycin-induced nephropathy. *China J. Chin. Mater. Med.* **2012**, *37*, 2926–2934.
39. Mao, Z.-M.; Shen, S.-M.; Wan, Y.; Sun, W.; Chen, H.-L.; Huang, M.-M.; Yang, J.-J.; Wu, W.; Tang, H.-T.; Tang, R.-M. Huangkui capsule attenuates renal fibrosis in diabetic nephropathy rats through regulating oxidative stress and p38MAPK/Akt pathways, compared to α-lipoic acid. *J. Ethnopharmacol.* **2015**, *173*, 256–265. [CrossRef] [PubMed]
40. Cai, H.; Su, S.; Qian, D.-W.; Guo, S.; Tao, W.; Cong, X.D.; Tang, R.; Duan, J. Renal protective effect and action mechanism of Huangkui capsule and its main five flavonoids. *J. Ethnopharmacol.* **2017**, *206*, 152–159. [CrossRef] [PubMed]
41. Panneerselvam, K.; Ramachandran, S.; Sabitha, V.; Naveen, K.R. Antidiabetic and antihyperlipidemic potential of *Abelmoschus esculentus* (L.) Moench. in streptozotocin-induced diabetic rats. *J. Pharm. Bioallied Sci.* **2011**, *3*, 397–402. [CrossRef] [PubMed]
42. Sabitha, V.; Panneerselvam, K.; Ramachandran, S. In vitro α-glucosidase and α-amylase enzyme inhibitory effects in aqueous extracts of *Abelmoscus esculentus* (L.) Moench. *Asian Pac. J.Trop. Biomed.* **2012**, *2*, S162–S164. [CrossRef]
43. Lu, Y.; Demleitner, M.F.; Song, L.; Rychlik, M.; Huang, D. Oligomeric proanthocyanidins are the active compounds in *Abelmoschus esculentus* Moench for its α-amylase and α-glucosidase inhibition activity. *J. Funct. Foods* **2016**, *20*, 463–471. [CrossRef]
44. Zhang, T.; Xiang, J.; Zheng, G.; Yan, R.; Min, X. Preliminary characterization and anti-hyperglycemic activity of a pectic polysaccharide from okra (*Abelmoschus esculentus* (L.) Moench.). *J. Funct. Foods* **2018**, *41*, 19–24. [CrossRef]
45. Gemede, H.F.; Haki, G.D.; Beyene, F.; Rakshit, S.K.; Woldegiorgis, A.Z. Indigenous Ethiopian okra (*Abelmoschus esculentus*) mucilage: A novel ingredient with functional and antioxidant properties. *Food Sci. Nutr.* **2018**, *6*, 563–571. [CrossRef]
46. Wang, K.; Li, M.; Wen, X.; Chen, X.; He, Z.; Ni, Y. Optimization of ultrasound-assisted extraction of okra (*Abelmoschus esculentus* (L.) Moench.) polysaccharides based on response surface methodology and antioxidant activity. *Int. J. Biol. Macromol.* **2018**, *114*, 1056–1063. [CrossRef] [PubMed]
47. Williamson, G.; Kay, C.D.; Crozier, A. The Bioavailability, Transport, and Bioactivity of Dietary Flavonoids: A Review from a Historical Perspective. *Compr. Rev. Food Sci. Food Saf.* **2018**, *17*, 1054–1112. [CrossRef] [PubMed]
48. Shen, D.-D.; Li, X.; Qin, Y.-L.; Li, M.-T.; Han, Q.-H.; Zhou, J.; Lin, S.; Zhao, L.; Zhang, Q.; Qin, W.; et al. Physicochemical properties, phenolic profiles, antioxidant capacities, and inhibitory effects on digestive enzymes of okra (*Abelmoschus esculentus*) fruit at different maturation stages. *J. Food Sci. Technol.* **2019**, *56*, 1275–1286. [CrossRef]
49. Gemede, H.F.; Ratta, N.; Haki, G.D.; Woldegiorgis, A.Z.; Beyene, F. Nutritional Quality and Health Benefits of Okra (*Abelmoschus esculentus*): A Review. *J. Food Process. Technol.* **2015**, *6*. [CrossRef]
50. Gonzales, G.B.; Smagghe, G.; Grootaert, C.; Zotti, M.J.; Raes, K.; Van Camp, J. Flavonoid interactions during digestion, absorption, distribution and metabolism: A sequential structure–activity/property relationship-based approach in the study of bioavailability and bioactivity. *Drug Metab. Rev.* **2015**, *47*, 175–190. [CrossRef]
51. Xia, F.; Zhong, Y.; Li, M.; Chang, Q.; Liao, Y.; Shi, Z.; Pan, R.-L. Antioxidant and Anti-Fatigue Constituents of Okra. *Nutrients* **2015**, *7*, 8846–8858. [CrossRef]
52. Arapitsas, P. Identification and quantification of polyphenolic compounds from okra seeds and skins. *Food Chem.* **2008**, *110*, 1041–1045. [CrossRef]
53. Ansari, N.M.; Houlihan, L.; Hussain, B.; Pieroni, A. Antioxidant activity of five vegetables traditionally consumed by South-Asian migrants in Bradford, Yorkshire, UK. *Phytother. Res.* **2005**, *19*, 907–911. [CrossRef]
54. Khomsug, P.; Thongjaroenbuangam, W.; Pakdeenarong, N.; Suttajit, M.; Chantiratikul, P. Antioxidative activities and phenolic content of extracts from okra (*Abelmoschus esculentus* L.). *Res. J. Biol. Sci.* **2010**, *5*, 310–313. [CrossRef]
55. Yuan, K.; Liao, H.; Dong, W.; Shi, X.; Liu, H. Analysis and comparison of the active components and antioxidant activities of extracts from *Abelmoschus esculentus* L. *Pharmacogn. Mag.* **2012**, *8*, 156–161. [CrossRef]
56. Nguyen, N.H.; Ta, Q.T.H.; Pham, Q.T.; Luong, T.N.H.; Phung, V.T.; Duong, T.-H.; Van Giau, V. Anticancer Activity of Novel Plant Extracts and Compounds from *Adenosma bracteosum* (Bonati) in Human Lung and Liver Cancer Cells. *Molecules* **2020**, *25*, 2912. [CrossRef] [PubMed]
57. Tavares-Carreón, F.; De la Torre-Zavala, S.; Arocha-Garza, H.F.; Souza, V.; Galán-Wong, L.J.; Avilés-Arnaut, H. In vitro anticancer activity of methanolic extract of *Granulocystopsis* sp., a microalgae from an oligotrophic oasis in the Chihuahuan desert. *PeerJ* **2020**, *8*, e8686. [CrossRef] [PubMed]

58. National Cancer Institute. Understanding Cancer. 2020. Available online: https://www.cancer.gov/about-cancer/understanding (accessed on 12 November 2020).
59. Vo, T.T.L.; Jang, W.-J.; Jeong, C.-H. Leukotriene A4 hydrolase: An emerging target of natural products for cancer chemoprevention and chemotherapy. *Ann. N. Y. Acad. Sci.* **2018**, *1431*, 3–13. [CrossRef] [PubMed]
60. Reddy, M.N.; Adnan, M.; Alreshidi, M.M.; Saeed, M.; Patel, M. Evaluation of Anticancer, Antibacterial and Antioxidant Properties of a Medicinally Treasured Fern Tectariacoadunata with its Phytoconstituents Analysis by HR-LCMS. *Anti-Cancer Agents Med. Chem.* **2020**, *20*, 1845–1856. [CrossRef]
61. Bhatt, M.; Patel, M.; Adnan, M.; Reddy, M.N. Anti-Metastatic Effects of Lupeol via the Inhibition of MAPK/ERK Pathway in Lung Cancer. *Anti-Cancer Agents Med. Chem.* **2020**, *21*, 201–206. [CrossRef]
62. Adetuyi, F.; Ibrahim, T. Effect of Fermentation Time on the Phenolic, Flavonoid and Vitamin C Contents and Antioxidant Activities of Okra (*Abelmoschus esculentus*) Seeds. *Niger. Food J.* **2014**, *32*, 128–137. [CrossRef]
63. Lin, Y.; Lu, M.-F.; Liao, H.-B.; Li, Y.-X.; Han, W.; Yuan, K. Content determination of the flavonoids in the different parts and different species of *Abelmoschus esculentus* L. by reversed phase-high performance liquid chromatograph and colorimetric method. *Pharmacogn. Mag.* **2014**, *10*, 278–284. [CrossRef]
64. Deng, Y.; Li, S.; Wang, M.; Chen, X.; Tian, L.; Wang, L.; Yang, W.; Chen, L.; He, F.; Yin, W. Flavonoid-rich extracts from okra flowers exert antitumor activity in colorectal cancer through induction of mitochondrial dysfunction-associated apoptosis, senescence and autophagy. *Food Funct.* **2020**, *11*, 10448–10466. [CrossRef]
65. Chaemsawang, W.; Prasongchean, W.; Papadopoulos, K.I.; Ritthidej, G.; Sukrong, S.; Wattanaarsakit, P. The Effect of Okra (*Abelmoschus esculentus* (L.) Moench) Seed Extract on Human Cancer Cell Lines Delivered in Its Native Form and Loaded in Polymeric Micelles. *Int. J. Biomater.* **2019**, *2019*, 9404383. [CrossRef]
66. Ping, M.-H. Hyperin Controls the Development and Therapy of Gastric Cancer via Regulating Wnt/β-Catenin Signaling. *Cancer Manag. Res.* **2020**, *12*, 11773–11782. [CrossRef]
67. Hong, C.-E.; Park, A.-K.; Lyu, S.-Y. Synergistic anticancer effects of lectin and doxorubicin in breast cancer cells. *Mol. Cell. Biochem.* **2014**, *394*, 225–235. [CrossRef] [PubMed]
68. Yau, T.; Dan, X.; Ng, C.C.W.; Ng, T.B. Lectins with Potential for Anti-Cancer Therapy. *Molecules* **2015**, *20*, 3791–3810. [CrossRef] [PubMed]
69. Figueiroa, E.D.O.; Da Cunha, C.R.A.; Albuquerque, P.B.; De Paula, R.A.; Aranda-Souza, M.; Da Silva, L.C.N.; Zagmignan, A.; Carneiro-Da-Cunha, M.G.; Da Silva, L.C.N.; Correia, M.T.D.S. Lectin-Carbohydrate Interactions: Implications for the Development of New Anticancer Agents. *Curr. Med. Chem.* **2017**, *24*, 3667–3680. [CrossRef]
70. Monte, L.G.; Santi-Gadelha, T.; Reis, L.B.; Braganhol, E.; Prietsch, R.F.; Dellagostin, O.A.; Lacerda, E.R.R.; Gadelha, C.A.A.; Conceição, F.R.; Pinto, L.S. Lectin of *Abelmoschus esculentus* (okra) promotes selective antitumor effects in human breast cancer cells. *Biotechnol. Lett.* **2013**, *36*, 461–469. [CrossRef]
71. Sharma, N.; Sabyasachi, S. *Immunomodulatory Potential of Phytochemicals: Recent Updates*; Springer Nature: Singapore, 2019.
72. Hosseinzade, A.; Sadeghi, O.; Biregani, A.N.; Soukhtehzari, S.; Brandt, G.S.; Esmaillzadeh, A. Immunomodulatory Effects of Flavonoids: Possible Induction of T CD4+ Regulatory Cells Through Suppression of mTOR Pathway Signaling Activity. *Front. Immunol.* **2019**, *10*, 51. [CrossRef]
73. Soares, G.S.F.; Assreuy, A.M.S.; Gadelha, C.A.A.; Gomes, V.M.; Delatorre, P.; Simões, R.C.; Cavada, B.S.; Leite, J.F.; Nagano, C.S.; Pinto, N.V.; et al. Purification and Biological Activities of Abelmoschus esculentus Seed Lectin. *Protein J.* **2012**, *31*, 674–680. [CrossRef]
74. Ortaç, D.; Cemek, M.; Karaca, T.; Büyükokuroğlu, M.E.; Özdemir, Z.; Kocaman, A.T.; Göneş, S. In vivo anti-ulcerogenic effect of okra (*Abelmoschus esculentus*) on ethanol-induced acute gastric mucosal lesions. *Pharm. Biol.* **2018**, *56*, 165–175. [CrossRef]
75. Sheu, S.-C.; Lai, M.-H. Composition analysis and immuno-modulatory effect of okra (*Abelmoschus esculentus* L.) extract. *Food Chem.* **2012**, *134*, 1906–1911. [CrossRef]
76. Chen, H.; Jiao, H.; Cheng, Y.; Xu, K.; Jia, X.; Shi, Q.; Guo, S.; Wang, M.; Du, L.; Wang, F. In Vitro and In Vivo Immunomodulatory Activity of Okra (*Abelmoschus esculentus* L.) Polysaccharides. *J. Med. Food* **2016**, *19*, 253–265. [CrossRef]
77. De Carvalho, C.C.C.R.; Cruz, P.A.; Da Fonseca, M.M.R.; Xavier-Filho, L. Antibacterial properties of the extract of *Abelmoschus esculentus*. *Biotechnol. Bioprocess Eng.* **2011**, *16*, 971–977. [CrossRef]
78. Lengsfeld, C.; Titgemeyer, F.; Faller, G.; Hensel, A. Glycosylated compounds from okra inhibit adhesion of *Helicobacter pylori* to human gastric mucosa. *J. Agric. Food Chem.* **2004**, *52*, 1495–1503. [CrossRef] [PubMed]
79. Fan, S.; Zhang, Y.; Sun, Q.; Yu, L.; Li, M.; Zheng, B.; Wu, X.; Yang, B.; Li, Y.; Huang, C. Extract of okra lowers blood glucose and serum lipids in high-fat diet-induced obese C57BL/6 mice. *J. Nutr. Biochem.* **2014**, *25*, 702–709. [CrossRef] [PubMed]
80. Liu, J.; Zhao, Y.; Wu, Q.; John, A.; Jiang, Y.; Yang, J.; Liu, H.; Yang, B. Structure characterisation of polysaccharides in vegetable "okra" and evaluation of hypoglycemic activity. *Food Chem.* **2018**, *242*, 211–216. [CrossRef]
81. Damodaran, D.; Jeyakani, J.; Chauhan, A.; Kumar, N.; Chandra, N.; Surolia, A. CancerLectinDB: A database of lectins relevant to cancer. *Glycoconj. J.* **2007**, *25*, 191–198. [CrossRef] [PubMed]
82. Pienta, K.J.; Nailk, H.; Akhtar, A.; Yamazaki, K.; Replogle, T.S.; Lehr, J.; Donat, T.L.; Tait, L.; Hogan, V.; Raz, A. Inhibition of Spontaneous Metastasis in a Rat Prostate Cancer Model by Oral Administration of Modified Citrus Pectin. *JNCIJ. Natl. Cancer Inst.* **1995**, *87*, 348–353. [CrossRef]

83. Nangia-Makker, P.; Nakahara, S.; Hogan, V.; Raz, A. Galectin-3 in apoptosis, a novel therapeutic target. *J. Bioenerg. Biomembr.* **2007**, *39*, 79–84. [CrossRef]
84. Olano-Martin, E.; Rimbach, G.H.; Gibson, G.R.; Rastall, R.A. Pectin and pectic-oligosaccharides induce apoptosis in in vitro human colonic adenocarcinoma cells. *Anticancer Res.* **2003**, *23*, 341–346.
85. Vayssade, M.; Sengkhamparn, N.; Verhoef, R.; Delaigue, C.; Goundiam, O.; Vigneron, P.; Voragen, A.G.; Schols, H.A.; Nagel, M.-D. Antiproliferative and proapoptotic actions of okra pectin on B16F10 melanoma cells. *Phytother. Res.* **2009**, *24*, 982–989. [CrossRef]
86. Soma Das, G.N.; Ghosh, L.K. Okra and its various applications in Drug Delivery, Food Technology, Health Care and Pharmacological Aspects—A Review. *J. Pharm. Sci. Res.* **2019**, *11*, 2139–2147.
87. Geng, S.; Liu, Y.; Ma, H.; Chen, C. Extraction and Antioxidant Activity of Phenolic Compounds from Okra Flowers. *Trop. J. Pharm. Res.* **2015**, *14*, 807. [CrossRef]
88. Hu, L.; Yu, W.; Li, Y.; Prasad, N.; Tang, Z. Antioxidant Activity of Extract and Its Major Constituents from Okra Seed on Rat Hepatocytes Injured by Carbon Tetrachloride. *BioMed Res. Int.* **2014**, *2014*, 341291. [CrossRef] [PubMed]
89. Hurrell, R.F. Influence of Vegetable Protein Sources on Trace Element and Mineral Bioavailability. *J. Nutr.* **2003**, *133*, 2973S–2977S. [CrossRef] [PubMed]
90. Faruque, A.; Ashraf, S.A.; Azad, A.A.; Prasad, P.B. Production and development of nutraceuticals using *Bacillus subtilis* NCIM 2708 under solid statefermentation by response surface methodology. *Eur. Sci. J.* **2013**, *9*. [CrossRef]
91. Subrahmanyam, G.V.; Sushma, M.; Alekya, A.; Neeraja, C.; Harsha, H.S.S.; Ravindra, J. Antidiabetic Activity of *Abelmoschus esculentus* Fruit Extract. *Int. J. Res. Pharm. Chem.* **2011**, *1*, 17–20.
92. Thöle, C.; Brandt, S.; Ahmed, N.; Hensel, A. Acetylated Rhamnogalacturonans from Immature Fruits of *Abelmoschus esculentus* Inhibit the Adhesion of *Helicobacter pylori* to Human Gastric Cells by Interaction with Outer Membrane Proteins. *Molecules* **2015**, *20*, 16770–16787. [CrossRef]
93. Messing, J.; Thöle, C.; Niehues, M.; Shevtsova, A.; Glocker, E.; Borén, T.; Hensel, A. Antiadhesive Properties of *Abelmoschus esculentus* (Okra) Immature Fruit Extract against *Helicobacter pylori* Adhesion. *PLoS ONE* **2014**, *9*, e84836. [CrossRef]
94. Dubey, P.; Mishra, S. A Review on: Diabetes and Okra (*Abelmoschus esculentus*). *J. Med. Plants Stud.* **2017**, *5*, 23–26.
95. Abi, I.; Abi, L.; Ladan, M.J. Hypoglycaemic Effect of *Abelmoschus Esculentus* Extracts in Alloxan-Induced Diabetic Wistar Rats. *Endocrinol. Diabetes Res.* **2017**, *3*. [CrossRef]
96. Yaradua, I.; Ibrahim, M.; Matazu, K.I.; Nasir, A.; Matazu, N.U.; Zainab, A.S.; Abdul Rahman, M.B.; Bilbis, L.; Abbas, A.Y. Antidiabetic Activity of *Abelmoschus esculentus* (Ex-Maradi Okra) Fruit in Alloxan-induced Diabetic Rats. *Niger. J. Biochem. Mol. Biol.* **2017**, *32*, 44–52.
97. Abbas, A.; Muhammad, I.; Abdulrahman, M.B.; Bilbis, L.S.; Saidu, Y.; Onu, A. Possible Antidiabetic Mechanism of Action of Ex-maradi Okra Fruit Variety (*Abelmoscus esculentus*) on Alloxan Induced Diabetic Rats. *Niger. J. Basic Appl. Sci.* **2018**, *25*, 101. [CrossRef]
98. Majd, N.E.; Tabandeh, M.R.; Shahriari, A.; Soleimani, Z. Okra (*Abelmoscus esculentus*) Improved Islets Structure, and Down-Regulated PPARs Gene Expression in Pancreas of High-Fat Diet and Streptozotocin-Induced Diabetic Rats. *Cell J.* **2018**, *20*, 31–40. [CrossRef]
99. Nguekouo, P.T.; Mbaveng, A.T.; Kengne, A.P.N.; Woumbo, C.Y.; Tekou, F.A.; Oben, J.E. Effect of boiling and roasting on the antidiabetic activity of *Abelmoschus esculentus* (Okra) fruits and seeds in type 2 diabetic rats. *J. Food Biochem.* **2018**, *42*, e12669. [CrossRef]
100. Hossen, M.A.; Jahan, I.; Mamun, M.A.A.; Sakir, J.A.M.S.; Shamimuzzaman, M.; Uddin, M.J.; Haque, M.E. CNS depressant and analgesic activities of Okra (*Abelmoschus esculentus* Linn.). *Mol. Clin. Pharmacol.* **2013**, *4*, 44–52.
101. Wang, H.; Chen, G.; Ren, D.; Yang, S.-T. Hypolipidemic Activity of Okra is Mediated through Inhibition of Lipogenesis and Upregulation of Cholesterol Degradation. *Phytother. Res.* **2013**, *28*, 268–273. [CrossRef]
102. Medina-Remón, A.; Casas, R.; Tresserra-Rimbau, A.; Ros, E.; González, M.M.; Fitó, M.; Corella, D.; Salas-Salvadó, J.; Raventós, R.M.L.; Estruch, R.; et al. Polyphenol intake from a Mediterranean diet decreases inflammatory biomarkers related to atherosclerosis: Asubstudy of the PREDIMED trial. *Br. J. Clin. Pharmacol.* **2016**, *83*, 114–128. [CrossRef]
103. Doreddula, S.K.; Bonam, S.R.; Gaddam, D.P.; Desu, B.S.R.; RamaRao, N.; Pandy, V. Phytochemical Analysis, Antioxidant, Antistress, and Nootropic Activities of Aqueous and Methanolic Seed Extracts of Ladies Finger (*Abelmoschus esculentus* L.) in Mice. *Sci. World J.* **2014**, *2014*, 519848. [CrossRef]
104. Adnan, M.; Patel, M.; Deshpande, S.; Alreshidi, M.; Siddiqui, A.J.; Reddy, M.N.; Emira, N.; De Feo, V. Effect of *Adiantumphilippense* Extract on Biofilm Formation, Adhesion with Its Antibacterial Activities Against Foodborne Pathogens, and Characterization of Bioactive Metabolites: An in vitro-in silico Approach. *Front. Microbiol.* **2020**, *11*, 823. [CrossRef]
105. Ashraf, S.A.; Adnan, M.; Patel, M.; Siddiqui, A.J.; Sachidanandan, M.; Snoussi, M.; Hadi, S.S. Fish-Based Bioactives as Potent Nutraceuticals: Exploring the Therapeutic Perspective of Sustainable Food from the Sea. *Mar. Drugs* **2020**, *18*, 265. [CrossRef]
106. Ahmad, F.; Ashraf, S.A.; Ahmad, F.A.; Ansari, J.A.; Siddiquee, R.A. Nutraceutical Market and its Regulation. *Am. J. Food Technol.* **2011**, *6*, 342–347. [CrossRef]
107. Ashraf, S.A.; ElKhalifa, A.E.O.; Siddiqui, A.J.; Patel, M.; AwadElkareem, A.M.; Snoussi, M.; Ashraf, M.S.; Adnan, M.; Hadi, S.S. Cordycepin for Health and Wellbeing: A Potent Bioactive Metabolite of an Entomopathogenic Cordyceps Medicinal Fungus and Its Nutraceutical and Therapeutic Potential. *Molecules* **2020**, *25*, 2735. [CrossRef]

108. Roy, A.; Shrivastava, S.L.; Mandal, S.M. Functional properties of Okra *Abelmoschus esculentus* L. (Moench): Traditional claims and scientific evidences. *Plant Sci. Today* **2014**, *1*, 121–130. [CrossRef]
109. Zaharuddin, N.D.; Noordin, M.I.; Kadivar, A. The Use of *Hibiscus esculentus* (Okra) Gum in Sustaining the Release of Propranolol Hydrochloride in a Solid Oral Dosage Form. *BioMed Res. Int.* **2014**, *2014*, 735891. [CrossRef] [PubMed]
110. Dwyer, J.T.; Coates, P.M.; Smith, M.J. Dietary Supplements: Regulatory Challenges and Research Resources. *Nutrients* **2018**, *10*, 41. [CrossRef] [PubMed]
111. Petropoulos, S.A.; Fernandes, A.; Barros, L.; Ferreira, I.C.F.R. Chemical composition, nutritional value and antioxidant properties of Mediterranean okra genotypes in relation to harvest stage. *Food Chem.* **2018**, *242*, 466–474. [CrossRef] [PubMed]
112. Institute of Medicine Standing (US) Committee on the Scientific Evaluation of Dietary Reference Intakes and its Panel on Folate, Other B Vitamins, and Choline. The National Academies Collection: Reports funded by National Institutes of Health. In *Dietary Reference Intakes for Thiamin, Riboflavin, Niacin, Vitamin B(6), Folate, Vitamin B(12), Pantothenic Acid, Biotin, and Choline*; National Academies Press: Washington, DC, USA, 1998.
113. Gasperi, V.; Sibilano, M.; Savini, I.; Catani, M.V. Niacin in the Central Nervous System: An Update of Biological Aspects and Clinical Applications. *Int. J. Mol. Sci.* **2019**, *20*, 974. [CrossRef] [PubMed]
114. Bakre, L.G.; Jaiyeoba, K.T. Effects of drying methods on the physicochemical and compressional characteristics of Okra powder and the release properties of its metronidazole tablet formulation. *Arch. Pharmacal Res.* **2009**, *32*, 259–267. [CrossRef] [PubMed]
115. Jeong, W.-S.; Kong, A.-N.T. Biological Properties of Monomeric and Polymeric Catechins: Green Tea Catechins and Procyanidins. *Pharm. Biol.* **2004**, *42* (Suppl. S1), 84–93. [CrossRef]
116. Adelakun, O.; Oyelade, O.; Ade-Omowaye, B.; Adeyemi, I.; Van De Venter, M. Chemical composition and the antioxidative properties of Nigerian Okra Seed (*Abelmoschus esculentus* Moench) Flour. *Food Chem. Toxicol.* **2009**, *47*, 1123–1126. [CrossRef]
117. Diwan, A.; Ninawe, A.; Harke, S. Gene editing (CRISPR-Cas) technology and fisheries sector. *Can. J. Biotechnol.* **2017**, *1*, 65–72. [CrossRef]
118. Adelakun, O.E.; Oyelade, O.J. Chemical and Antioxidant Properties of Okra (*Abelmoschus esculentus* Moench) Seed. *Nuts Seeds Health Dis. Prev.* **2011**, 841–846. [CrossRef]
119. Adetuyi, F.; Ajala, L.; Ibrahim, T. Effect of the addition of defatted okra seed (*Abelmoschus esculentus*) flour on the chemical composition, functional properties and Zn bioavailability of plantain (*Musa paradisiacal* Linn) flour. *J. Microbiol. Biotechnol. Food Sci.* **2012**, *2*, 69–82.
120. Jarret, R.L.; Wang, M.L.; Levy, I.J. Seed Oil and Fatty Acid Content in Okra (*Abelmoschus esculentus*) and Related Species. *J. Agric. Food Chem.* **2011**, *59*, 4019–4024. [CrossRef] [PubMed]
121. Agostoni, C.; Moreno, L.; Shamir, R. Palmitic Acid and Health: Introduction. *Crit. Rev. Food Sci. Nutr.* **2015**, *56*, 1941–1942. [CrossRef] [PubMed]
122. Dong, Z.; Zhang, J.; Tian, K.-W.; Pan, W.-J.; Wei, Z.-J. The Fatty Oil from Okra Seed: Supercritical Carbon Dioxide Extraction, Composition and Antioxidant Activity. *Curr. Top. Nutraceutical Res.* **2014**, *12*, 75–84.
123. López-Huertas, E. Health effects of oleic acid and long chain omega-3 fatty acids (EPA and DHA) enriched milks. A review of intervention studies. *Pharmacol. Res.* **2010**, *61*, 200–207. [CrossRef]
124. Jandacek, R.J. Linoleic Acid: A Nutritional Quandary. *Healthcare* **2017**, *5*, 25. [CrossRef]
125. Steyn, N.P.; Mchiza, Z.; Hill, J.; Davids, Y.D.; Venter, I.; Hinrichsen, E.; Opperman, M.; Rumbelow, J.; Jacobs, P. Nutritional contribution of street foods to the diet of people in developing countries: A systematic review. *Public Health Nutr.* **2013**, *17*, 1363–1374. [CrossRef]
126. Slavin, J.; Carlson, J. Carbohydrates. *Adv. Nutr.* **2014**, *5*, 760–761. [CrossRef]
127. Xiao, J.B.; Capanoglu, E.; Jassbi, A.R.; Miron, A. Advance on the FlavonoidC-glycosides and Health Benefits. *Crit. Rev. Food Sci. Nutr.* **2015**, *56*, S29–S45. [CrossRef]
128. Sunilson, J.J.; Jayaraj, P.; Mohan, M.; Kumari, A.G.; Varatharajan, R. Antioxidant and hepatoprotective effect of the roots of *Hibiscus esculentus* Linn. *Int. J. Green Pharm.* **2008**, *2*, 200. [CrossRef]
129. Shenkin, A. Micronutrients in health and disease. *Postgrad. Med. J.* **2006**, *82*, 559–567. [CrossRef] [PubMed]
130. Idris, S.; Yisa, J.; Itodo, A. Proximate and mineral composition of the leaves of *Abelmoschus esculentus*. *Int. J. Trop. Agric. Food Syst.* **2010**, *3*. [CrossRef]
131. Caluête, M.; Souza, L.D.; Ferreira, E.D.S.; França, A.; Gadelha, C.D.A.; Aquino, J.S.; Santi-Gadelha, T. Nutritional, antinutritional and phytochemical status of okra leaves (*Abelmoschus esculentus*) subjected to different processes. *Afr. J. Biotechnol.* **2015**, *14*, 683–687.
132. McCutcheon, A.; Ellis, S.; Hancock, R.; Towers, G. Antifungal screening of medicinal plants of British Columbian native peoples. *J. Ethnopharmacol.* **1994**, *44*, 157–169. [CrossRef]
133. Kumar, S. Physicochemical, Phytochemical and toxicity studies on gum and mucilage from plant *Abelmoschus esculentus*. *J. Phytopharm.* **2014**, *3*, 200–203.
134. Muhammad, I.; Matazu, I.K.; Yaradua, I.A.; Yau, S.; Nasir, A.; Bilbis, S.L.; Abbas, Y.A. Development of Okra-Based Antidiabetic Nutraceutical Formulation from Abelmoschus esculentus (L.) Moench (Ex-maradi Variety). *Trop. J. Nat. Prod. Res.* **2018**, *2*, 80–86. [CrossRef]
135. Reddy, M.T.; Kadiyala, H.; Mutyala, G.; Hameedunnisa, B. Heterosis for Yield and Yield Components in Okra (*Abelmoschus esculentus* (L.) Moench. *Chil. J. Agric. Res.* **2012**, *72*, 316–325. [CrossRef]

136. Dan-Dan, R.; Gu, C. Inhibition Effect of Okra Polysaccharides on Proliferation of Human Cancer Cell Lines. *Food Sci.* **2010**, *31*, 353–356. [CrossRef]
137. M.R. Future, Seeds Market Research Report. 2020. Available online: https://www.marketresearchfuture.com/reports/okra-seeds-market-7715 (accessed on 11 October 2020).
138. Wire, G.N. Okra Seeds Market Report Insights and Industry Analysis by Category (Conventional and Organic) and Region, Competitive Market Size, Share, Trends, and Forecast, 2018–2023. 2020. Available online: https://www.globenewswire.com/news-release/2019/07/25/1887743/0/en/Okra-Seeds-Market-Size-to-Reach-USD-352-7-Million-by-2023-at-9-8-CAGR-Predicts-Market-Research-Future.html (accessed on 15 October 2020).
139. WBOC. Okra Seeds Market—Global Countries Data, 2020. Top Leading Countries, Companies, Consumption, Drivers, Trends, Forces Analysis, Revenue, Market Size & Growth, Global Forecast 2025. 2020. Available online: https://www.wboc.com/story/42977060/okra-seeds-market-2020-regional-analysis-with-top-countries-data-trends-definition-share-market-size-and-forecast-report-by-2026 (accessed on 2 December 2020).
140. Uebelhack, R.; Bongartz, U.; Seibt, S.; Bothe, G.; Chong, P.-W.; De Costa, P.; Wszelaki, N. Double-Blind, Randomized, Three-Armed, Placebo-Controlled, Clinical Investigation to Evaluate the Benefit and Tolerability of Two Dosages of IQP-AE-103 in Reducing Body Weight in Overweight and Moderately Obese Subjects. *J. Obes.* **2019**, *2019*, 3412952. [CrossRef]

Article

New Freeze-Dried Andean Blueberry Juice Powders for Potential Application as Functional Food Ingredients: Effect of Maltodextrin on Bioactive and Morphological Features

Mauren Estupiñan-Amaya [1], Carlos Alberto Fuenmayor [2] and Alex López-Córdoba [1,*]

1 Facultad Seccional Duitama, Escuela de Administración de Empresas Agropecuarias, Universidad Pedagógica y Tecnológica de Colombia, Carrera 18 con Calle 22 Duitama, Boyaca 150461, Colombia; maurenest01@gmail.com
2 Instituto de Ciencia y Tecnología de Alimentos (ICTA), Universidad Nacional de Colombia, Av. Carrera 30 # 45-03, Bogota 111321, Colombia; cafuenmayorb@unal.edu.co
* Correspondence: alex.lopez01@uptc.edu.co; Tel.: +57-8-7604100

Academic Editors: Severina Pacifico and Simona Piccolella
Received: 29 October 2020; Accepted: 23 November 2020; Published: 30 November 2020

Abstract: Andean blueberry (*Vaccinium meridionale* Swartz) fruits are an underutilized source of anthocyanins and other valuable bioactive phytochemicals. The purpose of this work was to obtain Andean blueberry juice powders via freeze-drying processing and evaluate the effect of maltodextrin as a drying aid on their physicochemical, technological, microstructural, and bioactive characteristics. Andean blueberry juices were mixed with variable proportions of maltodextrin (20–50%); freeze-dried; and characterized in terms of their tristimulus color, Fourier transform infrared spectra (FTIR), moisture content, water activity, morphology, water solubility, flow properties, total polyphenols and anthocyanins content, and DPPH•-scavenging capacity. The powders obtained presented suitable characteristics in terms of their water activity (<0.5), solubility (>90%), and bioactive compound recovery (>70% for total phenolics, and >60% for total monomeric anthocyanins), with antioxidant activities up to 4 mg equivalent of gallic acid/g of dry matter. Although an increased content of maltodextrin resulted in lower concentrations of phytochemicals, as expected, it also favored an increased % recovery (over 90% of total phenolics at the highest maltodextrin proportion) and improved their flow properties. Freeze-dried juice powders are a potential alternative for the stabilization and value addition of this fruit as a new source of functionality for processed foods.

Keywords: anthocyanins; antioxidant activity; bioactive compounds; colorants; fruit juices; polyphenols; wild blueberry

1. Introduction

The use of natural ingredients has received widespread attention in recent years due to its high demand in different industrial fields such as food, pharmaceutics, and cosmetics [1]. In particular, in the food industry there is a growing interest in the extraction, characterization, and stabilization of new natural ingredients that can be further incorporated into functional foods.

The genus *Vaccinium* (family Ericaceae) comprises about 450 species, known for their high content of phytochemicals [2]. Andean blueberry (*Vaccinium meridionale* Swartz) is a wild shrub with few commercial exploitations that grows in the Andean region of South America at 2300–3300 m above sea level (m.a.s.l.) [3]. Several studies have reported that Andean blueberry fruits are a rich source of bioactive compounds, such as anthocyanins, flavonoids, and phenolic acids, which have been associated with antioxidant, anticarcinogenic, and anti-inflammatory properties [4–6].

Andean blueberry fruits are commonly marketed either fresh or processed into jellies and jams, with a very low supply of derived value-added products [4]. Andean blueberry juice is very attractive to small-scale food processors of rural areas because it is easy to produce and has adequate sensory properties [7]. Indeed, several studies had reported that Andean blueberry juice has a high potential for use as an antioxidant additive or functional ingredient in foods because it is rich in polyphenols and has a high antioxidant capacity [3,4]. However, as with most fruit juices Andean blueberry juices are susceptible to microbial spoilage and degradation due to their high moisture content. Moreover, the anthocyanin content and antioxidant capacity of Andean blueberry juice are significantly reduced during storage [3,8]. Therefore, it is necessary to develop strategies to increase their storage life.

The dehydration of fruit juices is a promising approach to obtain highly stable dried powders that are more resistant to microbial and oxidative degradation, light in weight, and readily soluble; furthermore, it enables room temperature storage over longer periods [9]. Several drying techniques are available for the production of food powders at an industrial scale, with freeze drying and spray-drying being the most successful methods for fruit juice powder production [10].

Freeze drying is a low-temperature dehydration process that removes ice or frozen solvent through sublimation and has the advantages of being flexible, straightforward, and easily scalable [11,12]. Additionally, freeze-drying is suitable for the processing of heat-sensitive active compounds, because substances are not exposed to high temperatures, unlike in conventional air-drying and spray-drying [12,13]. Nonetheless, fruit juices are generally difficult to dry due to their high content of sugars and organic acids, which make the dried products extremely hygroscopic, adhesive, and very susceptible to degradation during storage. Therefore, food-grade drying aids (carriers), such as maltodextrin and gum Arabic, have been added to spray-dried or freeze-dried fruit juices [9,14].

Maltodextrins are polysaccharides obtained by the partial enzymatic or acid hydrolysis of starch that consist of β-D-glucose units linked mainly by glycosidic bonds α(1→4). These low-cost carbohydrate polymers feature good film-forming properties, a high water solubility, and a neutral taste and aroma [10]. It has been reported that maltodextrins allow obtaining juice powders with a good stability against oxidation, ease of handling, improved solubility, and extended shelf life [10]. Several studies have been carried out to evaluate the optimal drying conditions for different fruit juices. Lachowicz et al. [15] studied the effect of maltodextrin and inulin on the protection of natural antioxidants in powders made of Saskatoon berry fruit, juice, and pomace, finding that the freeze-drying process using these wall materials led to the highest content of polyphenolic compounds and antioxidant activity. Pudziuvelyte et al. [16] encapsulated *Elsholtzia ciliata ethanolic* extract by freeze-drying using skim milk, sodium caseinate, gum Arabic, maltodextrin, β-cyclodextrin, and resistant maltodextrin, alone or in mixtures of two or four encapsulants. The highest value of encapsulation efficiency of phenolic compounds was obtained for powders prepared using sodium caseinate alone or in a mixture with resistant maltodextrin and maltodextrin.

The aim of the current work was to develop freeze-dried Andean blueberry juice powders for potential applications as ingredients with bioactive characteristics. The effect of the addition of different maltodextrin concentrations (20–50%) on the moisture content, water activity, water solubility, bulk density, flow properties, color attributes, morphology, polyphenols and anthocyanins content, and antioxidant capacity of the juice powders was evaluated. To the best of our knowledge, this is the first report on powders from freeze-dried Andean blueberry juice.

2. Results and Discussion

2.1. Andean Blueberry Juice Properties

The physicochemical properties of the Andean blueberry juice used in this work are presented in Table 1. The juice was characterized by a relatively high content of soluble solids (13.27 °Brix) and high concentrations of antioxidants, both total phenolic compounds and anthocyanins, which are expected to be more stable at the mildly acidic pH of this product (2.91) [17]. In fact, the juice presented a

DPPH radical scavenging capacity of 19.1 mg GAE/g dw. These characteristics, along with its deep ruby color, similar to that of Tannat or Merlot wines, and sweet characteristic taste highlight that this product is suitable for consumption as fresh juice with antioxidant features or for addition as a functional ingredient. The composition of berry fruits and juices depends on the cultivar, maturity stage, and agro-climatic conditions. In particular, the characteristics of Andean blueberry could be more variable because it grows as a wild shrub. According to previous reports, Andean blueberry fruits at maturity stages have a soluble solids content ranging between 12 and 18 °Brix, a pH between 2.5 and 3.0, a dry solid content between 17% and 23%, total monomeric anthocyanins between ~329 ± 28 mg cyd-3-glu/100 g, and phenolic compounds ~758.6 ± 62.3 mg gallic acid equivalent/100 g [3,18]. Franco et al. found similar values to those reported in this study (Table 1) when they evaluated the physicochemical properties of Andean blueberry nectar [19]. Casati et al. [20] evaluated the physicochemical characteristics from berries juices cultivated in Argentina (blueberry, elderberry, blackcurrant, and maqui berry), finding soluble solids contents ranging between 9.0 and 14.8 °Brix, a pH between 3.4 and 4.2, a water activity between 0.983 and 0.989, total polyphenol contents between 2970 and 9340, and total monomeric anthocyanins contents between 288 and 1795.4 mg cyd-3-glu/L.

Table 1. Physicochemical properties of the Andean blueberry juice.

Physicochemical Property	Value
Soluble solids content (°Brix)	13.27 ± 0.05
Dry solid content (%)	11.6 ± 0.3
pH	2.91 ± 0.07
Water activity	0.97 ± 0.01
Color coordinates (CIELAB)	$L^* = 22.7 \pm 0.4$
	$a^* = 22.5 \pm 0.4$
	$b^* = 7.8 \pm 0.5$
	$h = 19.2 \pm 1.9$
	$c = 23.2 \pm 0.8$
Total polyphenol content (mg GAE/L)	2032.5 ± 41.7
Monomeric anthocyanin content (mg cyd-3-glu/L)	371.5 ± 20.1
Antioxidant capacity (mg GAE/ g dw)	19.1 ± 0.3

mg GAE: milligrams equivalents of gallic acid; g dw: grams of dry matter; mg cyd-3-glu: milligrams of cyanidin 3-glucoside.

2.2. Physicochemical and Morphological Properties of the Juice Freeze-Dried Powders

Mixtures of the extracted blueberry juices and maltodextrin with no further ingredients were prepared and subjected to freeze drying processing to obtain juice powders. The general appearance of the juice powders obtained with the addition of maltodextrin is shown in Figure 1; unlike the maltodextrin-free freeze-dried juice (Figure S1, Supplementary Material), these powders were solid, easily manageable, and did not adhere to solid surfaces.

Figure 1. Images of freeze-dried powders obtained using different maltodextrin (MD) concentrations.

The CIELAB color coordinates of Andean blueberry juice powders are given in Table 2. Compared to the deep ruby color of the pure juice, the colors of the powders were dark fuchsia pink. The lightness of the Andean blueberry juice powders increased when the maltodextrin concentration increased from 20% to 40% (i.e., a lighter color). Above this material amount, a slight decrease in this parameter was observed (Table 2). Other authors have also described a higher lightness after the addition of higher maltodextrin amounts [21].

Table 2. Color parameter of Andean blueberry juice freeze-dried powders obtained with different concentrations of maltodextrin.

Maltodextrin Concentration (%)	L*	a*	b*	h	Chroma
20	47.2 ± 0.8 [a]	42.3 ± 0.5 [a]	2.6 ± 0.4 [a]	3.6 ± 0.5 [a]	42.3 ± 0.5 [a]
30	54.2 ± 0.6 [b]	36.6 ± 0.3 [b]	1.4 ± 0.1 [b]	2.2 ± 0.2 [b]	36.6 ± 0.3 [b]
40	56.7 ± 1.1 [c]	35.6 ± 0.6 [b]	3.4 ± 0.4 [a]	5.5 ± 0.6 [c]	35.8 ± 0.6 [b]
50	52.7 ± 1.7 [b]	38.0 ± 2.7 [b]	2.9 ± 0.4 [a]	4.3 ± 0.7 [b]	38.1 ± 2.7 [b]

Different letters within the same column indicate significant differences ($p < 0.05$).

Hue angle (h) and chroma are a very important color attributes that characterize the perception and the purity and intensity of the color, respectively. All the powders showed low h values (i.e., closest to the angle for red (0°)), as expected in this type of products (Table 2).

The samples with maltodextrin at 20% showed the higher values of chroma, indicating the highest intensity of color. Above this concentration, non-significant differences between the chroma values of the samples were observed (Table 2). As expected, the samples with a higher Andean blueberry juice concentration (maltodextrin at 20%) showed the higher values of coordinate a* (indicating redness) (Table 2).

The moisture content and the water activity of the Andean blueberry freeze-dried powders ranged from 4.0% to 9.0% and from 0.2 to 0.5, respectively (Table 3). The longer shelf life of dried products is closely related to lower moisture content and water activity. It has been reported that, in dried food powders with a low water activity ($a_w < 0.6$), no microbial proliferation occurs, and the product could be considered fully stable in that respect [22]. Similar results were obtained for sumac extract powders by Caliskan et al. [14].

Table 3. Moisture content, water activity, and water solubility of Andean blueberry juice freeze-dried powders obtained with different concentrations of maltodextrin.

Maltodextrin Concentration (%)	Moisture Content (%)	Water Activity (a_w)	Water Solubility (%)
20	6.1 ± 0.4 [a]	0.31 ± 0.03 [a]	94.6 ± 0.4 [a]
30	4.3 ± 0.1 [b]	0.27 ± 0.01 [a]	93.2 ± 0.9 [a]
40	5.4 ± 0.1 [a,b]	0.41 ± 0.05 [b]	92.8 ± 0.8 [a]
50	8.6 ± 0.3 [c]	0.52 ± 0.01 [c]	91.1 ± 0.5 [a]

Different letters within the same column indicate significant differences ($p < 0.05$).

The microstructure of the encapsulated dried fruit extracts is relevant to their water reconstitution behavior, flowability, and other techno-functional characteristics; although it is mainly dependent on the type of encapsulant and the drying technique, it can vary according to the extract-encapsulant interactions and ratio [23]. Figure 2 shows SEM images of the freeze-dried Andean blueberry juice powders obtained using different concentrations of maltodextrin. All the images show the typical morphology of freeze-dried powders, with an irregular glassy shape [12,16]. In this case, the microstructure of the powders appeared to be mainly defined by the characteristic crystallinity of the maltodextrin, even at lower (20%) encapsulant concentrations. This indicates that the thermodynamic compatibility between maltodextrin and the Andean blueberry juice solids allows for obtaining amorphous but microscopically homogeneous encapsulated materials via freeze-drying, and suggests that the macroscopical and techno-functional properties of these powders will be defined by those of the encapsulant, thus enhancing their manageability as a powdery food ingredient. The observed differences in the particle sizes might be attributed to the grounding of the freezing-dried cakes [24]. González-Ortega et al. [25] encapsulated olive leaf extract by freeze-drying, reporting that a porous cake was obtained due to the sublimation of the ice, giving rise to a structure made of a glassy matrix containing air cells whose size and shape depended on the processing conditions used and the composition of the initial system.

To evaluate the effect of freeze drying on the spectral characteristic of the juices and possible interactions between the Andean blueberry juice solids and maltodextrin at the processing conditions, the absorbance spectra of the powders in the mid-infrared region of electromagnetic radiation were recorded. Figure 3 shows the infrared spectra (4000–500 cm^{-1}) of the Andean blueberry juice powders obtained with different concentrations of maltodextrin. The spectra of the Andean blueberry juice and the maltodextrin are shown for comparison. Andean blueberry juices featured characteristic bands at 1712 cm^{-1} corresponding to –C=O bonds, and at 1630 cm^{-1} and 1521 cm^{-1} previously associated with the C=C vibrations of polyphenolic compounds from anthocyanin-rich berry extracts [26]. The IR spectra of all Andean blueberry juice powders showed the characteristic bands of maltodextrin at 3300 cm^{-1} (O–H stretching), 2905 cm^{-1} (C–H_2 asymmetric stretching), 1641 cm^{-1} (free carboxyl groups), 1150 cm^{-1} (C–O stretching), 1005 cm^{-1} (C–O stretching), and 929 cm^{-1} (C–O–C stretching of glycosidic bonds; CH_2 out of plane bending) [27]. The characteristic absorption bands of Andean blueberry juice were also detected in the juice in the freeze-dried powders (Figure 3).

The absorbance of the bands located at 1630, 1521, 1410, and 1024 cm^{-1}, associated with phenolic compounds and in particular with the presence of anthocyanins [26], were quantitatively correlated with the juice content of the powders, which indicates that FTIR-ATR measurements could be used as fast technique to assess the actual juice content in this type of ingredients. The absence of bands unrelated to either maltodextrin or Andean blueberry juice suggest that maltodextrin was an inert wall material with no observable chemical interactions with the Andean blueberry juices.

2.3. Technological Features of the Juice Freeze-Dried Powders

Table 4 shows the flow properties of Andean blueberry juice freeze-dried powders obtained with different concentrations of maltodextrin. The bulk density of the freeze-dried powders increased with the increase in the maltodextrin concentration—i.e., the samples with lower maltodextrin concentrations showed higher cohesiveness. The results obtained were similar to those determined for freeze-dried powders containing cinnamon infusions (536–554 kg × m^{-3}) [28] and sea buckthorn juice (512.7 kg × m^{-3}) [21]. All the powders showed an increase in their bulk density due to the tapping suggesting the presence of attractive forces and friction [29]. On the other hand, the samples with a maltodextrin concentration at 40% and 50% showed a slight decrease in the Hausner ratio and a lower Carr index and angle of repose than the powders with 20% and 30% of wall material, indicating better flow properties (Table 4). In general, higher maltodextrin proportions improved the flowability of the freeze-dried juices. Similar observations were reported by Caliskan et al. [14] when analyzing the effect of different amounts of maltodextrin addition on the powder properties of freeze-dried sumac extract powders.

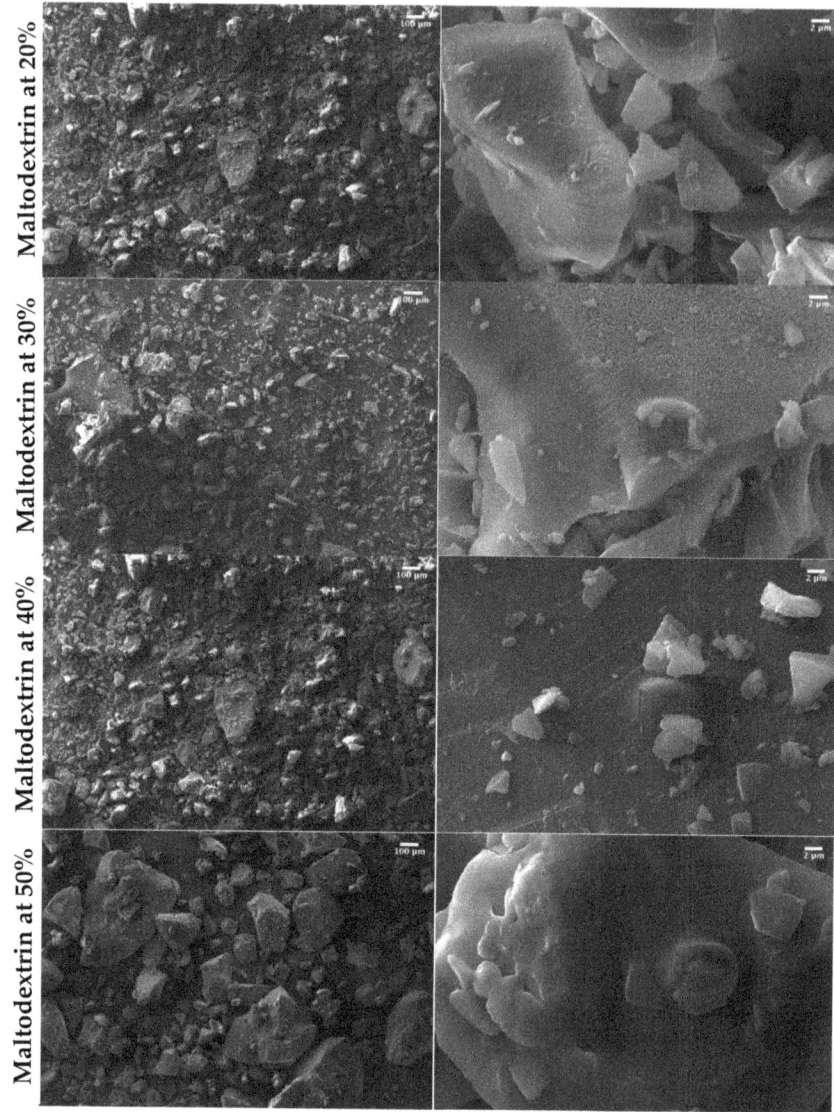

Figure 2. Scanning electron microscopy (SEM) images of freeze-dried powders obtained using different wall materials and mixtures.

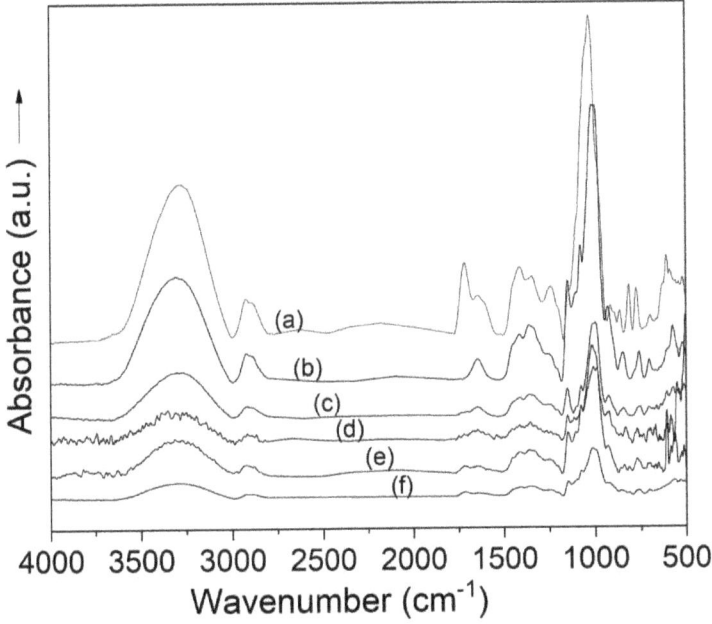

Figure 3. FTIR spectra of Andean blueberry juice (**a**); maltodextrin (**b**); and Andean blueberry juice powders with maltodextrin at 20% (**c**), 30% (**d**), 40% (**e**), and 50% (**f**).

Table 4. Flow properties of Andean blueberry juice freeze-dried powders obtained with different concentrations of maltodextrin.

Maltodextrin Concentration (%)	Bulk Density kg × m⁻³	Tapped Density kg × m⁻³	Hausner Ratio	Carr Index (%)	Angle of Repose (°)
20	470 ± 24 [a]	545 ± 44 [a]	1.2 ± 0.1 [a,b]	20.7 ± 2.6 [a]	35.4 ± 0.3 [a]
30	502 ± 20 [a]	615 ± 17 [b]	1.2 ± 0.1 [a]	15.6 ± 0.4 [b]	37.0 ± 0.5 [a]
40	585 ± 35 [b]	674 ± 39 [c]	1.1 ± 0.1 [a,b]	11.6 ± 0.7 [c]	36.3 ± 1.1 [a]
50	595 ± 41 [b]	650 ± 43 [b,c]	1.1 ± 0.1 [b]	6.1 ± 0.2 [d]	27.0 ± 0.9 [b]

Different letters within the same column indicate significant differences ($p < 0.05$).

The water solubility of the freeze-dried powders is significant for its incorporation in food systems [14]. All the freeze-dried Andean blueberry juice powders showed a similar water solubility—i.e., close to 93%—regardless of the maltodextrin concentration added (Table 3). This behavior can be attributed to the high solubility of maltodextrin in water and is in accordance with the microstructure similarity of the powders obtained at different encapsulant proportions in the SEM observations. Franceschinis et al. [30] reported a water solubility (%) of almost 100% for powders from blackberry juice obtained by freeze and spray-drying. Several authors have reported that the water solubility of freeze-dried powders depends on the morphology, the particle size, the inter-particle voids of powders, and the properties of juice and carrier agents [29].

2.4. Bioactive Characteristics of Andean Blueberry Powders with Different Maltodextrin Additions

The determination of the active compound content of freeze-dried fruit juice powders is important for estimating the powder amount necessary to reach a determined active compounds level in a food formulation. Figure 4 shows the total polyphenol content and total monomeric anthocyanin of freeze-dried Andean blueberry juice powders with different concentrations of maltodextrin. In general, a decrease in the total polyphenol content and the total monomeric anthocyanin of the powders was observed as the maltodextrin concentration in the formulations increased—i.e., as the juice amount in the powders decreased. The content of total phenolic compound ranged between 2.9 and 7.0 mg GAE/g of dry matter, while the monomeric anthocyanins content ranged between 0.19 and 0.60 mg cyd-3-glu/g of dry matter. The results obtained in this study were close to the ones previously reported by Casati et al. [20] for freeze-dried blueberry (*Vaccinium corymbosum*) powders: phenolic contents of 7.69 mg GAE/g and total monomeric anthocyanins content of 0.74 mg/g.

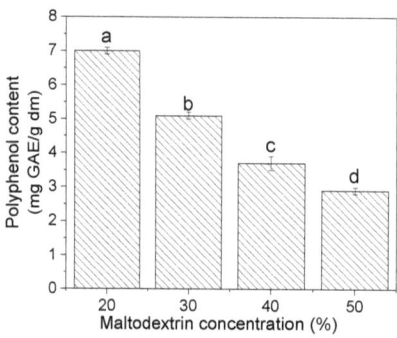
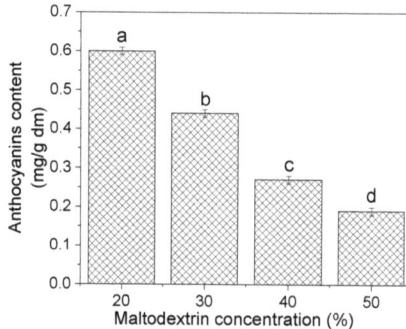

Figure 4. Total polyphenol content and total monomeric anthocyanin of freeze-dried Andean blueberry juice powders with different concentrations of maltodextrin. Values with different superscript letters are significantly different ($p < 0.05$).

On the other hand, the retention of phenolic compounds of the freeze-dried powders was significantly improved with increasing the maltodextrin concentration in the formulations (Figure 5). In all cases, retention percentages of phenolic compounds greater than 70% were obtained, with a higher recovery in the powders with 50% of maltodextrin. The percentage of recovery of monomeric anthocyanins increased significantly when the maltodextrin concentration increased from 20% to 30% (Figure 5). However, above this concentration a significant decrease in anthocyanin retention was observed. Romero-González et al. [31], when working with maqui juice freeze-dried powders, reported that the anthocyanin efficiency values decreased at a higher proportion of added polysaccharides (maltodextrin, gum Arabic, inulin, and their blends). Fraceschinis et al. [30], when working with freeze-dried blackberry powders, reported percentages of the retention of polyphenols and anthocyanins of 73% and 75%, respectively.

Figure 6 shows the antioxidant activity of the freeze-dried powders obtained using different maltodextrin (MD) concentrations. The DPPH•-scavenging activity of the freeze-dried powders increased as the maltodextrin concentration in the formulation decreased. In this sense, a high correlation ($R^2 = 0,99$) between the polyphenol content and the antioxidant activity of the powders was obtained, indicating a strong influence of the phenolic compound on this parameter. This correlation between polyphenol content and antioxidant capacity was also observed by Garrido-Makinistian et al. [32] in maqui powders obtained by spray-drying.

Garzón et al. reported that the high antioxidant activity of Andean blueberry fruits could be due to the high concentration and the chemical structure of its phenolic compounds. They detected in this fruit through high-performance liquid chromatography with photodiode array detection (HPLC-DAD) and HPLC-electrospray ionization tandem mass spectrometry (ESI-MS/MS) the presence of bioactive compounds with a strong antioxidant activity, such as monoglucosides of cyanidin and delphinidin, chlorogenic acid, and quercetin [4].

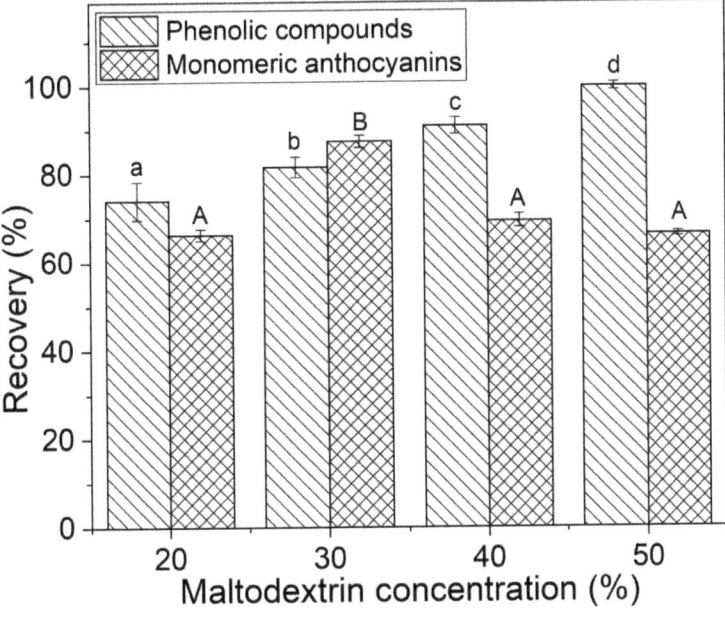

Figure 5. Recovery (%) of the total polyphenol content and total monomeric anthocyanin in freeze-dried Andean blueberry juice powders with different concentrations of maltodextrin. Values with different superscript letters are significantly different ($p < 0.05$).

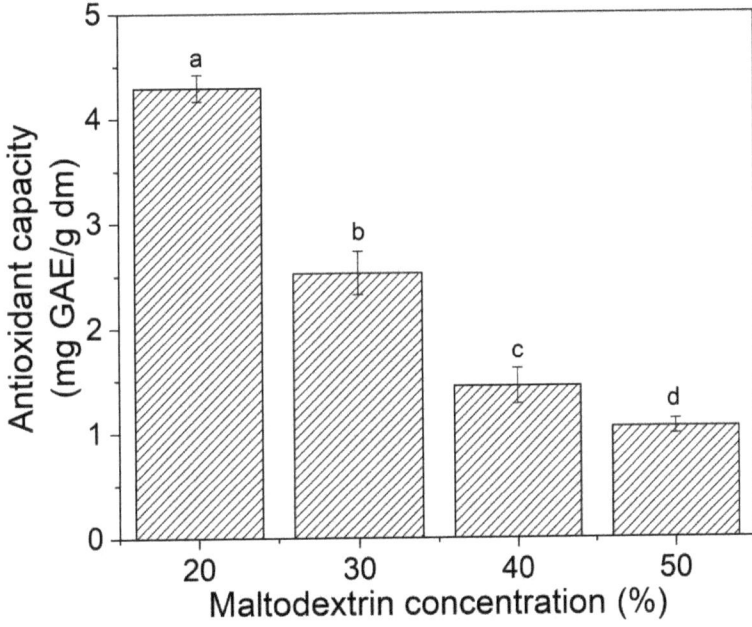

Figure 6. Antioxidant activity (DPPH• scavenging capacity) of the freeze-dried powders obtained using different maltodextrin (MD) concentrations. Values with different superscript letters are significantly different ($p < 0.05$).

3. Materials and Methods

3.1. Materials

Andean blueberries (*Vaccinium meridionale* Swart) maturity stage 5 (100% dark purple) were obtained in Ráquira (Boyacá, Colombia) at 2150 m.a.s.l. The berries were examined previous to its use to separate fruits with physical, mechanical, or microbial damage. The fruits were washed and disinfected with a 100 mg L^{-1} of chlorine solution.

Maltodextrin (MD) dextrose equivalent (DE) 18–22 from Tecnas S.A. (Medellín, Colombia) was used as a carrier. Folin–Ciocalteu reagent was purchased from Panreac (Barcelona, Spain) and gallic acid was purchased from Merck (Darmstadt, Germany). 2,2-diphenyl-1-picrylhydrazyl (DPPH•) reagent was purchased from Sigma-Aldrich (St. Louis, MO, USA). All the chemicals used were of analytical grade.

3.2. Preparation of Andean Bluebery Juice and Freeze-Drying Formulations

Andean blueberry juices were squeezed from crushed fruits using a juice press extractor and vacuum filtered through Whatman paper N °1.

Four formulations for freeze drying were obtained blending the juice (J) with maltodextrin (MD) in the following ratios (J:MD): 80:20 (MD20), 70:30 (MD30), 60:40 (MD40), and 50:50 (MD50). Maltodextrin was dissolved in the Andean blueberry juices under constant stirring using a EUROSTAR 20 vertical agitator at 800 rpm (IKA, Staufen, Germany). The homogenized formulations were poured on aluminum trays for the freezing drying process.

In preliminary experiments, the freeze drying of pure juices (100:0) and its blend with maltodextrin at 10% (90:10) was also assayed, but these formulations did not lead to stable juice powders (Table S1, Supplementary Material).

3.3. Freeze Drying

All the formulations were frozen at −20 °C for 24 h and then dried using a BUCHI Lyovapor L-200 freeze dryer (Flawil, Switzerland). It was operated at −55 °C at a chamber pressure of 0.1 mbar for 48 h. The freeze-drying cakes were grounded to obtain powders and stored in hermetic flask until use.

3.4. pH and Soluble Solids Content

The pH was assessed using a digital pH meter (Oakton Instruments, Vernon Hills, IL, USA) (AOAC 981.12). The soluble solids content was measured in the fruit juice using an Atago refractometer model PR 101 (Atago CO., Tokyo, Japan) and expressed as °Brix (AOAC 932.12).

3.5. Fourier Transform Infrared Spectroscopy (FTIR)

FTIR analysis was performed using FT/IR-4100 equipment (JASCO, Hachioji, Tokyo, Japan) equipped with a diamond single reflection attenuated total reflectance (ATR) module. Portions of the samples (approximately 10 mg) were placed on the ATR accessory and analyzed under reflectance mode. A total of 24 spectra per sample were acquired with 24 scans per spectrum with a spectral resolution of 4 cm^{-1} in the spectral interval 4000–450 cm^{-1}. The measured spectra were recorded and pre-treated with the built-in procedures for water elimination using the software Spectra Manager (v.2.7, JASCO Hachioji, Tokyo, Japan).

3.6. Color

Color was measured using a tristimulus Minolta colorimeter (Konica-Minolta CR-10, Osaka, Japan) and was reported in CIELab parameters (L*, a* and b* values), where L* was used to denote lightness, a* redness (+) and greenness (−), and b* yellowness (+) and blueness (−).

Hue angle (*h*) and chroma values were calculated using the following equations:

$$h = \tan^{-1}\left(\frac{b^*}{a^*}\right), \tag{1}$$

$$\text{Croma} = \left[a^{*2} + b^{*2}\right]^{1/2} \tag{2}$$

3.7. Morphological Analysis

The morphological characteristics of the freeze-dried powders were investigated by Scanning Electron Microscopy (SEM) using a ZEISS EVO MA10 microscope (Carl Zeiss SMT Ltd., Cambridge, UK). The powders were attached to stubs, coated with a layer of gold, and examined using an acceleration voltage of 20 kV.

3.8. Moisture Content and Water Activity

Humidity content (%) was measured gravimetrically, drying the freeze-dried powders in an oven at 105 °C until they reached a constant weight (AOAC, 1998). Values of water activity (a_w) were determined using an AquaLab Serie 3 TE (Pullman, WA, USA) apparatus (AOAC, 1998).

3.9. Water-Solubility

Water solubility was determined by blending 1 g of the freeze-dried powders with 100 mL of distilled water at room temperature with continuous stirring at 1000 rpm for 5 min (IKA RT5 magnetic stirring, Staufen, Germany). The rehydrated juice was centrifuged at 1500 rpm for 5 min and the supernatant was dried at 105 °C until constant weight. The dry weight was used to calculate solubility as a percentage.

3.10. Total Polyhenols Content

The total polyphenols content of Andean blueberry juice and powders was determined by the Folin–Ciocalteu method [33]. Briefly, 400 µL of reconstituted juice was mixed with 2 mL of Folin–Ciocalteu reagent (1:10 diluted). Then, 1.6 mL of sodium carbonate (7% *w/v*) was added to each sample. After 30 min, the absorbance was measured at 760 nm using a spectrophotometer (X-ma 1200 Human Corporation, Loughborough, UK). The results were expressed as gallic acid equivalents (GAE) per gram of dry solids.

3.11. Total Monomeric Anthocyanins Content

The total monomeric anthocyanins content of Andean blueberry juice and powders was measured by the pH differential method [34]. Absorbencies were read at 520 and 700 nm. The anthocyanin concentration was calculated and expressed as cyanidin 3-glucoside (cyd-3-glu) using an extinction coefficient (ε) of 26,900 L cm^{-1} × mol^{-1} and a molecular weight of 449.2 g/mol.

3.12. Active Compounds Recovery (%)

The recovery (%) of phenolic compounds and monomeric anthocyanins after the freeze drying process was calculated with the following equation [35]:

$$\text{Recovery (\%)} = \frac{L_c}{L_0} \times 100. \tag{3}$$

where L_c and L_0 are the total phenolic or monomeric anthocyanin content of the freeze-dried powders and the infeed dispersion, respectively.

3.13. DPPH•-Scavenging Activity

Antioxidant activity was tested as described in Brand-Williams et al. [36]. A volume of 100 µL of each reconstituted juice was mixed with 3.9 mL of 1,1-diphenyl- 2-picrylhydrazyl (DPPH•) ethanol solution (25 mg DPPH•/L). Absorbance was determined at 515 nm until the reaction reached a plateau. A calibration curve was performed using gallic acid as a standard. The results were expressed as the mg GAE per gram of dry solids.

3.14. Flow Properties

The loose bulk density was determined by pouring a known mass of freeze-dried powders delivered freely by gravity into a measuring cylinder and calculated by dividing the mass by the bulk volume. The tapped bulk density was calculated from the weight of powder and the volume occupied in the cylinder after being hand tapped until a constant value was reached [29,37].

The Hausner ratio and the compressibility index were estimated according to the procedures presented in López Córdoba et al. [37]

The angle of repose was determined by pouring a known mass of freeze-dried powder through a funnel located at a fixed height on a graph paper flat horizontal surface and measuring the height (h) and radius (r) of the conical pile formed. The tangent of the angle of repose is given by the h/r ratio [38].

3.15. Statistical Analysis

The statistical analysis was performed using the Minitab v.16 statistical software (State College, PA, USA). Analysis of variance (ANOVA) and Tukey's pairwise comparisons were carried out using a level of 95% confidence. The experiments were performed at least in triplicate, and the data were reported as means ± standard deviations.

4. Conclusions

Maltodextrin was found a suitable wall material for the stabilization of Andean blueberry juice via freeze-drying because it allowed obtaining powders with good handling properties and desirable features, such as low water activity and a high water solubility. Besides this, maltodextrin is an inert wall material that does not have chemical interactions with Andean blueberry juices.

The amount of maltodextrin used in the formulations significantly affected the lightness, the bulk density, the polyphenol and monomeric anthocyanins content, and the antioxidant activity of the freeze-dried Andean blueberry juice powders. The products with maltodextrin at 30% and 50% showed higher monomeric anthocyanin and polyphenol recoveries, respectively. The antioxidant activity of the freeze-dried Andean blueberry juice was highly correlated with its content of phenolic compounds. The produced powders could be potentially employed as functional ingredients for the formulation of new value-added foods.

Supplementary Materials: The following are available online, Figure S1: Images of the maltodextrin-free freeze dried-powder, Table S1: Physicochemical properties of the maltodextrin-free freeze dried-powder.

Author Contributions: Conceptualization was devised by A.L.-C.; Methodology, Validation, and Formal Analysis were carried out by M.E.-A., C.A.F., and A.L.-C.; Investigation, Resources, Data curation, Writing Original Draft Preparation were performed by A.L.-C.; Writing—Review and Editing was performed by C.A.F. and A.L.-C.; Project administration and Funding acquisition were performed by A.L.-C. All authors have read and agreed to the published version of the manuscript.

Funding: This research was funded by Minciencias, the programa Colombia BIO, and the Gobernación de Boyacá through the Fondo de Ciencia, Tecnología e Innovación del Sistema General de Regalías, managed by the Fondo "Francisco José de Caldas" (project 66038. Conv. 827-2018). Furthermore, A.L.-C. would like to thank the Universidad Pedagógica y Tecnológica de Colombia (UPTC) for their financial support.

Acknowledgments: The authors wish to thank the staff of the Institute of Food Science and Technology (ICTA) of the National University of Colombia, in particular Cristina Lizarazo for her kind technical support in the FTIR-ATR measurements.

Conflicts of Interest: The authors declare no conflict of interest.

References

1. Da Silva, B.V.; Barreira, J.C.; Oliveira, M.B.P. Natural phytochemicals and probiotics as bioactive ingredients for functional foods: Extraction, biochemistry and protected-delivery technologies. *Trends Food Sci. Technol.* **2016**, *50*, 144–158. [CrossRef]
2. Song, G.-Q.; Hancock, J.F. Vaccinium. In *Wild Crop Relatives: Genomic and Breeding Resources*; Kole, C., Ed.; Springer: Berlin/Heidelberg, Germany, 2011; pp. 197–221, ISBN 978-3-642-16057-8.
3. Celis, M.E.M.; Franco Tobón, Y.N.; Agudelo, C.; Arango, S.S.; Rojano, B. Andean Berry (Vaccinium meridionale Swartz). In *Fruit and Vegetable Phytochemicals: Chemestry and Human Health*, 2nd ed.; Yahia, E.M., Ed.; John Wiley & Sons Ltd.: Hoboken, NJ, USA, 2017; Volume 2, pp. 869–882.
4. Garzón, G.A.; Narváez, C.E.; Riedl, K.M.; Schwartz, S.J. Chemical composition, anthocyanins, non-anthocyanin phenolics and antioxidant activity of wild bilberry (Vaccinium meridionale Swartz) from Colombia. *Food Chem.* **2010**, *122*, 980–986. [CrossRef]
5. González, M.; Samudio, I.; Sequeda-Castañeda, L.G.; Celis, C.; Iglesias, J.; Morales, L. Cytotoxic and antioxidant capacity of extracts from Vaccinium meridionale Swartz (Ericaceae) in transformed leukemic cell lines. *J. Appl. Pharm. Sci.* **2017**, *7*, 24–30. [CrossRef]
6. Maldonado-Celis, M.E.; Arango-Varela, S.S.; Rojano, B.A. Free radical scavenging capacity and cytotoxic and antiproliferative effects of *Vaccinium meridionale* Sw. agains colon cancer cell lines. *Rev. Cuba. Plantas Med.* **2014**, *19*, 172–184.
7. Agudelo, C.D.; Ceballos, N.; Gómez-García, A.; Maldonado-Celis, M.E. Andean Berry (Vaccinium meridionale Swartz) Juice improves plasma antioxidant capacity and IL-6 levels in healthy people with dietary risk factors for colorectal cancer. *J. Berry Res.* **2018**, *8*, 251–261. [CrossRef]
8. Celli, G.B.; Dibazar, R.; Ghanem, A.; Brooks, M.S.-L. Degradation kinetics of anthocyanins in freeze-dried microencapsulates from lowbush blueberries (Vaccinium angustifolium Aiton) and prediction of shelf-life. *Dry. Technol.* **2016**, *34*, 1175–1184. [CrossRef]
9. Shishir, M.R.I.; Chen, W. Trends of spray drying: A critical review on drying of fruit and vegetable juices. *Trends Food Sci. Technol.* **2017**, *65*, 49–67. [CrossRef]
10. Nicoletti Telis, V.R.; Martínez-Navarrete, N. Biopolymers Used as Drying Aids in Spray-Drying and Freeze-Drying of Fruit Juices and Pulps. In *Biopolymer Engineering in Food Processing*; Nicoletti Telis, V.R., Ed.; CRC Press: Sound Parkway, NW, USA, 2012.
11. Delshadi, R.; Bahrami, A.; Tafti, A.G.; Barba, F.J.; Williams, L.L. Micro and nano-encapsulation of vegetable and essential oils to develop functional food products with improved nutritional profiles. *Trends Food Sci. Technol.* **2020**, *104*, 72–83. [CrossRef]
12. Fredes, C.; Becerra, C.; Parada, J.; Robert, P. The Microencapsulation of Maqui (*Aristotelia chilensis* (Mol.) Stuntz) Juice by Spray-Drying and Freeze-Drying Produces Powders with Similar Anthocyanin Stability and Bioaccessibility. *Molecules* **2018**, *23*, 1227. [CrossRef]
13. Aprodu, I.; Milea, Ş.A.; Anghel, R.-M.; Enachi, E.; Barbu, V.; Crăciunescu, O.; Râpeanu, G.; Bahrim, G.E.; Oancea, A.; Stănciuc, N. New Functional Ingredients Based on Microencapsulation of Aqueous Anthocyanin-Rich Extracts Derived from Black Rice (*Oryza sativa* L.). *Molecules* **2019**, *24*, 3389. [CrossRef]
14. Caliskan, G.; Dirim, S.N. The effect of different drying processes and the amounts of maltodextrin addition on the powder properties of sumac extract powders. *Powder Technol.* **2016**, *287*, 308–314. [CrossRef]
15. Lachowicz, S.; Michalska-Ciechanowska, A.; Oszmiański, J. The Impact of Maltodextrin and Inulin on the Protection of Natural Antioxidants in Powders Made of Saskatoon Berry Fruit, Juice, and Pomace as Functional Food Ingredients. *Molecules* **2020**, *25*, 1805. [CrossRef] [PubMed]
16. Pudziuvelyte, L.; Marksa, M.; Sosnowska, K.; Winnicka, K.; Morkuniene, R.; Bernatoniene, J. Freeze-Drying Technique for Microencapsulation of Elsholtzia ciliata Ethanolic Extract Using Different Coating Materials. *Molecules* **2020**, *25*, 2237. [CrossRef] [PubMed]
17. Fossen, T.; Cabrita, L.; Andersen, O.M. Colour and stability of pure anthocyanins influenced by pH including the alkaline region. *Food Chem.* **1998**, *63*, 435–440. [CrossRef]

18. Garzón, G.A.; Soto, C.Y.; López-R, M.; Riedl, K.M.; Browmiller, C.R.; Howard, L. Phenolic profile, in vitro antimicrobial activity and antioxidant capacity of Vaccinium meridionale Swartz pomace. *Heliyon* **2020**, *6*, e03845. [CrossRef]
19. Franco Tobon, Y.N.; Rojano, B.A.; Arbeláez Alzate, A.F.; Saavedra Morales, D.M.; Celis Maldonado, M.E. Efecto del tiempo de almacenamiento sobre las características fisicoquímicas, antioxidantes y antiproliferativa de néctar de agraz (Vaccinium meridionale Swartz). *Arch. Latinoam. Nutr.* **2016**, *66*, 261–271. (In Spanish)
20. Casati, C.B.; Baeza, R.; Sánchez, V. Physicochemical properties and bioactive compounds content in encapsulated freeze-dried powders obtained from blueberry, elderberry, blackcurrant and maqui berry. *J. Berry Res.* **2019**, *9*, 431–447. [CrossRef]
21. Tkacz, K.; Wojdyło, A.; Michalska-Ciechanowska, A.; Turkiewicz, I.P.; Lech, K.; Nowicka, P. Influence Carrier Agents, Drying Methods, Storage Time on Physico-Chemical Properties and Bioactive Potential of Encapsulated Sea Buckthorn Juice Powders. *Molecules* **2020**, *25*, 3801. [CrossRef]
22. Tapia, M.S.; Alzamora, S.M.; Chirife, J. Effects of Water Activity (aw) on Microbial Stability: As a Hurdle in Food Preservation. *Water Act. Foods* **2007**, 239–271.
23. Sarabandi, K.; Peighambardoust, S.H.; Sadeghi Mahoonak, A.R.; Samaei, S.P. Effect of different carriers on microstructure and physical characteristics of spray dried apple juice concentrate. *J. Food Sci. Technol.* **2018**, *55*, 3098–3109. [CrossRef]
24. Khazaei, K.M.; Jafari, S.M.; Ghorbani, M.; Hemmati Kakhki, A. Application of maltodextrin and gum Arabic in microencapsulation of saffron petal's anthocyanins and evaluating their storage stability and color. *Carbohydr. Polym.* **2014**, *105*, 57–62. [CrossRef] [PubMed]
25. González-Ortega, R.; Faieta, M.; Di Mattia, C.D.; Valbonetti, L.; Pittia, P. Microencapsulation of olive leaf extract by freeze-drying: Effect of carrier composition on process efficiency and technological properties of the powders. *J. Food Eng.* **2020**, *285*, 110089. [CrossRef]
26. Alzate-Arbeláez, A.F.; Dorta, E.; López-Alarcón, C.; Cortés, F.B.; Rojano, B.A. Immobilization of Andean berry (Vaccinium meridionale) polyphenols on nanocellulose isolated from banana residues: A natural food additive with antioxidant properties. *Food Chem.* **2019**, *294*, 503–517. [CrossRef] [PubMed]
27. Ballesteros, L.F.; Ramirez, M.J.; Orrego, C.E.; Teixeira, J.A.; Mussatto, S.I. Encapsulation of antioxidant phenolic compounds extracted from spent coffee grounds by freeze-drying and spray-drying using different coating materials. *Food Chem.* **2017**, *237*, 623–631. [CrossRef] [PubMed]
28. Santiago-Adame, R.; Medina-Torres, L.; Gallegos-Infante, J.A.; Calderas, F.; González-Laredo, R.F.; Rocha-Guzmán, N.E.; Ochoa-Martínez, L.A.; Bernad-Bernad, M.J. Spray drying-microencapsulation of cinnamon infusions (Cinnamomum zeylanicum) with maltodextrin. *LWT Food Sci. Technol.* **2015**, *64*, 571–577. [CrossRef]
29. López-Córdoba, A.; Goyanes, S. Food Powder Properties. *Ref. Modul. Food Sci.* **2017**, 1–7. [CrossRef]
30. Franceschinis, L.; Salvatori, D.M.; Sosa, N.; Schebor, C. Physical and Functional Properties of Blackberry Freeze- and Spray-Dried Powders. *Dry. Technol.* **2014**, *32*, 197–207. [CrossRef]
31. Romero-González, J.; Shun Ah-Hen, K.; Lemus-Mondaca, R.; Muñoz-Fariña, O. Total phenolics, anthocyanin profile and antioxidant activity of maqui, Aristotelia chilensis (Mol.) Stuntz, berries extract in freeze-dried polysaccharides microcapsules. *Food Chem.* **2020**, *313*, 126115. [CrossRef]
32. Garrido Makinistian, F.; Sette, P.; Gallo, L.; Bucalá, V.; Salvatori, D. Optimized aqueous extracts of maqui (Aristotelia chilensis) suitable for powder production. *J. Food Sci. Technol.* **2019**, *56*, 3553–3560. [CrossRef]
33. Singleton, V.L.; Orthofer, R.; Lamuela-Raventos, R.M. Analysis of total phenols and other oxidation substrates and antioxidants by means of Folin-Ciocalteu reagent. In *Methods in Enzymology (Oxidants and Antioxidants, Part A)*; Lester Packer, L., Ed.; Academic Press: San Diego, CA, USA, 1999; Volume 299, pp. 152–178.
34. Lee, J.; Durst, R.W.; Wrolstad, R.E. Determination of Total Monomeric Anthocyanin Pigment Content of Fruit Juices, Beverages, Natural Colorants, and Wines by the pH Differential Method: Collaborative Study. *J. AOAC Int.* **2005**, *88*, 1269–1278. [CrossRef]
35. Fang, Z.; Bhandari, B. Effect of spray drying and storage on the stability of bayberry polyphenols. *Food Chem.* **2011**, *129*, 1139–1147. [CrossRef] [PubMed]
36. Brand-Williams, W.; Cuvelier, M.E.; Berset, C. Use of a free radical method to evaluate antioxidant activity. *LWT Food Sci. Technol.* **1995**, *28*, 25–30. [CrossRef]

37. López-Córdoba, A.; Deladino, L.; Agudelo-Mesa, L.; Martino, M. Yerba mate antioxidant powders obtained by co-crystallization: Stability during storage. *J. Food Eng.* **2014**, *124*, 158–165. [CrossRef]
38. Rattes, A.L.R.; Oliveira, W.P. Spray drying conditions and encapsulating composition effects on formation and properties of sodium diclofenac microparticles. *Powder Technol.* **2007**, *171*, 7–14. [CrossRef]

Sample Availability: Samples of the freeze-dried powders are available from the authors.

Publisher's Note: MDPI stays neutral with regard to jurisdictional claims in published maps and institutional affiliations.

© 2020 by the authors. Licensee MDPI, Basel, Switzerland. This article is an open access article distributed under the terms and conditions of the Creative Commons Attribution (CC BY) license (http://creativecommons.org/licenses/by/4.0/).

Article

Development of Antioxidant-Loaded Nanoliposomes Employing Lecithins with Different Purity Grades

Cristhian J. Yarce [1,2], Maria J. Alhajj [1], Julieth D. Sanchez [1], Jose Oñate-Garzón [3] and Constain H. Salamanca [1,2,*]

1. Laboratorio de Diseño y Formulación de Productos Químicos y Derivados, Departamento de Ciencias Farmacéuticas, Facultad de Ciencias Naturales, Universidad ICESI, Calle 18 No. 122-135, 760035 Cali, Colombia; cjyarce@icesi.edu.co (C.J.Y.); mariajoalhajj@hotmail.com (M.J.A.); julieths28@outlook.com (J.D.S.)
2. Centro de Ingredientes Naturales Especializados y Biotecnológicos-CINEB, Facultad de Ciencias Naturales, Universidad ICESI, Calle 18 No. 122-135, 760035 Cali, Colombia
3. Facultad de Ciencias Básicas, Programa de Microbiología, Universidad Santiago de Cali, Calle 5 No. 62-00, 760035 Cali, Colombia; jose.onate00@usc.edu.co
* Correspondence: chsm70@gmail.com

Academic Editors: Severina Pacifico and Simona Piccolella
Received: 20 October 2020; Accepted: 11 November 2020; Published: 16 November 2020

Abstract: This work focused on comparing the ability of lecithins with two purity grades regarding their performance in the development of nanoliposomes, as well as their ability to contain and release polar (trans-aconitic acid) and non-polar (quercetin) antioxidant compounds. First, the chemical characterization of both lecithins was carried out through infrared spectroscopy (FTIR), electrospray ionization mass spectrometry (ESI/MS), and modulated differential scanning calorimetry (mDSC). Second, nanoliposomes were prepared by the ethanol injection method and characterized by means of particle size, polydispersity, and zeta potential measurements. Third, the encapsulation efficiency and in vitro release profiles of antioxidants were evaluated. Finally, the antioxidant effect of quercetin and trans aconitic acid in the presence and absence of nanoliposomes was assessed through the oxygen radical absorbance capacity (ORAC) assay. The results showed that, although there are differences in the chemical composition between the two lecithins, these allow the development of nanoliposomes with very similar physicochemical features. Likewise, nanoliposomes elaborated with low purity grade lecithins favored the encapsulation and release of trans-aconitic acid (TAA), while the nanoliposomes made with high purity lecithins favored the encapsulation of quercetin (QCT) and modified its release. Regarding the antioxidant effect, the vehiculization of TAA and QCT in nanoliposomes led to an increase in the antioxidant capability, where QCT showed a sustained effect over time and TAA exhibited a rapidly decaying effect. Likewise, liposomal systems were also found to have a slight antioxidant effect.

Keywords: antioxidant effect; lectins; nanoliposomes; purity grade; quercetin; trans-aconitic acid

1. Introduction

Lecithins are a type of raw material widely used by different sectors, such as pharmaceutical, food, and cosmetic sectors, where they are employed for different purposes due to their great versatility [1]. Some of these uses correspond to their ability to act as emulsifiers, as well as humectants and suspension stabilizers [2]. However, their most important application lies in their ability to form nanoliposomes, which represent a very special type of self-assembling system that is very useful as a nanometric formulation vehicle for multiple compounds of interest in the previously mentioned sectors [3,4].

Lecithins are complex raw materials because they consist of a mixture of various types of phospholipids, where phosphatidylcholine (PC) is usually the main component [5]. However, they may also contain other types of phospholipids, such as phosphatidylethanolamines (PE), phosphatidylserines (PS), phosphatidylinositols (PI), phosphatidylglycerols (PG), and glyphospholipids (GLP), as well as other components, such as phosphatidic acid (PA) and cholesterol [6]. Therefore, depending on the obtention source (vegetable or animal), the type of extraction, and the purification methods, lecithins can be composed of different types and amounts of phospholipids and thus have a wide diversity of references with multiple purity grades and commercial prizes [7]. This diversity of references can, in some cases, lead to confusion or a lack of criteria for choosing the correct lecithin. In the case of the pharmaceutical sector, this choice is easier to make due to the intrinsic features of such products and their regulatory affairs, in which it is practically mandatory to use raw materials with the highest possible purity grade. In contrast, this situation is different in other sectors, such as foodstuff and cosmetic sectors, where lower purity grade lecithins can be used with less restrictions. Therefore, the question presented in this study is as follows: Can a low purity lecithin develop nanoliposomes, just like a high purity lecithin? To address this question, the work focused on (i) chemically characterizing two types of soybean lectins with different purity grades and commercial value; (ii) the development and characterization of nanoliposomes loaded with two model antioxidants, corresponding to quercetin [8] and trans-aconitic acid [9]; and (iii) determining the encapsulation efficiency and in vitro release profiles of antioxidant compounds from nanoliposomal vehicles.

2. Results and Discussion

2.1. Lecithin Characterization

Firstly, it is important to highlight that for the comparison of the purity grade to be consistent, it is necessary that lecithins come from the same origin and therefore have a similar phospholipid composition. If this was not the case, the results would be meaningless and may not be reproduced. Considering this, the results discussed as the conclusions issued in this manuscript correspond to lecithins from a similar plant source (soybean). In the case of low purity lecithin, this was found to be a brown fluid with a sticky consistency. In contrast, the high purity lecithin was shown to be a yellowish granular waxy solid that could be easily handled. Such physical appearance characteristics are a very important factor to consider in foodstuff raw materials, since, depending on their consistency, various stages of product development, such as dispensing, weighing, and post-production cleaning, may be easier to carry out. Therefore, low purity lecithin has a disadvantage compared to high purity lecithin, which is easier to handle. Additionally, the color of the raw materials is another aspect to be considered, since the appearance of these plays an important role in the sensory characteristics of the product to be developed, where dark colors are more difficult to mask, which may be another disadvantage of low purity lecithin versus high purity lecithin. Consequently, the physical appearance of these lecithins is largely determined by their own chemical composition. In this way, and according to the respective technical data sheets, low purity lecithin has a distribution of phospholipids of around 90%, where ~50% is phosphatidylcholine, ~30% are inositol phosphatides, and ~10% is phosphatidylethanolamine; the remaining 10% corresponds to several impurities [5,10]. In contrast, high purity lecithin consists of 97% phospholipids, where ~92% is phosphatidylcholine, ~3% is lysophosphatidylcholine, and ~2% are other phosphatides, with the remainder also corresponding to compounds other than phospholipids [5].

2.1.1. FTIR Characterization

Figure 1A shows the FTIR spectra for low and high purity lecithins and also indicates the chemical structures of the two phospholipids that, according to the literature, are present in a greater proportion in lecithins [5]. In the case of Lp-SBL, signals were found at 3384, 3010, 2924, and 2854 cm^{-1}. These signals correspond to the stretching vibrations of the OH groups from (i) the carboxylic acid function of the fatty acids and the saccharides present in glyphospholipids; (ii) the amino group present in some phospholipids, such as phosphatidylcholine; and (iii) the symmetric and asymmetric tension vibrations of the CH_2 groups present in fatty acids. In the case of Hp-SBL, the signals were observed at 3334, 3009, 2924, and 2854 cm^{-1}. Moreover, the OH band amplitude and intensity suggest that low purity lecithin has a higher content of saccharides. On the other hand, the bands of 1744 cm^{-1} for Lp-SBL and 1739 cm^{-1} for Hp-SBL correspond to the stretching vibration of the carbonyl bond (C=O). However, the signal shifting suggests that the substitution for carbonyl is different in each raw material, which is due to differences in the length and proportion of substituent fatty acids in phospholipids. Likewise, the multiple bands that appear for both raw materials between 1238 and 1064 cm^{-1} correspond to the phosphate groups in the phospholipid chains. Finally, the wide shapes of the signals presented between 1064 and 1165 cm^{-1} in low purity lecithin are representative of the asymmetric tensions of OH groups present in sugars, which is consistent with a higher amount of saccharide molecules in that lecithin.

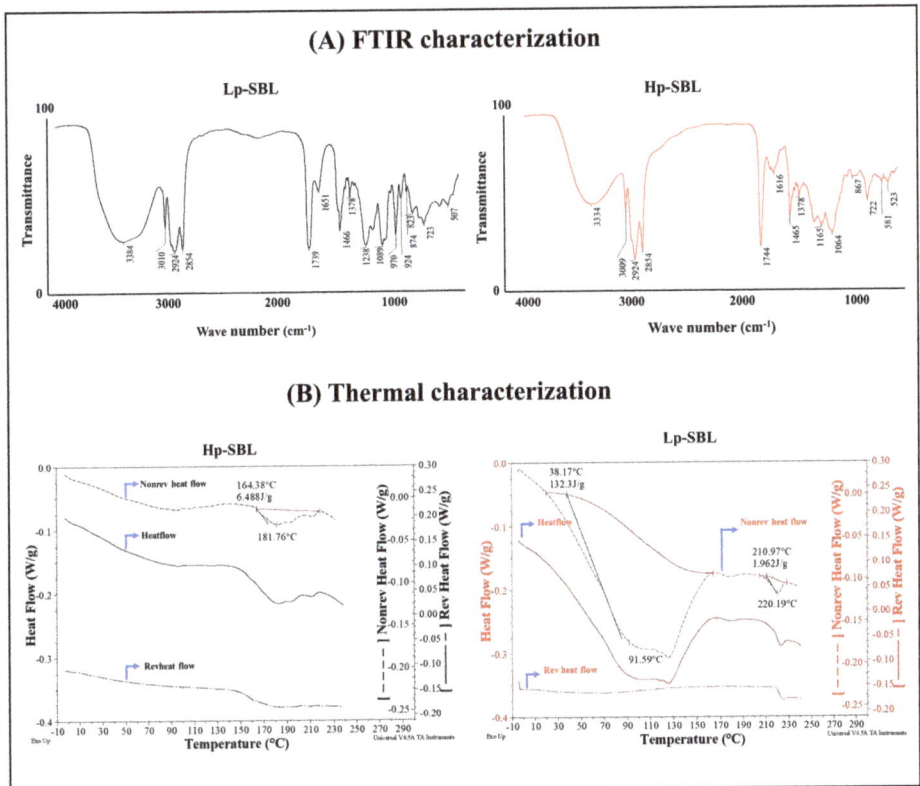

Figure 1. (**A**) Infrared spectroscopy (FTIR) signals for low purity grade (Lp-SBL) and high purity grade (Hp-SBL) soybean lecithins. (**B**) Modulated differential scanning calorimetry (mDSC) signals for low purity grade (Lp-SBL) and high purity grade (Hp-SBL) soybean lecithins. The solid line corresponds to the total heat flow, the short dashed line corresponds to the non-reversible heat flow, and the broken dashed line corresponds to the reversible heat flow.

2.1.2. Thermal Analysis

The results of the thermal characterization performed are shown in Figure 1B. In the case of low purity lecithin, it is possible to observe two thermal transitions. The first one had an onset at approximately 38.2 °C, a transition peak at around 91.6 °C, and an enthalpy value of 132.3 J/g.

This behavior can be observed in the total heat flow signal (solid line) and in the non-reversible heat flow signal (short dashed line), but not in the reversible heat flow signal (dashed line). This indicates that there is an endothermic-type transition caused by the kinetics of the molecules inside the material [11–13]. It leads to a reorganization and presents multiple rearrangements in their physical structure, suggesting a possible loss of volatile substances and water present in the Lp-SBL raw material. Consequently, according to the enthalpy of the transition, the energy expenditure required to remove volatile substances and water was relatively high, due to the contribution of the heat of water vaporization [12–17]. The DSC results for Lp-SBL are consistent with the mass spectra, where representative signs of mass loss due to the dehydration of saccharide-like molecules were observed.

On the other hand, in the case of high purity lecithin, the total heat flow signal presents an endothermic-type transition, with an onset at around 164.4 °C, a peak at 181.8 °C, and an enthalpy of 6.5 J/g. These signals are consistent for the reversible heat flux and the non-reversible heat flux. This shows that the high purity lecithin is a consistent material and that its composition is defined by compounds that operate around approximate temperature values. Furthermore, due to the relatively low enthalpy of the transition, it establishes the heat flow through the material being transmitted homogeneously, which requires a low energy expenditure for the fusion of phospholipids in Hp-SBL. Furthermore, for this raw material, there is no sign of water loss or signs due to the rearrangement of the material, as is the case for Lp-SBL. Finally, all these results are reliable in terms of the fluidity and physical appearance of the Lp-SBL, where it is necessary to remove the water and other volatile components prior to the fusion of the phospholipids, which is a transition that is displaced with an onset of 211.0 °C, with a peak at 220.2 °C and an enthalpy of 1.96 J/g.

2.1.3. Electrospray Ionization Mass Spectrometry (MS/ES$^-$/ES$^+$)

Figure 2 shows the mass spectra recorded in negative (ESI$^-$) and positive (ESI$^+$) ion mode. The signals are discussed together for both Lp-SBL and Hp-SBL. In the case of the mass spectrum in ESI$^-$ negative ionization mode, (Figure 2A left), the [M + H] ion is observed for phosphatidylcholine substituted with linoleic acid (C18:2) and palmitic acid (C16:0), at an m/z ratio of 758 Da. Additionally, an ion with an m/z of 833 Da, corresponding to phosphatidyl inositol, with substitutions of linoleic acid and oleic acid (C18:1), is observed. Conversely, in the case of the mass spectrum registered in positive ESI$^+$ mode (Figure 2A right), the presence of a molecular adduct of potassium for phosphatidylcholine, which is the ion [M + K] with an m/z ratio, can be found at 797 Da. In addition, the ion corresponding to the dehydrated form of [GLP] can be found at an m/z ratio of 832 Da. It is important to highlight that, in the case of low purity lecithin, the abundance between [M + H] and [GLP] is inverted regarding the abundance of these ions in high purity lecithin. Consequently, a greater abundance is observed for the phosphatidyl inositol signal than for the phosphatidylcholine signal, which is consistent with a higher presence of saccharides in the low purity lecithin.

On the other hand, Figure 2B shows the estimated fragmentation patterns in negative ionization mode for phosphatidylcholine and inositol glycophospholipid, respectively. It can be observed that the molecular ion [M + H] for phosphatidylcholine suffers a loss of fraction (1) of m/z 279 Da, which corresponds to the loss of the linoleic acid substituent of phosphatidylcholine, which leads to the generation of 479 Da m/fragment of z (3). In addition, a loss of m/z of 255 Da, corresponding to the palmitic acid substituent and a fragment (4) of the m/z ratio of 714 Da, is formed from the rupture of the nitrogen-bound methyl groups of the choline fraction. In Figure 3B, a water molecule from the phosphatidyl inositol [GLP] glycophospholipid is lost and forms a fragment with an m/z ratio of 831 Da. These processes occur in both low purity lecithin and high purity lecithin. Furthermore, Figure 3C shows the fragmentation patterns for phosphatidylcholine and inositol glyphospholipids in positive ionization mode. In ESI$^+$, a molecular adduct for phosphatidylcholine [M + K] is presented, which suffers losses of the substituent fatty acids in the phospholipid (1) and the methyl groups bounded to nitrogen in the choline fraction (4). In the case of [GLP], a loss of the dehydrated inositol molecule (5) and the substituent fatty acids in the phospholipid (2) can be observed. Accordingly, it is important to highlight that the results suggest that the ions present in the low purity lecithin are also present in high purity lecithin. Therefore, phospholipids such as phosphatidylcholine and phosphatidyl inositol are present in both raw materials and it is thus possible to obtain the advantages of these compounds when using lecithins for a process of transformation or the subsequent preparation of a food or cosmetic product. However, it is estimated that the phosphatidylcholine:phosphatidyl inositol ratio is lower for Lp-SBL compared to Hp-SBL. Furthermore, it should be recognized that other phospholipids, such as phosphatidylethanolamine and phosphatidylserine, may also be present in the raw materials [10], but their signals were not observed, because they could be overlapped with the abundance of ions from other compounds in lecithins.

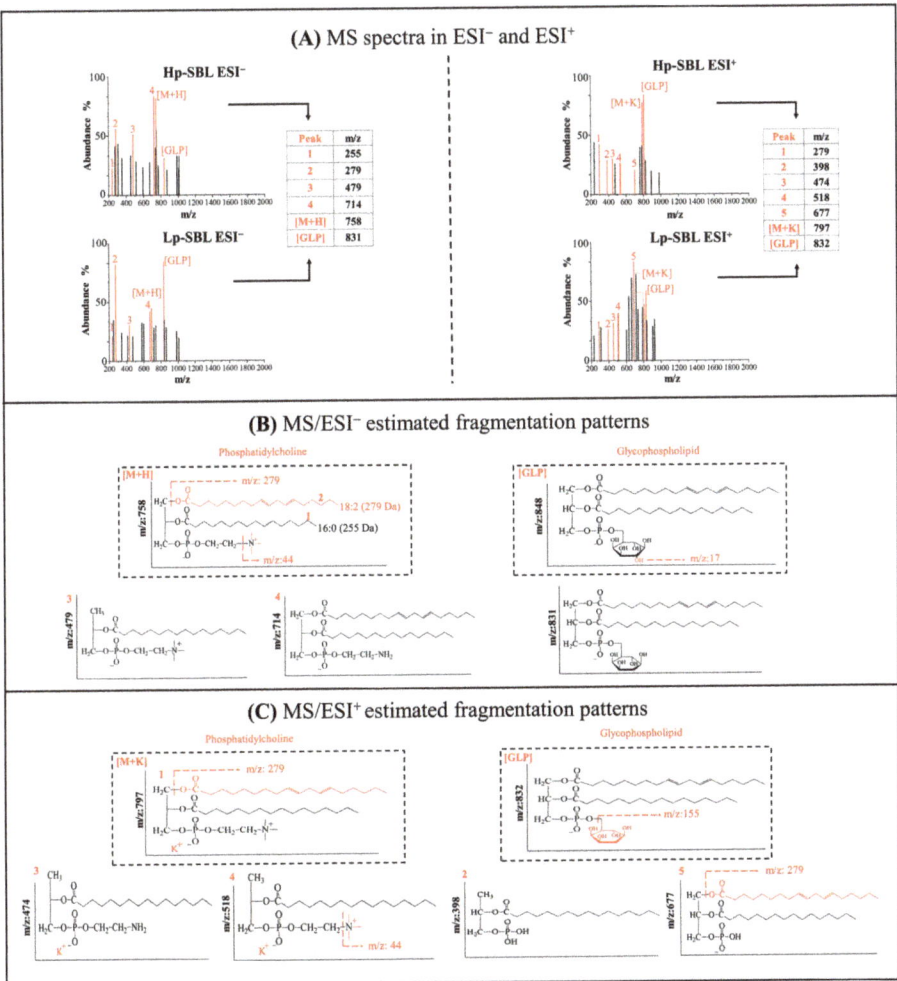

Figure 2. (**A**) Mass spectrometry (MS) spectra in total ion scanning mode for positive ion (ESI$^-$) and negative ion (ESI$^+$) modes for high purity lecithin (Hp-SBL) and low purity lecithin (Lp-SBL). (**B**) MS/ESI$^-$ estimated fragmentation patterns. (**C**) MS/ESI$^+$ estimated fragmentation patterns.

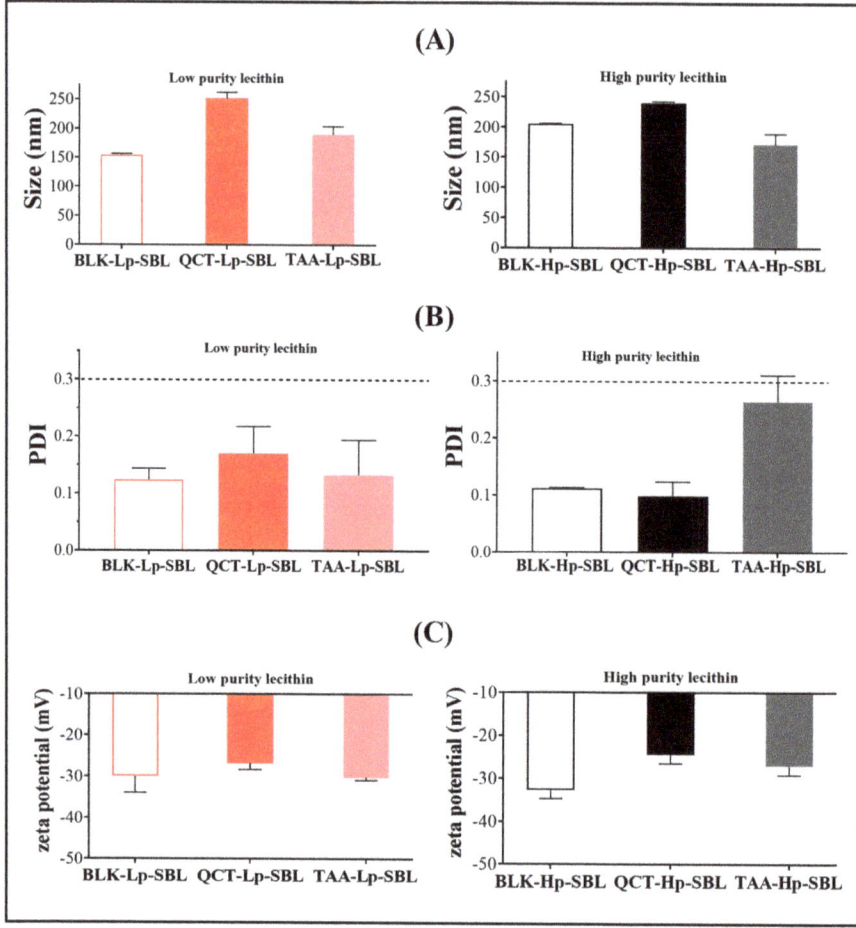

Figure 3. (**A**) Size, (**B**) polydispersity index (PDI), and (**C**) zeta potential for liposomes prepared with high purity lecithin (Hp-SBL) and low purity lecithin (Lp-SBL). BLK corresponds to unloaded nanoliposomes. Data are reported as the mean ± SD and n = 3.

2.2. Development of Nanoliposomes

2.2.1. Physicochemical Characterization

Figure 3 shows the results of the particle size, polydispersity index, and zeta potential for nanoliposomes prepared with low purity (Lp-SBL) and high purity (Hp-SBL) soybean lecithins. In the case of nanoliposomes prepared with Lp-SBL, the particle size of non-loaded nanoliposomes (BLK-Lp) was around 150 nm, while the particle sizes of the nanoliposomes loaded with QCT (QCT-Lp) and TAA (TAA-Lp) were ~250 and ~200 nm, respectively. In contrast, for the nanoliposomes prepared with Hp-SBL, the size of non-loaded nanoliposomes (BLK-Hp) was around 200 nm, while the sizes of the nanoliposomes loaded with QCT (QCT-Hp) and TAA (TAA-Hp) were ~250 and ~180 nm, respectively (Figure 3A). These differences in particle size can be explained considering the composition of phospholipids in both lectins and specifically, in those that consist of greater polar phospholipid amounts (inositol, saccharides, etc.), such as Lp-SBL, and that lead to a higher compaction of the aqueous liposomal core. Likewise, such a compaction effect can also be observed when the nanoliposomes are loaded with TAA, which is a very polar molecule [18–22]. On the other hand, it is important to highlight that the particle size reached in the liposomes prepared with both lecithins is on a nanometric scale (150–250 nm), which favors the permeation process in many biological membranes and is, in fact, an interesting characteristic for the development of functional food products [23–25]. Regarding the polydispersity values (Figure 3B), it was found that, regardless of the lecithin used, it always tends to form nanoliposomes (loaded and unloaded) with a low particle size distribution (PDI < 0.3). This result is very interesting, since it suggests that both lecithins can be used to produce nanoliposomes of a regular size, which is essential to guaranteeing the uniformity of the content inside the nanoliposomes [4,26,27]. In relation to the zeta potential (Figure 3C), this parameter showed values of around −30 mV in all of the nanosystems obtained, regardless of the lecithin used. These negative zeta potential values can be explained considering that liposomes consist of a small portion of fatty acids (oleic acid, linoleic acid, etc.), which can be slightly ionized in the aqueous medium. In the same way, it can be considered that the autoprotolysis effect of water, where some hydroxy anions are generated and located in the liposome-aqueous medium interface, leads to an increase in the negative zeta potential [28]. Therefore, this result suggests that nanoliposomes could have an adequate physical stability, because of electrostatic repulsion that prevent interparticle aggregation [29–31]. Furthermore, the systems loaded with TAA lead to more negative zeta potential values. This result can be explained by considering that TAA is a polycarboxylic acid (carboxylic acid-carboxylate), which can also be adsorbed in the interfacial zone (a fraction), increasing the anionic charge in this zone and therefore, the zeta potential.

2.2.2. Encapsulation and In Vitro Release of Antioxidant Compounds

Figure 4A shows the results of the encapsulation efficiency (EE) of antioxidant compounds (QCT and TAA) from nanoliposomes prepared with low purity (Lp-SBL) and high purity (Hp-SBL) soybean lecithins. In the case of nanoliposomes loaded with QCT and TAA and prepared with Lp-SBL, the results of the encapsulation efficiency were around 88% and 57%, respectively. For the nanoliposomes loaded with QCT and TAA and prepared with Hp-SBL, the results of the encapsulation efficiency were around 99% and 27%, respectively. These results suggest that the Lp-SBL favored the encapsulation of polar compounds, while the Hp-SBL favored non-polar compound encapsulation. This result may be explained considering the differences between the lecithins' composition, where Lp-SBL displayed a higher amount of glycophospholipids and sugars, which could interact with polar compounds such as TAA through hydrogen bond interactions. Another study also suggested that some small molecules, such as caffeine, are mainly located in the solvation layer adjacent to the liposomal lipid bilayer interface [32]. On the contrary, the Hp-SBL exhibited a composition of phospholipids with a non-polar character and where the encapsulation of QCT (non-polar) could possibly take place by a process such as micellar solubilization.

On the other hand, Figure 4B shows the in vitro release of antioxidant compounds (QCT and TAA) from nanoliposomes prepared with low purity (Lp-SBL) and high purity (Hp-SBL) soybean lecithins. In the case of QCT, it is noteworthy that the release profiles display a lag time of 240 min for nanoliposomes made with Lp-SBL and a maximum released amount of 15%. In comparison, for nanoliposomes elaborated with Hp-SBL, the lag time was 60 min, and the maximum released amount was 40%. These results are very interesting, since they show that the nanoliposomal vehicle considerably affects the QCT release, which can be easily appreciated when these are compared against the control (QCT alone, i.e., without a nanoliposomal vehicle), which is faster and exhibits a greater amount (99%). This result is very consistent, considering that the QCT is encapsulated within the hydrophobic pseudo-phase formed by the lamellar structures of the hydrocarbon chains of phospholipids. Likewise, this result supports the previous results of the encapsulation efficiency, where the QCT is encapsulated in high amounts in both nanoliposomal systems (88–99%), describing a greater affinity for the vehicle than the aqueous medium and therefore, its release is slow and controlled, as reflected in their respective lag times. On the contrary, the TAA-loaded nanoliposomes showed a faster release profile, regardless of the type of lecithin used. In this way, the TAA release from nanoliposomes described a similar behavior to that shown by the control (TAA alone), where rapid diffusion was observed. Regarding the control, it could be seen that around 80% of the TAA went through the bi-compartmental system in the first 5 min, and then remained almost constant, suggesting that the material balance had been reached. In the case of the TAA loaded in the nanoliposomes, it was found that the release was lower (~60%), which indicates that most of the TAA is in the interfacial zone and not within the liposomal aqueous core and therefore, the release is practically immediate. Likewise, the difference in the amount of TAA between the control and the nanoliposomal systems suggests that there is a fraction of TAA inside the nanoliposome and to achieve a higher released amount, a longer time or other external conditions would possibly be required. Similar results have also been reported, where it has been described that, depending on the molecule polarity, the encapsulation and release mechanisms may vary [32–37].

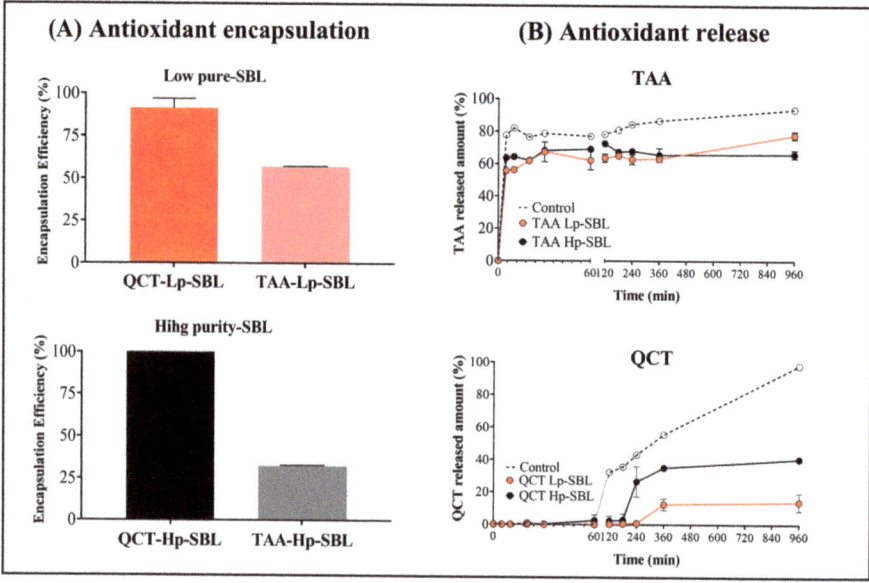

Figure 4. (**A**) Encapsulation efficiency and (**B**) release profiles of antioxidant (TAA and QCT) liposomes prepared with high purity lecithin (Hp-SBL) and low purity lecithin (Lp-SBL). QCT is quercetin and TAA is trans-aconitic acid. Data are reported as the mean ± SD and n = 3.

2.3. Antioxidant Effect Assay

The ORAC assay results for pure QCT and TAA, as well as loaded in nanoliposomes, are shown in Figure 5. In the case of Trolox® (standard), QCT, and TAA antioxidants, the increase in their concentrations prevented fluorescence quenching of the probe in different ways. In the case of pure QCT, a similar trend to that described by the standard antioxidant was observed, while TAA exhibited a different behavior. The maximum antioxidant effect was reached at concentrations of 30.5, 40.5, and 140 µg/mL for Trolox®, QCT, and TAA, respectively (Figure 5A). These results can be explained by considering the photo-physical mechanism that takes place in the ORAC assay, as well as analyzing the chemical structures of antioxidants (Figure 5B). In this context, fluorescein in aqueous medium emits fluorescent radiation at 520 nm, which remains practically unchanged over time. Then, the addition of the AAPH reagent leads to the oxidation of fluorescein by the reactive oxygen species (ROS) generated during the homolytic cleavage of such reagent [38,39]. Consequently, the oxidation in fluorescein leads to a decay of the fluorescence intensity over time (quenching) [38–40]. Therefore, the addition of an antioxidant establishes competition with fluorescein in the oxidation process and thus, when the antioxidant interacts with the ROS species, fluorescence decay is avoided, which is interpreted as an antioxidant effect. In these molecules, the phenyl substituent presents a thermodynamic equilibrium between phenyl and phenolate species, where the phenolate form is the one that reacts with the ROS species [8,38,39,41,42]. On the contrary, the antioxidant effect of TAA is due to the alteration of the phenyl-phenolate thermodynamic balance in fluorescein, leading to a slight predominance of the neutral form (R-phenyl), which is less reactive against ROS species. Moreover, the antioxidant effect was achieved at high concentrations (140 µg/mL), being moderate. However, it is important to note that TAA does not have oxidizable groups [9,43], whereas the Trolox® and QCT antioxidants do and therefore, TAA's ability to avoid fluorescence decay is more limited.

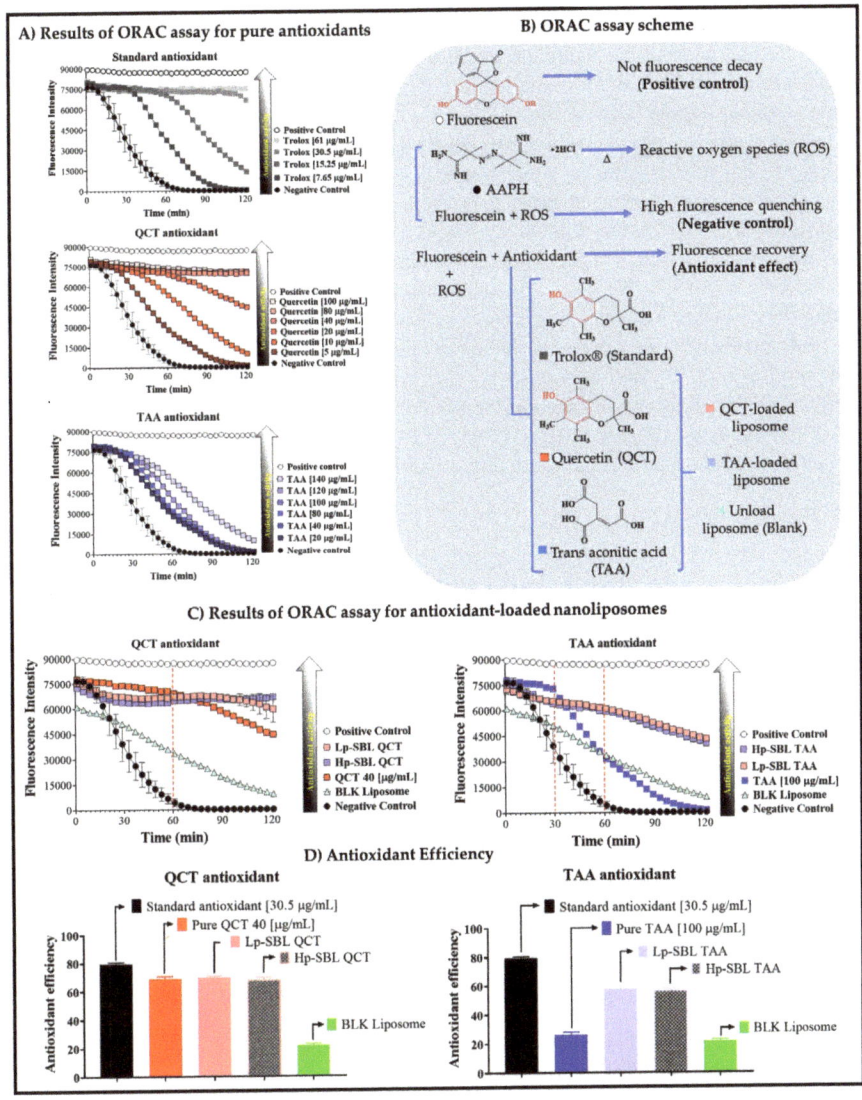

Figure 5. Results of the oxygen radical absorbance capacity (ORAC) assay. (**A**) Pure antioxidants. (**B**) ORAC method scheme. (**C**) Antioxidant-loaded nanoliposome. (**D**) Antioxidant efficiency. Data are reported as the mean ± SD and n = 3.

Regarding nanoliposomes loaded with QCT and TAA (Figure 5C), the antioxidant effect changes, depending on the type of compound. In the case of pure QCT, slight fluorescence decay of the probe was observed after 30 min, while QCT-loaded nanoliposomes displayed a sustained fluorescence effect. This result is consistent with that previously obtained in the QCT release profile from liposomes, where a sustained release was observed. Therefore, the use of nanoliposomes extends the antioxidant effect of QCT over time. On the contrary, TAA presented a considerable change when it was loaded in nanoliposomes, leading to a remarkable recovery of fluorescence. In the case of pure TAA, the antioxidant effect was very short and after 30 min, the probe fluorescence was considerably quenched. However, when TAA was loaded on nanoliposomes, there was a slight recovery of the fluorescent emission, suggesting a slight increase in the antioxidant effect. However, this recovery was less than that obtained with Trolox® and QCT antioxidants. This result is consistent considering that TAA does not have the phenyl substituent, which is involved in the oxidation by ROS species.

Regarding the antioxidant efficiency (Figure 5D), the results showed that pure QCT and that loaded in the nanoliposomes have practically the same antioxidant efficiency of around 70%, being very similar to that shown by the standard antioxidant. Nevertheless, the most interesting result is the sustained antioxidant effect over time, which was achieved when QCT was loaded inside the liposomes. In relation to TAA, the antioxidant efficiency was <30%, which is lower than the values for Trolox® and QCT antioxidants. However, this antioxidant efficiency was significantly improved when TAA was loaded in the nanoliposomes, resulting in an antioxidant efficiency value of around 60%. Regarding the type of lecithins used for the formation of nanoliposomes, no significant changes were observed concerning the antioxidant effect. In the case of Lp-SBL-QCT and Hp-SBL-QCT systems, as well as Lp-SBL-TAA and Hp-SBL-TAA systems, the antioxidant efficiencies were very similar to each other, with values of around 70% and 60%, respectively. Likewise, it was observed that both liposomal systems also exhibit a slight antioxidant effect of around 20%. These results are very interesting because they show that nanoliposomes formed with low purity lecithins can provide a similar antioxidant effect to that provided with nanoliposomes formed with high purity lecithins.

3. Material and Methods

3.1. Materials

Low purity grade lecithin (Lp-SBL) was obtained from Farmacia-Drogueria San Jorge Ltd.a (Cali, Colombia), whereas high purity grade lecithin (Hp-SBL) (Epikuron 200™, Mw = 786 g/mol) was purchased from Cargill Corporation (Wayzata, MN, USA). Both lectins claim to come from the same plant source (soybean) and were used as received. The phospholipid 1,2-dioleoyl-sn-glycero-3-phosphoethanolamine (DOPE, Mw = 744.03 g/mol) and cholesterol (Mw = 386 g/mol) were purchased from Avanti Polar Lipids (Alabaster, AL, USA). Quercetin (QCT), trans-aconitic acid (TAA), and Ethanol USP grade were purchased from Sigma-Aldrich (St. Louis, MO, USA) and ultrapure water was supplied by an Elix Essential Millipore® purification system, with a mean conductivity value of ~0.050 µScm. Methanol Lichrosolv™ mass spectrometry grade was obtained from Sigma-Aldrich (St. Louis, MO, USA). Regarding the antioxidant activity assay, the analytical reagents employed were fluorescein sodium salt, 2,2'-Azobis(2-methylpropionamidine) dihydrochloride (AAPH), 6-Hydroxy-2,5,7,8-tetramethylchromane-2-carboxylic acid (Trolox), potassium phosphate monobasic, and potassium phosphate dibasic, which were purchased from Sigma-Aldrich (Merck KGaA, Darmstadt, Germany)

3.2. Chemical Characterization of Lecithins

3.2.1. Infrared Spectroscopy (FTIR) Characterization

The structural characterization of lecithins was performed in an FT-infrared spectrometer coupled to an attenuated total reflectance (ATR) instrument (Nicolet 6700, Thermo Fisher Scientific,

Waltham, MA, USA). The spectra were recorded by using an attenuated total reflectance Smart iTR™ accessory, where the spectra of both lectins were compared.

3.2.2. Thermal Analysis

Thermal studies were carried out on a Q2000 differential scanning calorimeter (DSC; TA Instruments, New Castle, DE, USA) calibrated with indium T_m = 155.78 °C and ΔH_m = 28.71 J/g. Therefore, a modulated heating–cooling cycle from −10 °C (263.15 K) to 250 °C (523.15 K) at a heating rate of 5 °C/min was used. It was applied a ±0.5 °C modulation each 40 s. Approximately 10 mg of each sample was placed on a hermetic crucible with a lid, and an empty hermetic crucible was used as a reference.

3.2.3. Electrospray Ionization Mass Spectrometry

A methanolic solution of each respective lecithin was prepared at a concentration of 1 mg/mL, which was injected by direct infusion (flow of 3 µL/min) into a simple quadrupole mass spectrometer coupled to an electrospray ionization source (SQD2/ESI, Waters Corporation, Milford, MA, USA). The spectra were obtained in total ion scanning mode for positive ions (ESI$^+$) and negative ions (ESI$^-$), in a range of 200 to 2000 m/z. The test conditions for both ionization modes were as follows: Desolvation gas flow, 550 L/h; desolvation temperature, 500 °C; source temperature, 150 °C; extraction voltage, 3 V; cone voltage, 40 V; and capillary voltage 2.69 kV. The samples were injected in triplicate.

3.3. Development of Nanoliposomes

3.3.1. Preparation by the Ethanol Injection Method

The nanoliposomes were prepared based on a sequential process defined in several steps, depending on the antioxidant [44]. In the case of QCT-loaded nanoliposomes, they were prepared as follows: (i) Dispersion of phospholipids in organic phase: Ethanolic solutions of lecithin (1.3 mg/mL), cholesterol (0.64 mg/mL), DOPE (1.23 mg/mL), and QCT (350.8 µg/mL) were prepared. From those solutions, volumes of 30, 11.5, 30, and 28.5 µL were taken, respectively; (ii) mixture with aqueous phase: 100 µL of organic phase was slowly added to 100 µL of ultrapure water, and the solution was then stirred for 1 min and left to rest for 10 min; and (iii) Nanoliposome formation: The resulting mixture between the organic phase and aqueous media was diluted in 800 µL of ultra-pure water. This process led to nanoliposome formation with a final QCT concentration of 35 µg/mL. In the case of TTA-loaded nanoliposomes, these were prepared as follows: (i) Dispersion of phospholipids in organic phase: Ethanolic solutions of lecithin (1.3 mg/mL), cholesterol (0.64 mg/mL), and DOPE (1.23 mg/mL) were prepared. From those solutions, volumes of 42.4, 15.2, and 42.4 µL were taken, respectively. (ii) Mixture with aqueous phase: 100 µL of organic phase was slowly added to 100 µL of aqueous media trans-aconitic acid solution with a 1000 µg/mL concentration (Ultra-pure water), and the sample was then stirred for 1 min and left to rest for 10 min. The nanoliposome formation was conducted similarly to QCT-loaded nanoliposomes, where the TAA concentration in nanoliposomes was 100 µg/mL. Liposomes were purified by means of the ultrafiltration/centrifugation technique. An aliquot of each nanoliposome dispersion was transferred into an ultrafiltration tube (VWR, modified polyethersulfone-PES 10 kDa, 500 µL, diameter: 0.96 cm) and centrifuged (MIKRO 185, Hettich Lab Technology, Tuttlingen, Germany) at 10,000 rpm (1075 RFC) for 6 min. Each nanoliposomal system was prepared in triplicate at room temperature (25 °C).

3.3.2. Physicochemical Characterization

The particle size and zeta potential were determined using a Zetasizer nano ZSP (Malvern Instrument, Worcestershire, UK) with a red He/Ne laser (633 nm). The particle size was measured using dynamic light scattering (DLS) with an angle of scattering of 173° at 25 °C, in a quartz flow cell (ZEN0023), whereas the zeta potential was measured using a disposable folded capillary cell

(DTS1070). This instrument reports the particle size as the mean particle diameter (z-average), with the polydispersity index (PDI) ranging from 0 (monodisperse) to 1 (very broad distribution). All measurements were performed in triplicate after an appropriate dilution (~5:5000, v/v) of the liposome suspension in ultra-pure water and are reported as the mean and standard deviation of measurements made from freshly prepared liposomal dispersions.

3.4. Encapsulation and In Vitro Release of Antioxidant Compounds

3.4.1. Antioxidant Encapsulation Efficiency (EE)

The EE of QCT and TAA was assessed using the ultrafiltration/centrifugation technique. An aliquot of each nanoliposome dispersion was transferred into an ultrafiltration tube (VWR, modified polyethersulfone-PES 10 kDa, 500 µL, diameter: 0.96 cm) and centrifuged (MIKRO 185, Hettich Lab Technology, Tuttlingen, Germany) at 10,000 rpm (1075 RFC) for 6 min. For QCT quantification, an aliquot of the filtrate obtained in each system was obtained and evaluated with a microplate reader (Synergy, H1. Microplate reader, Biotek, Winooski, VT, USA). The amount of quercetin was determined by interpolation from a calibration curve that was previously prepared at the following concentrations, using ultra-pure water as the solvent: 1, 3, 10, 30, and 300 µg/mL. For the quantification of TAA, an aliquot of the filtrate obtained in each system was obtained and evaluated via HPLC (Lachrom elite, Merck, Darmstadt, Germany) equipped with a photo diode array (PDA) detector and an automatic sampling system. The mobile phase consisted of acetonitrile and water with a pH of 2.5 (10:90), and the flow rate was 0.8 mL/min. Separation was achieved using a 50 mm × 4.6 mm, Zorbax Eclipse XDB-C18 (Agilent technologies, Santa Clara, CA, USA) reversed-phase column, with an average particle size of 1.8 µm, keeping the column at 25 °C. The column effluent was monitored at 270 nm, and the chromatographic data analysis was performed with EZChrome software (Agilent technologies, Santa Clara, CA, USA). The amount of TAA was determined by interpolation from a calibration curve that was previously prepared at the following concentrations, using ultra-pure water as the solvent: 5, 10, 20, 40, 80, and 100 µg/mL. Finally, the amount of QCT and TAA loaded inside the nanoliposomes was calculated using the following equation:

$$EE = 100 - \left[\frac{Q_t - Q_s}{Q_t} \times 100\right] \tag{1}$$

where EE, Q_t, and Q_s correspond to the encapsulation efficiency, initial total amount of bioactive compound, and amount of bioactive compound in the filtrate, respectively.

3.4.2. In Vitro Antioxidant Release

The in vitro release of QCT and TAA from nanoliposomes was assessed by the dialysis method, using 5 mL of phosphate buffer with a pH of 7.0 and 150 mM aqueous medium under sink conditions. Therefore, volumes of 500 µL of the nanoliposomes were placed into a dialysis tube (VWR, modified polyethersulfone-PES cut-off 10 kDa, 500 µL, diameter: 0.96 cm) in triplicate and dialyzed at 37 °C for 16 h with constant stirring in an incubated orbital shaker (Inkubator 1000 with Unimax 1010, Heidolph Instruments, Schwalbach, Germany). Subsequently, the samples were taken from the external medium at intervals of 0, 5, 10, 20, 30, 60, 120, 180, 240, 360, and 996 min. The quantification of QCT and TAA was performed as described in the encapsulation efficiency section.

3.5. Antioxidant Effect Assay

The antioxidant activity was determined by the oxygen radical absorbance capacity (ORAC) assay [45,46], which is a method based on an evaluation of the ability of a compound to prevent the fluorescence quenching mediated by the AAPH reagent. For this, fluorescein (a fluorescent probe) and AAPH solutions were prepared in PBS (pH: 7.0) at concentrations of 0.02 mg/mL and 59.8 mg/mL, respectively. In contrast, the antioxidant compounds were made in PBS (pH: 7.0) at concentrations of

7.65, 15.25, 30.5, and 61 µg/mL for the standard antioxidant (Trolox®); 5, 10, 20, 40, 80, and 100 µg/mL for QCT; and 20, 40, 80, 100, 120, and 140 µg/mL for TAA. The evaluation of the fluorescent decay for fluorescein was conducted using a Synergy H1 microplate reader (Biotek, Winooski, VT, USA), where excitation and emission wavelengths of 485 and 520 nm were used, respectively, at 37 °C. Measurements were carried out in triplicate every 3 min for 2 h, and the data obtained from the fluorescent vs. time curves are reported as the average antioxidant efficiency (AE) of the antioxidant compound. This parameter is defined as the area under the fluorescent decay curve (AUC) recorded at a particular time in relation to the rectangular area (R) described by 100% of fluorescent emission of pure fluorescein (AUC of the positive control) at the same time. Therefore, the antioxidant efficiency can be calculated from:

$$AE = \frac{AUC}{R} \times 100\% \quad (2)$$

Finally, the pure fluorescein solution was named as the positive control because fluorescence decay does not take place. On the contrary, the fluorescein + AAPH solution was labeled as the negative control, because AAPH forms reactive oxygen species (ROS) that lead to a high quenching of the fluorescent probe. Similarly, Trolox® and unloaded nanoliposomes were labeled as the standard antioxidant and blank, respectively.

3.6. Statistical Analysis

The data were tabulated and analyzed using Microsoft Excel and Graph Pad Prism, respectively. The homogeneity of variance in the data was analyzed using Bartlett's test. Statistical comparisons were conducted using a one-way ANOVA. The Bonferroni post-hoc test was used to determine significant differences between the two independent groups. A confidence level of 95% was adopted. Data are expressed as the mean ± standard deviation.

4. Conclusions

In this study, it was established that low and high purity lecithins from a similar plant source (soybean) show differences and similarities, which can mean both advantages and disadvantages for their use as raw materials. First, the physical appearance is a determining factor for obtaining adequate handling and organoleptic characteristics at different stages of the product life cycle. Accordingly, the physical handling of a material such as low purity degree lecithin presents some difficulties compared to a material such as high purity degree lecithin. However, the physicochemical characterization of both materials using instrumental techniques such as FTIR, MS, and DSC, indicated that the chemical compositions of the two lecithins are very similar. Therefore, phospholipids such as phosphatidylcholine and phosphatidyl inositol were the main constituents of lecithins in this work. Regarding the ability to form nanoliposomal systems with adequate physicochemical characteristics (particle sizes < 300 nm, PDI < 0.3 nm, and zeta potential values of ~−30 mV), it was found that both lectins allow the preparation of these types of soft nanometric vehicles in a simple way. On the other hand, it was established that lecithins with a low purity grade consist of a greater amount of polar phospholipids, which tend to mainly encapsulate trans-aconitic acid (TAA). In contrast, lecithins with a high purity degree and which mainly consist of non-polar phospholipids tend to encapsulate a higher amount of quercetin (QCT). It was also established that the release of antioxidant compounds from nanoliposomal systems depends on their polarity and the way that they are encapsulated. In the case of the TAA, this is mainly located in the nanoliposome-water interface, where its release is very fast and around 60%. On the contrary, QCT release is slow and occurs at smaller quantities (15–40%), which is explained by the specific location of QCT within the lamellar structure of the nanoliposomes, where its affinity for such pseudo phase is great and therefore, its release is limited. It was demonstrated that low purity lecithins represent a viable alternative in terms of costs–benefits for obtaining innovative products for application to the food and cosmetic sectors. On the other hand, it was found that QCT presents a high antioxidant efficiency, which is sustained over time, describing a behavior very similar to the standard. On the contrary, it was found that TAA has a low antioxidant efficiency that can be increased when it is loaded with nanoliposomes. Finally, it was found that nanoliposomes formed with low purity lecithins can provide an antioxidant effect equal to that provided with nanoliposomes formed with high purity lecithins.

Author Contributions: M.J.A. and J.D.S. mainly performed the test of development and physical characterization of nanoliposomes. C.J.Y. is mainly responsible for the physicochemical characterization of lecithins, the encapsulation efficiency, the in vitro release assays, and the test of antioxidant effect evaluation. J.O.-G. is mainly responsible for validation and C.H.S. is mainly responsible for supervision, data analysis, validation, and final manuscript preparation. All authors have read and agreed to the published version of the manuscript.

Funding: This research was funded by the internal grant of Icesi University No. CA0413107.

Acknowledgments: The authors thank the Icesi University for the funding provided for the development of the study and Santiago de Cali University for supporting the APC.

Conflicts of Interest: The authors declare no conflict of interests.

Abbreviations

QCT	Quercetin
TAA	Trans-aconitic acid
Lp-SBL	low purity-soybean lecithin
Hp-SBL	low purity-soybean lecithin
PC	phosphatidylcholine
PE	phosphatidylethanolamines
PS	phosphatidylserines
PI	phosphatidylinositols
PG	phosphatidylglycerols
GPL	glyphospholipids
PA	phosphatidic acid

References

1. List, G.R. Soybean Lecithin: Food, Industrial Uses, and Other Applications. In *Polar Lipids: Biology, Chemistry, and Technology*; ACOS PRESS: Urbana, IL, USA, 2015; ISBN 9781630670450.
2. Gunstone, F.D. *Phospholipid Technology and Applications*; Oily Press: Bridgwater, UK, 2008; ISBN 9780955251221.
3. Klang, V.; Valenta, C. Lecithin-based nanoemulsions. *J. Drug Deliv. Sci. Technol.* **2011**, *21*, 55–76. [CrossRef]
4. Mozafari, M.R.; Johnson, C.; Hatziantoniou, S.; Demetzos, C. Nanoliposomes and their applications in food nanotechnology. *J. Liposome Res.* **2008**, *18*, 309–327. [CrossRef] [PubMed]
5. Scholfield, C.R. Composition of soybean lecithin. *J. Am. Oil Chem. Soc.* **1981**, *58*, 889–892. [CrossRef]
6. Palacios, L.E.; Wang, T. Egg-yolk lipid fractionation and lecithin characterization. *JAOCS J. Am. Oil Chem. Soc.* **2005**, *82*, 571–578. [CrossRef]
7. Szuhaj, B.F. Lecithin production and utilization. *J. Am. Oil Chem. Soc.* **1983**, *60*, 306–309. [CrossRef]
8. Lesjak, M.; Beara, I.; Simin, N.; Pintać, D.; Majkić, T.; Bekvalac, K.; Orčić, D.; Mimica-Dukić, N. Antioxidant and anti-inflammatory activities of quercetin and its derivatives. *J. Funct. Foods* **2018**, *40*, 68–75. [CrossRef]
9. Piang-Siong, W.; De Caro, P.; Marvilliers, A.; Chasseray, X.; Payet, B.; Shum Cheong Sing, A.; Illien, B. Contribution of trans-aconitic acid to DPPHrad scavenging ability in different media. *Food Chem.* **2017**, *214*, 447–452. [CrossRef]
10. Fernandes, G.D.; Alberici, R.M.; Pereira, G.G.; Cabral, E.C.; Eberlin, M.N.; Barrera-Arellano, D. Direct characterization of commercial lecithins by easy ambient sonic-spray ionization mass spectrometry. *Food Chem.* **2012**, *135*, 1855–1860. [CrossRef]
11. Verdonck, E.; Schaap, K.; Thomas, L.C. A discussion of the principles and applications of Modulated Temperature DSC (MTDSC). *Int. J. Pharm.* **1999**, *192*, 3–20. [CrossRef]
12. McPhillips, H.; Craig, D.Q.M.; Royall, P.G.; Hill, V.L. Characterisation of the glass transition of HPMC using modulated temperature differential scanning calorimetry. *Int. J. Pharm.* **1999**, *180*, 83–90. [CrossRef]

13. Linares, V.; Yarce, C.J.; Echeverri, J.D.; Galeano, E.; Salamanca, C.H. Relationship between degree of polymeric ionisation and hydrolytic degradation of Eudragit® E polymers under extreme acid conditions. *Polymers* **2019**, *11*, 1010. [CrossRef] [PubMed]
14. Sandoval, A.; Rodriguez, E.; Fernandez, A. Application of analysis by differential scanning calorimetry (DSC) for the characterization of the modifications of the starch. *Rev. Fac. MINAS* **2005**, *72*, 45–53.
15. Hatakeyama, T.; Iijima, M.; Hatakeyama, H. Role of bound water on structural change of water insoluble polysaccharides. *Food Hydrocoll.* **2016**, *53*, 62–68. [CrossRef]
16. Nakamur, K.; Minagaw, Y.; Hatakeyam, T.; Hatakeyama, H. DSC studies on bound water in carboxymethylcellulose-polylysine complexes. *Thermochim. Acta* **2004**, *416*, 135–140. [CrossRef]
17. Yarce, C.; Pineda, D.; Correa, C.; Salamanca, C. Relationship between Surface Properties and In Vitro Drug Release from a Compressed Matrix Containing an Amphiphilic Polymer Material. *Pharmaceuticals* **2016**, *9*, 34. [CrossRef]
18. Marianecci, C.; Petralito, S.; Rinaldi, F.; Hanieh, P.N.; Carafa, M. Some recent advances on liposomal and niosomal vesicular carriers. *J. Drug Deliv. Sci. Technol.* **2016**, *32*, 256–269. [CrossRef]
19. Rafiee, Z.; Barzegar, M.; Sahari, M.A.; Maherani, B. Nanoliposomal carriers for improvement the bioavailability of high—Valued phenolic compounds of pistachio green hull extract. *Food Chem.* **2017**, *220*, 115–122. [CrossRef]
20. Taylor, T.M.; Davidson, P.M.; Bruce, B.D.; Weiss, J. Liposomal nanocapsules in food science and agriculture. *Crit. Rev. Food Sci. Nutr.* **2005**, *45*, 587–605. [CrossRef]
21. Hsieh, Y.F.; Chen, T.L.; Wang, Y.T.; Chang, J.H.; Chang, H.M. Properties of liposomes prepared with various lipids. *J. Food Sci.* **2002**, *67*, 2808–2813. [CrossRef]
22. Raut, S.; Bhadoriya, S.S.; Uplanchiwar, V.; Mishra, V.; Gahane, A.; Jain, S.K. Lecithin organogel: A unique micellar system for the delivery of bioactive agents in the treatment of skin aging. *Acta Pharm. Sin. B* **2012**, *2*, 8–15. [CrossRef]
23. Hasanovic, A.; Hollick, C.; Fischinger, K.; Valenta, C. Improvement in physicochemical parameters of DPPC liposomes and increase in skin permeation of aciclovir and minoxidil by the addition of cationic polymers. *Eur. J. Pharm. Biopharm.* **2010**, *75*, 148–153. [CrossRef] [PubMed]
24. Refai, H.; Hassan, D.; Abdelmonem, R. Development and characterization of polymer-coated liposomes for vaginal delivery of sildenafil citrate. *Drug Deliv.* **2017**, *24*, 278–288. [CrossRef] [PubMed]
25. Elnaggar, Y.S.R.; El-Refaie, W.M.; El-Massik, M.A.; Abdallah, O.Y. Lecithin-based nanostructured gels for skin delivery: An update on state of art and recent applications. *J. Control. Release* **2014**, *180*, 10–24. [CrossRef] [PubMed]
26. Zhao, L.; Temelli, F.; Chen, L. Encapsulation of anthocyanin in liposomes using supercritical carbon dioxide: Effects of anthocyanin and sterol concentrations. *J. Funct. Foods* **2017**, *34*, 159–167. [CrossRef]
27. Laouini, A.; Jaafar-Maalej, C.; Limayem-Blouza, I.; Sfar, S.; Charcosset, C.; Fessi, H. Preparation, Characterization and Applications of Liposomes: State of the Art. *J. Colloid Sci. Biotechnol.* **2012**, *1*, 147–168. [CrossRef]
28. Marinova, K.G.; Alargova, R.G.; Denkov, N.D.; Velev, O.D.; Petsev, D.N.; Ivanov, I.B.; Borwankar, R.P. Charging of Oil–Water Interfaces Due to Spontaneous Adsorption of Hydroxyl Ions. *Langmuir* **1996**, *12*, 2045–2051. [CrossRef]
29. Barea, M.J.; Jenkins, M.J.; Gaber, M.H.; Bridson, R.H. Evaluation of liposomes coated with a pH responsive polymer. *Int. J. Pharm.* **2010**, *402*, 89–94. [CrossRef]
30. Sabín, J.; Prieto, G.; Ruso, J.M.; Hidalgo-Álvarez, R.; Sarmiento, F. Size and stability of liposomes: A possible role of hydration and osmotic forces. *Eur. Phys. J. E* **2006**, *20*, 401–408. [CrossRef]
31. Cantor, S.; Vargas, L.; Rojas, O.E.A.; Yarce, C.J.; Salamanca, C.H.; Oñate-Garzón, J. Evaluation of the antimicrobial activity of cationic peptides loaded in surface-modified nanoliposomes against foodborne bacteria. *Int. J. Mol. Sci.* **2019**, *20*, 680. [CrossRef]
32. Budai, L.; Kaszás, N.; Gróf, P.; Lenti, K.; Maghami, K.; Antal, I.; Klebovich, I.; Petrikovics, I.; Budai, M. Liposomes for topical use: A physico-chemical comparison of vesicles prepared from egg or soy lecithin. *Sci. Pharm.* **2013**, *81*, 1151–1166. [CrossRef]
33. Herman, A.; Herman, A.P. Caffeine's mechanisms of action and its cosmetic use. *Skin Pharmacol. Physiol.* **2012**, *26*, 8–14. [CrossRef]

34. Scherer, R.; Godoy, H.T. Antioxidant activity index (AAI) by the 2,2-diphenyl-1-picrylhydrazyl method. *Food Chem.* **2009**, *112*, 654–658. [CrossRef]
35. Li, M.; Du, C.; Guo, N.; Teng, Y.; Meng, X.; Sun, H.; Li, S.; Yu, P.; Galons, H. Composition design and medical application of liposomes. *Eur. J. Med. Chem.* **2019**, *164*, 640–653. [CrossRef] [PubMed]
36. Lombardo, D.; Kiselev, M.A.; Caccamo, M.T. Smart Nanoparticles for Drug Delivery Application: Development of Versatile Nanocarrier Platforms in Biotechnology and Nanomedicine. *J. Nanomater.* **2019**. [CrossRef]
37. Yadav, A.V.; Murthy, M.S.; Shete, A.S.; Sakhare, S. Stability aspects of liposomes. *Indian J. Pharm. Educ. Res.* **2011**, *45*, 402–413.
38. Kuti, J.O.; Konuru, H.B. Antioxidant Capacity and Phenolic Content in Leaf Extracts of Tree Spinach (*Cnidoscolus* spp.). *J. Agric. Food Chem.* **2004**, *52*, 117–121. [CrossRef]
39. Ehlenfeldt, M.K.; Prior, R.L. Oxygen radical absorbance capacity (ORAC) and phenolic and anthocyanin concentrations in fruit and leaf tissues of highbush blueberry. *J. Agric. Food Chem.* **2001**, *49*, 2222–2227. [CrossRef]
40. Güçlü, K.; Kibrisliŏglu, G.; Özyürek, M.; Apak, R. Development of a fluorescent probe for measurement of peroxyl radical scavenging activity in biological samples. *J. Agric. Food Chem.* **2014**, *62*, 1839–1845. [CrossRef]
41. Huang, D.; Boxin, O.U.; Prior, R.L. The chemistry behind antioxidant capacity assays. *J. Agric. Food Chem.* **2005**, *53*, 1841–1856. [CrossRef]
42. Prior, R.L. Oxygen radical absorbance capacity (ORAC): New horizons in relating dietary antioxidants/bioactives and health benefits. *J. Funct. Foods* **2015**, *18*, 797–810. [CrossRef]
43. Montoya, G.; Londono, J.; Cortes, P.; Izquierdo, O. Quantitation of trans-aconitic acid in different stages of the sugar-manufacturing process. *J. Agric. Food Chem.* **2014**, *62*, 8314–8318. [CrossRef] [PubMed]
44. Arévalo, L.M.; Yarce, C.J.; Oñate-Garzón, J.; Salamanca, C.H. Decrease of antimicrobial resistance through polyelectrolyte-coated nanoliposomes loaded with β-lactam drug. *Pharmaceuticals* **2019**, *12*, 1. [CrossRef] [PubMed]
45. Zheng, W.; Wang, S.Y. Antioxidant activity and phenolic compounds in selected herbs. *J. Agric. Food Chem.* **2001**, *49*, 5165–5170. [CrossRef] [PubMed]
46. Barros, R.G.C.; Andrade, J.K.S.; Denadai, M.; Nunes, M.L.; Narain, N. Evaluation of bioactive compounds potential and antioxidant activity in some Brazilian exotic fruit residues. *Food Res. Int.* **2017**, *102*, 84–92. [CrossRef] [PubMed]

Sample Availability: Samples of the compounds are available from the authors.

Publisher's Note: MDPI stays neutral with regard to jurisdictional claims in published maps and institutional affiliations.

© 2020 by the authors. Licensee MDPI, Basel, Switzerland. This article is an open access article distributed under the terms and conditions of the Creative Commons Attribution (CC BY) license (http://creativecommons.org/licenses/by/4.0/).

MDPI
St. Alban-Anlage 66
4052 Basel
Switzerland
Tel. +41 61 683 77 34
Fax +41 61 302 89 18
www.mdpi.com

Molecules Editorial Office
E-mail: molecules@mdpi.com
www.mdpi.com/journal/molecules

www.ingramcontent.com/pod-product-compliance
Lightning Source LLC
LaVergne TN
LVHW070145100526
838202LV00015B/1894